Ducks, Geese, and Swans of the World

Ducks, Geese, and Swans of the World

Paul A. Johnsgard

UNIVERSITY OF NEBRASKA PRESS
Lincoln and London

UNP Publishers on the Plains

Library of Congress Cataloging in Publication Data

Johnsgard, Paul A.
  Ducks, geese, and swans of the world.

  Bibliography: p. 387.
  Includes index.
  1. Anatidae.   I. Title.
QL696.A52J62   598.4′1   78–8920
ISBN 0–8032–0953–3

Contents

Illustrations

Color Plates

Preface

Of all the books that I have seen and wished that one day I could personally own, perhaps the ones that have most set me to daydreaming are the four-volume set *A Natural History of the Ducks,* by John C. Phillips. My first contact with this magnificent monograph was at Cornell University, where it could be seen in the rare book collection. I spent countless hours there during the late 1950s savoring the text and plates. At about that same time, the first volumes of Jean Delacour's *The Waterfowl of the World* were appearing, and these too gave me a chance to dream about such entrancing birds as torrent ducks, steamer ducks, musk ducks, and magpie geese. Since then it has been my good fortune to see all of these marvelous birds in life, and they have enriched my own life enormously. The waterfowl family is so rich in ecological diversity, in evolutionary lessons, and in behavioral complexities that I have never tired of either reading or writing about it.

Only two of my previous books, *Waterfowl: Their Biology and Natural History* and the earlier *Handbook of Waterfowl Behavior,* have dealt with the entire waterfowl family in a comprehensive fashion, and neither attempted to provide a systematic review of the biology of every species. Indeed, I have repeatedly thought about undertaking such an effort but each time have shrunk back at the thought of the huge number of species to be considered and of the vast literature that has accumulated on the group. Even in the 1920s it required four volumes for John C. Phillips to review the biology of the waterfowl family exclusive of the swans and geese, and Jean Delacour found that the same number of volumes was needed for his treatment of the entire family.

I decided that, in spite of the concessions that would need to be made, a shorter approach to the family Anatidae was possible, provided that a maximum condensation of text be attained, primarily through avoiding repetitous material such as separate accounts for each subspecies. I also decided to exclude taxonomic synonymies, extensive mensural data, and most avicultural information, all of which

can be found in Delacour's monograph, and likewise to reduce descriptive behavioral information to a minimum whenever it had already been summarized in my *Handbook of Waterfowl Behavior.* After considerable deliberation, it was decided to include plumage descriptions as an aid to identification but not to attempt to describe all molts and plumages. The "Natural History" section of each species account was considered to be the nuclear element, and an effort was made to include newer or at least different information from that summarized by Delacour. The "Status" and "Relationships" sections were added after the text was well underway, the former because of the increasing incidence of rare and endangered forms among the waterfowl, and the latter because more information on phyletic relationships is available now than in 1965, when I last dealt comprehensively with this problem in the *Handbook of Waterfowl Behavior.*

Except in a few instances, I have followed the taxonomic treatment used in the *Handbook,* both as to major taxonomic categories and sequences of species within such categories. The two exceptions constitute the recognition of special monotypic tribes for the Cape Barren goose and the torrent duck, since recent studies on both species have cast doubt on their presumed phyletic affinities and thus tribal recognition seemed to provide the best interim solution. The only other major change from the taxonomy used in the *Handbook* is the transfer of the genus *Thalassornis* from the stifftails to the whistling duck tribe. This action was based on my own studies after the publication of the *Handbook* and has received independent support from other investigators. A few changes have also been made in species limits and in subspecies recognized, and a moderate number of suggested changes in English vernacular names have been adopted. In part these have been to avoid the use of "common" for vernacular names, but also to allow for the consistent formation of distinctive vernacular names for each subspecies by simply adding an appropriate adjectival prefix. The group of four swans known as the "northern swans" has proven refractory to such treatment, partly because of their questionable taxonomic status and their still uncertain ancestral history. Thus, they have been given separate accounts and treated as if they are "good" species, in spite of the obvious fact that they are not. There are thus 148 separate species accounts in the text.

The writing of the text was begun in the spring of 1976 and continued through most of 1977, so that later literature was included only if it was regarded as of critical importance. Further, the number of literature citations was held to an absolute minimum, but one or more "Suggested Readings" for each species was included to provide additional access to the nearly endless literature on the waterfowl family. I would like to express my sincere appreciation to the School of Life Sciences for supporting this work and for allowing me the time needed to complete it.

In addition, I must again thank a number of individuals for their invaluable aid. These include virtually all of the staff of the Wildfowl Trust, but especially Sir Peter Scott, Dr. Geoffrey Matthews, and Dr. Janet Kear. I would also like to thank Jean

Delacour for encouraging me to undertake the book in spite of its potential competition with his own monograph. Not only have I relied extensively on the volumes by Delacour and Phillips but also on the regional volumes by H. J. Frith (Australia), G. Dementiev and N. Gladkov (U.S.S.R.), P. A. Clancey (South Africa), S. Ali and D. Ripley (India), and C. Tso-hsin (China) for various mensural and descriptive information, sometimes without citations to these sources. I also wish to thank David Skead for providing unpublished data on South African species.

The distribution maps have been based on a variety of sources, including the volumes just mentioned. The photographs are mostly of captive birds, taken at the Wildfowl Trust. All the photographs and line drawings are my own, although several of the drawings are based on published photographs taken by a variety of other photographers. In cases of rare or extinct species I have sometimes used photographs of live birds representing close relatives for proportions and postures, but virtually none of the drawings lack a documentary base.

xvi ❖ ❖ ❖

Introduction to the Family Anatidae

Inasmuch as the primary purpose of this book is to provide information on each of the species of the waterfowl family in a standardized format and easily accessible manner, it is important that the reader have some knowledge of the basis for my sequential organization of these species. A variety of attempts to provide a "natural" classification, or one that best reflects actual evolutionary relationships, of the family Anatidae have been made in recent years, with most of them being minor variations on a scheme first proposed by Jean Delacour and Ernst Mayr in 1945. In this landmark classification, emphasis was given to the association of species at the tribal level, rather than to the fragmentation of the family into a large number of subfamilies, as in earlier classifications.

My own behavioral studies of the family resulted in a proposed classification for the group in 1961, which was utilized in my subsequent books (1965a, 1968a) and has been subject to only minor modification since that time based on new information from my own and other studies. In brief, the family Anatidae is here regarded to be composed of 3 subfamilies, 13 tribes, 43 genera, and 148 recent species as follows:

Family Anatidae (ducks, geese, and swans)
  Subfamily Anseranatinae
    Tribe Anseranatini: Magpie goose (1 genus and species)
  Subfamily Anserinae
    Tribe Dendrocygnini: Whistling or tree ducks (2 genera, 9 species)
    Tribe Anserini: Swans and true geese (4 genera, 21 species)
    Tribe Cereopsini: Cape Barren goose (1 genus and species)
    Tribe Stictonettini: Freckled duck (1 genus and species)
  Subfamily Anatinae
    Tribe Tadornini: Sheldgeese and Shelducks (5 genera, 15 species)

    Tribe Tachyerini: Steamer ducks (1 genus, 3 species)
    Tribe Cairinini: Perching ducks (9 genera, 13 species)
    Tribe Merganettini: Torrent duck (1 genus and species)
    Tribe Anatini: Dabbling or surface-feeding ducks (4 genera, 39 species)
    Tribe Aythyini: Pochards (3 genera, 16 species)
    Tribe Mergini: Sea ducks (8 genera, 20 species)
    Tribe Oxyurini: Stiff-tailed ducks (3 genera, 8 species)

These groups are believed to be related to one another in the manner shown in figure 1, which indicates the probable relationships of the 13 tribes and 43 genera recognized in this book. Similar diagrams showing species relationships for each of the major tribes have been published earlier (Johnsgard, 1961a), and with relatively few more recent modifications still provide the basis for the sequence in which individual species are considered in this book. Behavioral and anatomical characteristics that provide the basis for the association of these species into tribes and subfamilies are those which are regarded as particularly significant, and thus it is worth reviewing such "emergent" characteristics before the species-by-species consideration of the entire family.

SUBFAMILY ANSERANATINAE

Tribe Anseranatini (Magpie Goose)

This subfamily is composed of a single or monotypic tribe, genus, and species, the magpie goose, which differs so much from the remainder of the family that some persons have argued that it might best be placed in a separate family. However, it also provides such important transitional characteristics between the more typical waterfowl and the South American screamers of the family Anhimidae that it is perhaps most useful to retain it in the Anatidae as a "land-

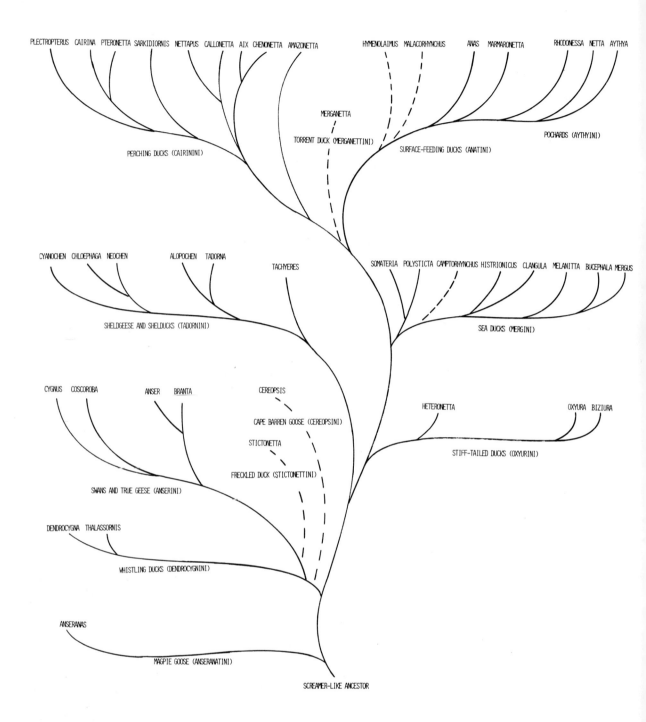

mark" form, against which the other species can usefully be measured.

Foremost among the magpie goose's peculiarities are its feet, which are only slightly webbed, and its unusually long hind toes, which are associated with the species' semiterrestrial adaptations and its perching behavior. Like the other Anatidae, the magpie goose molts its major flight feathers (primaries and secondaries) only once a year; but unlike virtually all of these other species, the magpie goose molts very gradually, with only a few feathers being absent at any one time, and thus the birds are continuously able to fly. There is no flightless period of increased vulnerability to predation, and the greatly prolonged period of molting is not a major disadvantage to this tropically adapted and nearly sedentary species.

A second unique anatomical feature of this species is the greatly elongated trachea in adults of both sexes, which in adult males may reach 150 centimeters. The magpie goose is unlike some swans in which the trachea is convoluted inside the sternum. Rather, in this species it penetrates the area between the breast muscles and the skin, in a manner comparable to that in certain species of Cracidae (Johnsgard, 1961b). As in the latter species, the resulting call is unusually low-pitched, and the elongated trachea probably functions as an effective resonator of low-frequency sounds. Other than this specialization, however, the anatomy of the sound-producing syrinx is relatively simple and gooselike.

A third remarkable feature of the magpie goose is related to its breeding behavior. The birds often form a trio consisting of a male and two females that drop their eggs in a single nest, with all three birds sharing the incubation responsibilities. This behavior, first observed in captive birds (Johnsgard, 1961c), was subsequently found to be typical also of wild individuals, and appears to be unique in the family. In addition, magpie geese are the only waterfowl that feed their young directly, the adults picking up edible materials in their bills and dropping them in front of their downy young. This behavior has its counterpart in the "tidbitting" behavior of many gallinaceous birds, but has not yet been reported from other Anatidae.

Subfamily Anserinae

The subfamily Anserinae includes the whistling duck, swans, true geese, and two "aberrant" species. Like the magpie goose, these species undergo only a single molt per year, which occurs after the breeding season and during which all feathers are replaced. However, the flight feathers are shed almost simultaneously, so that the birds are unable to fly for a period of from about three to six weeks. In all of the included species the unfeathered part of the lower leg, or tarsus—like that of the magpie goose—has an irregular network, or reticulated pattern, of scales throughout. Except in two semiterrestrial geese, however, the front toes are fully webbed. In all species

the sexes are essentially identical in plumage but often differ somewhat in body weight and, occasionally, in voice. Although the whistling ducks have relatively long hind toes, not even they perch as well as does the magpie goose, and the other species do not perch at all. All of the species lack iridescent coloration, even on the wings.

Tribe Dendrocygnini (Whistling Ducks)

The whistling ducks are a group of nine species, eight of which are readily placed in the single genus *Dendrocygna*, or "tree ducks." However, they are not primarily perching birds, and thus "whistling ducks" is a better vernacular designation than the frequently used "tree ducks." In nearly all species both sexes utter a clear, often multisyllabic whistle that readily identifies the bird as to its species and apparently provides important communication signals in these rather gregarious birds. Most of the species are tropical in distribution, and in all of them the pair bond seems to be permanent and potentially lifelong. Both sexes share equally in brood-rearing responsibilities, and in some species the male is known to participate in incubation as well. Unlike those of geese and swans, the downy young are distinctively patterned. Although the downy young of the white-backed duck differ somewhat from those of *Dendrocygna*, the species shares a sufficiently large number of behavioral and anatomical traits to be tentatively included in this tribe.

Tribe Anserini (Swans and True Geese)

The swans and true geese are moderately to extremely large waterfowl, which in common with the whistling ducks have plumage patterns that are alike in both sexes and lack iridescent coloration, and they also possess reticulated scale patterns on the tarsal surfaces. Most of the 20 species are found in the cooler parts of the Northern Hemisphere, the exceptions being 3 species of Southern Hemisphere swans. All are vegetarians, obtaining much of their food from terrestrial surface vegetation in the case of geese, and from subsurface aquatic vegetation in the case of swans. The patterns of the downy young tend to be pale and simple, without strong head or back patterning, and in most species the adult plumages are also fairly simple, with whites and blacks often predominating. The white plumage of most adult swans appears to be related to visibility needs associated with their high degree of territorial spacing. The most highly territorial swans are also the most strongly vocal ones, for the same reason. However, vocalizations of the sexes are very similar, and usually differ only in minor pitch characteristics. Most of the swans and geese are quite strongly migratory, but the insular Hawaiian goose not only has become nonmigratory but also has become semiterrestrial and has evolved reduced toe webbing as it has adapted to a mountainous existence on old lava fields.

Tribe Cereopsini (Cape Barren Goose)

Several recent studies have suggested that this unusual Australian gooselike bird is not very closely related to either the true geese or the sheldgeese, although it shares some traits with both groups. It has some unique traits, such as its unusually swollen bill, its adaptations to a terrestrial existence on the coastline and islands of the Bass Strait, and its very simple syringeal structure. Like the freckled duck, it is perhaps best considered as a survivor of a group transitional between the two major subfamilies of waterfowl that exist at the present time.

Tribe Stictonettini (Freckled Duck)

Even more than the Cape Barren goose, the freckled duck exhibits an assortment of anatomical traits that suggest affinities with the geese and swans, in spite of a fairly typical ducklike appearance and foraging behavior. Its unpatterned downy young, its remarkably primitive syringeal structure, and its reticulated tarsus all strongly argue for the position that this species is the sole survivor of a very ancient waterfowl lineage, with no near living relatives. A detailed study of its social behavior is greatly to be desired, for like the magpie goose, it provides an unequaled opportunity to gain insight into the evolutionary history of the waterfowl group through observation of a unique contemporary form.

Subfamily Anatinae

The third and largest subfamily, Anatinae, includes the sheldgeese, shelducks, and all of the typical ducks. It is subdivided into a varying number of tribes by different authorities; nine are recognized here. In most if not all of the species included in this subfamily there are two molts of the body feathers per year, and consequently there are usually two

distinct plumages, breeding (nuptial) and non-breeding (winter, or eclipse). In some species the breeding plumage of the male closely resembles that of the females, but more often the sexes have quite different breeding plumages, in conjunction with more frequently disrupted and renewed pair bonds. Likewise, the adult voices of the two sexes are normally very different in these species as the result of sexual differences in the structure of the syrinx and sometimes also the trachea. Unlike that in the previous subfamilies, the front surface of the lower tarsus has a linearly arranged (scutellated) scale pattern. Iridescent coloration is frequently present in the plumage, particularly among males, although on females it is usually restricted to the region of the secondaries of the wings. Such colorful patterns are called speculums and are usually species-specific.

In those species in which the sexes are quite different, or dimorphic, in size and appearance, the male is typically larger, more brilliantly patterned, and more aggressive. The plumages of juveniles and the nonbreeding plumages of males generally resemble that of the adult female, and lack both iridescent coloration on the body and the finely barred, or vermiculated, markings on the back, sides, and flanks that are typical of many of the species of this subfamily. The patterns of the downy young are often quite contrasting and distinctive, and usually include spotting and striping on the head and back. In no cases are males known to assist in incubation, but male participation in brood rearing is often variable within and between tribes, and is related primarily to such ecological factors as length and regularity of the breeding seasons.

Tribe Tadornini (Sheldgeese and Shelducks)

The most gooselike species of the subfamily Anatinae are the sheldgeese, which together with the closely related shelducks constitute the tribe Tadornini. This group of 14 species has a worldwide distribution except for North America. The typical sheldgeese are grazing birds, the ecological counterparts of the true geese, while the shelducks are mostly wading and dabbling birds, frequently feeding to a large extent on aquatic invertebrates. In all of the species the adult male has an enlarged bony bulla in the syrinx, which is evidently responsible for the whistling or whistlelike sounds that are important in sexual and aggressive displays. The females of all species have more reedy and typically gooselike or ducklike notes. In all species the males are larger than the females, and often differ appreciably from them in plumage. Although pair bonds are sometimes said to be permanent in the group, in some species at least they may be ruptured through interactions with outside males, and a premium is placed on male aggressiveness during sexual competition for mates. The wings are used in fighting and in most species they are strongly patterned, with white coverts and iridescent coloration on the secondaries or their greater coverts. The downy young are also typically patterned contrastingly in the shelducks and sheldgeese, a characteristic shared with various other cavity-nesting waterfowl.

Tribe Tachyerini (Steamer Ducks)

The South American steamer ducks have at times been included with the shelducks, but differ enough from them in structure and behavior to be regarded as a separate but closely related tribe. There are three species very similar in appearance; two are essentially flightless. All are found off the coasts of southern South America and the Falkland Islands, where they feed on mollusks and other marine invertebrates. The males differ slightly from females in their plumage and vocalizations, but in both sexes iridescent coloration is totally lacking and only a simple white speculum is present on the wings. Like the shelducks, however, they are highly aggressive during social display activities. The downy young to some extent resemble those of shelducks, but have less spotting and less contrasting coloration.

Tribe Cairinini (Perching Ducks)

This group of 13 species of primarily perching waterfowl is a rather heterogeneous assemblage of birds that are not easily characterized. In addition to being generally perching and cavity-nesting, they also have fairly wide, rounded wings, elongated tails, and sometimes also relatively long legs. Many of the species exhibit a great deal of iridescent coloration in their plumage, even among females, and in a few species nearly the entire body plumage is iridescent. In some of these brilliantly colored forms, such as the comb duck, muscovy duck, and the spur-winged goose, there is a great difference in the sizes of the sexes even though they may be very similar in plumage patterning. In spite of their brilliant plumages, males of only a few of the species exhibit definite eclipse plumages, probably in part because most of the forms are tropical, with

long or irregular breeding seasons. Pair bonding is generally seasonally established in the species that breed in temperate climates, while in the more tropical and seemingly more primitive species the pair bonds are weak and in a few cases apparently even nonexistent. The patterns of the downy young are typically well marked and characterized by white or yellow spots and stripes on a darker background; in many species they are not readily separable from the downy young of dabbling ducks. The two tribes are obviously closely related, and taxonomists are not agreed on which group some species such as the ringed teal, pink-eared duck, and Brazilian teal should be included in. However, to merge the tribes would make a comparatively large tribe and tend to obscure the well-defined differences in the behavior patterns and breeding biologies of these two groups.

Tribe Merganettini (Torrent Duck)

This remarkable stream-dwelling duck of the Andean mountains is certainly one of the most specialized of all waterfowl, and in part its anatomical and behavioral specializations have obscured its basic relationships, which appear to be with either the perching ducks or the dabbling ducks. In the absence of definitive evidence, it seems reasonable to maintain a separate tribe for this species, which exists as a series of relatively isolated populations between Venezuela and Tierra del Fuego that vary greatly in male plumage characteristics. Torrent ducks are essentially cavity- or ledge-nesting forms, although only a few nests have been found, and much still remains to be learned of their breeding biology.

Tribe Anatini (Dabbling Ducks)

The dabbling, or surface-feeding, ducks are, to judge from the number and abundance of the included species, the most successful of all waterfowl. This tribe includes all of the "puddle ducks" that constitute most of the important game species throughout the world. Of the 39 species making up the tribe, all but 3 can readily be placed in the single genus Anas. The tribe has a worldwide distribution, with some of the species occurring on several continents and having extensive transcontinental migration patterns. However, most of the species are temperate-breeding forms and are generally adapted to shallow, marshy habitats where food can be obtained from near the surface by dabbling or tipping-up. In most species the males have fairly elaborate and colorful breeding plumages, whereas the ground-nesting females are mostly cryptically patterned with buff and brown. In nearly all species both sexes have iridescent wing speculum markings, and in all species the trachea has an enlarged bulla at the syrinx. Pair bonds are generally reformed each year during the non-breeding season, but in some species with long breeding seasons the pair bonds are relatively permanent and the males remain with their offspring and help to protect them.

Tribe Aythyini (Pochards)

This tribe of primarily fresh-water diving ducks contains 16 species that collectively may be called pochards. The tribe has a nearly worldwide distribution, but only a few of its species have ranges that extend beyond a single continent. In addition to 15 species that are very similar in body proportions and diving adaptations, the apparently extinct pink-headed duck has anatomical characteristics that approach those of the pochards and it clearly should be included in this tribe. All of the typical pochards are adept at diving and possess large feet, with long outer toes and strongly lobed hind toes. Their legs are placed quite far apart and are situated farther back on the body than in other ducks, making the birds relatively awkward on land but improving their diving efficiency. They also have a heavier body size to wing-surface ratio, forcing them to run for some distance over the water when taking flight, in contrast to the "springing" takeoff of dabbling ducks. Although sexual dimorphism occurs to some extent in all species, male plumage patterns are generally not especially complex, and the wing speculum patterns are either lacking or limited to gray or white stripes. The females of pochards are usually rather uniformly brownish, and lack the very distinct patterning of female dabbling ducks. Nesting is usually done on land near water or, more commonly, in beds of emergent vegetation. The downy young tend to be weakly or obscurely patterned with shades of yellow and dark brown.

Tribe Mergini (Sea Ducks)

The sea duck tribe, which here includes the eiders, consists of 20 species that are all superb diving birds. They are found in fresh-water as well as marine habitats, and primarily have Northern Hemisphere distribution patterns. Except for the two isolated Southern Hemisphere species, all of the sea ducks have considerable sexual dimorphism in plumage. Males in breeding plumage are usually elaborately patterned, often with predominantly black and white markings, which probably serve very well to localize and identify these birds at considerable distances in their marine environments. However, iridescent coloration is generally limited to the head, and the wing speculum patterns, instead of being iridescent, are generally white and black. Nesting is usually done on fairly open shoreline or in grassy tundra, as in the eiders and long-tailed duck, or in cavities, under heavy brush, or in similar well-concealed locations, as in the goldeneyes, mergansers, and harlequin duck. Females of the ground-nesting forms approach the patterns found among female dabbling ducks in their cryptic coloration, and the downy young of these species are also rather brownish and obscurely patterned. Females of the hole-nesting species are more uniformly brownish and their ducklings are frequently contrastingly spotted with white and dark markings.

Tribe Oxyurini (Stiff-tailed Ducks)

The stiff-tailed ducks are characterized by several unusual features in addition to their long, stiffened tail feathers that serve as underwater rudders. Among other adaptations are their very large feet, which are placed so far back on the body that it is difficult for these birds to walk on land. The body feathers are small, numerous, and have a grebelike sheen, and the wings are so short that takeoff and flight is attained with difficulty. All of the species have relatively short and thick necks, which in males can be enlarged by the inflation of various internal structures during sexual display. Although all of the species exhibit some dimorphism of size or coloration, the plumage patterns are mostly shades of gray, ruddy brown, and black, with contrasting white markings largely limited to the head. Only one species, the masked duck, exhibits a distinct speculum, and this species is furthermore the only one that is able to attain flight easily. The downy plumage patterns are usually very similar to those of the adult females, which tend to be inconspicuously patterned with grays and browns. Non-breeding plumages of males also closely resemble those of the females, and during this period the ruddy coloration of the typical forms is largely lost, as is the unusual blue color of the male's bill. Two species that do not fit this general pattern are the Australian musk duck and the South American black-headed duck. The musk duck may readily be considered a typical stifftail that, under evolutionary pressures associated with a highly competitive mating system favoring strength and aggressiveness, has become remarkably large and has evolved bizarre male displays. However, the black-headed duck lacks many of the traits of the stiff-tailed ducks, and in many anatomical features more closely approaches the dabbling ducks. In addition to providing an apparent evolutionary link with that group, the black-headed duck is also the only species of Anatidae that seems to have become a total social parasite, abandoning its nest-building tendencies in favor of dropping its eggs in the nests of a variety of host species. In this sense it is one of the most specialized of all waterfowl species, although its parasitic adaptations are relatively primitive by comparison with those of some of the better-known species of social parasites.

Tribe Anseranatini (Magpie Goose)

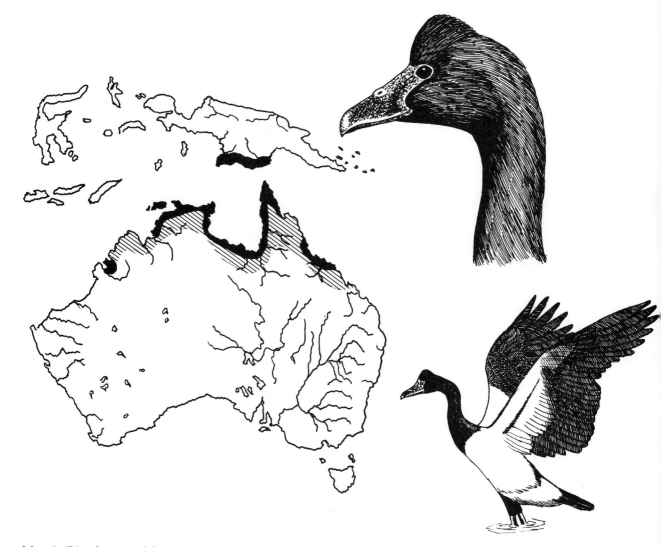

MAP 1. Distribution of the magpie goose, showing breeding (inked) and nomadic (hatched) ranges.

Magpie Goose

Anseranas semipalmata (Latham) 1798

Other vernacular names. Semipalmated goose, pied goose, black-and-white goose; Spaltfussgans (German); oie pie (French); gans overo o pintado (Spanish).

Subspecies and range. No subspecies recognized. The current breeding range includes much of northern Australia and southern New Guinea. See map 1.

Measurements and weights. (From Delacour, 1954–64, and Frith, 1967.) Folded wing: males, 368–450 mm; females, 356–418 mm. Culmen: males, 72–92 mm; females, 63–82 mm. Weights: males, 1,838–3,195 g (av. 2,766); females, 1,405–2,770 g (av. 2,071). Eggs: av. 80 x 54 mm, glossy white, 128 g.

Identification and field marks. Length 30–34" (75–85 cm). Plate 2. This is the only species of true waterfowl with only partially webbed feet. *Adults* of both sexes are black on the head, neck, wings, rump, and tail, and white elsewhere. The bill is long and straight, with a well-developed nail at the tip; as in swans, the head is featherless back to the eyes. *Females* resemble males but are somewhat smaller, lack an enlarged bony crown, and have a higher-pitched voice. Adult males have an elongated trachea that loops downward between the breast muscles and skin and can be felt through the skin; adult females and immature birds lack this

feature. *Immature* birds also have more mottled and grayish plumages than do adults. There are no seasonal variations in plumage.

In the field, this goose-sized bird can be readily recognized by its distinctive black and white plumage, its long legs, and its honking, gooselike call. In flight, the slow wing beat and rounded wing outline produce a somewhat vulturine appearance; in the wild the birds usually occur in family-sized groups or larger flocks.

NATURAL HISTORY

Habitat and foods. In Australia, magpie geese are particularly associated with the flood plains of tropical rivers within 50 miles of the coast, where several kinds of habitats occur. In permanent lagoons where the water is usually four feet deep or deeper, and where the dominant vegetation consists of lotus (*Nelumbo*), spike rush (*Eleocharis*), and water lilies (*Nymphaea*), the birds occur in great flocks, especially in the dry season, when they use these areas as refuges. In areas where the water is three to four feet deep and dries to damp soil in the dry season, tea trees (*Melaleuca*) grow and provide roosting sites. If a spike rush understory is present, the birds may also breed there. Low black-soil swamps dominated by spike rush and wild rice (*Orzya*) are primary breeding habitats, and the rice is also used for food. Higher soils that are rarely if ever flooded are covered by a variety of grasses and sedges, and are used as foraging areas by the geese when the plants are seeding (Frith, 1967).

The birds forage by bending down tall grasses with their feet to reach the seeds with the bill, by grazing, and by filtering beakfuls of mud. They also dig in the soil with their beaks to reach the bulblike roots of spike rush. Studies by Frith and Davies (1961) in the Northern Territory of Australia indicate that the birds are almost entirely vegetarians, with 70 percent of the food samples consisting of grass blades, the bulbs of spike rush, and the seeds of wild rice, wild millet (*Echinochloa*), *Paspalum*, and couch (*Cynodon*). Goslings consume the seeds of these same swamp grasses and water plantain (*Sesbania*), as well as some arthropods. The adults may bend down grasses for the young to reach the seed heads, and they may also bring up underwater food with the bill, if it is too deep for the young to reach. In captivity, adults often bring up submerged grain in this manner, after which they allow the young to feed on it as they release it from the bill (Johnsgard, 1961c).

Social behavior. Magpie geese are highly gregarious, and flocks may contain up to several thousand birds. Nesting is performed in colonies that vary greatly by locality and year but in one reported case averaged 135 nests per 100 acres in one year. The size of the nesting colony may vary from a few acres to several square miles. Although magpie geese mate for life and are always found in family groups, it is common for males to mate with two females. This has been reported to occur both in the wild (Frith, 1967) and in captivity (Johnsgard, 1961c). Kear (1973) suggests that this unique pair bonding arrangement carries the advantage of having more than two parents caring for the brood and is due to the need for rapid growth and early fledging in impermanent breeding habitats. After fledging, some flock movements occur, but Frith (1967) believes that these are usually only local ones. In some years, however, there are fairly extensive flock movements, especially in dry seasons as the geese leave their drying coastal swamps and move to inland lagoons. When extremely dry conditions occur, some geese may move as far south as southern Australia, and rarely they even reach Tasmania.

Reproductive biology. The nesting season begins with the first heavy rains of the wet season in northern Australia, which usually occur in October or November. The nest building occurs after the water depth in the swamp has become adequate. The nests consist of a large accumulation of vegetation that is gathered and trampled down by the geese. The nest has a deep cup at the top, but is not lined with down. It has been observed that in captivity nests are built off the ground by full-winged birds, sometimes in the tops of hedges or shrubs. All of the adults, usually a male and two females, help in building the nest, and several dummy nests may be built in advance of the one to be used. Usually one female begins to lay shortly in advance of the other, and each produces eggs at approximately 36-hour intervals. Wild clutches laid by a single female average 8.6 eggs, while those laid by two females averaged 9.4 eggs (Frith, 1967). A usual clutch size of 9 has been reported for captive birds (Kear, 1973). Incubation requires 28 days, and is shared by both sexes, with the male most often incubating at night (Johnsgard, 1961c). The young birds are actively fed by their parents, especially during the

first week, but occasionally feeding can be observed in goslings up to six weeks old. Fledging is said to require eleven weeks among wild birds; hand-reared birds have been reported to fledge in seven or eight weeks.

Status. Frith (1967) reports that although the number of magpie geese in the southern portions of Australia has greatly declined since early days, the species is still very numerous in the tropical areas. It can legally be hunted in the Northern Territory, but it is fully protected elsewhere from sport hunting. It has caused crop damage to rice in areas where it is most numerous, and perhaps the greatest danger to the species lies in the development of these tropical areas for grazing and agricultural purposes, which would destroy the specialized swamp vegetation on which the species depends for breeding.

Relationships. I have earlier (1961c) reviewed the numerous anatomical and behavioral characteristics of the magpie goose that distinguish it from other members of the Anatidae and suggest affiliations with the screamers (Anhimidae) and, more remotely, between the Anseriformes and the Galliformes. It is clear that subfamilial separation of the magpie goose is warranted, and that no other single species of the Anatidae provides such unique taxonomic and evolutionary interest. A recent study by Brush (1976) on the electrophoretic characteristics of feather proteins supports the position that the magpie goose is not closely related to any other waterfowl tribe and shows greater similarity to the crested screamer *(Chauna torquata)* than to other Anatidae.

Suggested readings. Frith, 1967; Johnsgard, 1961c.

4 ◈ ◈ ◈

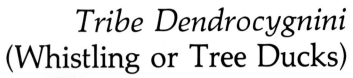

Tribe Dendrocygnini
(Whistling or Tree Ducks)

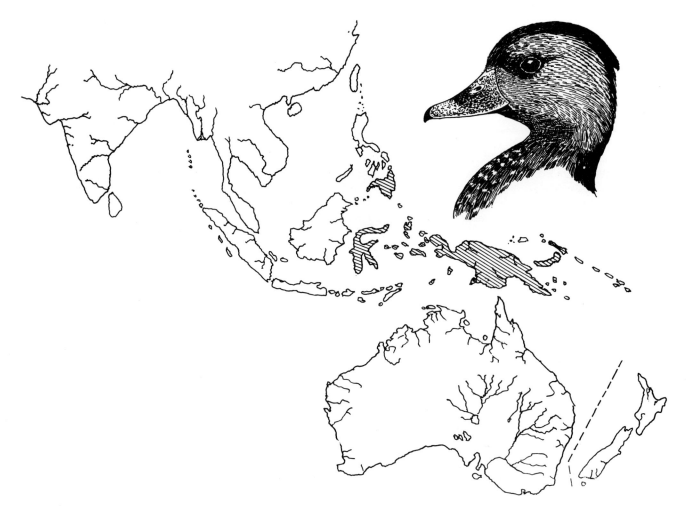

MAP 2. Residential or breeding distribution of the spotted whistling duck.

Drawing on preceding page: Spotted Whistling Duck

Spotted Whistling Duck

Dendrocygna guttata Schlegel 1866

Other vernacular names. Spotted tree duck; Tüpfel-pfeifgans (German); dendrocygne tacheté (French); pato silbador moteado (Spanish).

Subspecies and range. No subspecies recognized. Resident on Basilan, Mindanao (Philippines), Celebes, Buru, Ceram, Amboina, Tanimbar, Aru, and Kei islands, and in New Guinea and the Bismarck Archipelago (Delacour, 1954–64). See map 2.

Measurements and weights. Folded wing: adults of both sexes 212–23 mm. Culmen: adults of both sexes 41–46 mm. Weights: adults av. about 800 g (Lack, 1968). Eggs: av. 52 x 38 mm, white, 49 g.

Identification and field marks. Length 17–19″ (43–50 cm). Plate 3. This is a medium-sized whistling duck with a dark brown body, light gray face and throat, whitish belly, and fulvous flanks that have numerous rounded white spots surrounded by darker borders. The bill is reddish rather than black like that of the larger Cuban whistling duck, and the feet are tinged with red. *Females* cannot be readily distinguished from males. *Immatures* are generally duller, with flank feathers that are white, broadly edged with black, and the white spots on the sides drawn out into irregular streaks. There are no seasonal variations in plumage.

In the field, this species is likely to be confused only with the wandering whistling duck, which has an extensively overlapping range with this duck and often associates with it. The spotted whistling duck's darker body color and absence of tawny markings on the upper flanks will serve to distinguish it. The outermost primaries are deeply indented and produce a whirring sound in flight. It is one of the most silent of the whistling ducks, but some whistling calls have been reported, including a *whee'-ow* and a *whe-a-whew'-whew* (Johnsgard, 1965a).

NATURAL HISTORY

Habitat and foods. Little is known of this species in the wild, but in New Guinea it is said to inhabit marshes, lowlands, and lakes, with grassy waters its favorite habitat. It is said to feed on the seeds of various aquatic plants.

Social behavior. This duck is reportedly gregarious in the wild, with flock sizes ranging from a few up to several hundred birds. The birds regularly roost in dead trees near water at night, sometimes in flocks of hundreds. They often associate with wandering whistling ducks in New Guinea (Rand & Gilliard, 1967).

Reproductive biology. The nesting season in New Guinea is evidently long, as broods have been reported in March, a female with a formed egg has been collected in April, and nesting has also been reported in September. Although the breeding season is evidently prolonged, it is perhaps most likely to occur near the start of the wet season, during the austral spring that begins in September. In captivity the species has been bred only rarely, initially at the Wildfowl Trust. In 1959 a female spotted whistling duck laid 11 eggs in a wooden kennel (in the wild the species has been reported to nest in tree hollows), and the clutch was left for the parents to incubate. It is believed that, as in other species of whistling ducks, both sexes normally incubate. After a total of 31 days of incubation, the last 10 of which were completed in an electric incubator, 11 young hatched. This period is slightly longer than the 28–30-day incubation periods reported for other whistling ducks, and may have resulted from chilling between the time the parents left the nest on the 21st day and the onset of incubation in the electric incubator. The young were reared by a Cuban whistling duck foster mother after a few days of attempted hand-rearing, and a total of 7 ducklings were raised. Feathering of the ducklings began at 27 days and was completed in seven weeks (Johnstone, 1960).

Status. In spite of a relatively restricted range, the spotted whistling duck's status is apparently fairly secure. Rand and Gilliard (1967) reported it to be the most common and widespread duck in New Guinea, where, however, it is limited to lowland habitats.

Relationships. Although there are some superficial similarities in the adult plumage patterns of the spotted and Cuban whistling ducks, evidence from the patterning of the downy young and certain other similarities suggest that the nearest relative of this species is the plumed whistling duck (Johnsgard, 1965a).

Suggested readings. Delacour, 1954-64.

Plumed Whistling Duck

Dendrocygna eytoni (Eyton) 1838

Other vernacular names. Eyton whistling duck, plumed tree duck; grass whistle duck (Australia); Tüpfelpfeifgans (German); dendrocygne d'Eyton (French); pato silbador adornada (Spanish).

Subspecies and range. No subspecies recognized. Resident in Australia from New South Wales northward and westward to Cape York and the Fitzroy River, and in western Australia. See map 3.

Measurements and weights. (From Frith, 1967.) Folded wing: males, 222-42 mm; females, 215-45 mm. Culmen: males, 37-48 mm; females, 37-49 mm. Weight: males, 600-930 g (av. 788 g); females, 580-1,400 g (av. 792 g). Eggs: av. 48 x 37 mm, white, 40 g.

Identification and field marks. Length 16-18" (40-45 cm). Plate 4. This medium-sized whistling duck is generally light brown above, with yellow margins on the feathers, giving the bird a pale appearance, and with the abdomen, breast, foreneck, and throat pale brown to buff. The sides are chestnut brown, with vertical black barring, and the upper flank feathers are buffy yellow and greatly elongated, with black margins. The legs, feet, and bill are pink, the bill variably mottled with black. The iris is yellow rather than brown like that of other whistling ducks. *Females* cannot be outwardly distinguished from males, and *immatures* resemble adults but are paler and have narrower and less distinct barring on the sides, and their elongated flank plumes have broader blackish margins.

In the field, plumed whistling ducks can be recognized while standing or swimming in the water by their pale yellowish color and conspicuous flank plumes, which extend upward on each side to the top of the back or beyond. Both sexes utter shrill whistling notes, which vary from an extended twittering sound to a loud *wa-chew'* call. A whirring noise is also produced by the wings when in flight.

NATURAL HISTORY

Habitat and foods. This species of whistling duck is closely associated with tropical grasslands, where it forages on lagoon edges and in meadows and plains, but not in deep water. In this respect it differs considerably from the wandering whistling duck, with whose range its range extensively overlaps. It is closer in its habitat preferences to the Australian wood duck, which occurs on more temperate grasslands. Its relative independence from water also allows the bird to move well out into the arid plains, where only small pools may be found. The species is relatively nocturnal in its foraging behavior, spending the day in large roosting flocks near shorelines, and flying out in late afternoon or evening to foraging grounds that may be nearly 20 miles away. Attractive foraging areas are visited nightly until the food supply is exhausted, when the birds move to a new area. Foods consumed in the Northern Territory are almost entirely vegetation, during the wet season consisting of such grasses as wild millet (*Echinochloa*), couch (*Cynodon*), *Paspalum*, and some wild rice (*Oryza*). During the dry season the birds concentrate on foods available at the edges of marshes

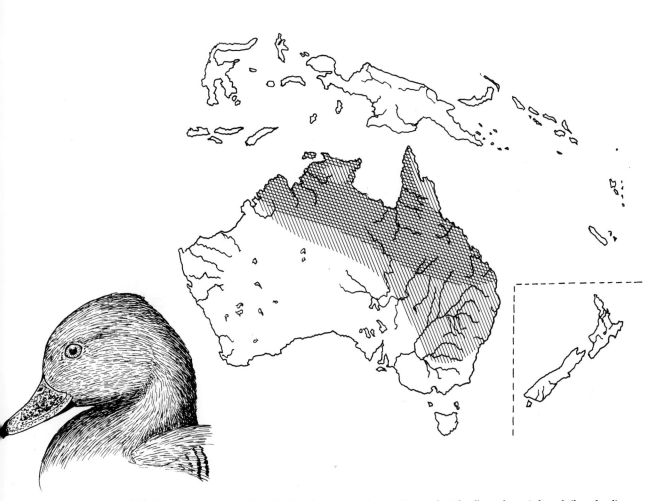

MAP 3. Distribution of the plumed whistling duck, showing primary (cross-hatched) and peripheral (hatched) breeding ranges.

and lagoons, such as spike rushes (*Eleocharis*) and sedges, according to Frith (1967). A study by H. J. Lavery in Queensland yielded similar results, with grasses being the most important food plant near the coast throughout the year, while sedges (especially *Fimbristylis*) were more important in inland locations, and particularly during the dry season.

Social behavior. Frith (1967) reports that although this species may sometimes be found in small groups, it is most often in very large flocks. As in all whistling ducks, pair bonds are strong and presumably permanent. The flocks thus consist mostly of mated pairs, and there is a good deal of aggressive behavior, especially at the start of the wet season. At that time display and associated pair-forming behavior is more frequent, and within a few weeks the flocks begin to disperse into shallow-water areas for breeding. In one study of banded birds it was found that dispersal from a dry-season concentration near Townsville

was multidirectional, with the birds moving both along the coast and inland, in a few cases to as far as the Murray River, some 1,200 miles away. Frith stated that the plumed whistling duck is a good deal more nomadic in its movements than the wandering whistling duck, even though the two species are often associated and may even breed in the same lagoons.

Reproductive biology. Plumed whistling ducks begin their breeding at the onset of the wet season; thus in Queensland and the Northern Territory nests are initiated throughout the period from February to April, but the greatest nesting activity takes place in February and March. The species may breed earlier, from August to October, in southern Australia. The nests are built on the ground, usually under grass or bush cover, and with a simple lining of vegetation, but no down. Clutch sizes reported in the wild have ranged from 8 to 14 eggs, and Johnstone (1970) reports a clutch of 10 to 12 eggs in captivity. The incubation period has been reported as 28 days by Lack (1968) and Frith (1967), and as 30 days by Johnstone (1970). After hatching, the young are led across the grasslands to water, which may be a mile or two away according to Frith. The fledging period is still unknown, but after the adult's postbreeding molt and the completion of fledging by the young, the birds begin to concentrate and move into dry-season habitats.

Status. Although no census data are available, the plumed whistling duck is obviously still an abundant species in Australia, at least in the northern portions. It is legally shot in the Northern Territory, but is not an important game bird, and few are killed. Frith suggests that the species has possibly been benefited by settlement and increased grazing, which provides an abundance of short grass cover and water tanks. There seems to be no reason for concern about its numbers.

Relationships. This species appears to be generally intermediate in its evolutionary affinities between the extreme represented by the spotted whistling duck and the central group represented by the wandering, fulvous, and lesser whistling ducks (Johnsgard, 1965a), with somewhat closer affinities to the latter than to the former.

Suggested readings. Frith, 1967.

Fulvous Whistling Duck

Dendrocygna bicolor (Vieillot) 1816

Other vernacular names. Fulvous tree duck, Sichelpfeifgans (German), dendrocygne à bec fauve (French), pato silbon and pichici colorado (Spanish).

Subspecies and range. No subspecies recognized. Currently breeds in southern Florida and southern Texas and Louisiana southward to Oaxaca and Tabasco, recently reaching Cuba and the Greater Antilles. In tropical South America it breeds from Colombia to the Guianas, and from Brazil southwest to Tucumán and southeast to Buenos Aires province. Breeds also in Africa south of the Sahara from Senegal and Ethiopia to Lake Ngami and Natal. Also resident on Madagascar, in India, and in Burma. See map 4.

Measurements and weights. Folded wing: males, 214–25 mm; females, 203–11 mm. Culmen: males, 45–48 mm; females, 44–47 mm. Weights: males, 621–755 g (av. 675 g); females, 631–739 g (av. 690 g) (J. Lynch, pers. comm.). Eggs: av. 53 x 38 mm, white, 50 g.

Identification and field marks. Length 18–21" (45–53 cm). This is a medium-sized whistling duck with a dark brown back, feathers which are tipped with tawny coloration, the face, breast, abdomen, and flanks all cinnamon, with no spotting or streaking except on the neck, which is buffy white streaked with dark brown. The upper flank feathers are elongated and creamy white, edged on the outer vane with blackish brown. The bill is blackish, and the feet and legs are bluish gray. *Females* differ from the males only in being slightly duller and smaller, and *juveniles* cannot easily be distinguished from adults, but the edges of their back feathers may average slightly darker. The upper wing coverts of young birds have little chestnut color, and their upper tail coverts are narrowly margined with brown.

In the field, fulvous whistling ducks might perhaps most readily be confused with wandering whistling ducks, but the ranges of these species do not overlap. In India and southeast Asia the fulvous and lesser whistling ducks occur together, and the smaller size and more grayish back of the latter species will serve to separate them. In flight, fulvous whistling ducks have the characteristic slow wingbeat and dangling-legs appearance typical of all members of this group, and their loud, whistled *wa-chew'* notes are commonly uttered in flight. In

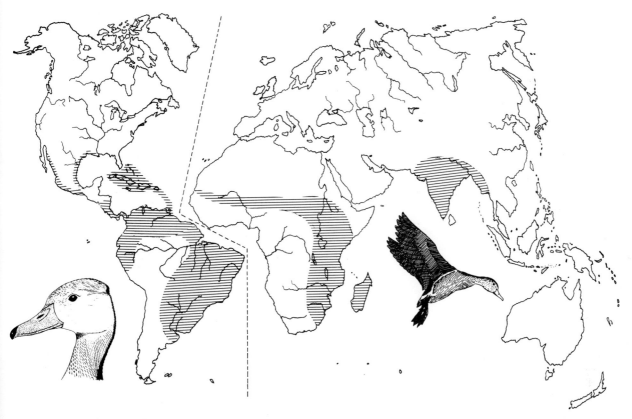

MAP 4. Distribution of the fulvous whistling duck.

Africa and South America the species often associates with the white-faced whistling duck, and in tropical America it may also be seen with the black-bellied whistling duck; the white markings on the face of the former and on the upper flanks of the latter will serve to separate them, as will the melodic and multisyllabic calls of these species.

NATURAL HISTORY

Habitat and foods. Clancey (1967) describes the habitat of this species in southern Africa as including fresh-water lakes, vleis, marshes, and swamps, including papyrus swamps, and the open portions of slowly flowing streams. Favored waters are those with a rich shoreline vegetation, banks of reeds, and floating plants such as water lilies. In the United States, nesting has occurred on fresh-water marshes where bulrushes or cattails grow interruptedly, and especially in weed-infested rice fields. The relationship between rice and the distribution of this species is quite clear in both Texas and Louisiana. Various studies in the United States have indicated that grass seeds of such types as wild millet (*Echinochloa*) and wild timothy (*Phleum*) are important foods, as are the seeds of broad-leaved herbs such as smartweeds (*Polygonum*) and sweet clover (*Melilotus*) (Johnsgard, 1975). Rice is apparently a major food in India, but the birds there have also been reported to eat various aquatic seeds, bulbs, leaf shoots, buds, grass, and rushes. There are few records of animal foods in the diet. The skull and bill structures of this species suggest that it feeds primarily by swimming and diving, as does the wandering whistling duck, whereas the black-bellied and plumed whistling ducks are structurally adapted for grazing (Rylander & Bolen, 1974). The birds dive surprisingly well, often remaining under water for up to 15 seconds.

Social behavior. Like all whistling ducks, this species is highly social, but it usually is not to be found in extremely large flocks. Flocks consist of mated pairs and families, and early studies in California indicated that social nesting was common there, with as many as fifty nests in an area about half a mile long by two hundred yards wide. More recent studies in Louisiana indicate much sparser nesting concentrations. During the non-breeding season these birds often associate with other whistling duck species and leave their roosting areas at dusk to feed. In many areas the nesting season is greatly extended, so that no single period of courtship appears evident. Courtship in this species, as in other whistling ducks, is prolonged and inconspicuous, and the only obvious aspect of pair formation is actual copulatory behavior. This occurs in water of swimming depth, with the birds facing each other and performing mutual head-dipping movements very much like normal bathing movements. The male then suddenly mounts the female, and after a brief period slips off to one side. The two birds then quickly rise in the water while treading water vigorously, and call while lifting the folded wing on the side opposite the partner. After treading, they bathe and preen extensively, often on shore (Johnsgard, 1965a). In the cooler parts of its range, as in the United States, movements out of the breeding area occur during winter, but most migrations are relatively short. In East Africa the birds also move out of the lakes in June, to return again in August, presumably after breeding.

Reproductive biology. The timing of breeding in this species is remarkably variable throughout its large range. In the United States it is a summer breeder, as it is in Argentina, where it nests between November and February. In India it nests from June to October, but mainly during July and August, while the nesting dates from Africa span nearly all of the months of the year. In southern Africa most of the nesting records are for the period between February and August. Birds in breeding condition have been reported in Senegal in January, and breeding on Madagascar has been reported for November, December, and April.

Although in most parts of its range, as in the United States and Argentina, nesting has been reported to occur in reed beds, rice fields, or beds of flag, nests have also been reported in tree crotches or natural hollows in a few areas, especially in India. Such tree locations are usually on small islands or overhanging water, and it seems much more characteristic of fulvous whistling ducks to build their nests in mats of aquatic vegetation that are trodden down to form a platform

above the water. Such nests are also usually concealed from above by overhanging grasses, and are lacking in down. The normal clutch size is from 8 to 16 eggs, averaging about 10, but multiple clutches, or dump nesting, are not infrequent, resulting in clutches of more than 20 eggs. Incubation is performed by both sexes, perhaps predominantly by the male, and the reported incubation periods range from 24 (Meanley & Meanley, 1959) or 26 (Johnstone, 1970) to 28 (Lack, 1968) days. Longer periods, as reported by Delacour (1954–64) are apparently not characteristic. The ducklings are guarded by both parents, and a fledging period of 63 days has been reported (Meanley & Meanley, 1959). In spite of a long nesting season in some areas, the birds are evidently single-brooded (Clancey, 1967).

Status. This species is still relatively abundant over most of its original range, which has retracted in some areas (California and Trinidad), but expanded in others (Cuba and the Greater Antilles). The species is hunted to some extent and is easily killed, but it is not a major sporting species anywhere. In some areas, such as Louisiana, it has been seriously affected by the coating of rice seeds with pesticides, but in general it does not appear to have suffered measurable inroads from man's activities.

Relationships. The fulvous whistling duck might well be regarded as the core of the genus *Dendrocygna*, with its nearest relatives being the wandering and lesser whistling ducks (Johnsgard, 1965a).

Suggested readings. Clancey, 1967; Johnsgard, 1975; McCartney, 1963.

Wandering Whistling Duck

Dendrocygna arcuata (Horsfield) 1837

Other vernacular names. Wandering tree duck, whistling teal, black-spotted tree duck; water whistle-duck (Australia); Wanderpfeifgans (German); dendrocygne à lunules (French); pato silbador errante (Spanish).

are virtually identical in appearance to males. *Juveniles* lack the broad pale edges on their back feathers, they are less reddish on the underparts, and there is little chestnut coloration on their lesser wing coverts and none on their median coverts, as is typical of adults.

In the field, wandering whistling ducks might most likely be confused with lesser whistling ducks; their range otherwise overlaps only with that of the spotted and plumed whistling ducks. Besides being larger than lesser whistling ducks, they have more conspicuous buff markings on the upper flanks, have whitish rather than chestnut upper tail coverts, and are darker on the back and above the eyes. The most frequently heard call is a rapid, descending whistle of five to seven notes, which is often uttered in flight, and the wings also produce a distinctive whistling noise.

NATURAL HISTORY

Habitat and foods. Frith (1967) reports that in Australia the favored habitats of this species consist of permanent water areas where aquatic foods are richest and can be obtained by diving. The birds are usually to be found in lagoons, billabongs, or flooded meadows, or sometimes on the edges of creeks and rivers. They dive into water up to ten feet deep, and also dabble along the water's edge, stripping seeds from plants within reach. Birds collected in the Northern Territory were found to have consumed primarily water lilies, including their seeds, buds, and leaves, and a variety of grasses typical of lagoons, such as *Paspalum*, rice (*Oryza*), couch (*Cynodon*), and wild millet (*Echinochloa*). Sedges were next most important, followed by a variety of other plants associated with water. Among birds collected in Queensland, grasses and various members of the gentian family were the two most important groups of food plants, followed by water lilies and sedges (Frith, 1967).

Social behavior. Like other members of this genus, wandering whistling ducks are highly gregarious, and their flocks consist of permanently mated pairs and families. During the nonbreeding dry season, flocks of one or two thousand birds may often be found on permanent water areas of the Northern Territory, especially in deep and permanent lagoons. Unlike the grazing-adapted plumed whis-

Subspecies and ranges. (See map 5.)
D. a. arcuata: East Indian wandering whistling duck. Resident in the Philippines, Borneo, Java, Bali, Sumba, Celebes, Timor, Roti.
D. a. australis: Australian wandering whistling duck. Resident in southern New Guinea and northern Australia.
D. a. pygmaea: Lesser wandering whistling duck. Resident on New Britain, and possibly also the Fiji Islands.

Measurements and weights. Folded wing: both sexes 173–222 mm. Culmen: both sexes 40–48 mm. Weights (for *australis*): males, 866–948 g (av. 741 g); females, 453–986 g (av. 732 g) (Frith, 1967). Eggs: av. 51 x 35 mm, white, 40 g.

Identification and field marks. Length 16–18″ (40–45 cm). This medium sized whistling duck has a dark brown mantle, with lighter brown edging on the back feathers, chestnut flanks and abdomen, conspicuous buffy yellow upper flank feathers with black and brown edges, and a brownish breast that is spotted with black markings. Unlike that of the fulvous whistling duck, its head is nearly black above the eyes, the neck streaking is lacking, and the bill is blackish rather than bluish gray. *Females*

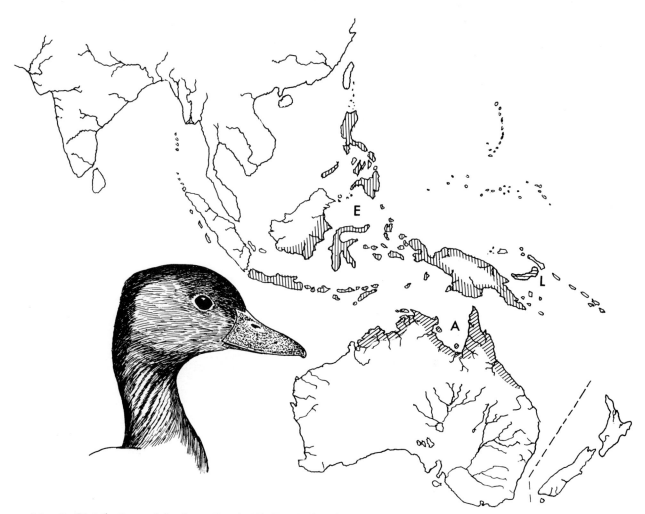

MAP 5. Distributions of the Australian ("A"), East Indian ("E"), and lesser ("L") wandering whistling ducks.

tling duck, this species does much foraging during the day, flocks frequently foraging as a group. As the wet season begins, these flocks break up and disperse to suitable nesting habitats provided by the filling swamps.

Reproductive biology. Because of the timing of breeding to coincide with the onset of the rainy season, most nesting in northern Australia occurs between December and April in Queensland, and between December and May in New Guinea (Frith, 1967). Records of breeding in the Philippines range from January to May, and there is a record of a duckling obtained in July in the Celebes. The nest is apparently always placed on the ground, either in grassy cover or under the protection of a bush. It may be placed quite close to water or some dis-

tance from it. Courtship and copulatory behavior appear to be identical with that of the fulvous whistling duck. Clutch-size information is scanty, but in general the clutch size has been reported to vary from 6 or 8 to about 15 eggs. Incubation periods have likewise been estimated at from 28 (Johnstone, 1970) to 30 (Lack, 1968) days. No doubt both sexes assist in incubation, as in other whistling ducks, and no down is present in the nest, presumably since it is never left unattended by at least one parent. Judging from ducklings that have been raised in captivity, the growth rate is fairly slow, with full feathering not occurring until the 10th week of life, and fledging at 12 to 13 weeks of age (Frith, 1967). Whether such a long prefledging period is also characteristic of wild birds remains to be determined, but ecological

14 ❖ ❖ ❖

pressures for rapid completion of breeding are not nearly so strong in this bird of permanent water areas as is with the more nomadic and opportunistically breeding species of Australian waterfowl.

Status. In Australia, wandering whistling ducks are reportedly still very abundant in tropical areas, and are not seriously affected by hunting or by agricultural practices. They are apparently also still quite abundant in the Philippine Islands. Doubtless the population in greatest jeopardy is the one that is apparently confined to New Britain Island; its status is unknown.

Relationships. The close anatomical and behavioral similarities of the wandering and fulvous whistling ducks, and their nonoverlapping distributions, indicate that these species are very closely related and should be considered a superspecies (Johnsgard, 1965a).

Suggested readings. Frith, 1967.

Lesser Whistling Duck

Dendrocygna javanica (Horsfield) 1821

Other vernacular names. Lesser whistling teal, lesser tree duck, Indian whistling duck; Zwergpfeifgans (German); dendrocygne de l'Inde (French); pato silbador de la India (Spanish).

Subspecies and range. No subspecies recognized. Resident from Pakistan and India eastward to the coast of southern China, Formosa, the Ryukyu Islands, Ceylon, the Andaman and Nicobar islands, the Malay Peninsula, Hainan, Indochina, the western half of Borneo, Sumatra, and Java. See map 6.

Measurements and weights. Folded wing: both sexes 170–204 mm. Culmen: both sexes 38–42 mm. Weights: about 450–600 g (Ali & Ripley, 1968). Eggs: av. 47 x 38 mm, white, 35 g.

Identification and field marks. Length 15–16" (38–40 cm). Plate 5. This smallest of the whistling ducks is generally pale brown and chestnut-colored, with chestnut upper tail coverts, light rufous-cinnamon underparts that lack breast spotting, a

pale buff face with a narrow yellow eye-ring, inconspicuous whitish flank feathers, and dark gray bill, leg, and foot coloration. *Females* are not distinguishable from males, and *juveniles* are duller-colored, with the back and scapular feathers margined with dingy fulvous instead of golden rufous coloration. The underparts are also a pale dull fulvous brown.

In the field, the small body size and chestnut rather than whitish upper tail coverts will serve to separate this species from the fulvous and wandering whistling ducks, with which it often associates. In flight a whistled *whi-wheee'* (or *"sea-sick"*) is uttered almost constantly, and the modified inner vane of the bird's outermost primaries is adapted for producing wing noise.

NATURAL HISTORY

Habitat and foods. In India and Pakistan this species is said to inhabit reedy and vegetation-covered tanks and jheels in plains country, or essentially the same habitats as used by the fulvous whistling duck, with which it often associates. In Ceylon it likewise favors lotus-covered tanks and swamps, especially if they are surrounded by jungle, but it avoids coastal lagoons and brackish estuaries. The presence of trees in or around the water is apparently important for roosting; treeless

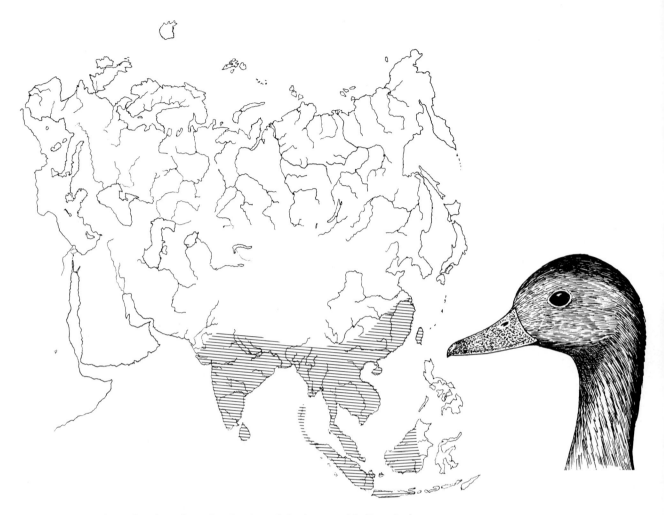

Map 6. Residential or breeding distribution of the lesser whistling duck.

waters that otherwise seem suitable are evidently avoided. Habitats favored are much the same as those used by cotton pygmy geese, and the species often associate. Wherever it is available, rice is the favored food, and the birds make dusk flights into rice paddies, as do fulvous whistling ducks. Young rice shoots are sometimes also grazed, as are some grasses in certain seasons. Reportedly this species eats some animal foods, most probably fresh-water snails, that makes its flesh less desirable for human consumption than that of other whistling ducks.

Social behavior. This species is highly social, with flock sizes that may number in the dozens but have also been reported to reach the thousands in India and Ceylon. One report in the early part of the twentieth century estimated flocks of hundreds of thousands. No doubt all flocks consist of pairs and families, as in other whistling ducks. A good deal of local movement is performed by these flocks as dry-season changes in water supplies dictate, and in some parts of India the bird is a local migrant. The breeding season is fairly prolonged, especially in southern parts of its range, and no specific period of pair formation has been reported. As in other whistling ducks and in geese, pair formation is extremely inconspicuous and probably a very prolonged process. Again as in most whistling ducks, copulatory behavior occurs in water in swimming depth, with the two birds close together and the male performing bill-dipping movements toward the female. In the few cases observed, only the male performed this display, while the female remained in an erect posture. The male then suddenly mounted the female and, after treading was completed, performed a vigorous step-dance, rising beside the female and raising the wing opposite her while treading water. The female performed the same display but with less pronounced wing lifting (Johnsgard, 1965a). Except for the Cuban and red-billed whistling duck, much the same behavior is characteristic of all the members of this genus.

Reproductive biology. In most of its range, the lesser whistling duck breeds during the rainy season. In India this activity is performed mainly during July and August but in rare instances occurs from mid-June to mid-October. Likewise in Pegu district of Lower Burma, nesting has been reported for July and August, and on the Andaman Islands during August and September. In Ceylon probably most breeding takes place during July and August, but breeding has also been reported during the drier months of December and January. On the very wet Malay Peninsula breeding is much more extended, lasting from August or September to January or February.

The nest of this species is apparently most often situated above the ground—in the natural hollow of a large tree or the fork of tree branches—the bird utilizes the old nests of kites, herons, or crows. Sometimes nests are found on the ground, among reeds and shrubs near a tank or jheel, in which case they consist of mounded herbaceous vegetation. Down is never present in the nest. Seven to 12 eggs are usually reported, rarely as many as 17, with 8 to 10 the most common number. The incubation period is probably 27 or 28 days (Lack, 1968; Johnstone, 1970), although Delacour (1954–64) has suggested it may be as long as 30 days and earlier estimates have been as short as 22 to 24 days. Both sexes are believed to incubate, and the male remains to help rear the brood. The fledging period is as yet unreported.

Status. This species has historically been one of the most common ducks in India and Ceylon and apparently it is still relatively common. Although it is hunted somewhat in Ceylon, it is not regarded in India as an important sporting bird because of its reportedly ill-tasting flesh and probably also its small size. Because it destroys rice, it may be locally hunted. However, there seems to be no current basis for concern about the species' status.

Relationships. Although the bird's behavior and plumage suggest affinities to the wandering as well as the fulvous whistling duck, details of its tracheal anatomy and its geographic distribution suggest somewhat closer affinities with the former species.

Suggested readings. Delacour, 1954–64.

White-faced Whistling Duck

Dendrocygna viduata (Linnaeus) 1766

Other vernacular names. White-faced tree duck; Witwenpfeifgans (German); dendrocygne à face blanche (French); pato silbador cara blanca or suiriri (Spanish).

Subspecies and range. No subspecies recognized. Resident and local migrant in tropical South America south to the Argentine Chaco, Paraguay, and Uruguay, and in Panama and Costa Rica. Also resident in Africa south of the Sahara south to Natal, and on Madagascar and the Comoro Islands. See map 7.

Measurements and weights. (Measurements from Clancey, 1967.) Folded wing: both sexes 219–40 mm. Culmen: both sexes 46–53 mm. Weights: males, 637–735 g (av. 686 g); females, 502–820 g (av. 662 g). Eggs: av. 47 x 37 mm, white, 36 g.

Identification and field marks. Length 17–19″ (43–48 cm). Plate 1. This is the only species of whistling duck that has white on the front half of its head and throat, and black behind. The breast and upper back are chestnut, while the flanks and underparts are transversely barred with black and white. The back feathers are elongated and dark brown with buff edging, producing a linear striping effect rather than crosswise barring as in other whistling ducks. The tail and tail coverts are blackish, the feet are bluish gray, and the bill is black with a lighter subterminal band. *Females* are identical in appearance to males, and *juveniles* are grayish buffy white on the face and front of the throat, while the back of the head is brownish black. The chestnut breast coloration is less rich in young birds, and also is less extensive.

In the field, the white face markings of this species are the best field mark. The most typical in-flight call is a three-note *wee-a-whew* (or *"su-ri-ri,"* its Spanish name), while the fulvous whistling duck

has a two-note call and the black-bellied a five- to seven-note call. In flight, it does not exhibit the buff upper tail coverts of the fulvous or the white wing markings of the black-bellied duck.

Natural History

Habitat and foods. In South Africa, this species' habitats are diverse and consist of fresh-water lakes, vleis, dams, reservoirs, marshes, swamps, and pans on flood plains, and, locally, sewage farms. Areas with mud or sandbars for roosting are preferred. Evidently it shows no tendency to inhabit wooded areas, and is sometimes found in large open waters. However, it is also frequently found on small water areas with rich shoreline vegetation, or on lakes with wide swamp margins. In South America the species likewise occupies a wide diversity of habitats, including both forested and nonforested areas, and fresh or brackish waters. However, the birds do not perch, and relatively open-country habitats seem to be favored over forested ones, and fresh-water areas over salt or brackish ones. Their foods are likewise diverse,

and are commonly obtained by diving. They evidently include invertebrates such as aquatic insects, mollusks, and crustaceans, as well as vegetable material such as aquatic seeds and sometimes rice. In Africa the birds often wade and dabble in shallow waters on grass and aquatic plant seeds and on aquatic invertebrates (Clancey, 1967).

Social behavior. This species is highly social, with flock sizes often numbering in the hundreds in favored habitats. During daylight hours of the non-breeding season the birds roost on banks, spending much time preening themselves and each other; mutual preening is highly developed in this species. Foraging is done primarily at night, so that there is a good deal of nocturnal flying. Nomadic movements are prevalent in some parts of Africa, as in Zambia, and seasonally there are large numbers on large lakes such as Lake Rudolf and the lakes of southern Nyasaland (Malawi). Where the two species exist together, white-faced and fulvous whistling ducks are often found in association. Pairforming behavior is inconspicuous and probably greatly prolonged, in conjunction with the permanent pair bonds and extended breeding season.

Map 7. Residential or breeding distribution of the white-faced whistling duck.

Mutual preening is probably an important part of pair-forming and pair-maintaining behavior. Copulation occurs in water of swimming depth, and in three instances observed by the author it was always preceded by bill-dipping and cheek-rolling movements by the male, while the female occasionally bill-dipped. Upending in the water by both birds was seen once. In all three instances copulation was followed by a typical step-dance with wing raising by both birds.

Reproductive biology. The time of breeding in South America is relatively poorly known, but considerable information is available for Africa. At the southern end of its range the species breeds from October through March. In Zambia the largest number of breeding records are for January and February, but they extend from December through June. Similar breeding times are reported for the southern Congo, but as one moves north to Kenya, the Sudan, and Chad, most breeding records are for the period July through August, and Senegalese records are for September and October. In Madagascar it is reported to nest from September to November and again in January and February. The nest site is a depression in dry ground or in reed beds over water. Few or no feathers are present in the nest. The usual clutch size is from 6 to 12 eggs, with an average of about 10. Incubation, which is performed by both sexes, requires 26 (Lack, 1968) to 28 (Johnstone, 1970) days. Both sexes tend the brood, and the ducklings are often kept hidden among water lilies and reeds (Clancey, 1967). The period to fledging has been estimated to be two months (Clark, 1976).

Status. The species is probably not in danger in any part of its range. It is extremely common in Africa, especially along the coast and on the larger lakes. Its broad ecological tolerances probably assure its continued abundance for the foreseeable future.

Relationships. In many respects, but particularly in its plumage patterns, the white-faced whistling duck appears to be relatively isolated from the other species of *Dendrocygna*. Probably its nearest, if not extremely close, relative is the black-bellied whistling duck (Johnsgard, 1965a).

Suggested readings. Clancey, 1967.

Cuban Whistling Duck

Dendrocygna arborea (Linnaeus) 1758

Other vernacular names. Black-billed whistling duck, Antillean tree duck, West Indian tree duck; Kubapfeifgans (German); dendrocygne à bec noir (French); pato silbador pico negro (Spanish).

Subspecies and range. No subspecies recognized. Resident in the Bahama Islands, the Greater Antilles, and the northern Lesser Antilles. See map 8.

Measurements and weights. Folded wing: both sexes 230–70 mm. Culmen: both sexes 45–53 mm. Weights: adult females av. 1,150 g (Lack, 1968). Eggs: av. 55 x 40 mm, white, 65 g.

Identification and field marks. Length 19–23" (48–58cm). This is the largest of the whistling ducks, and the only one with wing measurements in excess of 250 mm. The adults are dark umber brown, with a dark brown face and brown neck streaks that grade into a whitish foreneck and throat. The breast is a mottled brown, grading on the flanks into irregularly patterned black and white spotting and striping, the upper flank feathers tending toward stripes and the lower ones toward variably connected white spots. Spotting also occurs on the under tail coverts; the tail and upper tail coverts are blackish. The bill is black and the legs and feet are dark gray. *Females* do not differ noticeably from males, and *juveniles* are duller, with less distinctive spotting.

In the field, this species' restricted range makes it likely to be confused only with the fulvous and black-bellied whistling ducks. It is larger and more uniformly darker than either of those species, although its multisyllabic flight call of about five notes resembles that of the black-bellied duck. In flight, the Cuban whistling duck exhibits a pale gray upper wing area that corresponds to the white area exhibited by the black-bellied whistling duck.

NATURAL HISTORY

Habitat and foods. Cuban whistling ducks are often associated with forested swamps such as mangrove swamps, where they spend much of the daylight hours. They seem to be relatively terrestrial birds, at least in captivity, and spend a good deal of time on dry land. They also perch very well, and frequently can be seen perching in palm trees. Diving for food does not appear to have been reported. Most reports indicate that their favorite food is the fruit of the royal palm (*Roystonia*), and they regularly make dusk flights from their roosting areas to stands of these trees. They have also been reported to consume rice, sorghum grain in the milk stage, grass seeds, berries, and small fruit. Few instances of feeding on animal materials have been reported (Phillips, 1922–26).

Social behavior. Very little specific information on social behavior is available. The birds are evidently found in flocks throughout the year, and doubtless pair bonds are permanent in this species. Owing to the species' insular distribution, movements must be of a local nature only. Pair-forming behavior has not been specifically noted, and the copulatory behavior of this species differs from that of all other whistling ducks with the exception of the black-bellied in that it occurs while the birds are standing on ground or at the edge of water. Both sexes repeated making drinking movements, after which treading occurs. Afterward, both sexes call loudly with bills upraised and feathers ruffled, and the male parades rather stiffly around the female (Johnsgard, 1965a).

Reproductive biology. The period of breeding of this species is not yet well documented. Delacour (1954–64) indicates that in general the birds breed from June to October, and possibly later, while in Puerto Rico breeding reportedly occurs from October to December. Nesting sites are evidently very variable. They have been reported in palm trees, in tree branches or tree cavities, in clusters of bromeliads, in dry places among bushes, and among the roots of overturned trees in swamps. It would seem that the species is quite flexible in its range of acceptable nest sites. The clutch size is reportedly from 6 to 10 (Johnstone, 1970) or 8 to 12 (Delacour, 1954–64) eggs, or typically about 11 (Lack, 1968). The incubation period is generally reported to be 30 days and incubation no doubt is performed by both sexes. The fledging period is still unreported.

Status. Doubtless the population of this species has declined seriously in recent decades, particularly since the introduction of the mongoose into several islands where the birds used to be common, as on Jamaica, Puerto Rico, and Cuba. No estimates of the population are available, but it is clear that this species may eventually reach a threatened status.

Relationships. In spite of this species' distinctive adult plumage pattern, with some slight similarities to that of the spotted whistling ducks, its downy plumage and particularly its copulatory behavior pattern strongly indicate a close relationship to the black-bellied whistling duck (Johnsgard, 1965a; Bolen & Rylander, 1973).

Suggested readings. Phillips, 1922–26.

Black-bellied Whistling Duck

Dendrocygna autumnalis (Linnaeus) 1758

Other vernacular names. Red-billed whistling duck, black-bellied tree duck, gray-breasted tree duck; Herbstpfeifgans (German); dendrocygne à bec rouge (French); pato silbador pico rojo or pichichi comun (Spanish).

Subspecies and ranges. (See map 8.)
  D. a. autumnalis: Northern black-bellied whistling duck. Breeds from southeastern Texas, Sinaloa, and Nuevo León south to the Canal Zone. Migratory at northern end of range, otherwise resident.

D. a. discolor: Southern black-bellied whistling duck. Breeds from eastern Panama and northern South America southward to Guayaquil on the west and northern Argentina on the east. Also resident in Trinidad and occasional in some of the southern Lesser Antilles, Virgin Islands, and Puerto Rico.

Measurements and weights. Folded wing: males, 233–48 mm; females, 229–47 mm. Culmen: males, 49–56 mm; females, 51–56 mm. Weights: males, 680–907 g (av. 816); females, 650–1,020 g (av. 839) (Bolen, 1964). Eggs: av. 50 x 39 mm, white, 44 g.

Identification and field marks. Length 19–21" (48–53 cm). Plate 6. This is the only species of whistling duck with a bright pink bill, which grades to gray at the nail and to yellow behind the nostrils. The tail, upper tail coverts, flanks, and underparts are black, the breast and lower neck are chestnut to grayish (the latter in *discolor* only), and the back feathers are medium chestnut with tawny edging. The face is grayish, with a white eye-ring, and the uncrested crown is brownish. Unlike that of other whistling ducks, the upper wing surface is extensively white, including most of the coverts and the basal portions of the primaries. The feet and legs are flesh-colored. *Females* are not distinguishable from males, but *juveniles* have grayish feet and bills, and are generally duller and more obscurely patterned than adults, with bellies that are grayish white and marked with cross-barring. *In the field,* the strong contrast between the white upper wing surface and the black sides is evident and appears as a white stripe extending from near the tail to the bend of the wing. In flight the flashing white upper side of the wing contrasts strongly with the blackish underside, and the distinctive five- to seven-note call is uttered very frequently. The Spanish name "pichichi" is indicative of the usual flight call, which sounds like *wha-chew', whe-whe-whew.*

NATURAL HISTORY

Habitat and foods. The black-bellied whistling duck perches to a greater extent than does the fulvous, and thus is rarely found very far from trees. Tropical lagoons are a favorite habitat. In Mexico it breeds primarily along the tropical coasts, but sometimes also in the temperate uplands. It is

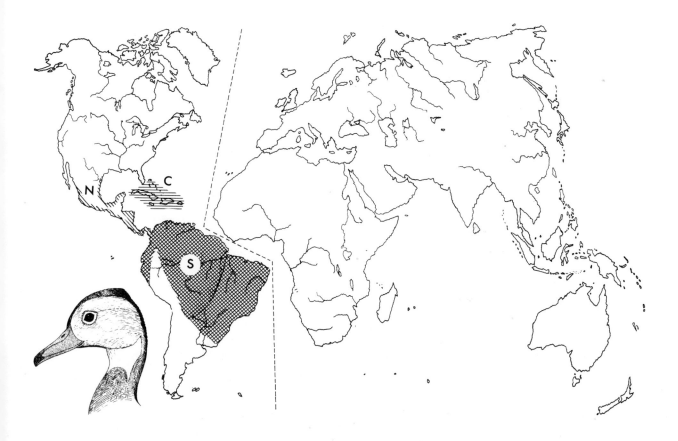

MAP 8. Breeding distributions of the northern ("N"), southern ("S"), and Cuban ("C") whistling ducks.

sometimes attracted to areas where corn is raised or rice culture occurs, and foraging flights to grassy pastures are also common. The black-bellied whistling duck walks in a more erect stance than does the fulvous, in association with its more terrestrial way of life. Likewise, its bill structure has been modified for terrestrial grazing in a number of respects, while that of the fulvous duck remains better adapted for sieve-foraging in water. (Rylander & Bolen, 1970, 1974). Studies in North America indicate that about 90 percent of the species' food is plant material, with sorghum grain and Bermuda grass *(Cynodon)* predominant items (Bolen & Forsyth, 1967). The intake of animal foods is very limited and confined largely to mollusks and insects. Deep waters are avoided, and areas with mud flats or sandbars for feeding and loafing are preferred.

Social behavior. This highly gregarious species often occurs in flocks of up to several thousand birds where it is common, and flocking is typical most of the year. The pair bonds are permanent, and banding studies have confirmed that the same birds are often paired in successive years (Bolen, 1971). In conjunction with the avoidance of deep water, copulation occurs in very shallow water or while the birds are standing at the shoreline. One or both members of the pair perform drinking movements before treading, and it is followed by mutual calling and partial wing lifting by the male. Family and flock movements are not well documented, but at least at the northern end of their range in Texas these birds are migratory, moving an unknown distance into Mexico for the winter. Even as far south as Guatemala there appears to be an influx of birds from Mexico during the colder months.

Reproductive biology. The nesting period for this species in Texas is from June through July, as it presumably also is in Mexico. There are no definite nesting records from Panama, but a laying female was collected in June. Nesting in Surinam occurs in

July and August, with ducklings seen as late as December, and nesting in Trinidad has been reported in November. Nesting in Brazil is probably more prolonged than the few records (July, September) of breeding or birds in breeding condition would indicate. The birds prefer to nest in cavities; of 20 nests studied in Texas, 17 were in trees and only 3 on the ground. Most of them were situated very close to water, and in the case of tree nests the cover below was usually herbaceous and a convenient perch was present. Ground nests consist of shallow baskets of woven grasses (Bolen et al., 1964). Clutch sizes are very variable as a result of dump-nesting, but usually range from about 12 to 16 eggs. Incubation is performed by both sexes, and estimates of its duration range from 26 to 31 days, with the longer periods probably more typical.

Status. The broad geographic range and association of this species with tropical lowland habitats that have not as yet been seriously affected by human activities make its status reasonably secure. It is not an important game species, but does cause local damage to corn or rice crops, which may result in control measures.

Relationships. The unusual adult plumage patterns and the distinctively brightly patterned downy young of this species set it well apart from nearly all other whistling ducks and most cause it to be regarded as an extreme variant in the genus *Dendrocygna* (Johnsgard, 1965a). Its nearest living relative is quite clearly the Cuban whistling duck (Johnsgard, 1965a; Bolen, 1973).

Suggesting readings. Johnsgard, 1975; Bolen, 1967.

White-backed Duck

Thalassornis leuconotus Eyton 1838

Other vernacular names. None in general English use. Weissrückenente (German); canard à dos blanc (French); pato lomo blanco (Spanish).

Subspecies and Range. (See map 9.)

T. l. leuconotus: African white-backed duck. Resident from eastern Nigeria and southern Ethiopia southward to the Cape, except for the Congo Basin.

T. l. insularis: Madagascan white-backed duck. Resident on Madagascar to elevations of 2,500 feet.

Measurements and weights. Folded wing: males 160–80 mm; females, 150–60 mm. Culmen: males, 38–45 mm; females, 35–42 mm. Weights: males, 650–790 g; females, 625–765 g. Eggs: Av. 68 x 48 mm, rich brown, 81 g.

Identification and field marks. Length 15–16″ (38–40 cm). Plate 7. This is an unusually short and thick-set duck, with legs and feet gray and placed well to the rear, and with a reticulated scale pattern as in *Dendrocygna*, rather than scutellated as in typical ducks. The bill is blackish, with yellow at the edges and below, and a large nail. Except for the lower back and upper tail coverts, which are broadly tipped with white, the head and body are predominantly shades of buff, tawny, and darker brown, all intermixed with extensive barring and spotting. There is a nearly white mark on the face behind the mandible (upper lores), otherwise the face and neck are buff, heavily spotted above with black. *Females* closely resemble males, and *immatures* are generally darker and less distinctively patterned, with the sides of the face and neck more heavily spotted with black.

In the field, the species' stubby shape somewhat resembles that of a stiff-tailed duck, but the neck is thinner and the tail is shorter. Like stifftails, it flies rarely, and thus its white back and upper tail coverts are usually not visible. In the air it often calls in the distinctive manner of whistling ducks, and on the water it sometimes also utters a clear whistling note. It dives well and forages in this manner.

MAP 9. Breeding distributions of the Madagascan ("M") and African ("A") white-backed ducks.

When taking off, it patters along the water some distance, in the manner of coots.

NATURAL HISTORY

Habitat and foods. The favored habitats of this species are quiet lagoons, lakes, or flood plains where there is an abundance of floating vegetation such as water lilies. Open waters are avoided, and on lakes it is mostly confined to shallow bays. It occupies much the same habitats as pygmy geese and Hottentot teal, and probably feeds on much the same foods as the former. Little has been written on its foods, which are believed to consist primarily of the seeds of aquatic plants as well as the leaves of water lilies. Clark (1969b) states that the bird dives to the bottom and sifts food from the mud and debris, often remaining under water for 20 to 30 seconds. Foraging times are from dawn to shortly after sunrise, and again from late afternoon to sundown; the birds are evidently not nocturnal as are most whistling ducks. During the middle of the day they are inactive and appear to be asleep.

Social behavior. In the wild, the birds are sociable, both resting and often foraging in groups. Flying is rarely undertaken for any distance, or at any great height off the water. There are some migratory movements associated with rainfall patterns; the birds are said to appear at the start of the rainy season in Rhodesia. They are usually to be found in pairs or larger groups, and the absence of conspicuous pair-forming behavior suggests that pair bonds are permanent, a view supported by the active role of the male in breeding. Some possible pair-forming behavior patterns that have been described consist of one bird (probably the male) swimming to another, calling, and the two birds swimming in parallel while calling (Johnsgard, 1965a). Another possible pair-forming or pair-maintaining display consisted of mutual bill dipping on the part of a presumed pair (Clark, 1969b). Aggressive chin lifting during calling has frequently been seen, as well as an aggressive head-back posture similar to that of whistling ducks. In extreme threat the bird will gape, spread its flank and scapular feathers, and hiss loudly while paddling the water vigorously (Johnsgard, 1967a). Precopulatory behavior consists of drinking movements on the part of the male as he approaches the female, followed by the female going prone in the water. After treading, the male utters a loud whistling note, and both sexes perform a step-dance in the manner of whistling ducks (Johnsgard, 1967a; Clark, 1969b).

Reproductive biology. Probably a prolonged period of breeding is typical of this species; Benson et al. (1971) list breeding records for nine months from December through August, with the largest numbers in February, April, and May. In South Africa the recorded nestings have mostly been from the period November through May, while Rhodesian records range from November through August (Clancey, 1967). Nests are built in reeds or papyrus beds, often in fairly deep water, with the nest near the edge of the reed bed. There is often a ramp leading to the nest, and it is usually concealed from above by overhanging reeds. Sometimes coot or grebe nests are used as a nest foundation. There may also be an approach channel between the nest and open water. The nest cup is not lined with down. The average clutch size is about 6 eggs (Lack, 1968), but up to 14 have been found in a single nest, probably because of multiple layings. The incubation period has been estimated to be as little as 26 days (Johnstone, 1970), but various eggs incubated under brood hens required from 29 to 33 days to hatch (Johnsgard, 1967a). It is now well established that the male strongly guards the nest and does much if not most of the incubation, as in typical whistling ducks. In both wild ducklings and those hatched in captivity it has been found that a favorite if not the only food consists of gnat larvae (*Chironomus*), which are sieved from the muddy bottoms of ponds (Kear, 1967).

Status. This species is locally abundant over its African range wherever water conditions are suitable, and appears to be fairly secure. It is considered inedible and is not hunted for food or sport. The status of the Madagascan race is unknown.

Relationships. The possible whistling duck affinities of this species were initially suggested by me in 1960, and later (1967a) supported by behavioral evidence. Anatomical evidence (Raikow, 1971) and analyses of duckling vocalizations (Kear, 1967) have provided additional support for this view. Most recently, Brush (1976) has analyzed the feather proteins of *Thalassornis* by electrophoretic methods and concluded that this hypothesis can be supported by his results.

Suggested readings. Johnsgard, 1967a; Clark, 1969b; Clancey, 1967.

Tribe Anserini (Swans and True Geese)

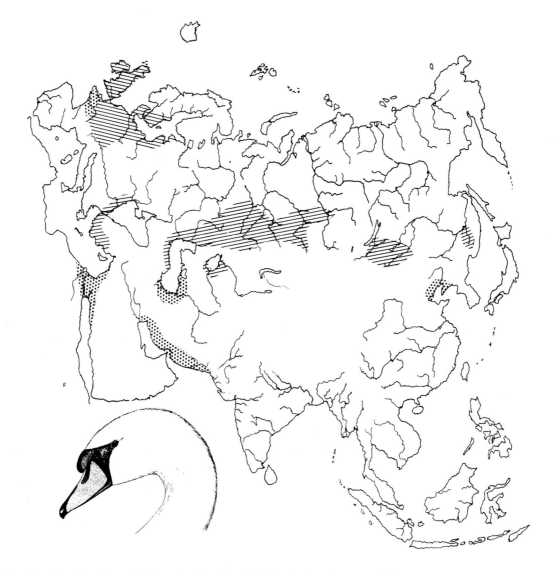

MAP 10. Breeding (hatching) and wintering (stippling) distributions of the mute swan, excluding introduced populations.

Drawing on preceding page: Trumpeter Swan

Mute Swan

Cygnus olor (Cmelin) 1789

Other vernacular names. White swan, Polish swan; Höckerschwan (German); cygne muet (French); cisne mudo (Spanish).

Subspecies and range. No subspecies recognized. Breeds under native wild conditions in southern Sweden, Denmark, northern German, Poland, and locally in Russia and Siberia; also in Asia Minor and Iran east through Afghanistan to Inner Mongolia; in winter to northern Africa, the Black Sea, northwestern India, and Korea. Breeds locally feral or semiferal in Great Britain, France, Holland, and Central Europe. Introduced and locally established in North America, mainly on Long Island and in northern Michigan. Also locally established in South Africa, Australia, and New Zealand. See Map 10.

Measurements and weights. (Mostly from Scott & Wildfowl Trust, 1972.) Folded wing: males, 589–622 mm; females, 540–96 mm. Culmen: males, 76–85 mm; females, 74–80 mm. Weights: males, 8.4–15.0 kg (av. 12.2 kg); females, 6.6–12.0 kg (av. 8.9 kg). Eggs: av. 115 x 75 mm, greenish blue, 340 g.

Identification and field marks. Length 50–61" (125–55 cm). Plate 8. *Adults* are entirely white in all postjuvenile plumages. The bill is orange, with black around the nostrils, the nail, and the edges of the mandible. The feet are black, except in the uncommon "Polish" color phase, which has fleshy-gray feet. *Females* are smaller (see measurements above) and have a less well developed knob over the bill. *Juveniles* exhibit a variable number of

brownish feathers which diminish with age (except in the Polish swan, which has a white juvenile plumage), and the knob over the bill remains small through the second year of life.

In the field, mute swans may be readily identified by their knobbed bill; their heavy neck, usually held in graceful curve; and their trait of swimming with the inner wing feathers raised, especially in males. In flight, the wings produce a loud "singing" noise, but apart from hissing and snorting sounds this species is relatively "mute."

NATURAL HISTORY

Habitat and foods. The habitats of this species are diverse and numerous, as a reflection of a long association with man and at least partial domestication. Probably, however, Temperate Zone marshes, slowly flowing rivers, and lake edges were its native habitats. Clean, weed-filled streams are preferred over larger, polluted rivers. Brackish or salt-water habitats are often used during nonbreeding periods. A variety of studies of foods taken in Europe and North America indicate that the leafy parts of a number of fresh-water or salt-water plants, such as pondweeds, muskgrass, eelgrass *(Zostera)*, and green algae *(Enteromorpha)*, are major sources of food where they are available. Also where it is available, waste grain is often consumed, and some grazing on terrestrial grasses near water is prevalent in many areas (Scott & Wildfowl Trust, 1972). Mute swans can reach underwater foods up to 50 centimeters below the surface by upending, but in common with other swans do not dive when foraging. A small amount of animal foods, including amphibians, worms, mollusks, and insects, may be taken when available, but these are minor parts of their diet.

Social behavior. Except in a few areas where local foraging conditions favor large concentrations, as on some coastal islands, mute swans are not notably social. Nevertheless, they may flock in groups of a thousand or more in summer molting areas, and similar flocking may also occur in the winter. In many parts of its range the mute swan is essentially sedentary, and in England, for example, studies of banded birds have revealed that most movements are less than 30 miles, and usually follow watercourses (Ogilvie, 1967). Probably the most extensive migrations occur in the breeding

birds of Siberia and Mongolia, from Lake Baikal east. It is presumably those birds that winter along the Pacific coast, from Korea south to the Yangtze Kiang, suggesting a migration of up to about 1,000 miles. Pairs and family units migrate together and remain together until about the end of the year, when at least in England the breeding adults begin to exhibit territorially. At that time the young birds remain in the winter flocks of nonbreeding birds, and they may remain together through the following summer and winter (Scott & Wildfowl Trust, 1972).

Pair-forming behavior occurs in the fall and winter, usually in the season before breeding initially. Among mute swans, initial breeding is most frequent in the third year, with some birds (mostly females) breeding when two years old and some not until four or older. Pair formation and pair bonding in this as in all typical swans occurs by mutual greeting ceremonies such as head turning, and bonds are firmly established by triumph ceremonies between members of a pair. This occurs after the male has threatened or attacked an "enemy" and returned to his mate or prospective mate with ruffled neck feathers and raised wings, calling while chin lifting. Precopulatory behavior may occur frequently among paired birds in winter

flocks, and consists of mutual head dipping and preening or rubbing movements along the flanks and back. After treading, both birds rise in the water breast to breast, calling and extending their necks and bills almost vertically for a brief moment, then subside in the water (Johnsgard, 1965a). Once formed, pair bonds are very strong, and so long as both members of a pair remain alive there is a low rate of mate changing. One study in England indicated that there was a 3 percent "divorce" rate among unsuccessful breeders or nonbreeders (Minton, 1968).

Reproductive biology. The nesting period of mute swans is in spring, which is generally March through June in its Northern Hemisphere range, and from September through January in New Zealand and Australia. Mute swans are highly territorial, although the size of the territory varies with breeding density. One study in Staffordshire, England, indicated a density of a pair per 2,000 hectares, but with territories limited to a few meters of stream-side locations. In a few areas of dense colonies, the nests may be located only a few meters apart. In England, nests are most often placed in or near standing water, less often beside running water, and least often in coastal situations.

Nests are large piles of herbaceous vegetation, built by both sexes, and often built on the previous year's nest, especially if it was a successful site. Eggs are deposited every other day until a clutch of 4 to 8 (usually 5 or 6) eggs is laid. Incubation is normally performed only by the female ("pen"), but the male ("cob") may occasionally take over for a time. The incubation period is usually 35–36 days, during which time the male rarely strays far from the nest. After hatching, both sexes attentively care for the young, frequently allowing them to ride on their backs. The rate of growth is very slow, and the fledging period is from 120 to 150 days, during which time the adults undergo their postnuptial molt. Thereafter, the family increases their food intake and fat reserves before leaving the breeding grounds (Kear, in Scott & Wildfowl Trust, 1972).

Status. This swan has in general benefited from man's influence, and has either been purposefully introduced or otherwise spread into new breeding areas in historical times. In many areas of Europe where the species was exterminated during the 1800s it is now reestablished and increasing.

Relationships. In spite of its similar geographic range and plumage characteristics, the mute swan is not closely related to the other Northern Hemisphere white swans, but instead its nearest relative is the black swan, as is indicated by a variety of behavioral characteristics (Johnsgard, 1965a).

Suggested readings. Scott & Wildfowl Trust, 1972.

Black Swan

Cygnus atratus (Latham) 1790

Other vernacular names. None in general English use. Trauerschwan (German); cygne noir (French); cisne negro (Spanish).

Subspecies and range. No subspecies recognized. The native breeding range includes Australia, except for the north and central areas, and Tasmania. Winters over most of its range. Introduced into New Zealand and well established on both islands. See map 11.

Measurements and weights. (Mainly from Frith, 1967.) Folded wing: males, 434–543 mm; females, 416–99 mm. Culmen: males, 57–79 mm; females, 56–72 mm. Weights: males, 4.6–8.75 kg (av. 6.27 kg); females, 3.7–7.2 kg (av. 5.1 kg). Eggs: av. 115 x 65 mm, pale green, 300 g.

Identification and field marks. Length 45–55" (115–40 cm). Plate 9. This is the only black swan; *adults* are uniformly dark brownish black, with somewhat paler underparts and white flight feathers. All the primaries and the outer secondaries are white, while the inner secondaries are white-tipped. The innermost wing feathers are strongly undulated. The bill is red to orange, with a whitish subterminal bar and nail; the iris is reddish (or sometimes whitish), and the feet and legs are black. *Females* are smaller and have a less brightly colored bill and iris. Also, they appear to have shorter neck feathers and a less ruffled back, which may simply be the result of variable feather lifting. *Juveniles* are a mottled grayish brown, with light-tipped feathers and a paler bill coloration.

In the field, the blackish plumage makes this species unmistakable for any other swan. In flight, the white flight feathers contrast strongly with the otherwise blackish coloration. The call of the adults is a rather weak bugling sound that does not carry great distances.

NATURAL HISTORY

Habitat and foods. The preferred habitat of black swans in Australia consists of large, permanent lakes, of either fresh or brackish water. Less often they are found on rivers and billabongs, and occasionally occur along the coast. Water depth is more important than water chemistry; abundant aquatic vegetation must be available to within three feet of the surface to provide a food source. Foraging is done primarily by upending in water, but sometimes the birds graze along the banks or stand in shallow water and dabble. Of the foods eaten by a sample of birds from New South Wales, all were of vegetable materials, except for a few items of animal origin that may have been accidently ingested with the plants. Over a third of the volume of food consisted of cattails, while the other materials that occurred in large quantities were algae, wild celery (*Vallisneria*), and pondweeds. A variety of grasses and their seeds also were part of

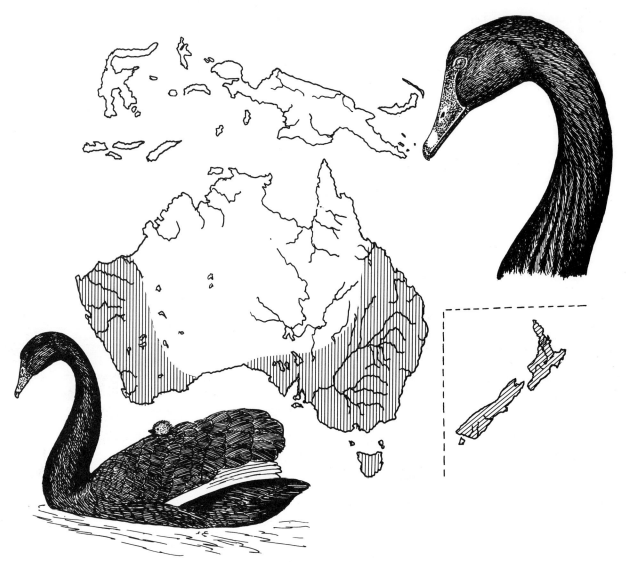

MAP 11. Native (vertical hatching) and introduced (diagonal hatching) distributions of the black swan.

the sample. Most of the food materials were of aquatic plants, with shoreline plants contributing only a very small percentage (Frith, 1967). Less is known of the foods consumed in New Zealand, but apparently the birds spend more time grazing in pastures there, and sometimes even strip and eat the leaves of willows.

Social behavior. The black swan is one of the more social of the swans, perhaps in part because of lack of definite territorial behavior, which favors the development of large breeding colonies as well as concentrations in the nonbreeding season. Many lakes in southern Australia regularly support from 5,000 to 15,000 swans, and in Lake Ellesmere of New Zealand an estimated 60,000 to 80,000 occur.

Such large flocks are of dynamic composition, but comprise the stable pair and family units of all swans. Further, the flocks are probably more mobile than is generally thought to be the case; banding studies of birds caught while molting in Australia indicate movements of up to several hundred miles in successive years. These studies indicate that most swans are essentially nomadic, moving widely as weather and availability of suitable habitat dictate. Pair formation presumably usually occurs during the second winter of life, although this has yet to be verified. As in the mute and other swans, pairs are formed and maintained by various mutual displays, particularly the triumph ceremony. This ceremony takes essentially the same form in the black swan as in the mute, with strong chin lifting, mutual calling, and ruffling of the neck feathers. Precopulatory behavior consists of mutual bathing movements but no preening, and after treading, the birds do not rise breast to breast, but rather extend their necks and heads vertically, call once, and then swim about in a partial circle before starting to bathe.

Reproductive biology. The timing of breeding in black swans is fairly variable. In its native range of Australia the birds breed in the rainy season of February to May in Queensland, and from June through August in western Australia, during that area's moister winter. In the central and southern areas breeding may be timed according to local rains, or during June and July in more permanent water areas (Frith, 1967). In New Zealand most breeding occurs from August to November, and in some areas nesting also occurs in the austral autumn. Among captive birds in Europe, records of breeding show a spread from March through September, but about 70 percent of the records are for

April and May (Petzold, 1964). Nests are built either on land or in swamps, and also sometimes are placed on stumps, at the bases of trees, or in floating debris. In swamps of cattail they consist of large clumps of these plants up to five feet wide and three or four feet above the water surface. Ground nests are appreciably smaller. Frequently the previous year's site is used; in Australia it has been found that a site may be used as many as four times. In large colonies, the nests may be as close as a meter apart (Guiler, 1966), or practically touching each other. It has been suggested that the slow growth rate of the cygnets, their vegetarian diets, and perhaps their increased protection from predation have facilitated the evolution of colonial nesting in this species (Kear, in Scott & Wildfowl Trust, 1972). Both sexes help build the nest, and a clutch of 4 to 10 eggs (most often 5 or 6) is laid on an alternate-day basis. The male definitely helps incubate in this species, and under natural conditions the average incubation period is about 40 days, but it ranges from 35 to 45 days (Frith, 1967). The

young grow fairly slowly, at least as compared with the swans which nest in the Arctic, and 140 to 180 days elapse before the flight feathers are fully grown (Frith, 1967).

Status. The black swan's status in Australia is very favorable; it is fully protected and has recently extended its breeding range in Queensland. Some short hunting seasons have been needed in recent years to cope with local crop depredation problems in Victoria and also in Tasmania. Likewise, in New Zealand the species is locally very common, and around Lake Ellesmere its numbers must be controlled by disturbance and allowing commercial egg collecting.

Relationships. The black swan is clearly a close relative of the mute swan, in spite of the differences in adult plumage coloration (Johnsgard, 1965a). A possible reason for the evolution of a dark plumage may be a reduction in ecological needs for conspicuousness in conjunction with the effective advertisement and defense of a large territory, which seem to be a primary reason for whiteness in the other swan species.

Suggested readings. Frith, 1967; Scott & Wildfowl Trust, 1972; Wilmore, 1974.

Black-necked Swan

Cygnus melancoryphus (Molina) 1792

Other vernacular names. None in general English use. Schwarzhalsschwan (German); cygne à col noir (French); cisne de cuello negro (Spanish).

Subspecies and range. No subspecies recognized. Breeds in Paraguay, Uruguay, Argentina, Chile, Tierra del Fuego, and the Falkland Islands. Winters as far north as the Tropic of Capricorn, in Paraguay and the three southern provinces of Brazil. See map 12.

Measurements and weights. (Mainly from Scott & Wildfowl Trust, 1972). Folded wing: males, 435–50 mm; females, 400–415 mm. Culmen: males, 82–86 mm; females, 71–73 mm. Weights: males, 4.5–6.7 kg (av. 5.4 kg); females, 3.5–4.4 kg (av. 4.0 kg). Eggs: av. 105 x 65 mm, cream, 247 g.

Identification and field marks. Length 45–55" (115–40 cm). Plate 10. This swan is the only species that is entirely white except for a black head and neck. There is also a large red caruncle behind the bill, which is bluish gray with a paler nail, and the legs and feet are pink. *Females* are noticeably smaller than males, but have well-developed caruncles behind the bill. *Juvenile* birds exhibit a variable amount of brownish gray in their plumage, which is lost by the end of the first year, after which immatures may be distinguished by their smaller or absent caruncles.

In the field, the contrasting white and black body plumage is unmistakable; the coscoroba swan is the only other swan in the native range of this species. The black-necked swan's voice is a relatively weak and wheezy whistle, and does not carry great distances.

MAP 12. Breeding or residential (hatching) and wintering (stippling) distributions of the black-necked swan.

Habitat and foods. This is a species of fresh-water and brackish-water marshes, predominantly the former. Weller (1975a) found the birds on large marshes or marshy lake edges in southern Argentina. Near Buenos Aires he (1967a) found them in lakes with tule edges or in large pools in marshes. On the Falkland Islands the swans favor large waters (over 40 hectares) that may be either brackish estuaries or fresh, but they spend little time in kelp beds (Weller, 1972). They were found where algae (*Nitella*) and mud plantain (*Heteranthera*) occurred, presumably feeding on them both. However, very little is known of the specific foods taken by this species. It is rarely seen ashore, and thus must be almost entirely dependent on aquatic plant materials for its diet. Johnson (1965) reported that aquatic insects and fish spawn may sometimes also be eaten.

Social behavior. This is a relatively gregarious swan, with nonbreeding birds forming flocks of up to several thousand birds seasonally. Weller (1967a) found a flock of five or six thousand birds on an Argentine marsh in midsummer after the young had fledged, but generally the observed flock sizes are much smaller. Migratory movements in South America are no doubt fairly substantial, and the birds obviously can fly long distances, as attested by an occasional stray reaching the Juan Fernandez Islands, 400 miles off Chile's coast. Even on the Falkland Islands there are some seasonal movements evident (Cawkell & Hamilton, 1961). Probably most of the northward movement from central and southern Argentina occurs in March and April, after the year's young are well fledged. Formation of pairs almost certainly takes place in winter flocks; two-year-old birds often breed in captivity, but it is assumed that under natural conditions the birds probably begin breeding when approaching the end of their second year. Pair formation is achieved by the repeated use of the triumph ceremony; although this is the smallest of the swans it is among the most aggressive, and males of pairs in captivity are almost constantly threatening other birds, then returning to their mates and performing a chin-lifting triumph ceremony. Unlike the mute and black swans, black-necked swans to not raise their folded wings in threat, but extend their neck and head low over the water as they rush toward the opponent. Because of this aggression, it seems unlikely that coloniality is normal in this species, and most observers indicate a low breeding density. Cawkell and Hamilton (1961) reported six to eight pairs on a large lake in the Falkland Islands, and Weller (1967a) found one nest and saw several broods on a 100-hectare lake near Buenos Aires. Yet, a cluster of six nests (four empty) were once found within 18 meters of one another on the Falkland Islands.

Reproductive biology. Black-necked swans nest in the southern spring, and were the earliest marsh-nesting species studied by Weller (1967a) in east-central Argentina. He says that birds apparently started nesting there in July. Likewise in central Chile breeding begins in July and August, but farther south it may begin at least a month later. On the Falkland Islands nesting occurs from early August to mid-September. In Chile the clutch size varies from 4 to 8 eggs, and the nest is placed in thick reed beds around lake edges or lagoons; preferentially it is placed on small islets (Johnson, 1965). It is usually large and bulky, often built of rushes and partially floating. The male closely guards the nest and stands over it when his mate leaves to forage, but does not normally help incubate. The incubation period is usually 36 days, with the female leaving the nest during the evening hours every few days for foraging. Dominican gulls (*Larus marinus dominicanus*) are reportedly a serious egg predator in the Falkland Islands (Cawkell & Hamilton, 1961). The newly hatched cygnets are closely attended by both parents, and more time is spent carrying the young on the backs by both adults than is true of any of the other swans. Indeed, the male sometimes does the majority of such carrying of young. The strong degree of parental carrying in this species may be associated with the fact that these swans rarely come ashore, and thus terrestrial brooding is impossible (Kear, in Scott & Wildfowl Trust, 1972). There is no specific information on the length of time to fledging in this species, but the growth rate is obviously very slow, apparently second only to that of the black swan, which requires nearly six months to fledge. Weller (1967a) reported that in central Argentina young had fledged by early January, while hatching occurred in October, suggesting a fledging period of about 100 days. During this fledging period the adults undergo their annual postnuptial molt, and departure from the breeding grounds occurs shortly after the young are on the wing.

Status. In Chile, the black-necked swan has recently reoccupied a part of its range that had previously been eliminated as a result of local persecution, and in Argentina the swan is locally common in many marshy areas. Destruction of marshes, either by their removal with drainage canals or by use of the marshes as a runoff repository for excessive rainfall on adjacent agricultural lands, rather than specific hunting or other persecution of the species is the biggest threat to its long-term survival.

Relationships. Although it is relatively isolated from the others of this group, the black-necked swan is fairly clearly a member of the mute swan and black swan evolutionary complex. It differs in a number of behavioral ways from these two species, however, and except for the coscoroba is perhaps the most isolated in an evolutionary sense of all of the swans.

Suggested readings. Wilmore, 1974; Scott & Wildfowl Trust, 1972.

Trumpeter Swan

Cygnus buccinator Richardson 1831

Other vernacular names. None in general English use. Trompeterschwan (German); cygne trompette (French); cisne trompetero (Spanish).

Subspecies and range. No subspecies recognized (this population, however, is often regarded as a subspecies of *C. cygnus*). The current breeding ranges consist of boreal forest marsh habitats in southern Alaska, British Columbia, Alberta, and Wyoming, with local breeding on various wildlife refuges in several western states as a result of reintroduction. Limited southward migratory movements occur during winter, especially in the Alaskan population, which winters along coastal British Columbia. See map 13.

Measurements and weights. (Mainly from Scott & Wildfowl Trust, 1972). Folded wing: males, 545–680 mm; females, 604–36 mm. Culmen: males, 104–20 mm; females, 101–12 mm. Weights: males, 9.1–12.5 kg (av. 11.9 kg); females, 7.3–10.2 kg (av. 9.4 kg). Eggs: av. 118 x 76 mm, white, 325 g.

Identification and field marks. Length 60–72″ (150–80 cm). Plate 11. This swan's great size, its entirely white plumage, and the absence of any yellow on the bill should serve to identify it. The bill of adults usually measures at least 50 millimeters from the front edge of the nostril to the tip of the nail, and the length of the middle toe (excluding claw) is at least 135 mm, compared to 133 mm maximum in whistling swans (Banko, 1960). The trachea of trumpeter swans is convoluted inside the sternum, which has a rounded protrusion on its mid-dorsal surface, and the total length of the trachea from glottis to the bronchi averaged (in two cases) over 130 centimeters. In contrast, the sternum of an adult whistling swan, although penetrated by the trachea, lacks a mid-dorsal protrusion and the trachea measured (in one case) under 120 centimeters. *Females* are identical to males, averaging only slightly smaller in most measurements. *Juveniles* exhibit some gray feathers in their plumage up to about a year from hatching; thereafter immature birds are essentially identical to adults in appearance.

In the field, the large size of the bird and absence of yellow on the bill are good field marks, as is the loud, trumpeting and resonant call, which is usually a double-noted *ko-hoo* that may be heard for a mile or more under favorable conditions.

NATURAL HISTORY

Habitat and foods. Banko (1960) characterized the breeding habitat of trumpeter swans as stable waters that lack marked seasonal fluctuations; quiet waters of lakes, marshes, or sloughs that are not subject to current or constant wave effects; and shallow waters that allow for foraging on aquatic plants. Originally, the birds bred over a large area of North America's northern or western regions, including boreal forests, montane pine forests, the western edge of the eastern deciduous forest, the short-grass plains, and tall-grass prairies. Obviously the surrounding habitats are less important than the water characteristics, which must provide security, reasonable seclusion, and adequate foraging and nesting opportunities. The primary foods consist of the leaves and stems such aquatic plants as pondweeds and crowfoot *(Ranunculus)*, as well as the roots of pondweeds and tubers of arrowhead *(Sagittaria)*. Emergent plants of the shoreline are also consumed, and in late fall the seeds of sedges

or water lilies *(Nuphar)* become important. Cygnets forage to a large extent on animal life in their early weeks, especially on aquatic insects and mollusks.

Social behavior. Large flocks of trumpeter swans are now essentially a thing of the past; too few exist to facilitate such groupings. However, winter flocks on the limited open water of Montana's Red Rock Lakes Refuge do become substantial; Banko (1960) provided a photograph of a flock of 80 birds on a small spring in January. Groups consist of family and pairs, the family bonds probably persisting through at least two years, judging from a limited amount of information on banded birds. A good deal of aggressive behavior and mutual display occurs in wintering flocks, and much of it is concerned with pair formation and pair-bond maintenance. Pair bonds are probably usually formed in

MAP 13. Breeding or residential (hatched and inked) and wintering (stippling) distributions of the trumpeter swan.

the second winter of life, or possibly the third winter in some individuals. Banko (1960) reported that the greatest incidence of social display occurs in fall when new birds join others on the wintering grounds, and in spring just before dispersal to the breeding grounds. Triumph ceremonies, with attendant wing waving and calling by both birds as they partially rise in the water, are the most common of these displays; copulatory behavior is more likely to occur on the nesting territory just before egg laying. Precopulatory behavior involves mutual head dipping closely resembling bathing movements by both members of a pair, and treading is followed by loud calling, wing waving, and rising in the water in a manner closely resembling the triumph ceremony (Johnsgard, 1965a). Dispersal to the breeding grounds from the wintering areas may involve movements of only a few miles, as on the western refuges, or a migration of several hundred miles, in the case of the breeding population of southern Alaska. On the Copper River delta and the Kenai Peninsula spring arrival is in late March or April (Hansen et al., 1971). Because of their long fledging period, the birds must begin nesting at the earliest possible moment; thus on their Alaskan range they are among the earliest breeding birds to arrive and the last to leave, establishing territories and nest sites on muskrat houses while these locations are still largely snow-covered and iced in (Wilmore, 1974).

Reproductive biology. At Red Rock Lakes, muskrat houses are the most commonly chosen nest site, and territorial limits are likely to encompass large areas with many suitable nest sites. A nesting density of about 4.5 birds per square mile, or 142 acres per pair, is average, with favorable areas supporting up to one pair per 70 acres. Birds nesting on islands defend smaller areas than those nesting on shorelines, and a feeling of spatial isolation, rather than food supply, probably is the major determinant of territorial size. Territories may be occupied as early as February, although eggs are not laid until the end of April or early May. In Alaska egg laying is initiated in late April or early May, and incubation is begun before the end of May (Hansen et al., 1971). Nests are built by both sexes, usually on prior nest sites, and clutches of 4 to 8 but most commonly 5, eggs are laid. Eggs are laid on an alternate-day basis, and the nest is strongly guarded by both sexes. Incubation is normally performed only by the female, but a few instances of apparent incubation by males have been observed in captivity in this species. The normal incubation period is from 33 to 37 days, on the basis of observations in Alaska and Montana and records from captivity. During this time the eggs may be left unattended for varying periods, but their large size probably reduces the rate of heat loss and possible chilling. However, even when the nest is not covered by the female, the male closely watches it for possible danger, and little if any egg predation occurs under natural conditions.

The cygnets feed on a diet rich in invertebrates and grow rapidly, especially in the long days of the Alaskan breeding grounds. A fledging period of 100 to 120 days has been estimated as typical for the cygnets in the Red Rock Lakes area, while in Alaska it is from 84 to 105 days. Females usually begin their wing molt from a week to three weeks before the eggs hatch, while males of successfully breeding pairs may begin to molt either before or after hatching, but usually before. Molting in non-breeding birds seems to be earlier and more synchronized than in breeders, but in either case the birds remain flightless for about 30 days (Hansen et al., 1971). Fall departure from the Alaska breeding grounds occurs in October, only shortly after the cygnets have fledged.

Status. The trumpeter swan was removed from the list of endangered species only a few years ago, when it became apparent that the Alaskan breeding population was much larger than previously known and that the overall population was thus appreciably greater than estimated earlier. The best estimates of this population are by Hansen et al. (1971), who found nearly 3,000 swans in 1968 and probably missed several hundred during the aerial survey. There were also about 150 swans that summer in western Canada, and over 300 in the vicinity of Red Rock Lakes, Yellowstone National Park, and Grand Teton National Park. There are about 600 additional birds in other U.S. refuges and zoos, suggesting that over 4,000 existed in North America around 1970. Success in establishing breeding swan populations in several western refuges suggests that at least remnant populations of the species can be maintained over much of the original range where habitat and protection allow.

Relationships. The trumpeter, whooper, whistling, and Bewick swans collectively constitute a close-knit evolutionary complex called the northern swans (Johnsgard, 1974a), in which the species limits and internal relationships are still far from clear.

Suggested readings. Hansen et al., 1971; Banko, 1960.

Whooper Swan

Cygnus cygnus (Linnaeus) 1758

Other vernacular names. None in general English use. Singschwan (German); cygne sauvage (French); cisne gritón (Spanish).

Subspecies and range. No subspecies currently recognized (although the trumpeter swan is often regarded as such). The breeding range includes Iceland and Eurasia from northern Scandinavia eastward through Finland and northern Russia, and in northern Asia from Kolymsk and Anadyr to Kamchatka and the Commander Islands, and southward through the tiaga and scrub forest zone to the Russian Altai, the lower Amur Valley, and Sakhalin. Winters well to the south, in Great Britain and northwestern Europe and east to the western border of China. Also winters in Japan, Korea, and along the Pacific coast of China. See map 14.

Measurements and weights. (Mainly from Scott & Wildfowl Trust, 1972.) Folded wing: males, 590–640 mm; females, 581–609 mm. Culmen: males, 102–16 mm; females, 97–112 mm. Weights: males, 8.5–12.7 kg (av. 10.8 kg); females 7.5–8.7 kg (av. 8.1 kg). Eggs: av. 113 x 73 mm, white, 330 g.

Identification and field marks. Length 55–65" (140–65 cm). Plate 12. The whooper is the only one of the northern swans with the yellow on the bill reaching farther forward than the nostrils, and the only swan with a partially yellow bill that has a folded wing length in excess of 580 milimeters. In the adult plumage both sexes are entirely white,

with black legs and feet and a black bill except for the large yellow marking. *Females* cannot be externally distinguished from males. *Juveniles* have a variable number of grayish feathers mixed with the white feathers of the immature plumage, and their bills are pinkish at the base rather than yellow.

In the field, the extensively yellow bill is the best field mark, and the more trumpetlike and less musical call of this species helps to distinguish it from the smaller Bewick swan.

NATURAL HISTORY

Habitat and foods. The preferred breeding habitat is shallow fresh-water pools and lakes, and along slowly flowing rivers, primarily in the coniferous forest (tiaga) zone, but also in the birch forest zone and treeless plateaus, but rarely in tundra (Voous, 1960). On the breeding grounds the foods of adults probably consist mainly of the leaves, stems, and roots of aquatic plants, including algae. A considerable amount of grazing on shoreline and terrestrial vegetation is performed; the whooper swan is thought to consume higher proportions of terrestrial plants and animal materials than is true of the Bewick swan. In the wintering areas of Europe the birds have been found to consume such aquatic plants as pondweeds and similar leafy foods (*Zostera, Ruppia, Elodea*) that can readily be reached from the surface. In some areas they also graze on winter wheat, waste grain, turnips, and potatoes (Owen & Kear, in Scott & Wildfowl Trust, 1972).

Social behavior. Whooper swans are gregarious and form large flocks in the nonbreeding season; in Aberdeenshire, Scotland, flocks of 300 to 400 often may be seen foraging in agricultural fields. Fairly large winter flocks also occur locally in Japan in protected sites such as Hyoko reservoir (Lake Hyo), on the island of Honshu. These wintering flocks consist of firmly paired, courting, and sexually immature birds, including first-year birds still consorting with their parents. So far as is known, the process of pair formation and pair bonding in whooper swans is exactly as in the trumpeter swan, as is the behavior associated with copulation. First-year birds remain with their parents through the winter period and start back toward the breeding grounds with them; but since the birds arrive on the breeding grounds in pairs, the family bonds must be broken during the spring migration period. Whooper swans leave their European and Asian wintering grounds in late February or March, and may not arrive at their northernmost breeding areas until May.

Reproductive biology. Immediately upon the birds' arrival on their breeding grounds, or at most within two weeks of arrival, nest building begins. Nests are built either on dry ground or in reed beds, often so large and with so deep a cup that the top of the sitting female may be flush with the rim of the nest. The nest cup is extensively lined with down, and a clutch of from 4 to 7 eggs, most commonly 5 or 6, is laid. In the U.S.S.R., egg lay-ing occurs during May and June, and in Iceland it also normally occurs at that time. Undoubtedly the time at which the nesting sites become snow-free dictates the year-to-year onset of nesting in this species. The probable usual incubation period is 31 days, although some estimates have been as high as 36 or 40 days. Incubation is performed entirely by the female, but the male remains in very close attendance. Typically she leaves for a short time during the warmest part of each day to forage. The male also closely guards the cygnets, which in Iceland have been reported to fledge in as few as 60 days. This is a remarkably short fledging period, although the latitude of Iceland would allow for 24-hour foraging in midsummer. The young birds

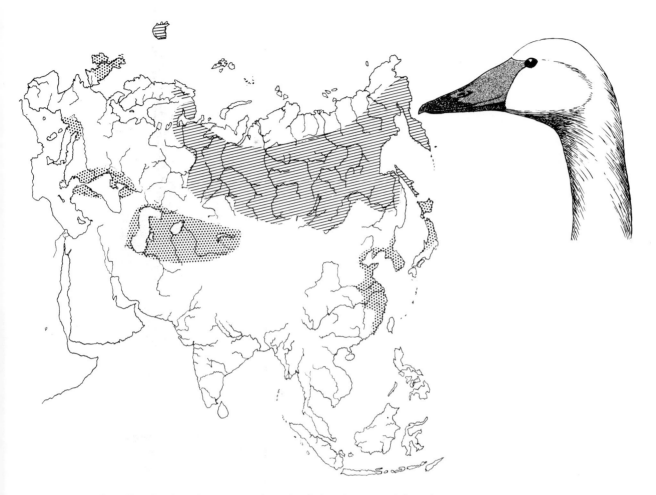

MAP 14. Breeding (hatched) and wintering (stippling) distributions of the whooper swan.

eat insect larvae, adult insects, and vegetation that grows on the water surface or just below it (Scott & Wildfowl Trust, 1972).

Status. The total world population of the whooper swan is certainly less than 100,000 birds, but nearly universal protection is given the species. The Icelandic breeding component is 5,000 to 6,000 birds, which winter either in Iceland or in Great Britain. Those breeding in Scandinavia and western Russia and wintering in northwestern Europe probably number about 14,000. The birds which breed farther east winter mainly on the Black and Caspian seas, and probably total at least 25,000. The Far Eastern breeding component that winters along the Pacific coast is least well documented, but close to 11,000 have been counted in a recent wintering census in Japan (Scott & Wildfowl Trust, 1972). At present there seems to be no real threat to this species' existence, at least on an overall basis.

Relationships. Although the whooper swan and trumpeter swan are now often regarded as conspecific (Delacour, 1954–65), this is not the only interpretation of relationships among the northern swans, and the safest present position is to consider them as separate species.

Suggested readings. Scott & Wildfowl Trust, 1972; Wilmore, 1974.

Whistling Swan

Cygnus columbianus (Ord) 1815

Other vernacular names. None in general English use. Pfeifschwan (German); cygne siffleur (French), cisne silbador (Spanish).

Subspecies and range. No subspecies recognized (although the whistling swan is often considered a subspecies of *C. bewickii*). Breeds in arctic tundra habitats of continental North America from western Alaska to Hudson Bay, and on Southampton, Banks, Victoria, and St. Lawrence Island. Winters principally along the Pacific coast from southern Alaska to California, and on the Atlantic coast from Chesapeake Bay to Currituck Sound. See map 15.

Measurements and weights. (Mainly from Scott & Wildfowl Trust, 1972.) Folded wing: males, 501–69 mm; females, 505–61 mm. Culmen: males, 97–107 mm; females, 92–106 mm. Weights: males, 4.7–9.6 kg (av. 7.1 kg);

females, 4.3–8.2 kg (av. 6.2 kg). Eggs: av. 110 x 73 mm, white, 280 g.

Identification and field marks. Length 48–58" (120–50 cm). The whistling swan is completely white in *adult* plumage, with black legs and feet and a bill that is typically entirely black except for a small yellow area in front of the eye. However, some whistling swans lack this yellow mark, and thus a bill length that is under 50 millimeters from the front of the nostrils to the tip of the bill is a better criterion for birds in the hand. *Females* are identical to males, and average slightly smaller in size. *Juveniles* possess some gray feathers for most of their first fall and winter of life, and their bills are mostly pinkish.

In the field, the yellow bill markings may not be evident at great distances, and thus the rather high-pitched, whistling, barklike call, *kow-wow,* is a better guide to identification. The neck of the whistling swan appears to be shorter and the bill profile somewhat more concave than those of the trumpeter swan.

Natural History

Habitat and foods. The whistling swan is associated with arctic tundra throughout its breeding range in North America, and thus is the ecological counterpart of the Bewick swan. Foods taken on the breeding grounds are not yet well studied, but in migration and wintering areas the birds usually feed extensively on such aquatic plants as wild celery (*Vallisneria*), wigeon grass, bulrushes, and pondweeds. The tubers of arrowhead (*Sagittaria*) are favored foods, and in brackish waters the birds may feed to some extent on mollusks, especially clams. Some grazing on agricultural lands is performed by wintering birds, which may consume grain and waste potatoes (Owen & Kear, in Scott & Wildfowl Trust, 1972).

Social behavior. Flock sizes in wintering areas and during migration are often large, and may number in the hundreds or even in the thousands, although the birds are strongly territorial and well scattered during the breeding season. The birds have a relatively long migratory route, often of more than 2,000 miles between wintering and breeding grounds, and counts made during spring in Wisconsin indicate that most flocks consist of units of up to as many as about 13 birds that remain together on local foraging flights. Thus, families

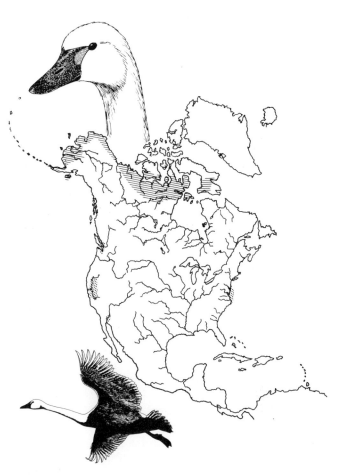

Map 15. Breeding (hatched) and wintering (stippling) distributions of the whistling swan.

and pairs are the obvious unit of substructure in whistling swan flocks. By the time they reach their breeding grounds, these flocks have broken up and the birds spread out widely over the tundra, often in densities of only about one or two pairs per square mile. Pair formation undoubtedly occurs in wintering flocks, presumably when the birds are in their second or third winter. There seems to be no record of captive whistling swans breeding before their fourth year, but it is uncertain that this is typical of wild birds. The triumph ceremonies and behavior associated with copulation are nearly the same in this species as in the trumpeter swan, although the speed of movements and associated vocalizations differ appreciably.

Reproductive biology. Arrival on the nesting grounds of western Alaska occurs in late May, and nesting is usually underway by the first of June. There is a high degree of synchrony of nest initia-

◈ ◈ ◈ 43

tion in individual areas, so that most hatching occurs over a period of only three or four days. Most pairs choose nest sites on the shore of a lake or a pond within 20 yards of water; somewhat fewer nests are built on islands or points of lakes, and even fewer on tundra or in other locations. Elevated sites, such as hummocks, are favorite nest sites and are also among the first areas to be snow-free in spring. There are marked year-to-year differences in average clutch sizes, but they range from about 3 to 5 eggs, with smaller average clutch sizes associated with late, cold springs. Incubation is normally performed only by the female, although a few observers have seen males sitting on and apparently incubating the eggs. An incubation period of 30 to 32 days is typical; in western Alaska most hatching occurs in late June and July. Most broods remain within 100 to 400 yards of their nest for some time, and the young grow at a very rapid rate, reaching weights of 11 to 12 pounds in 70 days, and fledging at about the same time, at between 60 and 75 days of age. In this area, the adults undergo their flightless period in July and early August, and nonbreeders begin to regain their power of flight in late August. About 85 days after the peak of hatching, families with fledged young begin to join the nonbreeders for the fall flight southward (Bellrose, 1976).

Status. The North American population of whistling swans probably consists of somewhat less than 100,000 birds, judging from federal surveys. About half of this population winters on the Atlantic and half on the Pacific coast. About 60 percent of the total breeding population nests in Alaska. There appears to have been a small increase in estimated numbers over the past 25 years (Bellrose, 1976). There have been a few limited experimental hunting seasons for whistling swans in recent years, but at present hunting cannot be said to have any effect on the population.

Relationships. The relationships of the northern swans have been discussed by me (Johnsgard, 1974a), and it is still uncertain whether the Bewick swan or the trumpeter swan is the whistling swan's nearest relative. The general consensus is that the Bewick and whistling swans are conspecific, but there are some conflicting data that do not support this position.

Suggested readings. Wilmore, 1974; Scott & Wildfowl Trust, 1972; Bellrose, 1976.

Bewick Swan

Cygnus bewickii Yarrell 1830

Other vernacular names. None in general English use. Zwergschwan (German); cygne de Bewick (French); cisne de Bewick (Spanish). The term "tundra swan" has been proposed by Palmer (1976) to designate the collective whistling swan–Bewick swan complex, on the assumption that they consist of a single species.

Subspecies and range. No subspecies recognized here, although an eastern race, *jankowskii*, is often recognized. The breeding range is from the Pechenga River, near the Fenno-Russian border, eastward along the north Siberian coast to about 160° east longitude, as well as Kolguev Island and southern Novaya Zemlya. Winters in two widely separated areas—Europe (primarily the Netherlands), and China, Japan, and Korea. See map 16.

Measurements and weights. (Mostly from Scott & Wildfowl Trust, 1972). Folded wing: males, 485–573 mm; females, 478–543 mm. Culmen: males, 82–108 mm; females, 75–100 mm. Weights: males, 4.9–7.8 kg (av. 6.4 kg); females, 3.4–6.4 kg (av. 5.0 kg). Eggs: av. 103 x 67 mm, white, 260 g.

Identification and field marks. Length 45–55" (115–40 cm). Plate 13. This is the smallest of the northern swans; it is the only one with a partially yellow bill that has a wing measurement of under 575 millimeters in adults. The yellow markings of the bill do not extend below and beyond the nostrils; as in the other northern swans, the feet and legs are black. *Females* are identical to males, and *juveniles* have mottled gray and white plumage for most of their first fall and winter of life. Older immature birds gradually become indistinguishable from adults.

In the field, the relatively small size and high-pitched musical call of this species provide useful field marks. The extent of yellow on the bill may be useful for close-range identification.

NATURAL HISTORY

Habitat and foods. The breeding habitat of this species consists of shallow tundra pools with abundant submerged vegetation and a luxuriant growth

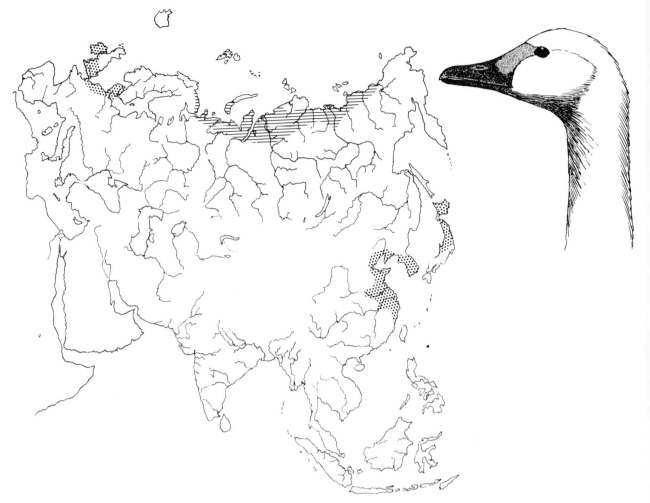

MAP 16. Breeding (hatched) and wintering (stippling) distributions of the Bewick swan.

of shoreline vegetation (Voous, 1960). Virtually no specific information on breeding-grounds foods is available, but in the U.S.S.R. the birds are reported to consume a variety of aquatic plants and grassy territorial plants. As with the other northern swans, the leaves, roots, and stems of pondweeds are favored foods on migration and in wintering areas, and the roots and rhizomes of eelgrass (*Zostera*) are also an important food. Various pasture grasses (*Glyceria, Agrostris,* etc.) are grazed extensively when they are wet or flooded in England, and sometimes grazing on grasses of drier pastures is also done (Owen & Kear, in Scott & Wildfowl Trust, 1972).

Social behavior. Probably more is known of the social behavior of this swan than any other of the northern swans, as a result of extensive observation of individually identified birds at the Wildfowl Trust in England. Through this work it is known that pair bonds are strong and permanent, without a single known case of "divorce" among hundreds of pairs studied. Birds which lose a mate may take up to three seasons to find a new one, but some establish new bonds much sooner. Further, family bonds persist in immature birds even up to the third winter of their lives, resulting in the association of up to four generations of related birds. Some tentative pairing may occur in birds as early as their second winter of life, but nearly all these bonds are broken by the following winter, and probably initial breeding does not occur in this species until the birds are nearly four years old (Scott & Wildfowl Trust, 1972). As in the other swans, pairs are formed and maintained by mutual displays such as the triumph ceremony; copulation probably plays little if any role in such pair-bond development as it rarely if ever occurs among

46 ❖ ❖ ❖

wintering flocks. Bewick swans winter at great distances, usually well over a thousand miles, from their breeding grounds, and the departure from England for the breeding grounds occurs in January and February. Arrival on the breeding grounds of Siberia may not occur until May, and is associated with the onset of thaws and the appearance of flowing water in rivers (Dementiev & Gladkov, 1967).

Reproductive biology. The swans spread out greatly on reaching their breeding grounds and select nest sites immediately. These sites are usually small hummocks on the tundra providing good visibility. The nests are constructed of tundra vegetation, especially sedges and mosses, with a lining of down. Old nest sites are frequently used again. The first eggs are laid on Novaya Zemlya at the end of May or early June, and clutch sizes are reportedly only 2 to 3 eggs there, although 3 to 5 is probably a normal range for the clutch. The female alone incubates, and the normal incubation period is 29 to 30 days, or slightly shorter than in the other northern swans. Likewise, a fledging period of only 45 days has been reported for this species, although this is difficult to accept in view of the appreciably longer fledging periods known to occur in the slightly larger whistling swan. Nonetheless, since the breeding territories are abandoned in September, the cygnets can be little more than 80 or 90 days old before they begin their thousand-mile southward migration (Dementiev & Gladkov, 1967).

Status. The population of Bewick swans that winters in northwestern Europe numbers about 6,000 to 7,000 birds, of which about half winter in the Netherlands, about 1,500 in England, 500 to 1,000 in Ireland, 700 in Denmark, and 300 in West Germany. The wintering populations have been stable in recent years, and wintering in England has increased appreciably with local protection and encouragement. Although the birds are protected everywhere, illegal shooting results in about 25 percent of the live birds carrying lead shot in their bodies. The restricted winter quarters for this population make it susceptible to future reductions, but in general the extensive tundra breeding grounds of the species are not now in danger (Hudson, 1975).

Relationships. The Bewick swan and whistling swan are at present regarded as conspecific by most authorities, but such an interpretation may be a considerable oversimplification of the evolutionary history of this group of swans. The small size of these two forms may simply reflect evolutionary convergence to comparable arctic tundra environments from separate stocks, rather than indicate conspecificity (Johnsgard, 1974a).

Suggested reading. Scott & Wildfowl Trust, 1972; Wilmore, 1974.

Coscoroba Swan

Coscoroba coscoroba (Molina) 1782

Other vernacular names. None in general English use. Koskorobaschwan (German); cygne coscoroba (French); cisne coscoroba (Spanish).

Subspecies and range. No subspecies recognized. Breeds in temperate South America from about 45° south latitude southward to Cape Horn and perhaps occasionally on the Falkland Islands. Winters variably northward, with some birds reaching central Chile, northern Argentina, Paraguay, Uruguay, and extreme southern Brazil. See map 17.

Measurements and weights. (Mostly from Scott & Wildfowl Trust, 1972.) Folded wing: both sexes 427–80 mm. Culmen: both sexes 65–70 mm; females only, 63–68 mm. Weights: males, 3.8–5.4 kg (av. 4.6 kg); females, 3.2–4.5 kg (av. 3.8 kg). Eggs: av. 91 x 63 mm, whitish cream, 185 g.

Identification and field marks. Length 35–45" (90–115 cm). Plate 14. Coscorobas are both swanlike and gooselike; the adult plumage is entirely white except for the distal portion of the primaries, which is black. The head is feathered in front of the eyes, as in geese, but the bill is ducklike in shape and bright pinkish red, while the feet and legs are fleshy pink. The iris is yellow to reddish in adult males. *Females* have a dark brown iris color, as do immatures. *Juvenile* birds have a distinctive brownish pattern on the head and upper parts, and their bill is grayish. Some gray or brownish feathers may persist, especially on the wings, until the second year.

In the field, coscoroba swans somewhat resemble large white domestic geese, but their black wing

tips (often scarcely visible except in flight) and bright pinkish red bills are distinctive. Both sexes have a loud *cos-cor-ooo* trumpeting call, uttered in flight and when on water.

NATURAL HISTORY

Habitat and foods. In Chile, the favored habitats of the coscoroba swan are the lagoons and swampy areas of the Magellanic region, where they are often found with black-necked swans. Weller (1967a) found them very common on a large fresh-water marsh with its shallow areas dominated by bulrushes and cutgrass *(Zizaniopsis)*. These birds normally feed by swimming or wading in shallow water and may rarely also come ashore to graze along the water's edge. These foraging traits, in conjunction with the ducklike bill structure, must effectively reduce competition with the black-necked swan for food. However, little is known specifically of their food; Johnson (1965) states that they eat various plants, aquatic insects, fish spawn, and apparently sometimes also small crustaceans.

Social behavior. Coscoroba swans seem most often to be seen in small numbers; rarely are flocks numbering in the hundreds ever mentioned. Weller (1967a) reported loafing groups of 60 to 100 along the shore of an Argentine lake; probably the species' greatest numbers are reached on various mid-summer molting grounds. It is presumed that these flocks consist primarily of paired birds, and that the pair bonds of this swan are permanent. Little is known definitely of sexual maturation rates; although captive birds have been known to breed at four years, it is assumed that the normal period to sexual maturity in the wild may be only two or perhaps at most three years. Unlike the typical swans, coscorobas do not have a triumph ceremony by which pair bonds can be established and maintained. A simple greeting ceremony of calling does exist and a threat display resembling that of the mute swan also occurs, but the means and timing of pair formation remain obscure. In further contrast to the other swans, copulation occurs while the birds stand in shallow water, after the male performs head dipping. Afterwards both birds call as the male partially raises his folded wings. In this respect and in lacking a triumph ceremony, the coscoroba swan strongly resembles some of the whistling ducks, such as the black-bellied (Johnsgard, 1965a).

Reproductive biology. The coscoroba swan nests in the austral spring; in Chile the breeding season extends from October to December (Johnson, 1965), while Gibson (1920) reported nesting in the vicinity of Buenos Aires during the period of June to November. Although normally the bird is a solitary nester, he mentioned finding 16 nests in an area measuring 400 square yards. Johnson (1965) describes the nest site as usually a bulky mound of soft vegetation placed among reed beds, in long grass close to water, or, if possible, on small islands. The nest cup is provided with an extensive down lining, and although only the female incubates the nest, it is closely tended by the male as well. The clutch size ranges from 4 to 7 eggs and probably averages about 6. It has been reported that the female usually leaves her nest to forage for only about an hour in the morning and afternoon during incubation, and before leaving carefully covers the eggs with down or even twigs. Incubation requires approximately 35 days, and all of the cygnets normally hatch on the same day. They differ from those of typical swans in being more distinctly patterned, with head markings slightly suggestive of those of whistling ducks. Adult cos-

coroba swans have never been seen carrying their young on their backs. They probably normally brood them on shore and thus back carrying is less likely than in the black-necked swan. The flightless period of the adults appears to be fairly variable; Weller (1975a) reported seemingly flightless birds on Tierra del Fuego in January or February, and elsewhere in Argentina they have been collected from mid-November to mid-April.

Status. The coscoroba swan is not extremely abundant anywhere in South America, but neither is it obviously declining. As with the black-necked swans, loss of its favored temperate marsh habitats would be the most serious threat to its continued survival. It is not particularly sought by hunters, although some trampling of nests by cattle probably occurs locally. It is apparently becoming more numerous on the Falkland Islands, where it was once regarded as only a rare visitor. However, it is still not a proven nester there.

Relationships. It is impossible to neatly pigeonhole this species taxonomically, since it exhibits an unusual combination of goose, swan, and whistling duck characteristics (Johnsgard, 1965a). Its skeletal anatomy suggests that it is a swan with some gooselike characteristics (Woolfenden, 1961), and except for the lack of a triumph ceremony, its behavior is generally swanlike. A study of its feather proteins by electrophoresis (Brush, 1976) indicates that the coscoroba swan should be retained within the tribe of geese and swans (Anserini), rather than being considered a link with the whistling ducks or an aberrant member of that group.

Suggested readings. Wilmore, 1974; Scott & Wildfowl Trust, 1972.

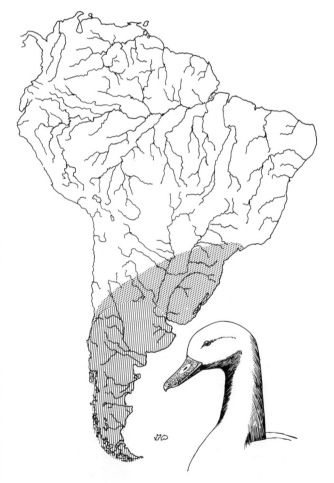

MAP 17. Breeding or residential (hatched) and wintering (stippling) distributions of the coscoroba swan.

Swan Goose

Anser cygnoides (Linnaeus) 1758

Other vernacular names. Chinese goose (domesticated form); Schwanengans (German); oie cygnoide (French); ganso-cisne (Spanish).

Subspecies and range. No subspecies recognized. The breeding range extends from the Ob and Tobol river drainages in south-central Siberia eastward to Kamchatka and the Commander Islands, south to

Central Asia and northern Mongolia. Winters in northern China and Japan. See map 18.

Measurements and weights. Folded wing: males, 450–60 mm; females, 375–440 mm. Culmen: males, 87–98 mm; females, 75–85 mm. Weights: males ca. 3,500 g; females, 2,850–3,450 g. (Kolbe, 1972). Eggs: av. 82 x 56 mm, white, 145 g.

Identification and field marks. Length 32–37" (81–94 cm). *Adults* have a dark chestnut crown, nape, and hind-neck stripe that extends from a narrow white band around the bill to the base of the neck; the rest of the head and neck is pale brown and sharply demarcated from the darker area. The mantle, scapulars, and flanks are ashy brown with buffy edges; the breast and abdomen are pale brown, grading to white behind the legs; and the under tail area is white. The tail is grayish brown with white tips and edging, and is bounded in front by a white bar on the upper tail coverts. The upper wing surface is mostly grayish brown, with whitish edging on the feathers. The bill is black, the iris is brown, and the legs and feet are reddish orange. *Females* differ from males in having shorter bills and necks, and are slightly smaller throughout. *Immatures* lack the white band around the base of the bill, lack conspicuous flank markings, and have dull, grayish brown upperparts.

In the field, the extremely long, black bill would serve to separate this species from the rather similar bean goose, which also lacks the distinctive two-toned head and neck of this species. The voice of this species closely resembles that of the other gray geese.

NATURAL HISTORY

Habitat and foods. Few specifics have been recorded on the ecology of this species; Dementiev and Gladkov (1967) state that it is always found in close association with rivers and lakes, but may be seen in mountains, valleys, or steppe terrain. In mountainous areas it is likely to be encountered in broad valleys, at the mouths of rivers, in forested lakes, marshes, or even in narrow ravines associated with swift currents. In steppes and plains the birds are likely to be encountered in broad river valleys, marshes, riverine and lacustrine meadows, and on fresh-water or brackish lakes with extensive emergent vegetation. Outside of the breeding season they are sometimes found far from water on barren steppes, and juvenile birds often seek out swampy and muddy lake shores or sandy shorelines of lakes and rivers. Like the other gray geese, the birds are exclusively vegetarians, and although little specific information is available, they are known to utilize sedges.

Social behavior. Flock sizes vary with the season, with spring flocks reportedly usually containing no more than from 20 to 40 birds, while fall flocks are sometimes appreciably larger. It is probable that the birds mature their second winter and at that time begin to form permanent pair bonds. At least under conditions of captivity the birds are most likely to breed initially in their second or third year (Ferguson, 1966). What little is known of the pair-forming behavior is suggestive that it differs in no appreciable way from that of the other true geese. Copulatory behavior is likewise the same, with the male swimming very high in the water before the female, cocking his tail, and repeatedly performing head-dipping movements that somewhat resemble foraging movements. These movements are accompanied by rather strong foot paddling, although the bird remains nearly in place. Less exaggerated movements of the same type are performed by the female until mounting occurs. When treading is completed, the male begins to open his wings as the female starts to call, and in this species both birds extend their necks to the maximum, point the bills nearly vertically, and partially open their wings and call loudly. The feet are paddled strongly at first, after which both birds settle back into the water and begin to bathe and preen (Johnsgard, 1965a). In the wild, pair formation probably occurs during the time the birds are on the wintering grounds, as courtship has not been observed in the breeding areas.

Reproductive biology. In the U.S.S.R., the birds arrive on their breeding areas in April, or even as late as May in some of the mountain plateaus of the trans-Baikal area. The nest site is typically in grasses and is lined with dry grass and down. In Sak-

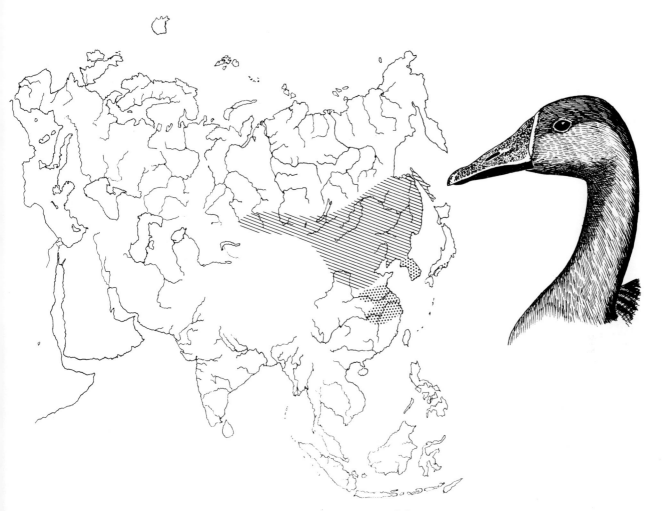

MAP 18. Breeding (hatched) and wintering (stippling) distributions of the swan goose.

halin, Gisenko and Mischin (1952) found numerous breeding pairs in a 25-kilometer stretch of river shoreline lined with heather (*Ledum*) and pines (*Pinus cembra*). The normal range in clutch sizes is from 5 to 8 eggs, but the majority of the nests contain either 5 or 6 eggs. Only the female incubates, but the male remains very near the nest and helps to defend it. In Mongolia, eggs have been seen as early as the last part of April, but May is probably the major month of nesting. The incubation period is 28 days. Goslings have been seen on the Amur River as early as mid-May, but in the more mountainous areas the broods hatch appreciably later. Brood sizes in the Amur are usually 5 to 6 goslings, but at times several broods unite and are jointly defended by 4 to 6 older birds. Immature and unsuccessful breeders tend to

flock and begin to undergo their molt well in advance of the successful breeders. Adults leading young begin to molt in July, and by the end of August the young birds have fledged and the adult birds have completed their postnuptial molts. At this time fall flocking begins, with most birds departing their breeding areas in September (Dementiev & Gladkov, 1967).

Status. Regrettably little is known of the status of this goose. Dementiev and Gladkov (1967) comment that although it is most common in the Amur River Basin, it has recently been declining there, and is generally rare or absent over much of its past range.

Relationships. This species seems to represent one extreme of the genus *Anser*, although there is no good behavioral justification for regarding it as a separate genus ("*Cygnopsis*"), as has often been done. Woolfenden (1961) has found anatomical evidence to support the inclusion of this species in the genus *Anser*.

Suggested readings. Dementiev & Gladkov, 1967; Uspenski, 1965.

Bean Goose

Anser fabalis (Latham) 1787

Other vernacular names. Pink-footed goose, Sushkin's goose; Saatgans and Kurzschnabelgans (German); oie des moissons and oie à bec court (French); genso de las habas and ganso de pies carmín (Spanish).

Subspecies and ranges. (See map 19.)

A. f. fabalis: Western bean goose. Breeds in wooded areas of northern Scandinavia and northern Russia, east to the Ural Mountains. Winters in Europe from Britain south to the Mediterranean and Black Sea.

A. f. johanseni: Johansen bean goose. Breeds in the wooded region of western Siberia, east to the Khatanga River. Winters in Iran, Turkestan, and western China.

A. f. middendorfi: Middendorf bean goose. Breeds in forests of eastern Siberia, from the Khatanga to the Kolima region and western Anadyrland. Winters in Japan and eastern China.

A. f. rossicus: Russian bean goose. Breeds in the tundra zone of Novaya Zemlya, and on the Yamal, Gyda, and Taimir peninsulas. Winters in Europe, Russia, western Siberia, Turkestan, and China.

A. f. serrirostris: Thick-billed bean goose. Breeds in the tundra zone from Khatanga to the Chukotsk Peninsula, Anadyrland, and Koryakland. Winters in Korea, China, and Japan.

A. f. brachyrhynchus: Pink-footed bean goose. Breeds in eastern Greenland, Iceland, and Spitsbergen. Winters in northwestern Europe. Sometimes considered a separate species.

Measurements and weights. Folded wing: both sexes 395–562 mm. Culmen: both sexes (*brachyrhynchus*) 37–54 mm; all other races 51–87 mm.

Weights: both sexes (*brachyrhynchus*) 2,750–3,500 g; both sexes (*fabalis*) 3,171–3,948 g. Eggs: av. 78–84 mm x 52–60 mm, white, 120–46 g.

Identification and field marks. Length 28–35" (71–89 cm). *Adults* of both sexes have a dark, chocolate brown head and neck, usually with no white behind the bill except in some instances when a narrow band may be present; a strongly furrowed neck; and a breast, flank, and underparts coloration that is mostly light brown, with darker flank markings and white edging at the top of the flanks. The rump and tail coverts are white, and the tail is grayish brown, with broad white tips and edging. The upper wing surface is brown, but the coverts and secondaries are broadly edged with whitish coloration. The legs and feet are bright yellow (most races) or pink (*brachyrhynchus*), the bill is black, with variably large yellow to pink markings that at times may include nearly the entire bill except for the nail. *Juveniles* resemble adults, but are generally duller, with less conspicuous light edging on the mantle feathers.

In the field, bean geese closely resemble white-fronted geese, but usually lack definite white forehead markings, and always lack uniformly pink bills and black breast markings. Even the pink-footed bean goose has a lower-pitched call, *ung-unk*, than that of the white-fronted goose, and the typical bean geese have even more reedy and bassoonlike notes that at times resemble the lowing of sheep.

NATURAL HISTORY

Habitat and foods. The several races of bean geese nest in a variety of habitats, with some of the subspecies associated with subarctic taiga lakes or subalpine forest river valleys. Others are more specifically confined to open tundra habitats, including moss-lichen, grassy, sedgy, and scrub tundra vegetation types, and even extending into rocky tundra virtually devoid of vegetation, as in the pink-footed race. Regardless of their race, bean geese are essentially vegetarians and consume grasses, sedges, mosses, herbs, and even some shrubby species. Wintering birds consume a wide variety of crop plants, including waste grain, abandoned potatoes, corn, and beans. (The latter accounts for the bird's English name; the Russian vernacular name means thresher or threshing-floor bird, referring to their appearance at threshing floors and plowlands, while the German name means corn or crop goose.) Berries, particularly blueberries, seem to be preferred foods for the

birds while on their breeding grounds. On their wintering grounds bean geese are rather more generalized foragers than white-fronted geese, and often make rather long flights from their roosting areas to their foraging grounds, which range from grasslands to hayfields and croplands (Palmer, 1976; Dementiev & Gladkov, 1967).

Social behavior. At least in the case of the pink-footed race, initial reproduction usually occurs as the birds approach three years of age, with a few two-year-olds attempting to breed. Almost certainly most birds mate by the time they are two years old, presumably during wintering-grounds associations. Wild birds are monogamous and pair bonds are thought to be permanent. It is known from captive birds that the aggressive postures of this species are exactly like those of the other *Anser* species and that the behavior associated with copulation is also the same (Johnsgard, 1965a).

Reproductive biology. The most intensively studied population of this species is that which breeds in

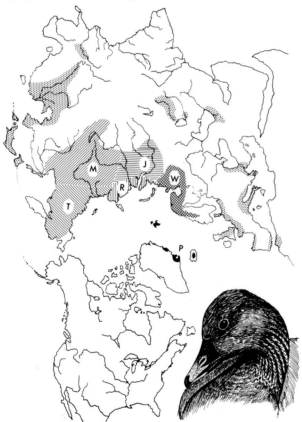

MAP 19. Breeding distributions of the pink-footed ("P"), thick-billed ("T"), Middendorf ("M"), Russian ("R"), Johansen ("J"), and western ("W") bean geese. The wintering distribution of all races is indicated by stippling.

Iceland and for many years has been the subject of intensive study by English biologists (Scott & Fisher, 1953; Scott et al., 1955). These birds typically arrive at their breeding areas in central Iceland in May and begin nesting almost immediately. Nests are generally built on dry and well-drained sites, often low frost mounds, that not only provide excellent visibility all around but usually also have, within about five meters of the nest itself, a second lookout perch available for the gander to use. Although the birds are colonial, nests are rarely placed closer than 15 meters apart. However, in areas of high nesting density of central Iceland, from 150 to 250 nests per square kilometer have been reported (Kerbes et al., 1971). Many nests are placed over sites used in earlier years, presumably by the same pairs, and the accumulations of manure, vegetation, and other materials gradually help to build up the nest site above its surroundings. Eggs are laid on a daily basis by the female, and full clutches are typically of 4 or 5 eggs, rarely as many as 7. Kerbes et al. (1971) reported a mean clutch size of 3.9 eggs for 312 nests of pink-footed bean geese; the literature from the U.S.S.R. suggests somewhat larger clutches for the larger "forest" forms of bean geese. Likewise, the incubation period of 26 days for the pink-footed race may be a day or two less than that characteristic of the larger and more southerly-breeding forms.

Once incubation begins, the female leaves her clutch only rarely, at most leaving once a day for foraging and defecation. There is a high degree of synchrony in hatching within goose colonies; and after leaving the nest, the young are rarely if ever taken back to it by their parents, and instead roost on the banks of streams and tarns at night. In some areas where they breed near the coast, the young are taken to salt water after about a month. The fledging period is probably between seven and eight weeks, which in Iceland means that they fledge in early August. Adults that do not breed molt in early July, and these birds are followed in about two weeks by those with broods. The flightless period is probably 26 to 28 days. Thus, both adults and young are flying by late August in Iceland. It is also known that a substantial number of Icelandic birds fly to northeastern Greenland to undergo their molt (Kerbes et al., 1971).

Status. Of the various races of bean geese, only the pink-footed race has a population that can be fairly accurately estimated. It has increased considerably in recent years, and during the winter of 1974–75 a peak count of 89,000 birds was obtained for Britain and Ireland. However, a substantial reduction occurred in 1975–76, after a disastrous breeding season (Ogilvie & Boyd, 1976). Perhaps only about 1,000 of these birds come from Greenland, while about 75 percent of the Icelandic population breeds in a very restricted area of the central highlands of Iceland, which is threatened by a possible hydroelectric development project. Those that breed in Spitsbergen winter in Denmark and the Low Countries, and in the late 1960s numbered about 12,000 to 15,000 (Kerbes et al., 1971). Bauer and Glutz (1969) provide a summary of wintering estimates for the western bean goose in various European countries, but little can be said of the population sizes of the other subspecies.

Relationships. Ploeger (1968) has discussed the evolutionary history of this species in terms of Pleistocene glaciation patterns. The nearest living relative of this species is probably the white-fronted goose, although the tundra-breeding forms are nearly ecological counterparts to the snow geese of North America. The forest-breeding birds come closer to the Canada goose in their habitat usage, and in general their foraging adaptations and behavior patterns seem to overlap considerably.

Suggested readings. Scott & Fisher, 1953; Scott et al., 1954.

White-fronted Goose

Anser albifrons (Scopoli) 1769

Other vernacular names. Specklebelly goose, tule goose, whitefront; Blessgans (German); oie rieuse (French); ganso frente blanca (Spanish).

Subspecies and ranges. (See map 20.)

A. a. albifrons: European white-fronted goose. Breeds in northern Russia and Siberia east to the Kolima River. Winters in England, along the coast of the North Sea, in the Mediterranean countries, and south of the Black and Caspian seas.

A. a. flavirostris: Greenland white-fronted goose. Breeds on the west coast of Greenland. Winters in Ireland, occasionally reaching England and the Atlantic coast of North America.

A. a. frontalis: Pacific white-fronted goose. Breeds in Alaska, on St. Lawrence Island, and in eastern Siberia, probably west to the Kolima Valley. Winters in the western United States, China, Japan, and northern Mexico.

A. a. gambelli: Gambel white-fronted goose. Breeding grounds probably in the Mackenzie Basin of Canada and adjacent western Alaska. Winters in Texas and Mexico.

A. a. elgasi: Tule white-fronted goose. Breeding grounds uncertain, presumed to be in the tiaga zone just south of the tundra in Alaska. Winters in the Sacramento Valley of central California.

Measurements and weights. Folded wing: males, 380–474 mm; females, 368–440 mm. Culmen: males, 42–63 mm; females, 42–62 mm. Weights: males (*albifrons*) 2,400–3,200 g; females 1,700–3,000 g (Kolbe, 1972). Males of *frontalis* and *elgasi* average 2,404 and 2,993 g, respectively, while females of the two races average 2,222 and 2,861 g (Nelson & Martin, 1953). Eggs: av. 76 x 54 mm, white, 125 g.

Identification and field marks. Length 26–34" (66–86 cm). *Adults* of both sexes have a grayish brown head and neck, except for a white forehead and anterior cheeks, bordered narrowly with blackish coloration. The neck feathers are vertically furrowed, and the lower neck, back, breast, and sides are sooty brown, narrowly edged with white, producing a slight barring. The tail coverts and sides of the body behind the flanks are white, the tail is brownish black tipped with white, and the abdomen and sides are variably blotched with black. The scapulars and tertials are grayish brown, tipped with white; the upper wing coverts are brownish gray (outwardly) or grayish brown and tipped with white, the white tips especially noticeable on the greater secondary coverts. The primaries and secondaries are bluish gray to black, while the underwing surface is slate gray. The bill is pinkish (orange-yellow in *flavirostris*), with a paler nail, grading to yellow (dorsally) and bluish (basally). The iris is brown, and the legs and feet are orange. *Immatures* in their first winter lack the black ventral markings and white facial patch, and have less conspicuous white markings on the flanks and upper surface. The abdomen is dull white with gray mottling, the bill is dull-colored, and the legs and feet are pale yellow. The white forehead is gradually acquired the first winter, but heavy black underpart markings are apparently not attained before the second year.

In the field, white-fronted geese are most readily identified by their distinctive high-pitched call, *leek-leek*, which resembles taunting laughter, and by the black belly markings of adults. The white forehead markings are also evident at considerable distances, and, except where the lesser white-front occurs, this feature helps identify the species. White-fronted geese also closely resemble graylag geese and the two often associate on wintering areas, but their voices as well as their forehead and abdominal markings are quite different.

NATURAL HISTORY

Habitat and foods. The breeding habitats of this widespread species are quite diverse, but at least in North America they typically consist of arctic vegetation of fairly low stature, near marshes, ponds, rivers, and lakes. Both coastal tundra and gently rolling upland tundra habitats are utilized, at times up to 700 feet above sea level. In addition, willow- and shrub-lined ponds and streams are used for nesting to a greater degree than is true of the other arctic-nesting geese. The birds are vegetarians throughout the year, and although few studies of summer foods have been undertaken, such plants as horsetails *(Equisetum)*, cotton grass *(Eriophorum)*, and grasses and herbs have been reported from summer specimens. During the fall and winter the birds concentrate on grasses and sedges, including such grain plants as wheat, rice, and barley, as well as many wild species including sedges, and the root-stalks of bulrushes and cattails. The birds often forage in company with Canada geese, and presumably eat much the same foods as that species. Winter habitats vary greatly, and range from coastal marshes to wet meadows and fresh-water marshes (Johnsgard, 1975; Dzubin et al., 1964).

Social behavior. Present evidence indicates that white-fronted geese form monogamous and permanent pair bonds during their second year of life, but probably normally breed for the first time when they are nearly three years old. Pair bonds are presumably formed over a rather long period during the second winter, and the limited observations on captive birds indicate that the associated behavior is virtually identical to that of the other "gray geese," with hostile encounters and associated triumph ceremony behavior serving to form pair bonds and maintain

MAP 20. Breeding distributions of the European ("E"), Pacific ("P"), Gambel ("G"), and Greenland ("Gr") white-fronted geese. The wintering distributions of all races is indicated by stippling.

family bonds (Boyd, 1953). Copulation is preceded by the usual mutual head-dipping behavior, and is followed by mutual calling and wing lifting (Johnsgard, 1965a). As the winter proceeds, the large flocks tend to break up into progressively smaller units consisting of groups of pairs and families that gradually spread out and become more inconspicuous as the spring migration progresses.

Reproductive biology. White-fronted geese arrive on their breeding areas only shortly after they become snow-free, and accompanying each pair are their young of the past year and sometimes also the two-year-old young, although the latter birds usually do not become territorial and soon leave the area. However, the yearlings remain close to their parents and provide important defence against predators and human intruders (Barry, 1966). Nesting pairs are usually widely scattered, and adults only rarely use

exactly the same nest site as in the previous year. The specific site selected is often on an incline or at the top of a low hill, with excellent visibility as well as grassy cover. The birds are very early nesters, normally laying their eggs at the rate of one per day, but at times skipping a day, until a clutch of 3 to 7 eggs is laid, with considerable yearly variations that seem to be related to the weather during the time of nesting. The incubation period is probably about 23 days on the average, but some estimates have been as low as 21 days and others as high as 28 days.

While jaegers and gulls are sometimes serious nest predators, the geese are large enough to be able normally to defend against these, and their nests are so scattered and the birds so well concealed that such predation is less likely to have an important effect than inclement weather during egg laying, incubation, and hatching. This is indicated by the large year-to-year variations in nesting success in this species. Within 24 hours of hatching, the goslings are led to water, and the broods are typically raised in areas with an abundance of sedge vegetation into which the birds can easily flee when disturbed. The fledging period is approximately 45 days for the geese in the Anderson River delta of the Northwest Territories of Canada (Barry, 1966), but there is also a somewhat questionable estimate of a five-week fledging period for the Greenland white-front, and an estimate of 55 to 65 days for the Yukon delta (Mickelson, 1973). The adults undergo a 35-day flightless period shortly after the young hatch, so that both adults and young attain flight at about the same time.

Status. Bellrose (1976) has reported that winter inventories between 1955 and 1974 indicate a North American population of about 200,000 white-fronted geese, while post-breeding season surveys in Alaska and Canada report about 250,000 and 55,000 birds, respectively. Most of these would be of the Pacific race; the Gambel white-fronted goose is relatively rare and Elgas (1970) believed that it may breed in the Old Crow Flats area of the Northwest Territories, which has about 1,800 birds in the post-breeding season (Bellrose, 1976). The tule goose's population is apparently even smaller, and its numbers are even more uncertain. S. R. Wilber (cited in Bellrose, 1976) estimated that about 250 birds of this form were present in 1955–56 among the 2,000 white-fronted geese wintering in the Suisun marshes of California, one of its few known wintering areas. It also winters in the Butte Creek Basin, near Marysville, and on the Sac-

❖ ❖ ❖ 57

ramento National Wildlife Refuge in small but unknown numbers. The total population may number about 1,500 birds (R. Elgas, pers. comm.). Estimates of the various components of the European population of white-fronted geese have been summarized by Bauer and Glutz (1968), who report that it is the most abundant species of goose in the Old World, with the European subspecies wintering primarily in Hungary and the Netherlands. Philippona (1972) estimated the total Eurasian population of the European white-fronted goose at about 300,000 birds, or 40,000 breeding pairs. The Greenland form numbers about 12,000 to 15,000 individuals wintering in Ireland, Scotland, and Wales. The size of the component of the Pacific white-fronted goose that winters in China, Japan, and elsewhere is unknown.

Relationships. This is the most widespread of all the species of *Anser* and is obviously closely related to the bean goose, the graylag goose, and especially the lesser white-fronted goose. Ploeger (1968) has discussed the possible role of glaciation patterns in the evolutionary history of these populations.

Suggested readings. Barry, 1966; Mickelson, 1973.

Lesser White-fronted Goose

Anser erythropus Linnaeus 1758

Other vernacular names. None in general English use. Zwerggans (German); oie naine (French); ganso frente blanca chico (Spanish).

Subspecies and range. No subspecies recognized. Breeds in subarctic areas from Scandinavia and Russian Lapland eastward through Siberia to the Kolima and Anadyr basin. Winters south to southern Europe, Egypt, Turkestan, northwestern India, China, and, rarely, Japan. See map 21.

Measurements and weights. Folded wing: males, 365–403 mm; females, 334–81 mm. Culmen: males, 42–52 mm; females, 25–35 mm. Weights: males, 1,440–2,300 g; females, 1,300–2,150 g (Kolbe, 1972). Eggs: av. 76 x 49 mm, white, 100 g.

Identification and field marks. Length 26–34″ (53–66 cm). Plate 15. *Adults* of both sexes have a dark brownish gray head, with a well-developed white forehead marking that extends almost to the top of the crown, and a bright yellow eye-ring. The furrowed neck, breast, abdomen, and flanks are light brown, with darker brown flank marking, irregular blotches of black on the abdomen and lower sides, and a narrow white upper flank edging. The rump is slate gray, the tail coverts are white, and the tail feathers are black with white tips and edging. The mantle and upper wing coverts are brown to brownish gray, with buff edging, and the flight feathers are brownish black. The iris is brown, the bill pink with a paler nail, and the legs and feet yellow. First-year *immatures* usually lack the white forehead markings and the black spotting underneath, and have black markings on the culmen.

In the field, the small size (it is the smallest of the "gray geese") and very high-pitched squeaking calls (*kow yu* or *kow-yu-yu*) help to separate this species from other geese. Not only is it appreciably smaller than the European white-fronted goose; its higher white forehead markings and the conspicuous yellow rings around the eyes are visible for considerable distances. Even first-year birds, which lack the white forehead markings and black belly spotting, have fairly bright eye-rings.

NATURAL HISTORY

Habitat and foods. Breeding habitats of this species include the tundra zone, parts of the northern woodlands, and especially brush-covered tundra of the scrubby timberline ecotone between tundra and subarctic or subalpine tiaga forest (Uspenski, 1965). In contrast to the white-fronted goose, it breeds not

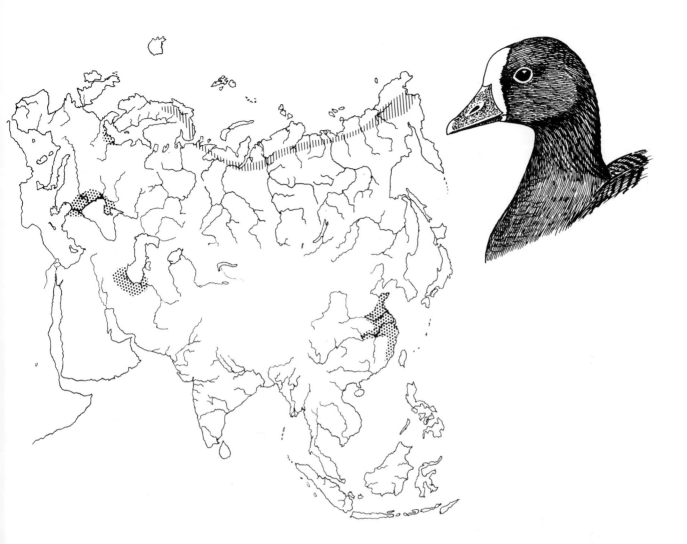

MAP 21. Breeding (hatched) and wintering (stippling) distributions of the lesser white-fronted goose.

only on tundra, but also on the lower parts of mountain streams, on mountain foothills, mountain lakes, and even on alpine precipices, often in thawing boggy areas or on stonefields. Like the larger species, it is a vegetarian, foraging in spring on freshly sprouting greenery. At least in Lapland, breeding birds also feed extensively on small arctic willows *(Salix herbacea)*. The very short bill of this species probably allows for extremely close cropping of low grasses, and on the wintering grounds the birds are sometimes found with the white-fronted goose on saltwort-dominated steppes and other semiarid terrain (Dementiev & Gladkov, 1967).

Social behavior. The period of time to sexual maturity has not been definitely established in this species, but it is probably less than two years, with nesting by females presumably often deferred until their third year. The sexual and aggressive postures and general social behavior patterns, including copulatory behavior, seem to be essentially like those of the white-fronted goose (Johnsgard, 1965a).

Reproductive biology. Lesser white-fronted geese return to their subarctic breeding grounds of the U.S.S.R. toward the end of May or the beginning of June, and as early as the first half of May in Lapland.

The nest sites are usually on newly thawed areas close to water, often on tundra, under dwarf birch vegetation, or in rocky areas on mountains. There is no indication that the birds are at all colonial or tend to nest in the vicinity of raptors or gulls. Clutches typically are of from 3 to 8 eggs, and most often consist of 4 or 5. Full clutches are found in Lapland about the beginning of June, while they are usually not completed until the middle or end of June in the U.S.S.R. The incubation period, based on information from captivity, is 25 days, and as in other geese incubation is performed by the female, with the male remaining close at hand throughout the entire period. Almost nothing is known of the birds' behavior during the fledging period, which requires about five weeks, but undoubtedly the adults also complete their own flightless period during this time. During the period of molt the birds often move to mountain lakes or the mouths of rivers with low accessibility to humans (Dementiev & Gladkov, 1967; Bauer & Glutz, 1969).

Status. There are no good numerical data from which to judge population trends in this species, but Uspenski (1965) estimated that its total population must number at least 100,000 individuals. Most of these birds now overwinter in the vicinity of the Caspian Sea, making this a critical area in terms of conservation of the species. Hudson (1975) indicated that this species has declined greatly in Finland and Scandinavia since 1940, and considered it endangered in Europe.

Relationships. This small goose is obviously a derivative of the larger species of white-fronted goose, but rather curiously has a more southerly breeding distribution, in spite of its seemingly greater adaptations to arctic breeding in terms of its short incubation and fledging periods. Ploeger (1968) suggests that it arose by isolation from other white-fronted goose stock by being separated on the Taimir Peninsula, where it adapted to a montane environment. Although the two species often associate in migration and in winter quarters, and some mixed pairs have been observed, no hybrids have been reported from the wild, suggesting that their isolating mechanisms are better developed than, for example, those of the snow and Ross geese of North America.

Suggested readings: Hudson, 1975; Owen, 1977.

Graylag Goose

Anser anser (Linnaeus) 1758

Other vernacular names. Gray goose; Graugans (German); oie cendrée (French); ansar (Spanish).

Subspecies and ranges. (See map 22.)

A. a. anser: Western graylag goose. Breeds in Iceland, Scotland, and from Scandinavia south to Austria, Yugoslavia, Macedonia, and Russia east to the Caucasus. Winters as far south as Spain, North Africa, Romania, Greece, Turkey, Iraq, and Iran.

A. a. rubrirostris: Eastern graylag goose. Breeds from west of the Urals and the lower Volga south to the Caspian Sea and east to northern Mongolia, Manchuria, and southeastern Siberia. Winters south to Asia Minor, India, and northern Indochina.

Measurements and weights. Folded wing: males, 445–82 mm; females, 416–70 mm. Culmen: males, 55–72 mm; females, 54–70 mm. Weights: males (both subspecies), 2,800–4,100 g; females, 2,500–3,800 g (Bauer & Glutz, 1968). Eggs: av. 85 x 58 mm, white, 160 g.

Identification and field marks. Length 30–35" (76–89 cm). Plate 16. *Adults* of both sexes are generally grayish brown throughout, including the head, which is uniformly brownish gray and normally lacks a white band behind the bill, although a small amount of white may be present. The head and neck are not markedly darker than the breast, although the neck is strongly furrowed, as in the bean goose. The underparts are often slightly mottled or spotted with dark brown, but do not have the black barring of the white-fronted goose. The flanks are somewhat vertically barred with darker brown and buff. The mantle feathers are dark brown, with broad buffy edging, and the rump is light gray. The tail feathers are dark brown, with white tips and edging, and the tail coverts are white, as is the entire posterior abdominal region. The upper wing surface varies from gray to dark brown, the coverts and secondaries being edged or tipped with buffy white. The iris is brown, with a pinkish eye-ring, the bill is orange *(anser)* or pink *(rubrirostris),* and the legs and feet are pink. First-year *immatures* resemble adults, but are generally less strongly patterned dorsally and lack spotting on the lower breast.

NATURAL HISTORY

Habitat and foods. In the U.S.S.R., the breeding habitat is said to consist of river flood plains, estuaries, and reedy bottomlands, as well as grassy bogs, damp meadows, and reed-lined fresh water lakes in grasslands. On the Outer Hebrides, where the species has retained one of its few indigenous breeding populations in the British Isles, graylags breed on small islets of various lochs that are surrounded by sedge- and heather-covered moorlands. There the birds feed primarily on grasses, supplemented in summer by marsh plants such as the roots of bulrushes, the leaves of duckweeds, and horsetails *(Equisetum)*, and such moorland plants as sedges and cotton grass *(Eriophorum)*. During the period of laying, grasses are nearly the only food eaten other than some clover *(Trifolium)*, but in late summer the birds feed on ripening grain crops. Likewise, in the U.S.S.R. the spring and summer foods are a variety of grass species plus aquatic plants such as pondweeds. The shoots and stems of such plants are apparently also important foods for goslings.

During fall migration and while in wintering areas the birds eat many cultivated cereal crops, and in coastal areas they also consume eelgrass *(Zostera)* or even the tubers or bulbs of steppe vegetation which they dig up from the ground (Dementiev & Gladkov, 1967; Newton & Kerbes, 1974).

Social behavior. Pair formation usually occurs when the birds are a year and a half old, although successful breeding may not occur until they are approaching three years of age. As in the other typical geese, the triumph ceremony plays an important role in forming and maintaining pair and family bonds. Fischer (1965) has studied its development with age and determined that it is derived from two elements, rolling and cackling. Cackling gradually develops out of calling while neck stretching in goslings, and

MAP 22. Breeding distributions of the eastern ("E") and western ("W") graylag geese. The wintering distributions of both races is indicated by stippling.

occurs in several intensities, related in part to the attachment drive of the individual. The other component, rolling, apparently reflects a conflict between the attachment drive and escape tendency, and the interplay of these two components results in the various forms of the triumph ceremony's performance. It serves a major role in maintaining the social structure of geese, through the cackling component's importance in keeping the family together and the rolling serving to maintain rank-order relationships and prevent attacks from developing within the group. Copulatory behavior takes the usual *Anser* form in this species, being preceded by mutual head-dipping behavior, and treading is terminated by both birds calling as the male stretches his head and neck almost vertically and strongly lifts his folded wings (Johnsgard, 1965a; Fischer, 1965).

Reproductive biology. In the Hebrides, mated pairs begin to occupy nesting islets several weeks before egg laying, which usually occurs in mid-April (Newton & Kerbes, 1974). In Scotland, egg laying occurs in late March or early April (Young, 1972). In both areas the same nest sites are used in successive years, with some in Scotland being used at least five successive years. There the nests are typically close to water and grazing fields, and in most cases are placed on wooded islands, presumably because of danger from foxes. However, nesting in association with gulls is not typical there. On favored islands the nests may be greatly concentrated; in one case 42 nests were present on an island of only 300 square meters, averaging only 11 meters apart. Although nesting in the Hebrides is also predominantly on islands, the nesting densities there are much lower. Nest building is performed by both sexes and usually requires from 3 to 6 days for completion. Laying is done at the rate of one egg per day, and clutches seem to vary considerably in size, from 2 to 12 eggs, but average between 5 and 6 eggs in each of these two areas. Clutches larger than 9 eggs presumably represent the efforts of two females, especially when the nests are placed close together, although broods of up to 12 goslings were reported by Young (1972).

Incubation by the females begins with the last or the penultimate egg and requires 28 to 29 days. Males remain by the nest throughout this period, standing sentinel, and undertake the major nest-defense responsibilities. Females leave their nests to defecate, and although they usually sleep on the nest, they may also forage some at night. In the New Hebrides, hatching success was related to the timing of the nesting season, with early nests having not only larger average clutch sizes but also a lower proportion of failures from predation and desertion. Predators there were apparently all avian rather than mammalian, consisting primarily of gulls and crows, and nesting success was in part related to effectiveness of concealment from overhead. Losses after hatching were apparently small in both study areas, with the sizes of well-grown broods averaging between three and four goslings. The fledging period is about 8 to 9 weeks, or appreciably longer than that for arctic-nesting geese, and certainly long enough to allow both parents to complete their own 35-day flightless period in the interim.

Status. Ogilvie and Boyd (1976) reviewed the trends in the Icelandic population of graylag geese that winter in Scotland and England, and indicated that an increase has been occurring since 1960, with over 60,000 geese wintering in the British Isles in the mid-1970s. By 1971, the Scottish breeding population slightly exceeded 1,000 birds, and there was a small additional population on Sutherland and Caithness, Outer Hebrides. The continental populations of Europe have been reviewed by Hudec and Rooth (1970) and Bauer and Glutz (1969); these in general are relatively small and scattered breeding groups. There seem to be no estimates available of the size of the population of the eastern graylag goose, but in general it seems to occupy a greatly disrupted and generally declining range (Dementiev & Gladkov, 1967).

Relationships. It seems realistic to regard the graylag goose as a central, or core, species in the genus *Anser*, lacking both the plumage and bill-structure

specializations seen in the swan goose, bar-headed goose, and snow goose, and basically adapted to temperate grassland breeding habitats. Its close relationships to the bean goose and white-fronted goose are evident in both plumage and behavior, and the graylag, bean goose, lesser white-fronted goose, and white-fronted goose constitute a series of ecological replacement forms extending across Eurasia from the grassland steppes of Central Asia to the high arctic tundra.

Suggested readings. Newton & Kerbes, 1974; Young, 1972.

Bar-headed Goose

Anser indicus Latham 1790

Other vernacular names. None in general English use. Streifengans (German); oie à tête barrée (French); ansar calvo (Spanish).

Subspecies and range. No subspecies recognized. Breeds in Asia on high mountain lakes from the Russian Altai and Tien Shans to Ladakh (India) and Koko Nor (China). Winters mainly in northern India from Sind to Assam and south to Mysore, and in northern Burma. See map 23.

Measurements and weights. Folded wing: males, 450–82 mm; females, 406–60 mm. Culmen: males, 48–63 mm; females, 47–55 mm. Weights: both sexes 2,000–3,000 g (Kolbe, 1972). Eggs: av. 82 x 55 mm, white, 141 g.

Identification and field marks. Length 28–30" (71–76 cm). *Adults* have heads that are entirely white except for a black bar extending from eye to eye around the occiput and a second black bar from the ears to the back of the head. A black stripe also extends down the back of the neck, and the foreneck is brown; these two areas are separated by a narrow white stripe down the side of the neck. The breast, flanks, and mantle are all silvery gray to brownish, with the feathers tipped with white, producing a barred effect on the breast and flanks. The rump and tail coverts are white, and the tail pale gray with white edging. The upper wing surface is mostly gray, with the tips of the flight feathers blackish. The iris is brown, the legs and feet yellow, and the bill yellow with a black

nail. *Females* are identical in plumage to males. *Juveniles* have a pale gray head and neck, with a dark brown line extending through the eyes, across the crown, and backwards down the neck. There is little or no indication of brown on the foreneck, and the legs and feet are greenish yellow.

In the field, the silvery gray body plumage, and the two black bars across the back of the head in adults, provide distinctive field marks. In flight the blackish tips of the flight feathers contrast with the otherwise white to silvery gray color of the wings and body. The voice of both sexes is a nasal honking reminiscent of an old car horn.

NATURAL HISTORY

Habitat and foods. Most of the breeding habitat of this species occurs on mountain plateaus having an elevation of 4,000 to 5,000 meters and bodies of water such as lakes, rivers, open tussocky bogs, or marshes. Both fresh-water and brackish impoundments are utilized, as are rivers with islands or tree-lined rivers, but the occurrence of cliffs nearby may be an important part of the habitat in at least some areas where nesting occurs on rocky outcrops (De-

mentiev & Gladkov, 1967). In Tibet, large, shallow lakes with broad belts of marsh and grassy islands provide the nesting habitat (Delacour, 1954–64). On their wintering grounds in India the geese are often found on large rivers, where they use large sandbars as safe resting places, flying out at night to croplands for foraging. Their foods there include grasses, tubers, grain crops, and the like (Ali & Ripley, 1968). In coastal areas they have been reported to consume seaweeds and such invertebrates as crustaceans, although it is most unlikely that such foods are a substantial part of their diet.

Social behavior. The bar-headed goose is unusually social, and even during the breeding season tends to cluster in groups of several pairs, or even larger colonies. In the winter, the birds occur in groups of as few as five or six birds to flocks of up to a hundred or more individuals. Sexual maturity probably occurs during the second winter of life; Würdinger (1970) has described the development of vocalizations and their role in the social life and pair-bonding process. The major aggressive displays of adults are a diagonal-neck posture and an erect posture, which are frequently alternated with an extremely intense forward display posture, with the head and neck held close to

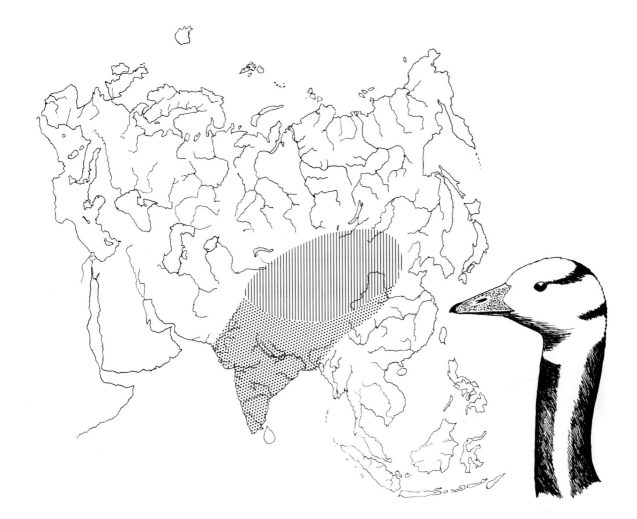

MAP 23. Breeding (hatching) and nonbreeding (stippling) distributions of the bar-headed goose.

the ground. During the erect posture the folded wings are often flicked alternately or spread maximally. After aggressive encounters the pair usually stand facing one another, calling simultaneously with their bills tilted nearly vertically upwards. Copulation is preceded by strong tail-cocking and head-dipping movements on the part of the male, while the comparable postures of the female are much less pronounced. The male usually spreads his wings fully as treading is completed, and both birds simultaneously call with their wings extended to the utmost (Johnsgard, 1965a).

Reproductive biology. The bar-headed geese are relatively late in leaving their wintering areas, and must migrate over the Himalaya range to their breeding grounds on the Asian plateaus before nesting.

Nesting in Ladakh, Kashmir, occurs during the end of May and June, and in Tibet at about the same time. Nest sites are quite varied, but probably islands in marshy lakes are preferred locations, with large colonies of birds sometimes developing at such locations. However, nests have also been found on rocky outcrops, in trees some four to six meters above the ground (presumably utilizing old nests of such species as ravens and hawks), and in boggy or swampy situations, where the nests consist of piled mosses and other vegetation. The clutch size among wild birds is usually only 4 eggs, with up to 8 being recorded. Incubation normally requires 27 days and is carried out entirely by the female. However, both sexes invariably defend the young, and sometimes several broods merge, an expected situation in view of the colonial nesting tendencies of this species.

Studies of captive birds indicate a normal fledging period of 53 days (Würdinger, 1973). In the Pamir Mountains of Central Asia, fledging of juvenile birds occurs in late September, by which time the adults have also completed their molts, and fall migrational movements get under way (Dementiev & Gladkov, 1967). Birds begin to arrive in West Pakistan and northwestern India about October or November, and are mostly gone again by the end of March (Ali & Ripley, 1968).

Status. Bar-headed geese are important sporting birds in India and eggs of this species are collected in some numbers in Tibet, but there seems to be no specific information on populations or population trends. Dementiev and Gladkov (1967) report that the species has declined greatly in the northern parts of its range (the Pamir and Tien Shan mountains), and remains common only in Tibet and Ladakh.

Relationships. Although obviously a member of the genus *Anser*, the bar-headed goose is one of several aberrant species whose relationships are rather obscure. Its juvenile plumage has some similarities to that of the snow goose, but the bill shapes and foraging behaviors of these two species are very divergent.

Suggested readings. Dementiev & Gladkov, 1967, Würdinger, 1973.

Snow Goose

Anser caerulescens (Linnaeus) 1758

Other vernacular names. Blue goose; Schneegans (German); oie des neiges (French); ganso die nieve (Spanish).

Subspecies and ranges. (See map 24.)
   A. c. caerulescens: Lesser snow (and "blue") goose. Breeds on Wrangel Island in Siberia, and locally in North America from the north coast of Alaska east to Baffin Island and south to the mouth of James Bay. Winters primarily in the central valley of California, along the Gulf Coast, and rarely in China and Japan.
   A. c. atlanticus: Greater snow goose. Breeds in Greenland and on Baffin, Devon, and probably Grinnell islands. Winters on the Atlantic coastline, mainly around Chesapeake Bay but south to North Carolina.

Measurements and weights. Folded wing: both sexes of *caerulescens*, 387–460 mm; of *atlanticus*, 425–85 mm. Culmen: both sexes of *caerulescens*, 50–62 mm; of *atlanticus*, 57–73 mm. Weights: males and females of *caerulescens* average 2,744 and 2,517 g, respectively, while those of *atlanticus* average 3,310 and 2,812 g (Johnsgard, 1975). Eggs: av. (*caerulescens*) 78 x 52 mm, white, 127 g.

Identification and field marks. Length 26–33″ (66–84 cm). Plate 17. *Adults* of both sexes occur in two color phases, with intermediates of all degrees. The "snow" phase is entirely white (often stained with rusty coloration in the face), except for black primaries, grayish primary coverts, and variably grayish alular feathers. The "blue" phase has only the head and upper neck pure white, with the lower neck, back, scapulars, breast, sides, and flanks grayish brown to blackish, with slight markings of cinnamon brown. The rump is gray to white, the tail coverts whitish with gray or brown mottling, and the lower abdomen white or ashy white, with this color extending variably forward in individuals of mixed genetic origin. The tail is silvery gray, margined and tipped with whitish coloration, and primaries and secondaries are black, with some inner secondaries (and the tertials) edged with white. The upper wing coverts are gray, with the larger one margined with white, and the underwing surface is gray, with the axillars and anterior coverts whitish. The iris is brown, the bill pink to carmine, with a whitish nail and a black streak along the sides of the mandibles, forming a "grinning patch." The legs and feet are dull red. *Immatures* of the snow phase resemble the adults, but during their first winter the head and neck are mottled with grayish coloration; the back and scapulars are ashy brown with white edging; the secondaries are dusky; and the bill, legs, and feet are all more dusky. First-winter immatures of the blue phase lack white on the head and have less elongated and pale gray greater secondary coverts and soft-part colors similar to those of white-phase birds. By the end of their first winter the heads of immature birds are white as in adults.

In the field, snow geese are likely to be confused only with Ross geese where that species also accurs (see Ross goose account for distinctions), and, in the blue phase, perhaps with white-fronted geese. The doglike barking call of snow geese helps to identify them at great distances, and the entirely white heads of the blue-phase birds helps to separate them from white-fronted geese.

Habitat and foods. The breeding habitats of lesser snow geese typically consist of low, grassy tundra, often on river flood plains or near ponds, lakes, or salt water, while greater snow geese are often found nesting where stony terrain meets wet and grassy tundra. Traditional winter habitats of snow geese have consisted of brackish or salt marshes with abundant emergent vegetation such as cord grass *(Spartina),* and other tall grasses and sedges that provide both food and cover. However, in recent decades lesser snow geese have gradually moved farther away from coastal marshes and into interior agricultural lands, where such crops as corn, rice, and pasture grasses provide abundant sources of food for them. Nevertheless, the birds have evolved and become highly adapted to grubbing out the roots and tubers of native aquatic plants, and for clipping off the tender shoots of grasses and sedges, thus reducing their competition with the more upland-adapted species of geese such as Canada and white-fronted geese (Johnsgard, 1975).

Social behavior. At least among the North American geese, snow geese are perhaps the most gregarious of all, and thus flocks numbering in the tens of thousands are common sights in major areas of concentration. This is in part a reflection of their communal nesting behavior and the strong degree of synchronization in migration arrival and departure times. These flocks are loosely bound entities, consisting basically of pairs and their offspring of the past year and sometimes of the past two years, resulting in a flock substructure unit from two to about ten birds. Snow geese pair for life, probably normally when less than two years of age, although many pairs are not successful nesters until their third year. Furthermore, the moderately high annual mortality rate of adults, estimated at from about 25 to 30 percent, results in a substantial rate of pair disruption and consequent need for re-forming of pair bonds in surviving birds. Among flocks of lesser snow geese that consist of both color phases, the predisposition for mating with birds of the same color phase as the individual bird's parents results in a high degree of assortative mating, and thus reduces the proportion of intermediate-plumage or heterozygotic offspring (Cooke & McNally, 1975). Pair-forming behavior is, as in the other geese, characterized by a considerable amount of threatening or chasing behavior, followed by a triumph ceremony of the potential pair. Copulatory behavior likewise follows the typical pattern of mutual head dipping during precopulatory stages, and mutual calling with tail cocking and wing lifting after treading is completed (Johnsgard, 1965a).

Reproductive biology. Snow geese form what are perhaps the largest and among the densest of all waterfowl nesting concentrations; it has been estimated that more than 100,000 nests were present in 1960 on Wrangel Island, the largest of all snow goose colonies. Under highly favorable conditions, the nests may be spaced so closely that each nesting territory averages only about 250 square feet. These colonies are usually single-species congregations, although snow geese sometimes also share nesting areas with brant geese and arctic-breeding Canada geese. Nesting behavior of snow goose colonies is highly synchronized, with females usually starting their nests as soon as the area becomes snow-free. The nests are slight heaps of vegetation in the grassy tundra which become progressively lined with down as the clutch is laid. Average clutch sizes are 4 or 5 eggs, laid on a daily basis. Only the female incubates, but the male is always nearby to help defend the nest. Once the 23-day incubation period is begun, the female leaves the nest only briefly, and normally does not leave at all during the late stages of incubation. Gulls and jaegers are usually the major predators of eggs and young; and shortly after hatching, the young are typically led away from the nest by both parents, moving to suitable foraging areas where the adults can undergo their molts in relative safety. The fledging period of the young is about 40 to 50 days, during which time the adults complete their own flightless period, so that with the onset of cold weather in late August or September the families are normally prepared to begin their migration southward (Johnsgard, 1974b, 1975).

Status. The two subspecies of snow geese differ appreciably in their numerical status, with the Atlantic coastal population of greater snow geese having once been threatened with extinction, but averaging over 50,000 birds during winter surveys in the period between 1955 and 1974. During the same period the lesser snow goose wintering population averaged nearly 1.3 million birds, wintering mainly in the lower Mississippi Valley and on the Gulf Coast. In general, all of the major breeding populations appear to be secure, but the constant threats of oil and mineral development in the arctic, the dangers of disease among the great concentrations of birds on migration and in wintering areas, and the sensitivity of all arctic-breeding geese to a series of disastrous breeding seasons when weather conditions are unfavorable requires that close attention be paid to the management of these flocks.

Relationships. The snow goose and its very close relative, the Ross goose, have at times been placed in a genus (*Chen*) distinct from the other "gray" geese, but this is impossible to justify on any anatomical or behavioral grounds. To be sure, there is no other species of *Anser* that provides an obvious link-form with the snow geese, but the snow geese almost certainly evolved as a result of the isolation of popula-

MAP 24. Breeding distributions of the lesser ("L") and greater ("G") snow goose. The wintering distributions of both races is indicated by stippling.

tions of Northern Hemisphere geese during the last major period of glaciation (Ploeger, 1968).

Suggested readings. Johnsgard, 1974b; Prevett, 1973.

Ross Goose

Anser rossi Cassin 1861

Other vernacular names. None in general English use. Zwergschneegans (German); oie de Ross (French); ganso de Ross (Spanish).

Subspecies and range. No subspecies recognized. Breeds in Canada, mainly in the Perry River region of the Northwest Territories, although local breeding has also been reported on Banks and Southampton islands, and on the McConnell River of Keewatin District. Winters mostly in central California, with increasing numbers recently occurring east to the Gulf Coast and rarely to the eastern states. See map 25.

Measurements and weights. Folded wing: males, 360–80 mm; females, 345–60 mm. Culmen: males, 40–46 mm; females, 37–40 mm. Weights: males and females average 1,315 and 1,224 g, respectively, with maxima of 1,633 and 1,542 g. Eggs: av. 70 x 47 mm, pink to white, 94 g.

Identification and field marks. Length 21–26″ (53–66 cm). *Adults* of both sexes are entirely white (sometimes stained with rusy coloration in the head region) except for the primaries, which are black, and the primary coverts, which are gray. The bill is red to pale purple, with a white nail and black edges but no definite "grinning patch," and with a warty bluish area at the base, especially in males. The iris is brown, and the feet and legs are deep pink. *Immatures* in their first winter are similar to adults, but are brownish gray on the head, back, and scapulars and have grayish to brownish rather than black outer flight feathers. The bill is initially brownish or greenish, gradually becoming pink, and the feet also change from greenish gray to pink during the first year. A few blue-phase birds, with measurements in the range of this species, have been collected recently in California, but their origin is unknown.

In the field, Ross geese might be confused with the larger snow geese, but the bill and neck are ap-

preciably shorter, and the birds are little larger than mallards. The call notes are also higher in pitch, but this is usually evident only when both species can be heard together.

NATURAL HISTORY

Habitat and foods. The breeding habitat of this arctic-nesting goose in the Perry River area is typically on islands in lakes that are too large for mammalian predators to swim across, and which are unlikely to have ice bridges present at the start of the nesting season. Nest cover in the form of rocks or low trees and shrubs is also an important habitat component (Ryder, 1967). Limited nesting also occurs on lake shores or along streams, but access by arctic foxes to such habitats makes them less desirable to these small and vulnerable geese. On their wintering grounds, Ross geese mingle with and use the same foraging habitats as small Canada geese, but tend to forage separately from snow geese. They feed there on various green grasses, whereas on the fall migration southward they depend heavily on waste grains such as wheat, barley, and rice.

Social behavior. Like the larger snow geese, Ross geese are highly gregarious, both on their nesting rounds and during migration. Flocks of several thousand birds are not uncommon, and nesting densities of birds in optimum habitats are even higher than those of snow geese. This is probably a reflection of the advantages of coloniality in reducing predation effects, and of the fairly few areas of relatively fox-free nesting habitats. Pair bonds are apparently strong, and are formed during the second year of life. However, only a few two-year-old females are successful breeders; presumably, competition for optimum nesting territories tends to delay most nesting efforts until the third year. Pair-forming behavior has not been observed on the nesting grounds, nor has copulation even been seen there, but it is known from observations of captive birds that Ross geese do not differ significantly from snow geese in these regards. Interestingly, in recent years an increasing number of wild hybrids between snow geese and Ross geese have been reported, and there has been an eastward movement of nesting Ross geese into breeding colonies of lesser snow geese. What effects this might have on the integrity of the much smaller gene pool of the Ross goose remains to be seen, but the consequences could be serious.

❖ ❖ ❖ 69

Reproductive biology. The studies by Ryder (1967, 1970) on the Ross goose provide the primary information on this species, which differs little from the snow goose in its general reproductive biology. Primarily, the differences seen attributable to this species' greater sensitivity to predation by arctic foxes, and also its vulnerability to gull and jaeger predation. Nest building is begun immediately after arrival on the breeding grounds, after the male has established a suitable territory. Nest spacing is very close, with the territories often averaging only about 100 square feet apiece, and in very favorable habitats the area may be as little as 50 square feet per nesting pair. Clutch sizes tend to be somewhat smaller than in snow geese, and average slightly under 4 eggs, with larger clutches in favorable nesting seasons and among those females that begin a few days earlier than the others. An interval of 1.5 days between eggs has been established, and very little down is placed in the nest until the laying of the penultimate egg. Only the female incubates, while the male remains close at hand to protect the nest. The average incubation period is 22 days, with the female fasting for most of this time. Ryder suggests that the low average clutch size of this species reflects the needs for efficient energy compartmentalization for nesting females, which undergo great stress at this time. Like the other arctic-nesting geese, females do not attempt to renest if their initial clutch is destroyed. After hatching, the male joins his mate and undertakes the major responsibility for the brood's defense, while the female typically leads them away from any source of danger. The fledging period is probably only slightly more than 40 days, and very often freezing weather sets in within 40 to 45 days following the peak of hatching. Thus, a critical feature of reproductive success is the efficiency with which the young can be brought to flight and the adults can complete their three- to four-week flightless period before the onset of freezing weather.

Status. Between the late 1950s and the early 1970s, the estimated wintering population of Ross geese was only about 23,000 birds. However, counts in the late 1960s on fall staging areas of Alberta and Saskatchewan, and on breeding grounds adjacent to the Queen Maud Gulf have indicated a population in excess of 30,000, suggesting that the number is currently increasing and the species is in no imminent danger. However, as with snow geese, the great tendencies for mass flocking and the reliance on a few breeding and wintering areas for the vast majority of the species make the Ross goose susceptible to rapid declines in population through disease, habitat disruption, and the like. Additionally, the increasing contacts and hybridization between Ross geese and snow geese add another dimension of uncertainty to the future of this species.

Relationships. The Ross goose is clearly a fairly recent offshoot of an ancestral snow goose type that probably differed little from the present-day form. The evolution of a smaller body size was most probably an adaptation to efficient breeding in areas with especially abbreviated breeding seasons, and perhaps also for the exploitation of foraging opportunities open to a small-billed goose.

Suggested readings. Ryder, 1967, 1970, 1972.

MAP 25. Breeding (inked) and wintering (stippling) distributions of the Ross goose.

Emperor Goose

Anser canagicus (Sewastianow) 1802

Other vernacular names. Beach goose; Kaisergans (German); oie empereur (French); ganso emperador (Spanish).

Subspecies and range. No subspecies recognized. Breeds in Alaska from the mouth of the Kuskokwim River to the north side of the Seward Peninsula, on St. Lawrence Island, and on the northeastern coast of Siberia. Winters along the Aleutian Islands and the Alaska Peninsula probably to Cook Inlet, with vagrants occurring farther south. See map 26.

Measurements and weights. Folded wing: males, 380–400 mm; females, 350–85 mm. Culmen: males and females 35–40 mm. Weights: males and females average 2,812 and 2,766 g, respectively, with maxima of 3,083 and 3,129 g (Nelson & Martin, 1953). Eggs: av. 76 x 52 mm, white, 120 g.

Identification and field marks. Length 26–30″ (66–89 cm). *Adults* of both sexes have the head and back of the neck white (often stained with rusty coloration on the face), and the chin, throat, and foreneck blackish, gradually merging with a grayish breast. The feathers of the back, scapulars, breast, sides, and flanks are silvery gray, with tips and subterminal black bars forming a scalloped pattern. The rump and tail covert feathers are white with gray bases. The primaries and the coverts are silvery to dusky gray, and the secondaries and greater coverts are blackish, with white edging. The other upper wing covers are gray, with black and white edging. The underwing surface is silvery gray. The bill is pink to flesh colored, with bluish coloration around the nostrils, a white nail, and a mostly black lower mandible. The iris is brown, and the legs and feet are bright yellow. *Immatures* in their first winter are like adults but have a duller plumage pattern, with gray mottling on the head and foreneck, and with brown rather than black barring on the back. The bill is black, and the legs and feet are olive brown.

In the field, the generally silvery gray body coloration, contrasting with a white head and a black foreneck, provides a distinctive combination for identification in this species' limited range. Like snow geese, the birds have a short neck and appear heavy-bodied when in flight, with strong and rapid wingbeats. Their usual calls consist of repeated *kla-ha* notes and an alarm note, *u-leegh.*

NATURAL HISTORY

Habitat and foods. In northeastern Siberia, the nesting habitats of emperor geese fall into two general types, grassy tundra associated with coastal

MAP 26. Breeding (hatched) and wintering (stippling) distributions of the emperor goose.

lagoons of brackish waters, and inland tundra of mosses and sedges associated with fresh-water ponds and lakes. The inland tundra, with an abundance of favored sedges for feeding and relative freedom from ice and flooding during the start of the breeding season, provides the most important breeding habitat for the geese (Kistchinski, 1971). In the Yukon delta the areas just inside the coastal fringe used by brant geese are also favored nesting habitats, where lowland hummocky or "pingo" tundra, sedge marshes, and tidal grasslands are all important. On their breeding grounds, the leaves of sedges and grasses are the primary food, but during the rest of the year, while the birds are at sea, they feed almost exclusively on such plants as eelgrass (Zostera), sea lettuce (Ulva) and to a very limited extent on barnacles. Goslings evidently initially feed on aquatic insects and marsh grasses and later begin to consume berries such as crowberries (Empetrum).

Social behavior. Emperor geese are not highly gregarious, but concentrate in large numbers in certain migratory and wintering areas, particularly in the area of Port Moller and Izembeck Bay at the tip of the Alaska Peninsula, where a total of as many as 120,000 birds have been reported among the flocks. The birds gather in large flocks on St. Lawrence Island during summer, when nonbreeding immatures numbering up to 20,000 birds concentrate there following molt migrations from the mainland. Early arrivals on the Alaskan breeding grounds are usually of small flocks, which soon scatter over the available habitats and nest in what might be called loose colonies, with nests averaging about 200 feet apart and

approxmately one nest per four acres (Eisenhauer & Frazer, 1972; Eisenhauer et al., 1971). Observations of captive birds suggest that the birds mate for life; and although data from the wild are not available, it is believed, on the basis of information from birds raised in captivity, that most do not nest until their third year. Likewise, the triumph ceremony behavior and copulatory behavior of emperor geese as observed in captivity is not significantly different from that of other typical geese. By the time the birds have reached their Alaskan breeding grounds in late May, the pair bonds are well established and females are ready to begin egg laying (Headley, 1967).

Reproductive biology. As noted above, nests are usually built in grass and sedge tundra, often on a slightly elevated hummock, and normally directly adjacent to water or at most no more than 75 feet from it. The clutch size averages slightly under 5 eggs, with some yearly variations that are evidently associated with prevailing weather and habitat conditions. Only the female incubates, and the average incubation period under wild conditions has been reported as 24 or 25 days, with extremes of 23 to 27 days. Gulls and jaegers are evidently the primary mortality factors in Alaska; observations in Siberia by Kistchinski led him to believe that the low insulation of the nest and poor resistance of the goslings to cold may limit the northward extension of the species' range. With the hatching of the eggs, the families usually move initially to the coast, where they may remain for a time before going to the larger rivers for molting. Molting areas may be as far as five miles from the nesting site (Eisenhauer et al., 1971). Evidently about 30 to 35 days are needed by the adults to complete their flightless molt, and during this time the young grow rapidly, requiring about 50 to 60 days to attain flight (Mickelson, 1973). Goslings as heavy as three pounds have been observed killed by glaucous gulls, but most gull predation occurs on younger goslings. In midsummer there is a substantial molt migration of nonbreeders to St. Lawrence Island, where they concentrate on the coast before undertaking the fall migration to winter quarters along the Aleutian Islands.

Status. The two breeding components of the emperor goose consist of an Alaskan segment that may number about 60,000 to 75,000 breeders and an equal number of nonbreeders during fall, and a much smaller but uncertain number of breeding birds in Siberia (Bellrose, 1976). Up to 2,000 nonbreeding birds gather on lagoons west of the Chukotsk Peninsula for a short time in summer; it may be imagined that a comparable number of breeding birds occur in the general area. The fact that about 90 percent of the Alaskan population of emperors nest in the Yukon delta, and thus well over half of the world's population occurs there, means that it is particularly important that the area be protected from disturbance or unnecessary development.

Relationships. Although in a few respects the emperor goose is one of the most aberrant of the "gray" geese in its maritime foraging behavior and its unusual adult plumage pattern, there is little evidence to support the contention that it should be generically separated (*"Philacte"*) from the other typical species of *Anser*. The surprisingly long fledging period of the young and their poor resistance to extreme cold suggest that the species must have evolved from a much more temperate-adapted form.

Suggested readings. Eisenhauer et al., 1971; Eisenhauer & Frazer, 1972; Eisenhauer & Kirkpatrick, 1977.

Hawaiian Goose

Branta sandvicensis (Vigors) 1833

Other vernacular names. Nene goose; Hawaiigans (German); bernache d'Hawaii (French); ganso hawaiano (Spanish).

Subspecies and range. No subspecies recognized. Native to the Hawaiian Islands, but now nearly extinct in the wild state and limited to the islands of Hawaii (Hawaii Volcanoes National Park) and Maui (the latter as a result of reintroduction in Haleakala National Park).

Measurements and weights. Folded wing: males, 372–78 mm; females, 350–68 mm. Culmen: males, 40–47 mm; females, 40–42 mm. Weights: 61 males averaged 2,212 g; 64 females averaged 1,923 g (J. Kear, pers. comm.). Eggs: av. 82 x 65 mm, white, 144 g.

Identification and field marks. Length 22–28" (56–71 cm). Plate 20. *Adults* have a black throat, face, crown, and hind neck, sharply set off from tawny buff on the sides of the head and the sides and front of the neck. The neck is deeply furrowed, with the darker feather bases producing a striped effect. The pale brown breast is separated from the neck by a blackish ring, and the underparts and flanks are mostly grayish with brown flank markings. The feathers of the upperparts are dark brown, with broad buffy edges. The tail and rump are black, while the upper and lower tail coverts are white. The upper wing coverts are patterned in the same manner as the back feathers, and the flight feathers are olive brown. The bill is black, the iris is brown, and the legs and feet are black. *Females* are nearly identical to males but are slightly smaller. *Juveniles* exhibit a dull version of the head and neck pattern, but the patterning is much duller, and the body coloration is more mottled.

In the field, the Hawaiian goose is not likely to be mistaken for any other species of waterfowl, since other geese occur as infrequent winter visitors. Hawaiian geese generally resemble Canada geese, but are more terrestrial, and the sides and front of the neck are tawny rather than black. The usual call is a low moaning sound, but the birds can also produce a more typical gooselike yelping call.

NATURAL HISTORY

Habitat and foods. On its native island of Hawaii, this goose inhabits the barren lowlands from sea level to 3,000 feet, and also occurs on the slopes of Mauna Loa and Hualalai between 3,000 and 9,000 feet. Most of this range is covered by lava flows with only scanty vegetation, but other parts of it include moist grasslands and open forests. Probably the birds originally spent the winter months in the lowlands, obtaining fresh green growth, and moved back up to the mountains during the breeding season to feed on herbs and berries, especially those of the ericad "ohelo" (*Vaccinium*). Studies by Baldwin (1947) based on the analysis of droppings indicated that the above-ground parts of a variety of grass species are consumed, as well as the seeds and green parts of sedges. The seeds, leaves, and stems of various forbs, especially certain composites (*Bidens, Cirsium, Gnaphalium, Hypochaeris*), are also main items of diet, and may provide both moisture and important nutrients. When foraging, the birds seem to pluck foods more than to peck for them and move about a good deal, probably covering several square miles in a single day. Almost no surface water occurs on the upland slopes, and thus plant materials that are high in moisture content, such as berries and green parts,

74 ◈ ◈ ◈

may be especially important to them. However, the birds are not dependent on any single plant for their food, and tend to consume whatever is abundant in any particular area.

Social behavior. The social unit of the Hawaiian goose is the family, and from the month of December through March the individual families are well separated as broods are being reared. However, starting in April these units begin to unite into larger flocks, and for about a month considerable wandering occurs before the birds move back to their summering areas. The summering period, lasting from late June through August, is the equivalent of the wintering period in temperate-zone waterfowl (Elder & Woodside, 1958), and presumably is the period of pair-bond formation. Like other geese, Hawaiian geese have a well-developed triumph ceremony, which is the apparent basis for the establishment and maintenance of pair bonds. Aggressive displays are very similar to those of Canada geese, and include a well-developed "sigmoid neck" posture, with the head and bill held close to the ground. However, this species is unlike all other true geese in that copulation occurs on dry ground, an obvious adaptation to the nearly water-free habitat. It is preceded by movements that are clearly derived from head-dipping movements, indicative of the secondary adaptation of the Hawaiian goose to an upland habitat. After treading, both birds call and assume a posture much like the postcopulatory posture of Canada geese and other *Branta* species (Johnsgard, 1965a). Over 80 percent of the females initially breed in their second year of life; a few yearling females may lay eggs but these have proven to be infertile (Berger, 1972).

Reproductive biology. Mainly during the months of August and September, the flocks that had summered together begin to break up and the individual pairs return to their nesting areas. Nesting is usually on "kipukas," which are vegetated areas surrounded by recent lava flows, and on the island of Hawaii nesting occurs at about 6,500 feet elevation. Probably the same territory is defended year after year, and Elder and Woodside (1958) reported that in 1957 at least six pairs used one small kipuka only a few acres in extent on the eastern slope of Mauna Loa. Eggs are laid over a fairly long period (October 29 to February 8) that centers on the winter solstice, and thus breeding occurs at the time of year when days are shortest, as opposed to arctic-nesting geese, which nest when day lengths are near the longest of the entire year. The clutch size is from 3 to 5 eggs and averages about 4.6, or somewhat smaller than that of most other true geese (Berger, 1972). Renesting can easily be induced in captive birds by the removal of the first clutch, and even three or four clutches can thus be obtained. However, Elder and Woodside found no definite evidence of renesting in the wild. Incubation is by the female and requires 29 days; during that time the male remains nearby and helps to defend the nest. A large number of introduced predators now occur on Hawaii, including feral cats, dogs, pigs, and mongooses, against most of which the geese are relatively defenseless. The young grow relatively slowly, and a period of 10 to 12 weeks is needed for them to attain flight. During this time the adults also undergo their flightless period of from 4 to 6 weeks, and until all of the birds are able to fly they are likely to remain within two or three miles of the original nesting site. After fledging, however, families begin to merge and form postbreeding flocks.

Status. The long and sad history of the Hawaiian goose has been told in detail by Baldwin (1945), and more recent accounts of efforts to preserve it have been detailed by Fisher et al. (1969) and Zimmerman (1974). The low ebb of the species' condition probably occurred about 1947, when only about 50 birds were thought to be present in the wild and in captivity. The raising of captive birds by the Wildfowl Trust in England and the Hawaiian Board of Agriculture's game farm at Pohakuloa gradually began to increase the known number of birds. The first releases of captive-raised birds back into the wild occurred in 1960, and since then over 1,000 birds have been released, with the majority on Hawaii and several hundred on Maui. There has been relatively good survival of these birds in the wild; some have been seen as long as 12 years after their original release. However, only about one-tenth of the broods that have been seen since 1965 have been from pairings in which both mates were hand-reared, suggesting lower reproductive success among these than among wild stock. It is now estimated that there may be at least 600 birds living under wild conditions in Hawaii, and possibly 200 on Maui (Zimmerman,

1974). In addition, there are several hundred birds in captivity, primarily at the Wildfowl Trust and at Pohakuloa.

Relationships. Most workers have agreed that the Hawaiian goose is probably an offshoot of *Branta* stock, perhaps derived from a Canada goose–like ancestor, and as such should be retained in that genus. Recently, however, Woolfenden (1961) found that a large number of its skeletal characteristics are different from those of *Anser* as well as *Branta,* and urged that the genus *Nesochen* be recognized for it. However, Miller's (1937) study of the leg musculature of this species indicated that only minor differences occurred between it and more typical geese, in spite of the terrestrial adaptations it had undergone.

Suggested readings. Elder & Woodside, 1958; Berger, 1972; Zimmerman, 1974.

Canada Goose

Branta canadensis (Linnaeus) 1758

Other vernacular names. Cackling goose, Canadian goose, honker, Hutchin's goose; Kanadagans (German); bernache du Canada (French); ganso del Canada (Spanish).

Subspecies and ranges. (See map 27.)

B. c. canadensis: Atlantic Canada goose. Breeds in eastern Labrador, Newfoundland, and on Anticosti and Magdalen islands.

B. c. interior: Hudson Bay Canada goose. Breeds in northern Quebec, Ontario, and eastern Manitoba north to Churchill.

B. c. maxima: Giant Canada goose. Originally bred on the Great Plains from Manitoba and Minnesota south to Kansas, Missouri, Tennessee, and Arkansas. Now largely limited to wildlife refuges where captive birds have been released. Closely similar to *moffitti,* and the two are considered by Palmer (1976) as a single form.

B. c. moffitti: Moffitt (Great Basin) Canada goose. Breeds in the Great Basin over a broad but discontinuous range, intergrading in southern Canada with *parvipes* to the north and with *maxima* to the east.

B. c. parvipes: Lesser Canada goose. Breeds from central Alaska eastward across northern Canada to Hudson Bay, and south to the Prairie Provinces, where it intergrades with *moffitti* and *maxima.*

B. c. taverneri: Taverner (Alaska) Canada goose. Breeds in the interior of Alaska from about the base of the Alaska Peninsula to the Mackenzie River delta, probably intergrading broadly with *parvipes,* and regarded by Palmer (1976) as inseparable from that form.

B. c. fulva: Vancouver Canada goose. Breeds along the coast and islands of British Columbia and southern Alaska north to Glacier Bay. North of Glacier Bay it is replaced by *occidentalis,* which Palmer (1976) does not consider separable from *fulva.*

B. c. occidentalis: Dusky Canada goose. Breeds along the Prince William Sound, Cook Inlet, and inland through the Copper River drainage east to Bering Glacier.

B. c. leucopareia: Aleutian Canada goose. Breeds on the Aleutian Islands; at present limited to

Buldir Island and perhaps also Amchitka where birds have been released.

B. c. asiatica: Bering Canada goose. Extinct; once bred on the Commander and Kurile islands. Not considered separable from *leucopareia* by Palmer (1976).

B. c. minima: Cackling Canada goose. Breeds along the western coast of Alaska from Nushagak Bay to the vicinity of Wainwright.

B. c. hutchinsii: Baffin Island (Richardson) Canada goose. Breeds on the coast of the Melville Peninsula, Southampton Island, Baffin Island, and western Greenland, west at least to southern Victoria Island, and perhaps to the Anderson River delta.

The wintering ranges of most subspecies greatly overlap, and cannot easily be summarized, but the major wintering areas of the species as a whole are shown on map 27. Canada geese have also been released and become established in such areas as England, New Zealand, Norway, and Sweden; most of these are of the largest (*moffitti* and *maxima*) races.

Measurements and weights. folded wing: both sexes of *maxima, moffitti, canadensis, fulva,* and *interior,* 410–550 mm; of *parvipes, occidentalis,* and *taverneri,* 356–460 mm; and of *asiatic, hutchinsii,* and *minima* 330–70 mm. Culmen: both sexes of these three groupings have ranges of 42–68 mm, 32–49 mm, and 26–39 mm, respectively. Weights: males of the first group average 3,809–6,523 g; females, 3,310–5,514 g. Males of the second group average 2,241–3,754 g, and females, 2,059–3,131 g. Males of the third group average 2,005–41 g; females, 1,360–856 g (Johnsgard, 1975). Egg measurements vary between subspecies from 74 x 49 to 87 x 60 mm, are white, and weigh from 91 *(minima)* to 163 *(canadensis)* g.

MAP 27. Breeding distributions of the Aleutian ("A"), Baffin Island ("B"), cackling ("C"), dusky ("D"), eastern ("E"), giant ("G"), Hudson Bay ("H"), lesser ("L"), Moffett ("M"), Taverner ("T"), and Vancouver ("V") Canada geese. An area of probable Moffett-giant intergrades is indicated by "M-G," and the wintering areas of all races is indicated by stippling. See text for introduced populations.

Identification and field marks. Length 22–43" (55–110 cm). *Adults* of both sexes have a black head and neck except for white cheek markings that extend from the throat to behind the eyes; rarely a white forehead patch may be present as well. The breast, anterior abdomen, and flanks vary from light grayish or nearly white to a barred dark chocolate brown, either merging with the black neck or separated from it by a narrow to broad white collar. The back and scapulars are darker brown, the feathers having lighter tips; the rump is blackish, and the tail is black or blackish brown. The tail coverts and posterior abdomen are white, as are the sides of the rump. The flight feathers are brown or blackish, the upper wing coverts are dark brown with lighter tips, and the underwing surface is grayish brown. The iris is brown, the bill black, and the legs and feet black with a greenish cast. *Immatures* in their first winter have duller and more blended tones of black and brown, and brownish rather than white cheeks, and the breast and flanks are mottled rather than barred.

In the field, the black "stocking" on the head and neck, with the contrasting white cheek patches, provides the best field mark. Barnacle geese are generally smaller and also have black extending down on their breast, and their white head markings extend over the forehead in front of the eyes. Calls vary in pitch from the "honking" characteristic of the larger races to high-pitched yelping sounds of the smallest forms.

NATURAL HISTORY

Habitat and foods. The large number of subspecies that breed across North America in habitats ranging from sagebrush semidesert to temperate rain forest and arctic tundra make a simple characterization of preferred habitats impossible. This species is the most adaptable of all North American geese, and has also successfully been introduced into new breeding habitats and climates in Great Britain, Iceland, and New Zealand. Wintering habitats are slightly less diverse, and in eastern North America the combination of extensive agricultural lands growing grain crops closely adjacent to estuarine or salt-water marshes where extensive beds of bulrush and cord grass provide both food and protection represents nearly ideal conditions. Likewise, in the midwestern states, refuges that provide both fresh-water marshes and abundant food in the form of corn or other grain crops have in recent years attracted increasingly larger numbers of Canada geese that in earlier times migrated all the way to the Gulf Coast (Johnsgard, 1975; Bellrose, 1976).

Social behavior. The strong family and pair bonding generally believed to be exhibited by Canada geese is almost legendary, and in general is an accurate assessment. However, enforced separation of a bird from its mate will result in the establishment of a new pair bond, usually during the next breeding season (Hanson, 1965). Potentially permanent pair bonds can be established surprisingly rapidly among experienced birds that are acquainted with one another, and initial pairing in one study occurred among two-year-old birds on the nesting grounds (Sherwood, 1965). At least some Canada geese, particularly of the larger subspecies, regularly attempt to breed as two-year-olds, and probably all birds normally attempt to breed by their third year. Immature birds retain social bonds with their broodmates and parents until well into their second year, and it is this strong familial bonding system that provides the basis for the social structuring of Canada goose flocks. Inter-pair and interfamilial aggressive behavior is typically followed by performance of the triumph ceremony, which facilitates the formation and maintenance of these social bonds. In addition, head-tossing behavior, special vocalizations, and distinctive plumage features such as the white upper tail coverts help to synchronize takeoff and facilitate in-flight coordination of pairs and families (Raveling, 1969).

There is a tendency in both sexes, but especially in females, to return to their natal homes when they begin nesting; such behavior tends to produce the localized adaptations and to promote rapid subspeciation, a condition that is particularly obvious among Canada geese. Breeding Canada geese tend to be relatively territorial and dispersed wherever the habitat allows for such spacing, but in particular the availability and distribution of suitable nesting sites seem to dictate the degree of sociality that is tolerated during the nesting season. Furthermore, the small arctic-breeding Canada geese, such as cackling geese, often nest in dense concentrations that may reach or even exceed 100 nests per square mile. Occasionally the larger Canada geese will tolerate great crowding of nests on island situations, or even in such unusual nest locations as on haystacks (Williams, 1967).

Reproductive biology. Inasmuch as there are great regional and population differences in the extent and timing of migration and the length of the potential breeding season, little can be said of the timing of breeding other than that it becomes more predictable and abbreviated as one moves toward the Arctic. Southern-nesting forms are prone to varied times of

nest initiation and to repeated renesting attempts; the most northerly ones are highly synchronized and exhibit no renesting behavior. However, clutch sizes vary little if at all across the entire nesting range, and average close to 5 eggs, with an egg-laying rate of about one per day. The subspecies do differ measurably in terms of egg size and incubation periods, with the larger forms having incubation periods of about 28 days and the smallest ones about 24 days. In all, however, only the female incubates, while the male remains on territory to help defend the female and nest. Most races of Canada geese are large enough to deal with the majority of predators, and thus losses from flooding or weather seem to be more important mortality factors than are nest predators. There are marked differences in the rate at which the goslings develop and attain flight, with the cackling Canada goose reportedly fledging in as few as 40 days and the giant Canada goose requiring from 64 to 86 days. Likewise, adults of the smallest forms of Canada goose are flightless for periods of from 20 to 30 days, while the larger races are flightless for 30 to 40 days. Adults of the arctic races typically become flightless when their goslings are between one and two weeks of age, while farther south the young may be about a month old before the adults attain their flightless condition. In some populations of Canada geese the nonbreeding segment of the population undertakes a molt migration of several hundred miles before undergoing their annual molt; some of the giant and Great Basin birds may actually fly close to 2,000 miles to the northern parts of the Hudson Bay region for this period (Bellrose, 1976; Johnsgard, 1975).

Status. In a general sense the Canada goose population is probably more secure at present than it has been in historical times. Winter surveys have suggested that the numbers have roughly doubled between the mid-1950s and the mid-1970s, and a fall population of about 3 million birds was estimated in 1974 (Bellrose, 1976). However, certain of the subspecies are by no means secure, and one subspecies native to the Aleutian Islands has been close to extinction and in 1977 totaled less than 2,000 birds (Springer et al., 1978). Another, the Bering Canada goose, is known only from a few museum specimens. Bellrose (1976) has provided fall or winter population estimates for the other subspecies or regional populations, the rarest of which is presumably the giant Canada goose. Because of disagreement as to its distinction from the similar and intergrading Great Basin Canada goose its numbers are impossible to assess with certainty, but Bellrose suggested that the "eastern prairie" population alone probably contained an average of 27,000 giant Canada geese in the period 1969–73, and additional birds of this size class occur in the "tall-grass prairie" flock.

Relationships. The Canada goose and barnacle goose occupy mutually exclusive breeding ranges and one might readily imagine them to be replacement forms, but they are probably not quite so closely related as a superficial comparison of them would suggest. Perhaps the Canada goose might best be thought of as the most generalized form of *Branta*, with the brant goose, the barnacle goose, and the Hawaiian goose all representing variably specialized relatives.

Suggested readings. Hanson, 1965; Williams, 1967; Bellrose, 1976.

Barnacle Goose

Branta leucopsis (Bechstein) 1803

Other vernacular names. None in general English use. Weisswangengans (German); bernaches nonnette (French); barnacla (Spanish).

Subspecies and range. No subspecies recognized. Breeds in northeastern Greenland, Spitsbergen, on the southern island of Novaya Zemlya, and on Vaigach Island. Winters in the British Isles and along the North Sea. See map 28.

Measurements and weights. Folded wing: males, 385–425 mm; females, 385–417 mm. Culmen: males, 27–32 mm; females, 28–31 mm. Weights: males (in March) 1,370–2,010 g (av. 1,672 g); females, 1,290–785 g (av. 1,499 g) (Bauer & Glutz, 1968). Eggs: av. 76 x 50 mm, white, 107 g.

Identification and field marks. Length 23–28″ (58–71 cm). Plate 18. *Adults* of both sexes have a white or buffy white head, except for a black stripe between the eye and bill, and a black crown which extends over the back of the neck and sides of the head to the base of the throat and continues downward to the back and lower breast, where it is sharply terminated. The back and scapulars are blackish and

silvery gray, and the rump and tail are black. The tail coverts are white, and the white extends forward on the abdomen to the black breast and laterally on the sides to merge with pale barred gray flanks. The flight feathers are silvery gray, and the upper wing coverts gray with white tips and a subterminal black bar, forming a scalloped pattern. The iris is brown, the bill black, and the legs and feet black. *Immatures* in the first winter of life are similar to adults, but have duller markings, a more grayish neck, and gray flecking in the white head pattern. When seen in flight, the upper wing surface of first-year birds appears darker than that of adults.

In the field, the strong black and white patterning of the head, neck, and breast of this goose is quite distinctive, and is likely to be confused only with that of Canada geese. In the few areas where Canada geese and barnacle geese might be seen together, the black breast and whitish flanks of the barnacle goose provides an excellent field mark. The calls of barnacle geese are rapidly repeated single-syllable barking notes, rather like those of a small dog.

NATURAL HISTORY

Habitat and foods. Barnacle geese are in general confined as breeding birds to arctic areas where cliffs or rocky slopes are located close to lakes, rivers, marshes, the upper portions of fjords, or even coastlines, but they also nest on level ground at times, in the vicinity of brant or eiders. For the rest of the year the birds are essentially maritime, foraging on tidal flats and in coastal marshes and adjoining grassy areas. They are essentially exclusively vegetarians as adults, and feed largely on grasses, sedges, and the leaves of various herbs and shrubs, including even the leaves, twigs, and catkins of arctic willow *(Salix arctica)*. In a sample of 14 birds obtained during winter in Eng-

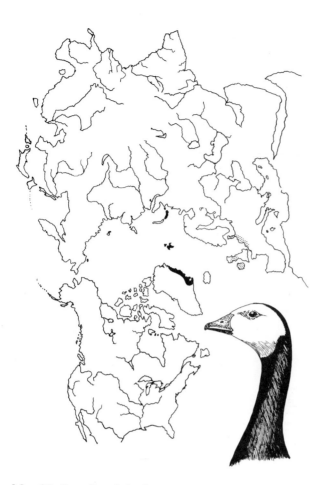

MAP 28. Breeding (inked) and wintering (stippling) distributions of the barnacle goose.

land, over 90 percent of the food remains consisted of grasses, and no animal materials were present (Palmer, 1976).

Social behavior. It is probable, but not certain, that barnacle geese mature at two years, but inasmuch as most other high-arctic geese usually do not nest successfully before their third year, this may also apply to this species. Records from captivity also indicate that few birds nest as two-year-olds, and most breed when three (Ferguson, 1966). Pair formation has not been studied in the wild, but observations in captivity indicate that the hostile behavior patterns and triumph ceremonies are nearly identical to those of Canada geese. The two species likewise perform preflight head-shaking movements that effectively expose their white cheek and throat areas. The precopulatory displays consist of mutual head-dipping movements, and although during the postcopulatory display the male raises his folded wings somewhat

more strongly than do Canada geese, the posturing is otherwise almost identical (Johnsgard, 1965). It is assumed that most pair-forming behavior occurs on the wintering grounds and the birds are well paired by the time of their arrival at the breeding areas.

Breeding biology. On Novaya Zemlya, these geese typically seek out rocky outcrops, ledges of steep cliffs, or the tops of low hills for nesting, which is done in scattered colonies that are highly conspicuous (Dementiev & Gladkov, 1967). Observations in northeastern Greenland by Ferns and Green (1975) indicate comparable nest sites, as on a nearly vertical basalt exposure about 50 meters high, where the birds chose flat and gently sloping ledges about 75 centimeters deep and one or two meters wide. Evidently nest sites are often used year after year and gradually accumulate nesting materials. The clutch size typically ranges from 4 to 6 eggs, but up to 9 have been reported. The inaccessibility of nests to most ground predators probably reduces nesting losses and is presumably the reason for the cliffside nesting; even the birds have difficulty in landing on the small ledges. Incubation is performed by the female alone and requires from 24 to 25 days. There is evidently a rather high incidence of nonnesting or unsuccessful nesting, judging from the low percentages of goslings reported both on the breeding grounds and in wintering areas; Ferns and Green (1975) noted average brood sizes of 2.7 to 2.8 young for single families as well as for amalgamated family parties, but only 11.1 percent young in the total population seen. They calculated that about 90 percent of the adult population either did not attempt to breed or failed in their efforts. The fledging period, based on captive birds, is 6 to 7 weeks, a short period but typical for arctic-breeding geese.

Status. Ogilvie and Boyd (1976) summarized population estimates of barnacle geese wintering in Britain, and reported that the Greenland component numbered about 24,000 birds in 1973, an increase from 8,300 in 1959. The population breeding on Spitsbergen also winters in Scotland; in 1962 it numbered about 2,300 birds and more recently was estimated at about 4,000 individuals. The population breeding on Novaya Zemlya and Vaigach Island consisted of about 20,000 individuals in 1959 and more recently of about 25,000. All of these figures would suggest a total world population of about 50,000 barnacle geese, making this species one of the rarer forms of true geese (Kumari, 1971). This, added to the seemingly low reproductive efficiency of barnacle geese, makes them vulnerable to rapid population declines in spite of apparent recent increases.

Relationships. Plumage and behavioral similarities suggest that the barnacle goose and Canada goose are close relatives (Johnsgard, 1965a). Ploeger (1968) discusses the evolutionary history of this species in a perspective of glaciation patterns.

Suggested readings. Ferns & Green, 1975; Jackson et al., 1974.

Brant

Branta bernicla (Linnaeus) 1758

Other vernacular names. Brent; Ringelgans (German); bernachea cravant (French); branta (Spanish).

Subspecies and ranges. (See map 29.)

B. b. bernicla: Dark-bellied (Russian) brant. Breeds in northern Europe and Asia from Novaya Zemlya to the Khatanga River, and on Franz Joseph Land and Kolguev Island. Winters along the coast of northwestern Europe.

B. b. orientalis: Pacific brant. Breeds in arctic Siberia from about the delta of the Lena River to the Anadyr basin, and on the adjacent arctic islands. Winters along the coast of Korea and China.

B. b. nigricans: Black brant. Breeds in eastern Siberia from the Anadyr River to the Bering Strait, and in North America from western Alaska to central northern Canada, where it intergrades with *hrota* on Prince Patrick Island and in the Perry River region. Winters mainly along the Pacific coast of North America.

B. b. hrota: Light-bellied (Atlantic) brant. Breeds in northern Canada from Prince Patrick Island and the Perry River area eastward to Baffin Island, and on Greenland and the adjacent arctic islands. Winters along the Atlantic coast of North America.

Measurements and weights. Folded wing: Males, 310–51 mm; females, 315–31 mm. Culmen: males, 29–37 mm; females, 31–51 mm. Weights: males, 1,400–2,250 g; females, 1,200–769 g. Eggs: av. 70–75 x 47–50 mm, white, 85 g.

Identification and field marks. Length 22–26" (55–66 cm). Plate 19. *Adults* of both sexes have the head, neck and upper breast black, except for a somewhat crescentic white patch on both sides of the upper neck, interrupted by parallel striations of the neck feathers, exposing the darker bases. The back and scapulars are dark brown, with inconspicuous lighter tips, and the rump is dusky centrally and white laterally. The tail coverts are white and elongated, nearly hiding the black tail, and the abdomen is whitish posteriorly, and white or gray anteriorly. The sides and flanks vary from nearly white (*hrota*) to grayish (*bernicla*) or black and white (*orientalis*), with some blackish barring usually evident toward the rear. The flight feathers are black, and the upper wing coverts are grayish brown to dark brown, with inconspicuous paler edging. The iris is brown, the bill is black, and the legs and feet are black. *Immatures* in their first winter are much like adults but have conspicuous white edges on the upper wing coverts, the scapulars, and the back. Young birds gradually acquire the white throat markings of the adult during their first winter of life.

In the field, the maritime habitat and tiny size of these geese set them apart from all others, and their general black and white appearance is unique. They swim with their hind quarters well raised, exposing the white tail coverts, and when in flight tend to be in undulating and irregular lines. Their calls consist of soft *ruk-ruk* or *r-r-r-ruk* sounds that do not carry very far.

NATURAL HISTORY

Habitat and foods. Throughout its breeding range the brant is associated with lowland coastal tundra never very far from the tidal zone. Generally the birds are associated with tundra-covered flats dissected by numerous tidal streams, only a few feet above high tide line, and in some areas the birds extend inland to tundra lakes with grassy islands or to grassy slopes of low mountains near the coast, but such areas are secondary. Summer foods are mainly the newly regenerating shoots of sedges that line tidal ponds, and the birds probably also feed on other grasses and broad-leaved herbs as well. The young birds eat insects, aquatic invertebrates, and the vegetative parts of pondweeds that occur in the tidal channels. As the young mature, they and the adults shift over to more maritime foods, consisting primarily of eelgrass (*Zostera*) and secondarily of

❖ ❖ ❖ 83

wigeon grass and sea lettuce (*Ulva*). A very limited amount of animal materials are at times consumed, perhaps in part accidentally with plant life, and including fish eggs, worms, snails, and amphipods. But in general the birds seem largely dependent on eelgrass wherever it is abundant, and the depletion of the North Atlantic eelgrass beds in the early 1930s was accompanied by a substantial population decline in the brant of that region as well as a shift to other food sources (Johnsgard, 1975; Bellrose, 1976).

Social behavior. Sexual maturity in these geese, as indicated by formation of pairs, is probably normally attained in the second winter of life, but evidently only a very small proportion of two-year-old females breed successfully, and probably many do not attempt to nest until they are nearly three years old. Pair formation apparently may even begin during the summer on the breeding grounds, but reaches its highest incidence in wintering areas, where hostility

and aerial chases become highly evident. Such flights are seemingly more evident in brant than in the larger geese, but as in the latter, the mechanism of pair formation seems to be through the use of the triumph ceremony by pairs or potential pairs. The triumph ceremony is apparently also used when a family member regains its own group (Jones & Jones, 1966). Copulation is preceded by mutual head-dipping movements but not the strong tail cocking typical of many geese in this situation. Afterward the male stretches his neck and calls, but no other obvious ritualized posturing is evident, judging from observations of captive birds (Johnsgard, 1965a).

Reproductive biology. Brant usually arrive in flocks that concentrate along the coast of their breeding grounds before the nesting areas are free of ice and snow, and pairs gradually leave these flocks to seek out suitable nesting sites, with the female leading such flights. In areas of dense concentration the birds are distinctly colonial nesters, with densities often approaching and sometimes exceeding one nest per acre. Presumably only the immediate nest site is defended in such situations, and it is usually located on a small islet or along the shoreline of a pond or stream, generally in rather low grass or sedge cover, where the dark-colored female remains rather conspicuous in spite of her small size. The male usually remains within 100 yards of the nest during egg laying, which usually requires only as many days as there are eggs in the clutch—typically 3 to 5 and usually 4. The female incubates for 22 to 25 days, keeping her head and neck low to the ground during much of this time, and the male continues to keep his vigil over the nest for the entire time, but rarely comes closer than 15 feet from the nest unless it is required for defense. There is a high degree of synchrony of egg laying, and thus hatching, in brant colonies; likewise, no renesting efforts are made by unsuccessful females. Storm tides often cause massive destruction of brant nests, and late winter snow storms can also cause extensive damage and desertion of nests (Barry, 1966). After the young are hatched, they are joined by the male, who assumes the major responsibility for brood defense. The family typically moves out into the tidal flats, where the young begin to feed on invertebrates. About 10 days after hatching, the adults begin their flightless molt, which lasts about 30 days, resulting in their resumption of flying abilities about the time the young complete their 40- to 50-day fledging period. The fall migration begins very soon afterward, and in at least

MAP 29. Breeding distributions of the dark-bellied ("D"), Pacific ("P"), black ("B"), and light-bellied ("L") brant geese. Wintering areas of all races is indicated by stippling.

some areas there is an earlier premolt migration by immature birds and unsuccessful breeders to favored areas, such as near Cape Halkett, Alaska (King, 1970).

Status. Bellrose (1976) reported that although the average winter population of light-bellied, or Atlantic, brant between 1955 and 1974 was 177,000, sharp declines reduced the number from 151,000 in 1971 to 42,000 in 1973, followed by an increase to nearly 90,000 in the two following years. This serious loss was largely the result of unfavorable nesting seasons, but in part resulted from overharvesting (Penkala et al., 1975). In contrast, the dark-bellied brant goose that winters along the coast of Great Britain and northern Europe increased from a rather frightening estimate of 16,500 birds in the mid-1950s to about 110,000 in the winter in 1975-76, in part apparently as the result of improved protection in the wintering areas (Ogilvie, 1976b). The black brant, which winters along the Pacific coast of North America, has evidently remained at a fairly constant level in the last 25 years, with an average wintering population estimate of 140,000 for the period 1951-74 (Bellrose, 1976). There are no numerical estimates available for the size of the Pacific brant population that winters along the coastlines of Korea and China.

Relationships. Ploeger (1968) has discussed the probable basis of subspeciation in the brant goose in connection with glacial events; most writers now accept the idea that only a single species of brant should be recognized. However, this species is well distinguished from all of the other extant *Branta* species, and it seems impossible to judge its nearest living relative. Its downy young most closely resemble those of the barnacle goose, which is likewise a rather marine-adapted and high-arctic species and is perhaps its nearest living relative (Johnsgard, 1965a).

Suggested readings. Barry, 1966; Mickelson, 1973; Einarsen, 1965.

Red-breasted Goose

Branta ruficollis (Pallas) 1769

Other vernacular names. None in general English use. Rothalsgans (German); bernache à cou roux (French); ganso de pecho rojo (Spanish).

Subspecies and range. No subspecies recognized. Breeds on the Siberian tundra from the Ob to the Khatanga. Winters at the west end of the Black Sea, the south end of the Caspian Sea, and occasionally at the north end of the Persian Gulf. See map 30.

Measurements and weights. Folded wing: males, 342-90 mm; females, 342-60 mm. Culmen: males, 24-26 mm; females, 24-25 mm. Weights: males, 1,315-625 g; females, 1,150 g (Bauer & Glutz, 1968). Eggs: 71 x 48 mm, cream-colored, 76 g.

Identification and field marks. Length 21-22" (53-55 cm). Plate 21. *Adults* of both sexes have a black head and hind neck except for an oval white patch between the eye and bill, and a larger bright chestnut patch behind and below the eye, this latter patch surrounded by white that extends down the side of the neck. The foreneck and breast are also reddish chestnut, and are separated from the black underparts and sides by a white stripe. The upper and rear flank feathers are broadly marked with white, and the rear underparts and upper tail coverts are also white. The mantle, upper wing surface, rump, and tail are all black, except for white tipping on the greater secondary coverts and the row of coverts immediately ahead of them. The iris is brown, and the bill, legs, and feet are black. *Immatures* in their first winter generally resemble adults, but are considerably duller, with white spotting present on the sides and mantle, and with the chestnut markings much weaker and less well demarcated.

In the field, the tiny size and unique chestnut head and breast markings make this species almost impossible to confuse with any other waterfowl, and the strong black and white pattern of the flanks can be seen at great distances. The birds utter loud, staccato calls that are almost always double-noted, *kik-wik,* or *kee-kwa,* something like the barking of a tiny dog.

NATURAL HISTORY

Habitat and foods. The surprisingly restricted breeding grounds of the red-breasted goose consist mostly of moss- and lichen-covered tundra and brush tundra, mainly on the Gyda Peninsula and the western part of the Taimir Peninsula. During the nesting season the birds seem to prefer the driest and highest areas of tundra and brush close to water. In the winter they often occupy low-lying regions around

MAP 30. Breeding (hatched) and wintering (stippling) distributions of the red-breasted goose.

lakes or reservoirs where halophytic herbs and ephemeral grasses are to be found (Dementiev & Gladkov, 1967; Uspenski, 1965). However, in Romania they have been seen inhabiting sprouting wheatfields and corn stubble in company with white-fronted geese (Scott, 1970). Throughout the year they are vegetarians, concentrating during the breeding season on various grasses and sedges, particularly cotton grass (*Eriophorum*), which sprouts at about the time they begin nesting. In their winter quarters they reportedly make daily flights to water holes to drink and obtain sand, then return to steppelike areas where they consume grasses and arid-adapted herbs (*Salicornia*).

Social behavior. The period to sexual maturity and usual time of initial breeding have not been established for wild birds, but in captivity 7 of 13 initial breeding records were for third-year birds, while the remainder were equally divided between the next two years (Ferguson, 1966). Thus, it is likely that little if any breeding occurs in two-year-old birds, even though they may well be mated before that time. In spite of their small size, red-breasted geese are highly vocal and relatively aggressive among themselves, and perhaps their unusually bold coloration is related to its role in emphasizing their threat displays, which in part involve a vibration of the "mane" feathers and rotary head shaking. Precopulatory behavior is also highly ritualized, and involves mutual head dipping alternated with a very erect attitude as the male faces the female. Postcopulatory behavior is also marked by an extreme raising of the wing, cocking of the tail, and stretching of the neck while calling (Johnsgard, 1965a).

Reproductive biology. These geese arrive on their breeding grounds in small flocks of 3 to 15 birds, usually during the first half of June, about the time that the tundra is starting to become free of snow and grasses are starting to sprout. The birds often nest in small colonies of 4 or 5 pairs, often on steep river banks, in shrub- or grass-covered gullies or ravines, or in small depressions on sloping ground. The nests are often quite conspicuous, and almost invariably the birds nest in close vicinity to peregrine falcons or rough-legged hawks, or even near colonies of larger gulls such as glaucous gulls or herring gulls. It has recently been reported (Krechmar & Leonovich, 1967) that 19 out of 22 peregrine nests had from 1 to 5 red-breasted geese pairs nesting nearby, from as close as 5 to as far as 300 feet away from the peregrine nest. This remarkable nesting association apparently evolved as an antifox adaptation, and the serious worldwide decline in peregrine populations in recent decades has been suggested as a possible explanation for the concurrent decline in red-breasted goose numbers (Ogilvie, 1976a). The clutch size is typically from 3 to 7 eggs, but rarely up to 9 have been reported, and probably 6 or 7 eggs would represent the normal clutch. Incubation is performed by the female alone, but the male remains nearby throughout the incubation period of 23 to 25 days. Hatching typically occurs by the end of July, and molting by adults must begin at about the same time, since most of the adults and juveniles are flying by the last third of August, suggesting a very short fledging period for the young (Dementiev & Gladkov, 1967; Uspenski, 1965).

Status. Apparently the red-breasted goose is quickly becoming a candidate for extinction. Hudson (1975) reported that the world's red-breasted goose population has declined from about 50,000 birds in 1956–57 to about 30,000 more recently, and that the major wintering areas have shifted to the west from Azerbaijan toward Romania. The numbers of birds found in traditional wintering areas and sanctuaries vary greatly, according to the severity of the weather, and Scott (1970) noted that although some 25,000 were present in Romania in the winter of 1968–69, in the following winter only about 3,000 to 4,000 were present in the same area. If Ogilvie (1976a) is correct in his pessimistic estimate of only about 15,000 birds, the species is in real danger and will have to be closely monitored in the future.

Relationships. The highly distinctive plumage and remarkably short bill of this species sets it well apart from the other *Branta* species, and its behavior provides no real clues to its probable relationships. It is seemingly the most isolated of all the "black geese," and probably diverged from the other *Branta* stock quite early (Ploeger, 1968).

Suggested readings. Scott, 1970; Hudson, 1975; Owen, 1977.

Tribe Cereopsini (Cape Barren Goose)

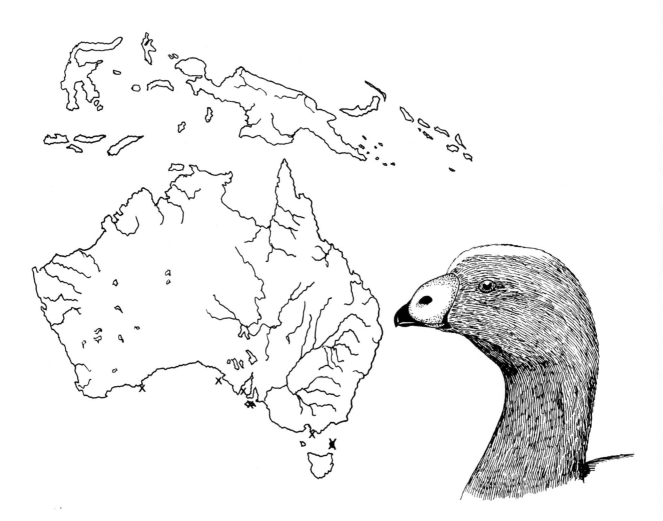

MAP 31. Locations of residential populations of the Cape Barren goose; major population indicated by large *X*, other populations represented by small *x*.

Cape Barren Goose

Cereopsis novaehollandiae Latham 1801

Other vernacular names. Cereopsis goose, pig goose; Hühnergans (German); céréopse (French); ganso del Cabo Barren (Spanish).

Subspecies and range. No subspecies recognized. Breeding currently limited to the islands off Australia's southern coast, from the Furneaux group in the east to the Recherche Archipelago to the west, with nonbreeding groups occurring on the adjacent mainland during summer. See map 31.

Measurements and weights. Folded wing: males, 450–90 mm. Culmen: males, 45–50 mm. Weights: both sexes range 3,170–6,800 g (males average 5,290 g; females average 3,770 g (D. Dorward, pers. comm.). Eggs: av. 78 x 55 mm, white, 137 g.

Identification and field marks. Length 30–39" (75–100 cm). Plate 22. The cereopsis has a unique short and cere-covered bill that is pale greenish yellow except for the nail and cutting edges, which are black. The plumage almost entirely dove gray, with the larger feathers having dark brown rounded or heart-shaped spots, especially on the larger wing coverts. The feet are black, and the legs vary from black to pink or carmine. *Females* are identical in appearance to males, and *juveniles* average slightly lighter than adults, but have heavier spotting on the wings. Their legs may vary from black to greenish in color.

In the field, Cape Barren geese are unmistakable; their limited range places them out of contact with all other gooselike birds. Their call is a piglike grunt, and males also have a louder honking or trumpetlike note.

NATURAL HISTORY

Habitat and foods. Cape Barren geese are limited at present to various small islands, where they inhabit such open areas as beaches, promontories, and grassy areas, but avoid scrub except during nesting. They also avoid entering water except when injured, molting, or as juveniles. Their bills are adapted for grazing, and about two-thirds of all their food is composed of grasses, of which a common tussock grass (*Poa*) is the most important. Many other grasses are consumed, and the leaves and seeds of some broad-leaved plants (*Trifolium, Medicago,* etc.) and sedges are also eaten in smaller quantities

(Frith, 1967). Cape Barren geese can survive where little or no fresh water is available, and are able to excrete a solution saltier than sea water through their salt glands (Marriott, 1970).

Social behavior. Cape Barren geese are fairly gregarious, particularly nonbreeders, which may form flocks of up to 300 individuals. The breeding birds are found in flocks only during the molting period in October, otherwise they remain in pairs. From November through January they forage near their territory, from February to April they are actively defending their territories, and the period of nesting and rearing of young extends from May to October. The maximum density of breeding birds seems to be from one pair per acre to about one pair per 1.5 acres. After the breeding season many birds leave their nesting islands, and at that time flocks appear on the mainland of Australia, sometimes containing 50 to 60 birds; rarely as many as 500 occur near Lakes Alexandrina and Albert in South Australia. Pair formation takes the form typical of geese, with a well-developed triumph ceremony between the mates or potential mates; in this respect the species is closer to the true geese than to the sheldgeese. As in geese, pair bonds are strong and permanent, and the male fiercely defends the territory and nest. Copulation occurs on dry land, a nearly unique situation among waterfowl. Thus there are many differences in behavior from that of other geese, including an absence of head-dipping movements. The male walks around the female, pecking at her back and seemingly trying to push her to the ground. After she goes prone, the male walks around her several times, flaps his wings, and mounts. Afterward there is a mutual calling ceremony much like the usual triumph ceremony (Johnsgard, 1965a).

Reproductive biology. On the Furneaux group, nest building by the birds may begin as early as May, but June is the more typical month for this. Nests are usually placed in bushes or beside rocks or tussocks of *Poa,* especially in areas where there is good visibility. Rarely, nests have been found in trees such as tea trees (*Leptospermum*) as high as 18 feet from the ground. Nests are usually widely spaced and are sometimes built beside old sites. The nearest that two nests have been found has been 13 yards apart; close spacing seems to be typical of scrubby habitats rather than open and grassy ones. Clutch sizes are usually from 3 to 6 eggs, with 5 being the most frequently encountered number. The nests are well covered with down when the female leaves them; so far as is known, only the female incubates. The incubation period lasts about 35 days; and for a time after hatching, the adults keep their young in the breeding territory, leading them to water only when disturbed. The young cluster into nursery flocks at the age of about six weeks and begin to fly about October. In captive-raised birds, the age of fledging has reportedly been 70 days (Veselovsky, 1973). October is also the period of molting by adults and subadults (Guiler, 1967).

Status. The major stronghold of the Cape Barren goose is the Furneaux group, where in 1967 Guiler reported that aerial surveys in the 1960s resulted in counts of from about 1,000 in 1960 to 1,600 in 1965.

There are no permanent human residents on these islands, and although some are grazed, there is relatively little disturbance. Barring major changes in human disturbance, the conservation of at least this relict population of geese appears to be feasible (Guiler, 1967). It is probable that, as of 1975, about 16,000 Cape Barren geese existed in the wild, and thus they must be watched carefully if the species is to be preserved (Pierce, 1975).

Relationships. Delacour's (1954–64) relegation of this species to the sheldgoose tribe has been discussed by me (Johnsgard, 1961a, 1965a) and I have argued that it should instead be regarded as a primitive and aberrant goose. Woolfenden (1961) has supported the evidence for anserine affinities and suggested the erection of a monotypic tribe for the species. The species' swanlike (but not gooselike) characteristics have also been supported by a study of the photoresponses of the species for breeding (Kear & Murton, 1973), and these authors have also indicated that the bird is primitive and close to the stem-line of anatid stock. For these reasons the recognition of the tribe Cereopsini would appear justified. However, Brush (1976) recently, on the basis of an analysis of its feather proteins, supported my early position that the species is allied to the geese and swans rather than the sheldgeese, and that tribal separation is not warranted.

Suggested readings. Frith, 1967; Guiler, 1967, 1974.

1. White-faced whistling duck, adult

↑ 2. Magpie goose, pair with male in foreground   ↓ 3. Spotted whistling duck, female and brood

↑ 4. Plumed whistling duck, pair   ↓ 5. Lesser whistling duck, pair

↑ 6. Black-bellied whistling duck, pair dabbling

↓ 7. African white-backed duck, adult male

↑ 8. Mute swan, male wing-flapping

↓ 9. Black swan, female and brood

↑ 10. Black-necked swan, pair carrying brood

↓ 11. Trumpeter swan, adult and cygnet

↑ 12. Whooper swan, pair and four juveniles

↓ 13. Bewick swan, adult and brood

↑ 14. Coscoroba swan, adult   ↓ 15. Lesser white-fronted goose, pair

↑ 16. Eastern graylag goose, female and brood   ↓ 17. Lesser snow goose, female and gosling

↑ 18. Barnacle goose, female and brood

↓ 19. Atlantic brant, pair

20. Hawaiian goose, adult male

21. Red-breasted goose, pair

↑ 22. Cereopsis goose, pair   ↓ 23. Andean goose, female and goslings

24. Magellan goose, pair with female in foreground

↑ 25. Ashy-headed sheldgoose, adult and goslings

↓ 26. Egyptian goose, pair

↑ 27. Cape shelduck, pair with male in foreground

↓ 28. Magellanic flightless steamer duck, pair

Tribe Stictonettini (Freckled Duck)

Map 32. Breeding or residential distributions of the freckled duck.

Freckled Duck

Stictonetta naevosa (Gould) 1840

Other vernacular names. Oatmeal duck, monkey duck; Pünktchenente (German); canard moucheté (French); pato manchado (Spanish).

Subspecies and range. No subspecies recognized. Regularly resident in only a limited area of south-central Australia in the Murray-Darling Basin, and in the southwestern portion of Western Australia. Elsewhere the birds are extremely local breeders or occur irregularly as vagrants. See map 32.

Measurements and weights. (Mainly from Frith, 1967.) Folded wing: males, 186–258 mm; females, 205–36 mm. Culmen: males, 50–59 mm; females, 46–53 mm. Weights: males, 747–1,130 g (av. 969 g); females, 691–985 g (av. 842 g). Eggs: av. 63 x 47 mm, cream or ivory, 66 g.

Identification and field marks. Length 20–22" (50–55 cm). The freckled duck is unique in several respects, including its "oatmeal" coloration of dark brown with small buff-colored freckles on the head and upperparts, which grade into a vermiculated pattern on the flanks and tail coverts. The bill is high at the base but unusually flattened toward the tip, with a large nail, and is generally dark gray, except in breeding males, which exhibit a bright orange red color at the base of the bill. The wings are brownish above and below, with coverts freckled like the back, and the legs and feet are slate gray. The front of the legs have a reticulated scale pattern, as in whistling ducks. *Females* resemble males, but are lighter in color and have a less contrasting pattern of freckles. *Juveniles* resemble females, but are much lighter in color, and their freckles are a deep buff color.

In the field, freckled ducks slightly resemble Australian black ducks, but are smaller and have a nearly uniformly dark head. The head shape is characteristic; a small crest makes the head appear almost triangular in profile from the side. The most commonly heard call is a flutelike note similar to that of a black swan. Frith (in Delacour, 1954–64) was erroneous in describing the flight as slow and bitternlike; it is almost as rapid as a mallard's.

NATURAL HISTORY

Habitat and foods. The favored breeding habitats of freckled ducks are heavily vegetated areas of fresh water, often a permanent swamp or a newly flooded creek. Permanent cattail (*Typha*) swamps or lignum swamps are typical inland habitats, while near the coast the birds favor tea tree swamps. Their foraging is done by filter-feeding at the surface, dabbling while immersing the bill or upending, and by wading along shore, dipping the bill to the bottom and filtering food through it in the manner of typical stifftails (*Oxyura*). The bill is well adapted in shape for this mode of feeding, and resembles that of stifftails. On the basis of samples of food taken from wild birds, it appears that algae are the most constant source of food, with seeds of such aquatic or shoreline plants as smartweeds (*Polygonum*) and docks (*Rumex*) also important. Many other kinds of plant seeds have been found in smaller quantities, and a small percentage of animal foods, mainly insects, has been reported (Frith et al., 1969).

Social behavior. During the nonbreeding season freckled ducks gather in small to moderately large flocks on lakes, lagoons, and billabongs. There they remain in fairly closely spaced groups, with little aggressive behavior evident. Whether they are mostly paired or unpaired is a matter of conjecture, but I observed almost no pair-forming activities in a group of nearly 200 birds just before the nesting season (Johnsgard, 1965b). The only social display I observed consisted of a rapid and extreme vertical neck stretching associated with gaping (and probably calling) on the part of two birds while closely facing each other. Although it was similar in form to the triumph ceremonies of some swans such as the black swan, I observed no prior aggressive behavior by the birds. I also observed no inciting on the part of females. Regrettably, no observations on copulatory behavior have yet been made. Frith (1967) kept the species in captivity for some time but paid little or no attention to their social behavior, and nobody else has had the opportunity to study the birds under these conditions. With the advent of local or general drought, the birds move about nomadically, sometimes for considerable distances, but there is apparently no general pattern of migratory movements evident for this species. On the other hand, during unusually wet years the birds are sedentary and the breeding season may be quite prolonged, from about July until January in most areas, while in normal years it probably occurs between September and December (Frith, 1967).

Reproductive biology. Only a limited amount of information is available on the reproduction of this unusual species. It seems that the breeding season is normally regular in this species and relatively late as compared with that of such birds as the black duck and the Australian white-eye, but the birds are able to breed at other times when flooding conditions provide a favorable situation. The observed nests of this species are few, and have consisted of bowl-shaped structures built of small lignum (*Meuhlenbeckia*) sticks and spike rush (*Eleocharis*) placed in flooded lignum bushes in about four feet of water. They are probably normally built close to the water level, but declining waters may cause them later to become elevated above it. A moderate amount of down is present as a nest lining, and unlike the male whistling ducks, the male of this species evidently does not help incubate. The clutch size ranges from 5 to about 10 eggs, and 7 is the number most often found. The probable incubation period is from 26 to 28 days (Frith, 1967), but Braithwaite (1976) estimated a 36-day period for one clutch. He also noted that only the female incubated, and that there was no indication of a strong pair bond during the incubation period, as no males were seen near the nests he observed. On the basis of information from captive-raised birds, it would seem that the fledging period is relatively long, requiring about nine weeks (Frith, 1967).

Status. The status of this species is of special significance in view of its remarkable evolutionary interest. Frith (1967) has indicated that the species' future depends on preservation of the relatively few permanent swamp habitats where it breeds. Many of these swamps in South Australia are now being drained, and such activities could spell disaster for this unique species.

Relationships. In 1960 I suggested that a special anserine tribe Stictonettini should be erected for the freckled duck, and after observing the species in life added (1965b) supporting observations on that point. Frith's (1964) description of the downy young as swanlike has provided additional support for this position. Most recently, Brush (1976) has analyzed the feather proteins of this species and found that in this regard the species is completely unlike the dabbling ducks; although its electrophoretic pattern is unique, it is somewhat closer to the goose and swan complex than to the whistling ducks. Likewise, studies of the wax esters secreted by the uropygial gland of waterfowl (Edkins & Hansen, 1972; Jacob & Glaser, 1975) indicate that these secretions in the freckled duck are similar to those of certain swans such as the mute swan.

Suggested readings. Frith, 1965; Johnsgard, 1965b; Braithwaite, 1976.

Tribe Tadornini (Sheldgeese and Shelducks)

MAP 33. Residential distribution of the blue-winged goose.

Drawing on preceding page: Male Magellan Goose

Blue-winged Goose

Cyanochen cyanopterus (Rüppell) 1854

Other vernacular names. Abyssinian blue-winged goose; Blaüflugelgans (German); bernache à ailes bleues (French); ganso alas azules de Abisinia (Spanish).

Subspecies and range. No subspecies recognized. Limited to the highlands (above 8,000 feet) of Ethiopia, primarily in the area between Lakes Zwai and Tana, but extending north to about 15° north latitude. See map 33.

Measurements and weights. Folded wing: males, 368–74 mm; females, 314–34 mm. Culmen: males, 32–33 mm; females, 30–31 mm. Weights: adult female, about 1,520 g (Lack, 1968). Eggs: av. 70 x 50 mm, cream, 85 g.

Identification and field marks. Length: 23–29″ (60–75 cm). *Adults* are predominantly grayish brown, with paler brown tones on the lower back and upper tail coverts, and the under tail coverts are white. The tail and primaries are black, and the secondaries are glossy green, while the upper wing coverts are a distinctive powder blue except for a small area at the bend of the wing, which is white, as are the underwing coverts. The bill, feet, and legs are black. *Females* are identical to males, but are considerably smaller (see wing measurements above). *Juvenile* birds are somewhat duller in color and pattern than adults.

In the field, no other gooselike bird occurs in the limited geographic range of this species with which it could be confused. The birds often walk and swim with their heads resting on the back, and the male often utters a rapidly repeated whistling note, while the female has a similar but somewhat harsher call.

NATURAL HISTORY

Habitat and foods. The preferred habitat of these birds is the highland rivers of Ethiopia, where the country is open and where meadows of short grass come down to the river banks. They also occur at the edges of swamps that are not overgrown with bushes or banks of reeds, but rarely if ever are they found in the middle of ponds or in deep water. Their major foods undoubtedly consist of grasses, sedges, and similar herbaceous vegetation of these areas, judging from their bill shape. However, the stomachs of some wild birds shot in the wild have revealed such animal life as worms, insects, insect larvae, and snails.

Social behavior. In the wild, blue-winged geese are usually found only in pairs, and seldom in larger groups. They apparently never move great distances, and even when frightened they do not fly very far. They are reportedly somewhat nocturnal under natural conditions. Nothing is known of seasonal movements, but these are likely to be small or lacking.

Reproductive biology. Almost no nests have been described from wild birds, although broods of downy young have been reported in May and June, and one nest containing 5 eggs was found in March. In captivity, nests have been constructed under bushes, in a clump of sedge, and in boxes buried in a bank. Nest-building and mating activities are evidently carried out at night, since almost none of these behavior patterns have been observed by aviculturists. It is known from captive birds that the usual clutch is from 4 to 9 eggs, perhaps averaging 7, and that only the female incubates. The incubation periods reported by various observers range from 30 to 34 days. The downy young are as brightly patterned as those of typical sheldgeese, and grow fairly rapidly. It has been reported that in captivity completion of feathering, except for the wing feathers, is achieved at about six weeks, although the fledging period is probably appreciably longer (Delacour, 1954–64; Kolbe, 1972). The period to reproductive maturity is believed to be two years, but specific data seem lacking.

Status. This species has the most restricted range of all of the sheldgeese, and almost nothing is known of

103

its actual population size, which must be relatively small.

Relationships. The blue-winged goose is a typical sheldgoose, whose closest relationships are quite clearly with the South American genus *Chloephaga.* In some ways the species is an ecological counterpart of the Andean goose.

Suggested readings. Delacour, 1954–64.

Andean Goose

Chloephaga melanoptera (Eyton) 1838

Other vernacular names. None in general English use. Andengans (German); bernache des Andes (French); guayata, ganso andino (Spanish).

Subspecies and ranges. No subspecies recognized. Resident in the Andes above 10,000 feet elevation from southern Peru and Bolivia to the latitude of Mendoza Province in Argentina and Nuble in Chile. See map 34.

Measurements and weights. Folded wing: males, 460–75 mm; females, 420–30 mm. Culmen: males, 38–43 mm; females, 34–37 mm. Weights: Both sexes range from 2,730 to 3,640 g (Kolbe, 1972). Eggs: av. 75 x 50 mm, cream, 131 g.

Identification and field marks. Length 29–32" (75–80 cm). Plate 23. This sheldgoose is the only one having the combination of a coral-red bill and orange legs and feet, and further, its plumage is mostly white except for a black tail. Some flight feathers (primaries and tertials) are blackish, and most of the greater secondary coverts are iridescent purple, while the other coverts and secondaries are white. The longer scapulars are blackish, while the shorter and more anterior ones are white with oblong blackish central markings. *Females* closely resemble males but are considerably smaller, while *juvenile* and *immature* birds are less pure white, and dark gray above rather than blackish.

In the field, since no other sheldgoose occurs in the high Andean habitats used by this species, it can be readily recognized by its gooselike shape and its black and white plumage pattern. The call of the male is a loud and shrill repeated whistling note, while that of the female is a grating growl.

NATURAL HISTORY

Habitat and foods. The habitat of the Andean goose consists of Andean lakes, marshes, and well-watered valleys, where an abundance of fresh green grass and similar herbaceous vegetation is to be found. No studies of the birds' foods under wild conditions have been undertaken, but they are probably entirely of such materials. For most of the year the birds remain at altitudes above 10,000 feet, but in the southern parts of their range they may temporarily move down into the damp meadows and marshes of Chile's central valley after winter snows in the mountains (Johnson, 1965).

Social behavior. During much of the year these birds are found in loose flocks, which are probably quite sedentary and rarely are forced to fly. As in other sheldgeese, the pair bonds are strong and, once formed, are probably permanent. It is believed that all sheldgeese mature in the second year of life, and thus courtship probably occurs when the birds are about one and a half years old. The species is relatively pugnacious, and the sexual behavior of the female consists of a well-developed inciting call and

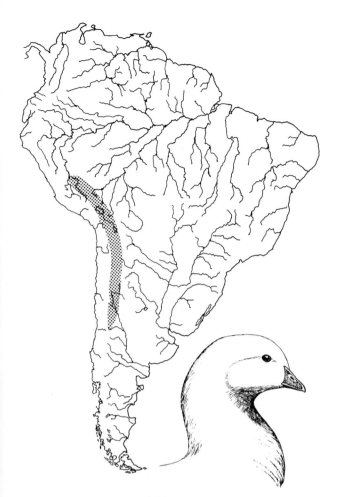

MAP 34. Residential distribution of the Andean goose.

posture. While uttering a repeated *gack* call, the female walks in high-stepping and erect posture, shaking her tail from side to side. The female often thus walks around her mate, which typically responds with a variety of aggressive and sexually oriented displays. The most conspicuous is a repeated whistling call, uttered with neck and head diagonally outstretched. This appears to be an aggressive call, and is often alternated with an erect and stereotyped march around the female, with neck feathers fluffed, the bill tilted downward and resting against the neck, the tail repeatedly shaking, and a flatulent *humm-pah* sound uttered. Males also have an aggressive wing-flapping display, as well as body-shaking and head-rolling movements that all closely resemble the normal comfort movements of this species (Johnsgard, 1965a).

Reproductive biology. At least in the southern parts of its range, the Andean goose's breeding season begins in November, with young being hatched in late December and early January. The nests are usually simple scrapes generally placed among sparse vegetation, and are often on hilly slopes overlooking water, or even on bare ground directly below snow line. Typically from 5 to 10 eggs are laid and are incubated entirely by the female, although the male remains close by and strongly defends the nest. The normal incubation period is 30 days. After the young hatch, they are led by the parents to the nearest available water; and when swimming, the male always takes the lead, followed by the brood and finally by the female (Johnson, 1965). Judging from observations of captive-raised birds, the young require about three months to become fully feathered, by which time they closely resemble the adults except for the scapular feathers, which are mostly brownish gray, with poorly defined black markings. In captivity at least, breeding usually does not occur in birds less than three years old, but the situation in the wild is not known.

Status. The fairly remote environments of this species probably keeps it well out of reach of most human persecution, and furthermore its habitats are ones unlikely to be seriously modified by man in the foreseeable future. Its population size is completely unknown, but there is no indication that the bird is currently in any danger.

Relationships. The entire genus *Chloephaga* is obviously a close-knit one, and many of this species' minor differences from the others must be related to its high-montane ecological adaptations. The evolution (or retention) of a conspicuous white plumage in the female is curious and deserves some attention.

Suggested readings. Johnson, 1965; Delacour, 1954–64.

Magellan Goose

Chloephaga picta (Gmelin) 1789

Other vernacular names. Upland goose; Magellangans (German); oie de Magellan (French); caiquen, ganso magallánico (Spanish).

Subspecies and ranges. (See map 35.)

C. p. picta: Lesser Magellan goose. Breeds from the latitude (36° south) of Talca in Chile and Río Negro in Argentina south to Tierra del Fuego; occurs north to Colchagua in Chile and Buenos Aires in Argentina during winter.

C. p. leucoptera: Greater Magellan goose. Resident on the Falkland Islands; introduced into South Georgia island but now extirpated.

Measurements and weights. (Both subspecies.) Folded wing: males, 395–462 mm; females, 380–425 mm. Culmen: males, 33–47 mm; females, 31–45 mm. Weights (of *picta*): males, 2,834 g; females, 2,721–3,200 g. Eggs: 74 x 50 mm, cream, 122 g.

Identification and field marks. Length 23–26″ (60–65 cm). Plate 28. This is the only large sheldgoose (folded wing 380 mm or longer) that exhibits black barring on the flanks. In females and males of some individuals of the lesser subspecies the barring is extensive, extending around the abdomen and up the neck, while in males of the greater subspecies and some lessers the barring is restricted to the flanks; the breast, abdomen, neck, and head are entirely white. The tail is black or black tipped with white; the upperparts are gray posteriorly and barred with black on the smaller scapulars; the wing has an iridescent green speculum formed by the greater secondary co-

verts; the secondaries and lesser coverts are white. The bill, legs, and feet are grayish black. *Females* have yellow legs and feet, and the head, neck, breast, and anterior flanks are overlain with a reddish cinnamon cast, but the black barring pattern remains evident. *Juveniles* and first-year immature males show dusky brown feathers in the head region.

In the field, this sheldgoose may be found in company with ruddy-headed and ashy-headed geese, which are both smaller and lack white heads. Females slightly resemble ruddy-headed geese, but lack that species' white eye-ring markings. In all sheldgeese the call of the male is a repeated whistling note, while the female produces a loud grating sound.

NATURAL HISTORY

Habitat and foods. This is a species of the semiarid, open grasslands of Patagonia, sometimes occurring relatively far from water. Except when flightless or with broods, the birds are completely terrestrial, and are found most often in cultivated pastures and grassy valleys. Grassy islands or stream shorelines are favored territorial and brood-rearing areas in the Falkland Islands, and larger green areas, often far from water, are used by nonbreeding flocks. Grassy areas near the sea are used by birds approaching their molt, so that they might escape to the sea when flightless. The leaf tips and seed heads of grasses such as meadow grasses (*Poa*) are favored foods of young and adult birds, and even very young birds apparently have an entirely vegetarian diet (Weller, 1972).

Social behavior. This species is the most abundant of all the sheldgeese, and some flocks must be relatively enormous. These flocks move around considerably according to the local food supplies, and strongly compete with sheep flocks. Where they are extremely abundant, their excrement may drive sheep away, and on some sheep *estancias* in Tierra del Fuego as many as 75,000 eggs have been reported destroyed in a single year. On the Falkland Islands, flocks of up to 100 birds are commonly seen. The existence of flocks of nonbreeders indicates that the period of sexual im-

MAP 35. Breeding or residential distribution of the greater ("G") and lesser ("L") Magellan goose. Wintering distribution of the greater Magellan goose is shown by stippling.

maturity is at least two years, and Weller (1972) suggests that a three-year period to maturity is more likely. Pair formation is achieved by the combination of female inciting behavior and aggressive responses of her potential mate. In captive birds it is confined to late winter and spring. Aggressive displays include whistling with the neck and head high and often are followed by a rapid running attack with the brilliantly patterned wings partially spread and ready to strike the opponent. One observed case of copulation occurred in shallow water and was preceded by mutual head-dipping movements and followed by apparently mutual calling and partial wing raising by the male (Johnsgard, 1965a).

Reproductive biology. In Chile, nesting occurs during November (Johnson, 1965), but in the Falkland Islands it may extend from early August to late November, with most activity between mid-September and late October (Woods, 1975). Chilean nests have been found scattered indiscriminately over the countryside but usually are near water, while in the Falkland Islands the nests are typically placed among ferns, "diddle-dee" (*Empetrum*), or white grass (*Cortaderia*). The usual clutch size is 5 to 8 eggs, and incubation by the female requires 30 days. Males remain near the nest at this time; and shortly after hatching, the young are led to water. Weller (1972) estimated a 9- to 10-week fledging period for this species on the Falkland Islands. By late December, adults have begun their molt and become flightless, and at that time the birds move near the seacoast for protection.

Status. In spite of tremendous persecution by sheep-growing interests, this species remains remarkably common over much of its range. The Falkland Island race is obviously much less numerous than the mainland form, and in some areas the young birds are caught and used for food. Together with the ruddy-headed goose, these birds are considered pests and may be killed at any time. They nevertheless remain relatively common.

Relationships. This species seems to be most clearly related to the kelp goose, judging from adult plumage patterns, but a more detailed anatomical or behavioral analysis of *Chloephaga* must be undertaken before intrageneric affinities become evident.

Suggested readings. Woods, 1975; Johnson, 1965.

Kelp Goose

Chloephaga hybrida (Molina) 1782

Other vernacular names. None in general English use. Tanggans (German); bernache antarctique (French); caranca, ganso del cachiguyo (Spanish).

Subspecies and ranges. (See map 36.)

C. h. hybrida: Patagonian kelp goose. Resident in Chile from Chiloé Islands south to Tierra del Fuego, and rare in Argentina from Santa Cruz and Chubut southward.

C. h. malvinarum: Falkland kelp goose. Resident on the Falkland Islands.

Measurements and weights. (Both subspecies.) Folded wing: males, 363–96 mm; females, 334–80 mm. Culmen: males, 35–40 mm; females, 35–40 mm. Weights: males, 2,607 g; females 2,041 g. Eggs: av. 73 x 53 mm, deep cream, 139 g.

Identification and field marks. Length 22–25″ (55–65 cm). *Male* kelp geese are the only entirely white waterfowl except for swans; and unlike any swans, they have yellow legs and feet and a short, black bill with a crimson spot on the culmen. *Females* have dark brown heads with a white eye-ring, heavily barred breasts and flanks, and a dark brown mantle, with a white tail and tail coverts. The wing pattern of females is like that of the other typical sheldgeese, rather than white as in males. *Juvenile* and immature males are rather like the female but have brown secondary coverts and dull greenish yellow legs and feet, and young females have dark upper tail coverts.

In the field, the entirely white plumage of the adult male is unmistakable, and females appear generally dark except for their white hindquarters. The call of the male is a repeated whistle, that of females a raucous snarling note.

NATURAL HISTORY

Habitat and foods. The habitat of kelp geese consists of rocky shorelines and shingle beaches. Occasionally the birds visit fresh-water ponds for bathing and drinking. Nesting also occurs near fresh-water areas less than a kilometer from the coast. Foraging is restricted largely to filamentous algae (*Enteromorpha*), sea lettuce (*Ulva*), and leafy algae (*Porphyra*) growing on tidal rocks. When nesting, adults also consume green grasses near the nests, and in winter the berries of "diddle-dee" (*Empetrum*) may be an impor-

tant food. Young birds apparently eat much the same foods as adults, with animal materials only a very minor component (Weller, 1972).

Social behavior. Kelp geese form moderately large flocks of nonbreeding and molting birds; Weller (1972) reported a group of 232 adults feeding at low tide in one bay, and Gladstone and Martell (1968) saw a flock of over 300 molting birds. The Falkland Island population is essentially sedentary, with only local movements during winter, leaving breeding areas in March and April, and returning in September and October. The continental population undertakes substantial northward migrations during winter, but the details are still largely unstudied. Pair formation presumably occurs in wintering flocks; Woods (1975) reports that display may be seen from September to late November. Observations by Gladstone and Martell indicate a behavior much like that of the Andean goose. Pair bonds are almost certainly permanent; Pettingill (1965) reported seeing birds in solitary pairs from February through October on the Falkland Islands. The period to reproductive maturity is still uncertain but must be at least two years.

Reproductive biology. Nesting gets under way on the Guaitecas Islands of southern Chile in November

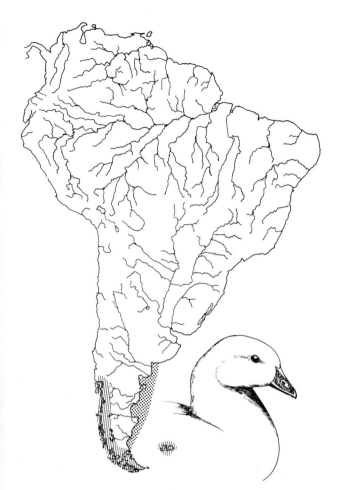

low population densities of the bird. Upon hatching, the goslings are quickly led to water and feed, during the low-tide periods of each day, mainly on algae of the filamentous (*Enteromorpha*) type. This restricted foraging period probably accounts for their slow growth rate; Weller (1972) estimated the fledging period at 12 to 13 weeks, compared to 9 or 10 in the upland goose. The flightless period of adults extends from late November through February on the Falkland Islands.

Status. This species is still reasonably common over most of its range, and is not in competition with human sheep-raising interests, as are the other sheldgeese.

Relationships. Apart from its dietary specializations, the kelp goose does not diverge very far from the other species in this genus. The entirely white plumage of the male, even including the greater secondary coverts, is of special ecological interest, and is in strong contrast to the dark, concealing coloration of the female. Most probably the species evolved from stock not very different from the modern Magellan goose.

Suggested readings. Gladstone & Martell, 1965; Pettingill, 1965.

MAP 36. Breeding or residential (hatched) and wintering (stippling) distributions of the kelp goose.

(Johnson, 1965) but begins appreciably earlier on the Falklands, with most nesting in late October and early November. Nests are almost always placed within ten yards of the high-tide line, often on low cliff ledges, in clumps of tall tussock grasses, or even under old planks on the beach. Ledges from four to eight feet up a cliff, with a cover of stunted tussock grass, are favorite locations (Gladstone & Martell, 1965). The nest is lined with vegetation and gray down, and the normal clutch size ranges from 3 to 7 eggs, with 6 the most commonly encountered number. The incubation period is 30 days, and the conspicuous male remains close by throughout the period, thus often revealing the nest's location. The adaptive significance of the male's white coloration is obscure, but may be related to its effectiveness as a dominance signal in territorial encounters. Territorial defense of limited shoreline food resources may be a significant factor in maintaining the fairly

Ashy-headed Sheldgoose

Chloephaga poliocephala Sclater 1857

Other vernacular names. None in general English use. Graukopfgans (German); bernache à tête grise (French); canquen, avutarda de cabeza gris (Spanish).

Subspecies and range. No subspecies recognized. Breeds from southern Magellanes Province of Chile, and southern Argentina (Santa Cruz) southward to Tierra del Fuego and the Falkland Islands. Winters north to Buenos Aires in Argentina but only to Colchagua in Chile. See map 37.

Measurements and weights. Folded wing: males, 355–80 mm; females, 335–40 mm. Culmen: males, 30–33 mm; females, 26–28 mm. Weights: males, 2,267 g; females, ca. 2,200 g. Eggs: av. 70 x 50 mm, pale buff, 89 g.

Identification and field marks. Length 20–22" (50–55 cm). Plate 25. This rather small sheldgoose is essentially tricolored, in both sexes, with a gray head (whitish eye-ring), a chestnut breast, and white sides and abdomen overlaid with vertical black barring. The rump and tail are black and the under tail coverts pale chestnut, while the mantle is brown. The bill is black, and the legs and feet are two-toned orange and black. The wings are as in the other species, with white secondaries and iridescent green secondary coverts, the other coverts being white. *Females* are smaller and have somewhat barred breasts, and *immature* birds are more extensively brownish and have brown rather than iridescent coverts.

In the field, this species is most likely to be confused with the ruddy-headed sheldgoose, but the grayish head color and pure white abdomen will serve to identify it. The calls of all sheldgeese are similar, consisting of repeated whistling notes in the males and grating calls by the females.

Natural History

Habitat and foods. Although the ashy-headed sheldgoose often associates with the open-country Magellan goose, it is relatively scarce where that species is most common, and is far more prevalent in wooded areas. In mountains it is most closely associated with swampy areas where rushes and associated plants occur amid small clearings in the forest (Johnson, 1965). Unlike the other *Chloephaga* sheldgeese, these birds perch readily in trees and normally nest among them. Since they often feed in company with Magellan geese, they presumably eat much the same grassy foods, but so far no studies have been carried out on this problem.

Social behavior. The relatively few descriptions of this bird in the wild indicate that it is usually to be found in small flocks outside the breeding season. Flock sizes of up to as many as 200 birds have been recorded. In southern Chile it is probably most abundant in the vicinity of Chiloé and the Guaitecas Islands to the south, becoming rare in the open grasslands of northern Tierra del Fuego and the southern part of Magellanes (Johnson, 1965). It presumably breeds at two or three years of age, and in captivity its social behavior shows all of the typical sheldgoose features. Captive birds tend to remain in pairs all year, and often threaten intruders.

Females are easily stimulated to incite other birds, and the male's normal response is an intensely performed alternation of extreme body erection and quick bowing movements of the bill nearly to the ground, accompanied by a huffing call (Johnsgard, 1965a). The same display serves as a kind of triumph ceremony following an attack or threat by the male, and no doubt is important in pair bonding. Behavior associated with copulation has not yet been described.

Reproductive biology. Very few nests of this species have been found, and only Johnson's (1965) report provides much information. On the Guaitecas Islands he found several nests during November, usually hidden in long grass and with an overhead half dome of the same material, abundantly lined with down. In the mountains of northern Aysen Province of Chile, he found the birds nesting abundantly in November, in the hollows of burned trunks and branches of large trees. By late November many nests had hatched, and the young were taken to the nearby lakes for rearing. The clutch size is typically 4 to 6 eggs, with a probable average of 5, and the eggs are somewhat darker than those of other *Chloephaga*. The incubation period is 30 days, and only the female is known to incubate. There is no information on the rate of growth of the young under wild conditions. After the fledging of the young and the completion of the postnuptial molt by adults, they begin to move

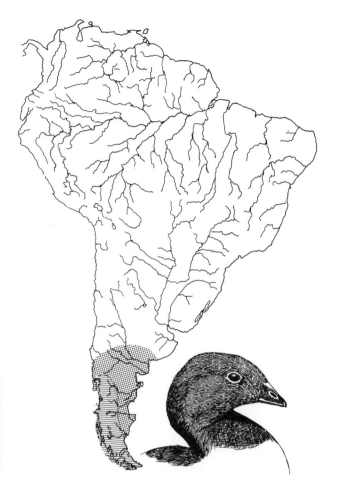

MAP 37. Breeding or residential (hatched) and wintering (stippling) distributions of the ashy-headed sheldgoose.

north, with the Argentine population arriving in Buenos Aires Province about the middle of March, seldom in flocks of more than 100 birds.

Status. This species is apparently more common in Chile than anywhere else in its range, and it is rare on the Falkland Islands and is also relatively scarce in Tierra del Fuego. It is not in direct conflict with humans over most of this range, but at times does feed on sheep pastures with Magellan geese and thus might be locally persecuted.

Relationships. Several adult plumage similarities between this species and the ruddy-headed sheldgoose support the view that they are very closely related to one another, and the downy young of the two species are also very similar. Likewise, the social display patterns are almost identical.

Suggested readings. Johnson, 1965.

Ruddy-headed Sheldgoose

Chloephaga rubidiceps Sclater 1860

Other vernacular names. None in general English use. Rotkopfgans (German); bernache á tête rousse (French); avutarda cabeza colorada (Spanish).

Subspecies and range. No subspecies recognized. Breeds in southernmost Argentina and Chile from southern Magellanes and Santa Cruz south to Tierra del Fuego, and on the Falkland Islands. Winters north in Argentina to about 37° south latitude. See map 38.

Measurements and weights. Folded wing: males, 330–50 mm; females, 310–20 mm. Culmen: males, 28–30 mm; females, 25–28 mm. Weights: both sexes ca. 2,000 g. Eggs: av. 65 x 48 mm, deep cream, 90 g.

Identification and field marks. Length 18–20″ (45–50 cm). This smallest of the sheldgeese has a bright chestnut-colored head and neck, with a white eye-ring, while the breast, abdomen and flanks are all finely barred black over a buff background, becoming chestnut on the under tail coverts. The rump and tail are black and the upper parts pale

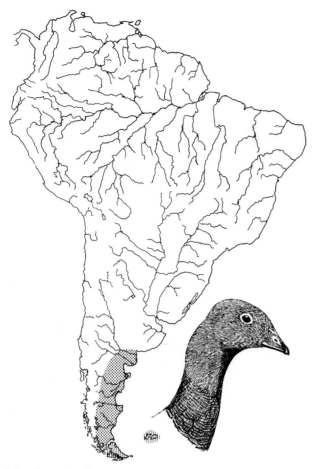

MAP 38. Breeding or residential (hatched) and wintering (stippling) distributions of the ruddy-headed sheldgoose.

grayish. The wing coloration is like that of the other sheldgeese, with an iridescent area formed by the greater secondary coverts, while the secondaries and the other coverts are white. The bill is black, and the legs and feet are bright orange, variably spotted with black. *Females* are slightly smaller than males (see folded wing measurement), but otherwise essentially identical. *Juveniles* and subadults through their first year have noniridescent greater secondary coverts.

In the field, this species most closely resembles the female Magellan goose and may be seen with it, but is much smaller and has a conspicuous eye-ring and finer barring on the body. Males utter a repeated whistling note and females a rasping quack.

NATURAL HISTORY

Habitat and foods. The ruddy-headed goose, like the Magellan goose, is characteristic of open plains and meadows, and is appreciably more restricted than the latter to coastal areas having a fine grass cover. The two species often feed together on the Falkland Islands, but the ruddy-headed goose is evidently more of a grubber than primarily a clipping, grazing bird, and thus consumes roots and entire small plants as well as leaves and seed heads (Weller, 1972). The extent to which it thus avoids direct foraging competition with the larger Magellan goose is unknown, but is of some interest.

Social behavior. Up until recent years, this species was one of the most abundant of the sheldgeese in Tierra del Fuego, and often occurred in quite large flocks. There they most often mix with ashy-headed geese, and less often with the much larger Magellan geese. The two smaller species more often are to be found in the vicinity of farms, while the Magellan geese are more generally distributed around the countryside. This population is strongly migratory, and is on the breeding grounds only from April to September, while on the Falkland Islands the birds are resident throughout the year. Scott (1954) believes that probably both ashy-headed and ruddy-headed sheldgeese have a relatively prolonged wing molt that avoids a completely flightless period; this has not been confirmed with wild birds, but my own observations of captive ruddy-headed sheldgeese support this view. The ruddy-headed sheldgoose is extremely pugnacious and aggressive in spite of its small size, often attacking larger birds, and males precede such attacks by repeatedly calling with whistled notes in a highly erect stance, holding the wings slightly away from the flanks and thus exposing the conspicuous white coverts. When displaying before the female in a triumph ceremony, the male rapidly alternates between this erect posture and one in which the bill almost touches the ground in a bowing display (Johnsgard, 1965a). These displays appear to be the basic pair-forming and pair-maintaining activities of the species. Copulatory behavior in this species has never been described, and it would be of interest to learn if it ever occurs on land.

Reproductive biology. The nesting season of this species in the Falkland Islands is from late September to early November; likewise, in Tierra del Fuego it is reported to nest in October and November. In Tierra del Fuego the nests are placed in the same situations as those of the Magellan goose, but can readily be distinguished by the cinnamon-colored down (Johnson, 1965). On the Falklands, the nests are usually well hidden in long grasses or rushes, are placed under rock outcrops, or may even be situated in old penguin burrows (Woods, 1975). From 5 to 8 eggs are normally laid, and incubation is performed by the female alone, while the male typically waits at the nearest pond, which may be a considerable distance from the nest. After an incubation period of 30 days, the newly hatched young are led to water and thereafter the family leads a semiaquatic life. There is no specific information available on the fledging period or growth rate of the goslings.

Status. This species has greatly decreased in numbers on Tierra del Fuego since the 1950s, and Weller (1975a) suggests that this might be the result of egg-destruction programs aimed primarily at the Magellan sheldgoose or because of the intentional introduction of foxes (*Dusicyon griseus*) into the islands. In any case, the species' present stronghold is the Falkland Islands, where it is numerous only in some areas of West Falkland (Woods, 1975). The continental population may be under 1,000 birds (A. E. Rumboll, unpublished m.s.).

Relationships. As indicated in the account of the ashy-headed goose, there is good evidence that these two species are very closely related to each other. However, the entire genus is one of great morphological and ecological similarities, and deserves a thorough analysis from these standpoints.

Suggested readings. Johnson, 1965; Woods, 1975.

Orinoco Goose

Neochen jubata (Spix) 1824

Other vernacular names. None in general English use. Orinokogans (German); oie de l'Orenoque (French); oca del Orinoco (Spanish).

Subspecies and range. No subspecies recognized. Resident in the Orinoco and Amazon basins, south to southern Amazonas, northern Mato Grosso, and São Paulo in Brazil, Paraguay, and Salta in Argentina. See map 39.

Measurements and weights. Folded wing: males, 315–33 mm; females, 300–310 mm. Culmen: males, 38–40 mm; females, 35–37 mm. Weights: females, 1,250 g. Eggs: av. 60 x 44 mm, cream, brownish, or pale greenish, 63 g.

Identification and field marks. Length 24–26″ (61–66 cm). The Orinoco goose has a head, neck, and upper breast of pale yellowish brown color, with the neck feathers noticeably elongated and slightly furrowed as in geese. The sides and abdomen are chestnut, paling to buff on the upper flanks, and with white under tail coverts. The tail and back are glossy black, except for chestnut scapulars that are broadly buff-tipped. The upper wing surface is mostly purplish black, except for iridescent green secondaries that have white on the basal portion of their outer vanes. The bill is red and black, and the legs and feet are bright salmon or pink. *Females* resemble males but are slightly smaller, and *juveniles* have washed-out colors and paler legs and feet.

In the field, the tropical forested and savannah habitats of this species separate it from all other sheldgeese, and its long, ruffled neck and chestnut flanks are also distinctive. Males utter a strong

whistling note, and females produce a harsh cackling sound.

Natural History

Habitat and foods. The Orinoco goose is a truly tropical goose, and is most often associated with forest-lined rivers or wet savannah areas. It is reportedly the commonest species of waterfowl on the Orinoco River, but is rarely found on the coast. Very little is known of the birds' foods under natural conditions, but a variety of such animal materials as butterfly larvae, aquatic insects, small mollusks and worms have been reported in addition to the plant materials that are presumably the species' primary dietary component. In captivity the birds graze on

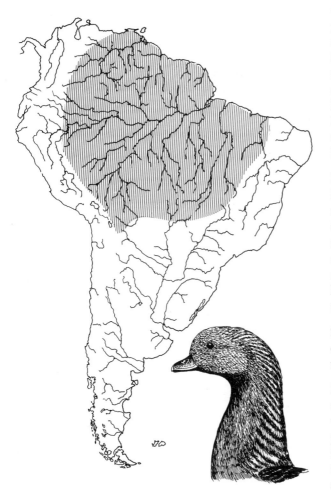

Map 39. Residential or breeding distribution of the Orinoco goose.

greenery to much the same degree as the more typical sheldgeese.

Social behavior. Orinoco geese apparently rarely occur in large flocks; postbreeding groups of from 5 to 20 individuals appear to be the largest aggregations that are normally encountered. No obvious migratory movements are undertaken. Rather, the birds are usually in pairs or family groups throughout the year. Delacour (1954–64) reports having seen, in December, paired birds spaced at distances of about a mile apart along Venezuelan rivers, where breeding is said to occur during the winter months. In Bolivia, families with well-grown young have been observed in September, supporting the idea that breeding occurs during the dry season (Kolbe, 1972). Pairforming behavior and territorial display are similar to that of the other sheldgeese, but one unusual feature is the high degree of social preening that occurs between birds, presumably normally between pair members. Male aggressive displays consist of a repeated whistling note uttered with the neck held diagonally outward and the feathers somewhat ruffled; as the male turns toward his inciting mate, he assumes a strongly erect position with head held far back and with one or both of his wings strongly lifted, and he utters a wheezy *wi-chuff* note (Johnsgard, 1965a). No observations on copulatory behavior are yet available.

Reproductive biology. Few nests of this species have been found in the wild, but they are evidently normally situated in hollow trees. The range of eggs in the clutch is from 6 to 10, probably averaging 8, and the nest is well lined with white down. Incubation requires about 30 days, and is performed by the female. Males rejoin the family when the goslings hatch, and closely guard them. No specific information on growth rates or fledging time is yet available.

Status. This species is apparently still fairly common over its relatively broad range, and its habitats are unlikely to be extensively modified in the near future.

Relationships. This species is unique among the sheldgeese in having black upper and under wing coverts, and in having a bicolor wing speculum composed of the secondary feathers. In a few respects, as in female inciting behavior, *Neochen* seems to bridge the evolutionary gap between the typical sheldgeese and the shelducks (Johnsgard, 1965a).

Suggested readings. Delacour, 1954–64; Phillips, 1923–26.

Egyptian Goose

Alopochen aegyptiacus (Linnaeus) 1766

Other vernacular names. Nile goose; Nilgans (German); oie d'Egypte (French); ganso de Egipto, oca del Nilo (Spanish).

Subspecies and range. No subspecies recognized. Resident in Africa south of the Sahara; also occurs throughout the entire Nile Valley. Occurs casually in southern Europe and northernmost Africa. Feral in parts of southeastern England. See map 40.

Measurements and weights. Folded wing: males, 400–430 mm; females, 350–85 mm. Culmen: males, 52–56 mm; females, 49–53 mm. Weights: males, 1,900–2,250 g; females, 1,500–1,800 g. Eggs: av. 68 x 50 mm, creamy white, 97 g.

Identification and field marks. Length 28–29" (71–73 cm). Plate 26. The *adult* plumage is predominantly grayish on the head, neck, breast, underparts, flanks, and back, with darker chocolate brown tones around the eyes, the nape, on the upper wing coverts, and as an irregular blotch on the lower breast. The primaries and tail feathers are black, as is the rump, while the secondaries are iridescent green and the upper coverts white except for a narrow black bar extending across the front of the greater secondary coverts. The bill, legs, and feet are pink. *Females* are virtually identical to males, but smaller. In *immatures* the white upper wing surface is tinged with sooty coloration, and the brown marks around the eyes and on the breast are lacking.

In the field, the species resembles a large shelduck rather than a goose, and might be confused with the ruddy shelduck, which is smaller and more uniformly buffy brown. Males have a loud, gusty breathing note, and females a loud cackling call. In

flight, the white upper and under wing coverts are very conspicuous.

NATURAL HISTORY

Habitat and foods. Nonforested environments of wide diversity, including fairly arid open plains, are this species' preferred habitats. However, the birds do perch at night in trees, and usually return to the same location each night. Clancey (1967) describes the habitats as including shores or flats bordering lakes, marches, vleis, dams, reservoirs, rivers, and other bodies of water. The diet of the adult is exclusively vegetable materials, particularly the shoots and seed heads of grasses and the shoots and flowers of herbs, as well as the corms of cyperus (*Cyperus*). The species often feeds in fields of sprouting grain crops, and sometimes does considerable damage there. Most foraging is done in early morning and again at dusk, but the birds also forage at night and rest during the hottest part of the day (Clancey, 1967).

Social behavior. This species is believed to form strong and potentially permanent pair bonds, and thus the basic social unit is the pair or family. In the wild, it may be seen in pairs, small parties, or flocks of 100 or more birds. Egyptian geese often associate with other large species of waterfowl, and sometimes are also seen with herons or storks. The birds fly well, and outside the breeding season vagrants may occur almost anywhere in northern Africa or southern Europe. However, they are not regularly migratory. The breeding season is very prolonged and irregular, and thus the period of pair formation is probably also quite prolonged. In captivity Egyptian geese do not breed before their second year, and presumably the same applies in the wild. During pair formation a great deal of inciting behavior is performed by the females, and intensive threatening or fighting behavior among males is typical. However, display may even occur in the absence of other males. The male response to female inciting is to stand extremely erect, utter a repeated breathing note, and suddenly flash open his wings, exposing white coverts. Copulation apparently occurs in shallow water, and probably is preceded by head-dipping movements by one or both birds. After treading, the male lifts his wing on the side opposite the female to a nearly vertical angle (Johnsgard, 1965a).

Reproductive biology. Nesting is extremely prolonged in some areas; in Zambia records extend

weeks longer than has been established for various species of *Tadorna* (Lack, 1968). The adults undergo their wing molt between May and July in South Africa, and are flightless for about 30 days (Clancey, 1967) or 40 days (D. Skead, pers. comm.).

Status. This species is relatively abundant over most of its range, and in some areas is considered an agricultural pest. It is considered by sportsmen to be a good gamebird, but has coarse flesh and at times is unpalatable. There should be no concern for its population status at present.

Relationships. In many respects the Egyptian goose might be regarded as a large shelduck, even though it is primarily a grazing bird. There can be little doubt that *Tadorna* represents the nearest living relative of *Alopochen*, and it is somewhat more distantly related to the other genera of sheldgeese.

Suggested readings. Clancey, 1967; Cramp & Simmons, 1977.

Map 40. Residential or breeding distribution of the Egyptian goose.

from January to November, but the largest number of records is from June through September. In South Africa the peak nesting activity occurs from July to October. In tropical Africa breeding occurs most of the year, while in the northern parts of the species' range it breeds from July to October (Senegal), July to September (Sudan), and July and August (Ethiopia). The nest site is extremely variable, ranging from burrows or holes in the ground to nests in trees or on cliffs, amid herbaceous ground cover, or on top of an old or incomplete nest of some other bird, particularly the hammerhead (*Scopus umbretta*). There are even records of the birds' nesting among colonies of cliff-nesting vultures. The nest is usually well constructed of available materials and lined with smoke-gray down by the female. Clutch sizes are quite variable, ranging from 6 to 12 eggs and averaging about 7. Incubation requires 30 days, and according to Clancey (1967) it is shared by both sexes, but this requires confirmation, as it is certainly not typical of captive birds. The fledging period of the young has been reported at 14 weeks, which seems incredibly long for a bird of this size, and from about 4 to 7

Ruddy Shelduck

Tadorna ferruginea (Pallas) 1764

Other vernacular names. Brahminy duck; Rostgans (German); casarca roux (French); oca colorada (Spanish).

Subspecies and range. No subspecies recognized. Breeds chiefly from southeastern Europe and the western Mediterranean lands north and east across Asia to Transbaikalia and the upper Amur, south to Iran, the Himalayas, and southeastern China.

Winters in the southern part of its breeding range and south to the Nile Valley, India, Burma, Thailand, and southern China. See map 41.

Measurements and weights. Folded wing: males, 365–87 mm; females, 340–55 mm. Culmen: males, 43–48 mm; females, 38–42 mm. Weights: males, 1,200–1,640 g; females, 925–1,500 g. Eggs: av. 68 x 46 mm, creamish color, 83 g.

Identification and field marks. Length 25–26" (63–66 cm). *Adults* of both sexes are almost entirely buffy brown to rust colored, with pale buff but not white faces, blackish primaries and tail, and white upper wing coverts and an iridescent green speculum on the secondaries. The bill, legs, and feet are blackish. *Females* lack the narrow black neck-ring of males and have a paler face. In *immatures* the plumage is paler, the white upper wing coverts are tinged with sooty coloration, and the head is white.

In the field, this species' strongly golden to rusty brown body coloration separates it from all other dabbling ducks in its range. It is relatively terrestrial and often found away from water. In flight, the white upper and under wings coverts contrast strongly with the other, darker tones of the wing and body. The female has a loud, nattering call, and that of the male is a loud *choor* or *cho-hoo'* note.

NATURAL HISTORY

Habitat and foods. Habitats used in the U.S.S.R. are quite varied, but include steppe lakes and rivers in hilly regions, especially where waters are salty or brackish. Although open country seems to be the preferred habitat, the birds also nest around alpine lakes and in the timberline belt of birches, but avoids tiaga forest. In some areas they reach elevations of

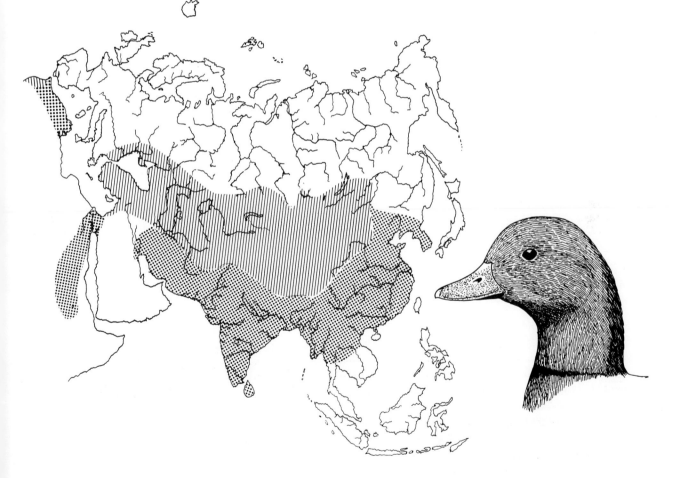

MAP 41. Residential or breeding (hatched) and wintering (stippling) distributions of the ruddy shelduck.

4,000 to 4,500 meters, but generally spend the fall and winter months in lowlands, where they concentrate on broad rivers, salty lakes, and lacustrine floodwaters. Spring and summer foods are mostly of vegetable matter, especially green plant shoots, while fall flocks may concentrate on millet, wheat, or other cereal grains. In the winter, sprouting greens are taken when available, but the birds also feed on garbage and sometimes even carrion (Dementiev & Gladkov, 1967). Much of the daytime hours are spent in resting, while feeding often occurs at night or during twilight hours.

Social behavior. Ruddy shelducks are usually found in pairs or family groups, and large flocks are probably limited to molting assemblages on certain U.S.S.R. lakes. Flocks of moderate to large size have also at times been seen on the wintering grounds in India, but the basic aggressiveness of this species probably tends to keep flock sizes small. Pair bonds are fairly strong and tend to be permanently held; two years are required to attain sexual maturity in this species. Pair formation consists of female inciting and male threat or attack behavior toward other individuals, in the usual manner of shelducks. Copulation probably normally occurs in water of swimming depth (but has also been reported on land) and is preceded by mutual head-dipping movements. After treading, both sexes call and the male slightly raises his folded wing on the side opposite the female (Johnsgard, 1965a). Migratory movements are well developed in this species, and generally involve southward flights of varying distances. However, at least until recently there was a regular flight of birds from Morocco northward to the coast of Spain for the winter season, the only known case of an African bird species wintering in Europe. Probably the major wintering grounds are in Persian Azerbaijan, where about 40,000 birds winter around Lake Rezaiyeh, and the primary breeding area is from the Caspian Sea eastward (Hudson, 1975).

Reproductive biology. In the U.S.S.R., breeding birds arrive already in pairs, and they often appear in the breeding areas before the lakes are ice-free. Egg dates in U.S.S.R. range from the last half of April through June, with a few records of newly hatched young in July. Nests are placed in hollow cavities, including the hollows of larches, at heights of up to ten meters, in ground burrows, in rocky cliff crevices, or even in ruined buildings. Like many other hole-nesting birds, females of this species may make a snakelike hissing sound when disturbed on the nest.

The clutch averages from 8 to 12 eggs, which are deposited daily, and incubation begins with the last egg. It requires from 27 to 29 days, and is carried out entirely by the female. However, the male warns of danger, and may make threatening flights toward intruders. Both adults closely tend the brood, which feeds in the shallows on aquatic insects, brine shrimps, or even locusts. Fledging has been reported to require 8.5 weeks in this species; in the U.S.S.R. recently fledged young have been seen from mid-July to late August. Postnuptial molts in adults begin about the time that the young become independent, or in August and September. The juvenile birds molt again in the winter, with the young males attaining their black neck-rings in February (Dementiev & Gladkov, 1967).

Status. In Europe this species is considered endangered, with both the population and the breeding range contracting. It is probable that fewer than 50 pairs breed outside of Russia in Europe; these may occur in Greece, Bulgaria, and Romania. In the northern part of Africa the birds breed primarily in Morocco, in the Atlas Mountains, and on the coastal desert south to Cape Juby. The causes of the European decline are still uncertain, but loss of habitat, shooting, and egg collecting may all be involved (Hudson, 1975).

Relationships. This species is obviously a close relative of the Cape shelduck, and the two may be regarded as constituting a superspecies (Johnsgard, 1965a).

Suggested readings. Dementiev & Gladkov, 1967; Cramp & Simmons, 1977.

Cape Shelduck

Tadorna cana (Gmelin) 1780

Other vernacular names. South African shelduck, gray-headed shelduck; Graukopfkasarka (German); casarca du Cap (French); oca Sud Africana (Spanish).

Subspecies and range. No subspecies recognized. Resident in Cape Province, Orange Free State, Transvaal, southeastern Botswana, and South-West Africa north to the highlands of Damaraland. See map 42.

Measurements and weights. Folded wing: males, 365–80 mm; females, 330–40 mm. Culmen: males, 46–51 mm; females, 43–45 mm. Weights: males average 1,758 g; females average 1,417 g. Eggs: av. ca. 70 x 50 mm, creamish, 83 g.

Identification and field marks. Length 24–26" (61–66 cm). Plate 27. *Adults* of this species have a clear golden brown to rust breast color, which becomes darker hazel and vermiculated with blackish lines on the flanks and anterior back. The longer scapulars, lower back, and upper tail coverts and tail are blackish and the under tail coverts are cinnamon. The head is uniform gray in males, but is extensively white around the eyes in females, and the upper and lower wing surfaces are patterned exactly as in the ruddy shelduck. *Females* exhibit the white face markings only when adults; *immature* birds of both sexes resemble males but are paler and duller throughout, with brownish edging on the upper wing coverts and with a relatively dull speculum.

In the field, Cape shelducks do not resemble any of the other ducks of southernmost Africa, and are the only ones with white upper and lower wing coverts except for the Egyptian goose, which is appreciably larger and more grayish overall. The call of the female is a loud and repeated nattering, while the male has low-pitched *korrr* and *ka-thoo'* notes.

NATURAL HISTORY

Habitat and foods. This species' preferred habitats consist of shallow fresh-water lakes, pan, dams, reservoirs, and pools in river courses, especially where exposed mud flats can be found. Feeding is done mostly while standing or swimming in shallow water, and the birds frequently consume such algae as *Zostera* and *Spirogyra*, as well as herbaceous greens and the heads of flowers. Some animal materials such as phyllopods (*Brachiopoda*) have also been reported as foods, and the birds have been known to enter food crops and do damage to them (Clancey, 1967).

Social behavior. Although essentially sedentary, these birds do at times, during the period October through February, gather in flocks of as many as 400 individuals on various water areas suitable for undergoing the flightless period during molting. At this time the birds spend their days on deep water, moving at night to shallow areas and banks for feeding.

In adult flocks the sex ratio is often strongly unbalanced in favor of females, which leads to a competition among females for mates. Nevertheless, it is believed that the species is basically monogamous, and that once pair bonds are formed they are relatively permanent. Displays consist of the usual combination of female inciting and male aggressive responses to intruding birds, especially other males. The male's two-noted call is, like that of related species, the one typically directed toward females, while the single-noted call is more directly a threat. As in the other shelducks, the conspicuous white wing coverts also make an effective threat display. The timing of pair formation is not clear, but it is certain that two years are required for the attainment of sexual maturity in shelducks. Copulatory behavior occurs in water of swimming depth, and is typically begun by head-dipping movements on the part of the male, which may be also performed by the female. Treading is followed by a strong wing-raising display by the male as he treads water in a high and erect posture. Both birds then begin bathing (Johnsgard, 1965a).

Reproductive biology. The breeding season of this species extends from the latter part of July until October or November. Nest sites are often at considerable distances from water, and may be in surprisingly arid and open country. The nest is often on the slope of a ridge or hill, at the end of a burrow. Frequently the burrow of an antbear, porcupine, or other mammal is used, and some nests have been located as far as 27 feet from the entrance of the hole, although a few have also been found among rocks. The nest is constructed primarily of the female's down and also more common in South-West Africa than previously believed, and this apparent change in status may be the result of dam and reservoir construction (Clancey, 1967).

Relationships. As noted in the last account, the ruddy shelduck and Cape shelduck are extremely closely related, and constitute a superspecies. The Cape shelduck is only slightly less closely related to the Australian and New Zealand species of shelducks.

Suggested readings. Clancey, 1967.

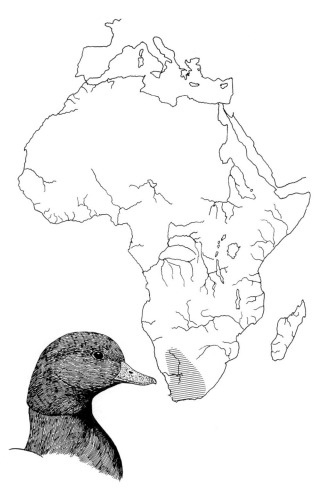

MAP 42. Breeding or residential distribution of the Cape shelduck.

feathers, but occasionally a small amount of vegetable material is also incorporated. While the clutch size in the wild usually ranges from 6 to 13 eggs, up to 15 have been reported. Only the female is known to incubate, but the male always remains fairly close to the nest site. Incubation requires 30 days, and shortly after hatching, the young emerge to be led to water by both parents. They are raised mainly in areas of shallow water and mud, retreating to deeper water only when danger requires. A surprisingly long 10-week fledging period has been reported. Further, there is a substantial molt migration to permanent water areas, where the 40-day flightless period is passed.

Status. Although this species has a relatively restricted range, it is fairly common in some areas, particularly in the Karroo regions of the Cape. It is

Australian Shelduck

Tadorna tadornoides (Jardine and Selby) 1828

Other vernacular names. Mountain duck, chestnut-breasted shelduck; Australische Kasarka (German); casarca d'Australie (French); oca Australiana (Spanish).

Subspecies and range. No subspecies recognized. Resident of southern Australia, including Tasmania, from the Southern Tablelands of New South Wales across Victoria and South Australia to Western Australia and north to Cape Leveque, with uncertain occurrence in the Australian Bight. See map 43.

Measurements and weights. Folded wing: males, 318–92 mm; females, 304–55 mm. Culmen: males, 44–49 mm; females, 38–45 mm. Weights: males, 990–1,980 g (av. 1,559 g); females, 878–1,850 (av. 1,291 g). Eggs: av. 68 x 49 mm, creamy white, 93 g.

Identification and field marks. Length 25–28" (63–71 cm). In *adults*, the head is largely or entirely black, the breast is cinnamon-colored, and the underparts, flanks, and mantle are a vermiculated brown and black. The tail, tail coverts, and primaries are black; the secondaries are iridescent green for most of their length, and the upper and under wing coverts are white. *Females* exhibit small white patches behind the bill and around the eyes (these may be continuous in some birds), and the breast is a brighter chestnut color. In *immatures* the upper wing coverts are flecked or margined with gray, and some white edging is present at the tips of the secondaries.

120 ❖ ❖ ❖

In the field, these birds are likely to be confused only with radjah shelducks or magpie geese; no other Australian waterfowl have white wing coverts. The chestnut or cinnamon breast color, as well as the major differences in body size, should serve to identify this species. Females have a loud and raucous call, often repeated, while the usual call of the male is a loud honk. His threat call is a strong *ho*, and the sexual display call sounds like *ho-poo*. In flight, a harsh *chank chank* call is frequent.

NATURAL HISTORY

Habitat and foods. The preferred habitat of this species consists of muddy shorelines of large brackish lakes or estuaries, but the birds also occupy fresh-water lakes, lagoons, and billabongs. In spite of their name "mountain duck," they do not prefer such habitats. On open plains, they show a preference for lightly timbered areas, but occasionally may be seen on small claypans and ground tanks. Although often found on salt water, the species apparently must drink fresh water, and often flies to fresh-water soaks for this purpose. Relatively few studies of food habits have been undertaken, but in a sample of 30, all but 3 contained both animal and vegetable materials. The plants most frequently found were clover (*Trifolium*) leaves and seeds, green algae (*Chlorophyceae*), couch grass (*Cynodon*), muskgrass, and azolla (*Azolla*). The commonest animal foods were insects, especially midges (*Chironomidae*) and water boatmen (*Corixa*), as well as cladocerans and other aquatic invertebrates (Frith, 1967).

Social behavior. Since these shelducks are found in pairs and family groups throughout the year, it is believed that pair bonds are potentially permanent. Some banding studies have supported this view, since pairs have been found together in the wild for several years, and at times have used the same tree

hole for breeding in several successive years. Sexual maturity is attained when the birds are nearly two years old, although temporary pair bonds may be formed by younger birds. Further, birds that fail to breed successfully often dissolve their pair bonds (Riggert, 1977). The only time that substantial flocking occurs is during the molting period, when flocks concentrate on lakes and estuaries. On Lake George, near Canberra, up to 2,000 birds gather between November and January, and similar concentrations have been seen in Tasmania, South Australia, and Western Australia. As winter approaches, these groups disperse to breeding grounds, over distances as great as 400 miles (Frith, 1967). Pair-forming and pair-maintaining displays in this species are essentially identical to those of the other shelducks already discussed, with the major differences being apparent in male vocalizations. Copulation occurs in water of swimming depth, and is preceded by alternated head dipping and calling in an erect posture by the male. After treading, both birds typically rise in the water, at least the female calling, while each vertically raises its wing on the side opposite its partner (Johnsgard, 1965a).

Reproductive biology. The relatively few available breeding records indicate that in Western Australia and perhaps South Australia eggs are laid between the middle of June and the end of September. In New South Wales breeding begins in July and may persist until November, while in Tasmania eggs are found in August and September. The favored nest site is a tree hole, often situated as high as 60 or 70 feet above the ground. In treeless areas the birds may nest in rabbit holes, or even on the ground surface, amid thick grass or other concealing vegetation. Riggert (1977) reports that on Rottnest Island, Western Australia, the birds nest in limestone crevices or burrows and the female may spend days seeking out a suitable site while her mate stands guard nearby. Breeding density is determined by the availability of areas suitable for brood territories, which require sources of fresh water, since newly hatched ducklings are unable to utilize salt water. The nests may be as far as a kilometer from the ocean or up to 9.5 kilometers from a salt lake, and the eggs are laid at approximately daily intervals. Various studies indicate that the clutch size is typically from 10 to 14 eggs. Incubation is performed entirely by the female, and requires from 30 to 32 days. During this time the male defends the brood territory and the pair have very little contact with each other. However, when the ducklings are two days old, they are led from the nest by both adults and taken to the brood territory. The fledging period varies from about 50 to 70 days, when the family groups break up. At about the time of fledging, or slightly before, the adults begin their molt, which includes a flightless period of 26 days. At that time the juveniles begin to flock, reaching their greatest concentrations in November. By December the young have dispersed from Rottnest Island.

Status. Frith (1967) reports that these birds are still numerous and are not extensively hunted over their entire range, partly because of their very poor table quality. In some salt-water areas, however, these birds are regularly hunted. Further, the species may do local damage to crops, especially sprouting pastures and fields of wheat and peas. At present, its conservation is not a serious problem.

Relationships. Probably the New Zealand shelduck is the species' closest living relative, as mentioned in the account of that bird.

Suggested readings. Frith, 1967; Riggert, 1977.

New Zealand Shelduck

Tadorna variegata (Gmelin) 1789

Other vernacular names. Paradise shelduck; Neuseeländische Kasarka (German); casarca de paradis (French); oca del paraíso (Spanish).

Subspecies and range. No subspecies recognized. Resident on New Zealand, occurring on both islands, but more common on South Island and absent from the northern parts of North Island. See map 43.

Measurements and weights. Folded wing: males, 365–80 mm; females, 325–55 mm. Culmen: males, 42–45 mm; females, 37–40 mm. Weights: females, 1,260–1,340 g. Eggs: av. 65 x 47 mm, creamish, 91 g.

Identification and field marks. Length 25–28″ (63–71 cm). The sexes are quite dimorphic in adults. *Males*

have greenish black heads; blackish breasts; gray to brownish vermiculated underparts, flanks, and mantle; and black rump and tail. The under tail coverts are bright chestnut; the upper and under wing coverts are white, while the secondaries are iridescent green and the primaries blackish. *Females* have entirely white heads as adults, while their body color changes seasonally from dark reddish chestnut (breeding) to grayish brown (nonbreeding) in general tone. Vermiculations are also present on the female, especially on the mantle. *Immatures* of both sexes resemble males, but young females show irregular white feathering on the head, which gradually increases in amount, and young males are tinged with brown on the head. In both, the margins of the upper wings coverts may be narrowly edged with dull fulvous brown.

In the field, this species is unlikely to be mistaken for any other New Zealand species of waterfowl; the female's white head and the conspicuous white wing coverts of both sexes make for easy identification. The female has a shrill, nattering call, while the male utters gutteral *horr* and *ha-hoo'* notes as threat and sexual display calls, respectively. (The Maori name "Putangi," or "wail of death," refers to the call of this species.)

NATURAL HISTORY

Habitat and foods. These shelducks are widespread in New Zealand, occupying both fresh-water and coastal areas. They are found along mountain streams, river beds, open grassy plains, and lowland downs, and around lakes and coastal waters.

Favored breeding habitats are the tussock flats in South Island's mountain areas. The bird is obviously highly adaptable and in these varied habitats no doubt consumes a wide variety of foods. Among those that have been reported are soft grasses and herbs, insects, and crustaceans (Oliver, 1955).

Social behavior. Like the other shelducks, this species is most often encountered in pairs, and perhaps more than the others is likely to be in such groups rather than in flocks, as the birds are highly aggressive at most times of the year. The only time that flocking on a large scale occurs is during the postbreeding molt, when fairly large flocks develop on various lakes and estuaries, such as those in south Westland. Other than local movements to molting and breeding areas, no major migratory patterns are evident in this species. Wintering birds occur on the Bay of Plenty, the Firth of Thames, and sometimes Manukau. The social pair-forming behavior of the New Zealand shelduck has the same general pattern as that in the other species, but it represents an extreme in the trends that are apparent in the group. The sexes differ more in overall size and in plumage dimorphism than any of the other shelducks, and this species is the only one to exhibit definite breeding and non-breeding, or eclipse, plumages. Further, unlike those of other waterfowl, the females exhibit these seasonal changes more than the male, a point related to the fact that it is the female among shelducks that takes the initiative in forming pair bonds. This is done through intensive inciting behavior toward males, which stimulates both aggressive and sexual responses in the males. The white head of the female is especially conspicuous during inciting, as it is lowered and moved from side to side rapidly. Copulatory behavior in this species apparently has not yet been adequately described (Johnsgard, 1965a).

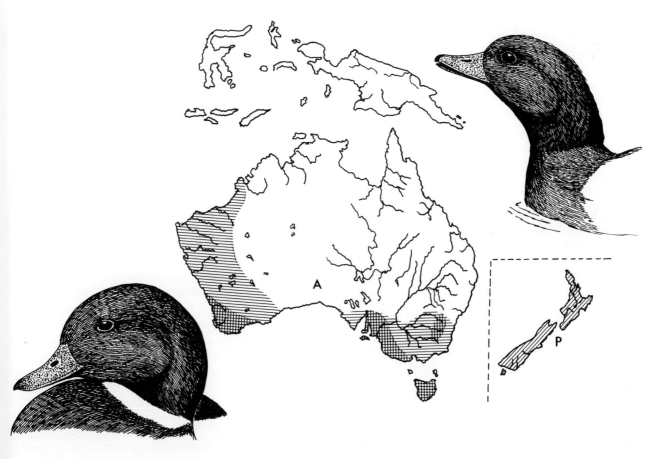

MAP 43. Breeding or residential distributions of the Australian ("A") and New Zealand or paradise ("P") shelducks. The primary range of the Australian shelduck is indicated by cross-hatching, and the peripheral range by horizontal hatching.

Reproductive biology. The normal breeding period for this species in New Zealand is from August to January; reportedly two broods are often reared during this prolonged period. Females seek out nest sites that may be a ground location well hidden by tussock grasses, in a rock crevice, on a cliff face far away from water, or in a tree cavity that may be 15 to 20 feet above the ground. The nest is typically well lined with down, and only the female incubates. The normal range in the clutch is from 5 to 11 eggs; 8 is the typical number. As in the other shelducks, the normal incubation period is 30 days, with the male remaining close to the nest to help protect it and the female from danger. After hatching, the young are immediately led to water and initially feed on insects and crustaceans (Oliver, 1955). The fledging period has not yet been established, but, judging from the other shelducks, is likely to require from about 7 to 10 weeks.

Status. This species is now absent from a few areas in which it once occurred before intensive settlement, but nonetheless it is still widespread and is even extending its range northward, largely as a result of liberations on the North Island. Although a game species, it is protected in some districts, and the short open season on it does not seem to be significant. It is probable that this species has been less seriously affected by man's activities than any other New Zealand waterfowl except the shoveler and gray teal (Williams, 1964).

Relationships. This species is obviously a very close relative of the Australian shelduck and the two perhaps should be regarded as a superspecies.

Suggested readings. Oliver, 1955.

Crested Shelduck

Tadorna cristata (Kuroda) 1917

Other vernacular names. Korean shelduck; Schopfkasarka (German); tadorna huppé (French); oca de Corea (Spanish).

Subspecies and range. No subspecies recognized. Presumably extinct; the species is known only from three extant specimens, of which two are from Korea and one from near Vladivostok.

Measurements. Folded wing: male, 320 mm; female, 310 mm. Culmen: male, 45 mm; female, 41.5 mm. Weights: unknown. Eggs: unknown.

Identification and field marks. Length ca. 25–28" (63–71 cm). The *adult male* is black on the crown of the head, nape, breast, upper tail coverts, tail, and primaries; the rest of the head is brownish black. The underparts, flanks, and mantle are dark gray with black vermiculations, while the coloration of the wing is like that of other shelducks. *Females* have a small white eye-ring, and the head and neck are also white except for the crown stripe and crest. Most of the rest of the body is dark brown with narrow white lines. *Immature* plumages are unknown.

DISCUSSION

This history of this species is one of the major mysteries in ornithology, inasmuch as so far only three extant specimens are known, and their validity has been doubted. The bird was first described in 1890 by W. Sclater, from a male specimen shot near Vladivostok and presumed to be a hybrid between the falcated duck and the ruddy shelduck. In 1917, N. Kuroda described a second, apparently female, specimen, taken near Fusan in Korea, and regarded it as a new genus and species of shelduck. Another bird, a male, was taken about the same time near Kunsan, Korea, and a fourth, another male, was shot in 1924 near Seoul. Reportedly three were shot from a flock of six birds seen in northwestern Korea in 1916, but none was preserved. As Kuroda has described, the species was apparently known to early Japanese aviculturists as the chosen-oski (Korean mandarin duck), and was allegedly imported regularly from Korea in the early 1700s. What appears to be this species was also figured on various old Chinese tapestries and paintings.

The only possible recent record of the Korean shelduck dates from 1964, when three shelducks bearing crests were seen among a group of harlequins in the Rimskii-Korsakov archipelago southwest of Vladivostok by two Russian observers. The drake reportedly had a pinkish bill and legs and was crested, which would eliminate the possibility of its being some other shelduck. This sighting has renewed hopes that the species may yet survive in small numbers, perhaps breeding in Russian Ussuriland (Fisher et al., 1969).

Suggested readings. Phillips, 1923–26; Fisher et al., 1969.

Northern (Common) Shelduck

Tadorna tadorna (Linnaeus) 1758

Other vernacular names. Sheldrake; Brandgans (German); tadorne de belom (French); oca común (Spanish).

Subspecies and range. No subspecies recognized. Breeds along the coasts of Norway, Sweden, Great Britain, France, and northern Europe east to Estonia, and locally around the shoreline of the Mediterranean. Also breeds inland from the Black and Caspian seas eastward on the saline lakes of Central Asia to eastern Siberia, Mongolia, and Tibet. Winters from the southern part of its breeding range to northern Africa, Iran, India, southern China, and Japan. See map 44.

Measurements and weights. Folded wing: males, 318–50 mm; females, 290–334 mm. Culmen: males, 52–60 mm; females, 44–54 mm. Weights: males, 980–1,450 g; females, 801–1,250 g. Eggs: av. 65 x 47 mm, creamy white, 78 g.

Identification and field marks. Length 24–25" (61–63 cm). *Adults* in breeding season have greenish black heads and necks, white breasts, and a broad chestnut brown band over the lower breast, while the underparts (except for a median blackish band), flanks, mantle, and tail coverts are white. The outer scapulars, primaries, and tail are black, and the secondaries are iridescent green. The upper and under wing coverts are white, while the tertials are brownish. The bill is bright red, and the legs and feet are pink. In the nonbreeding season the bill is less bright and less enlarged at the base, and the colors and patterning of the feathers are noticeably duller. *Females* are noticeably smaller than males, have white feathers between the eye and bill, and the bill is not enlarged basally. *Juveniles* of both sexes are mostly white on the underparts and dull black to gray dorsally, with no trace of the chestnut breast band. The head is blackish above and white on the cheeks and throat, and the upper wing coverts are whitish or dull gray. By midwinter young males assume their first adult plumage and can be distinguished from females by their solid blackish head color.

In the field, the strongly patterned white, chestnut, and black plumage combination is distinctive and virtually unmistakable with any other species. The loud whistling call of males and the raucous quacking calls of the females are frequent and can be heard for considerable distances.

NATURAL HISTORY

Habitat and foods. To an apparently greater degree than that of most other shelduck species, the northern shelduck's ecology is closely associated with the availability of small mollusks, especially the estuarine snail *Hydrobia.* To a much smaller extent, small fish and fish spawn, insects and insect larvae, and very small amounts of algae are consumed. Olney's (1965) summary of foods taken in the British Isles confirms the importance of *Hydrobia* in the species' diet. All of 46 specimens containing food remains had eaten this snail, regardless of locality or time of year. This mollusk is present in great quantities in estuarine and salt-marsh mud flats and in muddy sands, but does not extend into fresh-water environments. Bryant and Leng (1975) closely investigated the relationship between the abundance of *Hydrobia* and foraging activities of shelducks, and found a strong relationship between the intensity of feeding and the distribution and abundance of this food source. Feeding was done mainly on the flood tide and at times of high water, while the ebb tide and low-water periods were times of nonfeeding. The usual method of feeding was by standing on shore or in shallow water and probing for food, while head dipping or upending was done in water of swimming depth. Diving for food is not done by northern shelducks.

Social behavior. During most of the year northern shelducks are gregarious and occur in flocks of varying sizes. Pair formation occurs in wintering areas, and pair bonds are thought to be relatively permanent. Most birds reach maturity and presumably normally breed when two years old (Hori, 1964). Inciting by the female seems to play an important role in pair formation; and unlike any of the other shelducks, this species has a ritualized form of preening behind the wing (exposing the iridescent speculum) as an important part of pair-forming behavior (Johnsgard, 1965a). With the spring migration to their breeding areas, most pairs take up mutually exclusive feeding territories, although a part of the population remains together in a communal flock, presumably unable to establish territories within the area of favorable food supplies. The territories were

from 1,000 to 2,000 square meters in size in Young's (1970) study area, and were directly related to *Hydrobia* densities. A majority of the pairs reclaimed the same territories in subsequent years; and although the territories were defended primarily by males, the females had the stronger territorial attachment. When territorial males lost their mates, they soon abandoned their territories, while females whose mates had been shot quickly acquired new mates. Two of eight marked females whose mates had died or deserted them had different consorts in three years, and one changed mates three times in a single year. Since nesting does not occur on the territory, its role is probably to restrict the number of breeding pairs to the available food supply (Young, 1970), and perhaps it also helps to maintain an effective pair bond (Hori, 1969).

Reproductive biology. The studies of Hori (1964, 1969) on the breeding of northern shelducks in northern Kent, England, are unusually complete, and provide an excellent source of information. In that area,

MAP 44. Known European breeding areas (inked) and presumptive Asian breeding distribution (hatched) of the common shelduck; wintering distributions of both populations indicated by stippling.

most birds arrive in March, although intensive nest-site prospecting does not begin immediately. The nests are often used by the same pair repeatedly, and may be from 20 yards to two miles from the foraging territories established by the pair. Tree sites, hay or straw stacks, rabbit holes, and miscellaneous cavities are used about equally frequently, but open sites are used only rather rarely. There is no marked dispersion of nesting sites, and adjacent trees are sometimes used. Both members of the pair search for a nest site, but the female does so more actively. The male uses a variety of postures when directing his mate toward a prospective nest, but there is no observable indication to confirm that the female has chosen a particular site. Multiple use of a nest site by two or more females is not uncommon, judging from the number of clutches in excess of 12 eggs, but there was no indication of reduced hatching success in unusually large clutches. Sometimes two females may attempt to incubate the same clutch of eggs, and little or no antagonism occurs among breeding "communes."

The normal clutch size is about 9 eggs, but multiple clutches may contain from 14 to 30 (rarely 50) eggs. Incubation requires 30 days and is performed entirely by the female, with the male close at hand. Although no renesting occurs, pairs which have failed remain attached to other birds in their nesting "commune" and associate with them until the last eggs are hatched. As each clutch is hatched, the pair takes their young to brackish or salt water and have no further contact with other pairs. For the first two or three weeks the broods remain in family units, but larger groupings, or crèches, gradually develop, especially in broods hatched late. At times, groups of 100 or more ducklings may be seen together. They are cared for by their actual parents, and other adults, presumably failed breeders, also attempt to care for them. The fledging period is apparently between six and a half and eight weeks, at the end of which time the birds begin to disperse toward late-summer molting areas. Most of the British population migrates in July to the German coast of the

North Sea to undergo their postnuptial molt, but a limited number also move to Bridgwater Bay, Somerset (Eltringham & Boyd, 1960). The flightless period is from 25 to 31 days, with females averaging 27 days and males 29 days.

Status. The gathering of 90 percent or more of the European population in the North Sea molting areas provides an excellent basis for estimating population trends in this species, which appears to be on the increase. In recent years this population has exceeded 100,000, while the population that gathers in the Mediterranean and Black Sea areas numbers about 30,000 birds (Ogilvie, 1975). The population of the more easterly areas is still unknown.

Relationships. I have earlier suggested (Johnsgard, 1965a) that this species and the radjah shelduck represent rather isolated offshoots from the general group of shelducks, without especially close affinities to each other or to the other *Tadorna* forms. The juvenile plumage of the northern shelduck has some slight similarities to that of the Egyptian goose, and perhaps the bird provides an evolutionary link between *Alopochen* and *Tadorna*.

Suggested readings. Hori, 1964, 1969; Young, 1970.

Radjah Shelduck

Tadorna radjah (Garnot) 1828

Other vernacular names. Burdekin duck, white-headed shelduck; Radjahgans (German); casarca radjah (French); oca rajá (Spanish).

Subspecies and ranges. (See map 45.)

T. r. radjah: Black-backed radjah shelduck. Resident in the Moluccas, New Guinea, western Papua Islands, Fergusson Island, and the Aru Islands.

T. r. rufitergum: Red-backed radjah shelduck. Resident in northern and eastern tropical Australia, from the Fitzroy River to northern Queensland.

Measurements and weights. Folded wing: males, 260–68 mm; females, 246–98 mm. Culmen: males, 40–54 mm; females, 42–55 mm. Weights: males, 750–1,101 g (av. 750 g); females, 600–1,130 g (av. 839 g). Eggs: av. 60 x 42 mm, white, 59 g.

Identification and field marks. Length 20–24″ (51–61 cm). *Adults* of both sexes have white heads, breasts, underparts, and flanks, the white broken by a narrow blackish band around the lower breast. The upperparts from the scapulars back to the tail are generally black (brownish in *rufitergum*); the wing is white except for black primaries, white-tipped iridescent green secondaries, and a narrow black line formed by the greater secondary coverts. The bill, legs, and feet are pink, and the eyes of both sexes are uniquely white. *Females* are identical to males, and *immatures* resemble adults but the speculum is duller, the upper wing coverts are margined with blackish coloration, and the black band on the greater secondary coverts is broader.

In the field, the predominantly white coloration of this species sets it apart from all other shelducks and all other ducks of the area except for the much smaller pygmy geese. The sexes may sometimes be identified in the field by the female's rattling call and the male's wheezy whistle.

NATURAL HISTORY

Habitat and foods. According to Frith (1967), this species' preferred Australian habitat consists of brackish water, mud flats, and mangrove swamps in coastal areas. Fresh-water ponds are seemingly avoided at all times other than the dry season, and

the littoral strip extending inland about a mile or so from the coastline is most commonly utilized. In New Guinea the birds utilize very similar habitats, particularly mangrove and sago swamps, and the muddy banks and shoals of the larger rivers. Few data on foods are available; but of 21 samples from Australia, the predominant materials present were animal life, consisting almost entirely of mollusks. Some large insects were found, and the only plant materials present were sedge materials and small quantities of algae. Reportedly, the birds feed almost exclusively on land or in water only a few inches deep. They usually forage in pairs, with each pair maintaining its own feeding territory and visiting this territory each morning and evening (Frith, 1967).

Social behavior. Like other shelducks, this species has strong pair bonds, and thus flocks consist of pairs and family units that are the basic social unit. Concentrations of birds and moderately large flocks may develop during the dry season, when groups of up to 50 or 60 may occur locally. These are usually on rivers or permanent lagoons, particularly in the Northern Territory of Australia. The birds probably do not move very great distances between the dry and wet seasons, and probably most of the seasonal migrations are within rather than between river valley systems. With the start of the wet season in the Northern Territory, movements back toward the breeding grounds again occur. Breeding is timed to occur at the end of the wet season and early in the dry season, when floodwaters are receding and bare mud flats are exposed between the water and the fringe of pandanus. Radjah shelducks are unusually aggressive and highly territorial for their small size. Quite possibly the strongly white plumage is associated with this need for territorial advertisement. Frith (1967) reports that the average density of territories on the Adelaide River in one year was about one pair per one and three-quarters miles, a surprisingly low

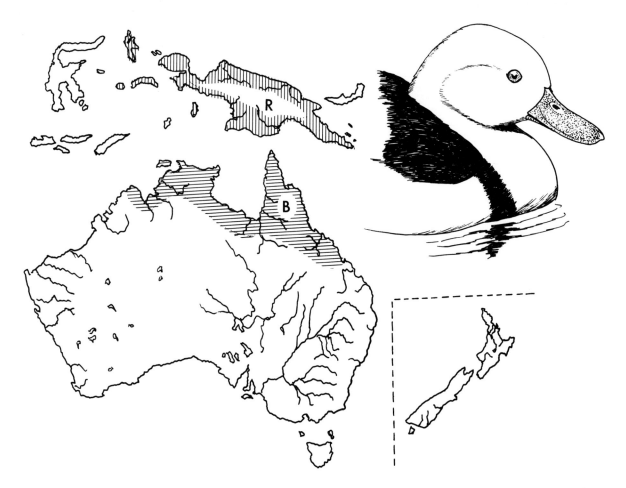

MAP 45. Breeding or residential distributions of the red-backed ("R") and black-backed ("B") radjah shelducks.

density for a bird of seemingly generalized foraging needs. The breeding territory includes both a foraging area such as a stretch of river bank and a suitable nesting site, and territorial establishment and defense is probably an important part of pair-bonding behavior. Observations on copulatory behavior are still inadequate, but it evidently occurs on water and preliminary displays involve head dipping by the male. The postcopulatory display is relatively weak, if the single observed instance was at all typical (Johnsgard, 1965a).

Reproductive biology. Nesting occurs in the Northern Territory between February and July, but most records are for May and June, while it has been reported to be considerably earlier (December to February) in Queensland. Apparently the yearly variations in the onset of the rainy season may affect the timing of reproduction. Nesting is most often done in trees, and both members of the pair may spend considerable time looking for a suitable location in a hollow limb or sprout. Nests are typically near water, and have only a little down present for lining. Few nests have been seen in the wild, and the clutch size is estimated to range from 6 to 12 eggs. Incubation is performed only by the female, and requires 30 days. After hatching, both parents lead the ducklings to water and in most cases probably raise the brood within the confines of their territory. As water areas dry up, the fledged young and adults may be forced to move to more permanent waters; otherwise they are likely to remain on the breeding territory until the next breeding season (Frith, 1967).

Status. This species is unusually vulnerable to destruction through hunting, and in many areas of Australia where the bird was once common it is now rare or absent. It is probable that more effective protection from hunting will have to be forthcoming if the species is to be saved from complete eradication (Frith, 1967).

Relationships. This shelduck seems to be the most atypical of the entire genus, and its precise relationships to the other species are not at all clear. Its bill shape is adapted for dabbling much like that of the northern species, and might reflect either real evolutionary affinities or simply ecological convergence to a comparable foraging niche.

Suggested readings. Frith, 1967.

Tribe Tachyerini (Steamer Ducks)

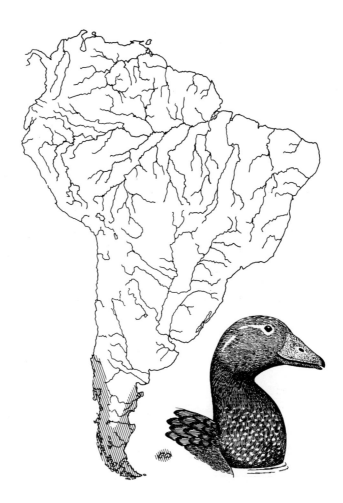

Map 46. Breeding or residential distribution of the flying steamer duck.

Drawing on preceding page: Falkland Flightless Steamer Duck (pair)

Flying Steamer Duck

Tachyeres patachonicus (King) 1830

Other vernacular names. Canvasback (Falkland Islands); Patagonische Dampfschiffente (German); canard-vapeur volant (French); pato quetru volador (Spanish).

Subspecies and range. No subspecies recognized. Breeds on the Falkland Islands and in southern South America from Tierra del Fuego north to about 48° south latitude in Argentina and to 37° south latitude in Chile. See map 46.

Measurements and weights. Folded wing: males, 287–317 mm; females, 276–301 mm. Culmen: males, 48–57 mm; females, 50–59 mm. Weights: males, 2,892–3,175 g (av. 3,073 g); females, 2,438–2,835 g (av. 2,616 g). Eggs: av. 76 x 52 mm, ivory, 115 g.

Identification and field marks. Length 26–28" (66–71 cm). *Adult males* are grayish-lead to wine-colored above, with a pale gray head except for a reddish throat and a whitish eye-ring and eye-stripe, and with white underparts. The bill is relatively small, is orange-yellow except around the nostrils, where it is bluish, and has a relatively narrow (under 15 mm) black nail. Overall, the plumage is somewhat more brownish than that of the other two species of this tribe, especially the Magellanic flightless form. *Females* are generally more brownish, especially on the head, which is almost uniformly brown except for a whitish eye-ring and eye-stripe. The bill is yellowish on the culmen, shading to bluish on the cutting edge. *Immature* males closely resemble

females in head coloration as well as in the color of the bill. *Juveniles* are generally grayish, lacking wine-colored margins on their breast and flank feathers.

In the field, the flying ability and more slender shape of this species helps to separate it from both of the flightless forms, although it prefers to "steam" over the water with flapping wings rather than to take flight. The calls of the males range from female-like "rasping grunts" to mechanical "ticking grunts" and whistlelike "sibilant grunts." The female calls are entirely low-pitched grunting sounds, often uttered repeatedly in long series.

NATURAL HISTORY

Habitat and foods. This species, to a greater degree than the flightless forms, is often found on fresh-water ponds as well as on the sea. It often frequents such ponds during the daylight hours, and may also occur on rivers or lakes some distance from the coast. So far as is known, its foods are similar to those of the flightless forms, but it consumes a lower proportion of thick-shelled mollusks and a correspondingly greater amount of crabs and crustaceans that occur in both salt- and fresh-water environments. During the winter the birds are found mainly on the coast, but during the summer they spread inland to breed over considerable distances. A favorite breeding ground on Tierra del Fuego is Mantu Lake, which is 28 miles from the coast (Murphy, 1936; Johnson, 1965).

Social behavior. Flocking in this species is largely limited to immature and molting birds. Small flocks of from 4 to 15 may still be commonly seen in June, while the major flightless period is probably somewhat earlier, in April. For a good part of the spring and summer the birds are in pairs on territories, and the breeding season in Chile is said to extend from November to January. However, even as early as June some females collected on the island of Chiloé had begun to show signs of gonadal growth. In the Falkland Islands the breeding season begins sooner, in October or possibly earlier. Moynihan's (1958) observations on display were made in Tierra del Fuego in November and December, or near the peak of the nesting season, and much of the hostile behavior he observed appeared to be associated with territorial defense. Most attacks he saw were by resident males on intruders, usually females, who flew immediately. Female displays included false drinking, vertical neck and head stretching, and grunting calls. Those of the male are basically similar, but grunting occurs in a graded series of intensities that evidently reflect differing degrees of attack and escape tendencies. Displays between mated pairs were relatively few and simple, and often took the same form as the mutual triumph ceremonies of geese and swans, which is understandable, considering the strong and seemingly permanent pair-bonding arrangement. The few copulations that were seen occurred after mutual or male-only bill-dipping behavior; and after treading, the two birds assumed an alert posture and swam away from each other, orienting the back of their heads toward one another (Moynihan, 1958).

Reproductive biology. The nesting behavior of this species is much like that of the other two, and at least in Chile the nests are typically placed on small islets surrounded by water, rather than along shorelines. They are usually very well hidden by grass or other vegetation, but at times are placed in exposed sites. The clutch size may range from 5 to 9 but is typically 7 eggs. Only the female incubates; but as in the other species, the male is never found far from the nest. Although the specific incubation period has not been established, it is probably between 30 and 40 days, judging from information on the Falkland flightless species. In Tierra del Fuego many of the young have hatched by December, but the period to fledging is evidently quite prolonged. Thus, birds with barely sprouted primaries but otherwise well feathered out have been collected at Ushuaia in April (Murphy, 1936; Johnson, 1965).

Status. The flying steamer duck is probably less common than either of its two flightless relatives over their common ranges, but confusions over identities make this difficult to ascertain with certainty. Nevertheless, there is no cause for concern over its status either on the mainland of South America or on the Falkland Islands.

Relationships. This species presumably is nearest the ancestral form that produced all of the extant steamer ducks. The entire group is apparently most closely related to the shelducks and sheldgeese, but the relationship nevertheless is somewhat obscure, and tribal separation seems warranted (Moynihan, 1958; Johnsgard, 1965a). Woolfenden (1961) believes that the birds should be regarded as aberrant dabbling ducks on the basis of skeletal anatomy, but there is so far no supporting evidence for this.

Suggested readings. Murphy, 1936; Weller, 1976.

136 ◈ ◈ ◈

Magellanic Flightless Steamer Duck

Tachyeres pteneres (Forster) 1844

Other vernacular names. Racehorse, seahorse, loggerhead; Magellan-Dampfschiffe (German); canard-vapeur de Patagonie; pato quetru novolador, pato vapor magallanico (Spanish).

Subspecies and range. No subspecies recognized. Resident in western South America from Chiloé Island south to Cape Horn and Staten Island. See map 47.

Measurements and weights. Folded wing: males, 260–88 mm; females, 255–71 mm. Culmen: males, 55–66 mm; females, 54–61 mm. Weights: males, 5,897–6,180 g (av. 6,039 g); females, 3,629–4,763 g (av. 4,111 g). Eggs: av. 82 x 56 mm, ivory, 167 g.

Identification and field marks. Length 29–33″ (74–84 cm). Plate 28. This is the largest of the steamer ducks. *Adult males* have a bluish gray to whitish head, with reddish brown lower cheeks and throat, a body pattern of bluish gray scales with darker slate edges, and

white underparts. The wing pattern, with a large white speculum formed by the secondaries and their greater coverts, is like that of the other species. The massive bill is bright orange, with a large and wide (over 15 mm) black nail, and the legs and feet are mostly yellow, with some black on the underside of the feet. *Females* are considerably smaller than males, more wine-colored on the back, with a more brownish head and more yellowish bill. *Juveniles* are entirely gray and lack wine-colored margins on the breast and flank feathers.

In the field, its very large size and nearly flightless condition help separate this species from the flying steamer duck. It also has a substantially heavier bill and nail than that of the flying steamer duck; and the white speculum of the flightless species is limited to the outer six or seven secondaries, while in the flying

steamer duck it extends to the tenth or twelfth secondary (Murphy, 1936). The calls of this species include a shrill, repeated *qu-ie-u-ll* and a quick *kek* note, while the female produces croaking sounds.

NATURAL HISTORY

Habitat and foods. These birds are found along rocky shorelines and offshore kelp beds, where they forage on a variety of invertebrate forms. Johnson (1965) found mytilid mussels in some, and states that mollusks, crustaceans, and small fish are their major items of diet. They forage at or near high tide periods, and as the tide goes out they resort to preening or sleeping on rocks. Chitons, limpets, snails, and crustaceans have all been reported in the stomachs of this species, and probably most of the foods are obtained by diving in the relatively shallow kelp beds. The birds probably rarely move far inland except during nesting, but apparently in some areas make daily trips to fresh-water springs for drinking, as has been suggested by various observers (Murphy, 1936).

Social behavior. Except for young and nonbreeding birds, flocking in this species is nonexistent. Flocks of nearly one hundred birds have been reported in mid-January in extreme southern Chile, at the time that broods are out. Even when molting, however, the birds are usually to be found in pairs or only small flocks. During the winter months they sometimes associate with flying steamer ducks. The courtship period in Chiloé has been estimated to begin in August or September, and nearly all of the broods are hatched by December in that area (Murphy, 1936). Although it is almost certain that the hostile and sexual displays of this species are very much like those of the flying steamer duck, no actual detailed observations on this point are yet available. I (1965a) have observed a very few examples of display in captive individuals of this species, and the posturing and calling of both sexes appear to be nearly the same as Moynihan's (1958) account of these patterns in the flying steamer duck. Apparent inciting behavior by the female, called stretching, and a variety of male calls that are progressively more sibilant and variably associated with tail cocking and head lifting are the most common displays. Both sexes also perform false drinking frequently.

Reproductive biology. Nesting is probably largely confined to the period September through December, but few actual egg dates are available. Nests are often

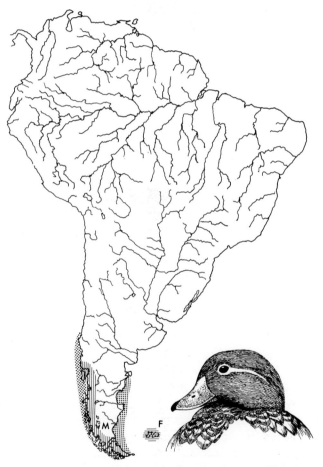

MAP 47. Residential range of the Falkland flightless steamer duck ("F"), and breeding ("M") and wintering (stippling) distributions of the Magellanic flightless steamer duck.

placed quite close to water, rarely more than 300 yards from it according to Murphy (1936). He states that they are usually under shrubbery or otherwise well concealed. Johnson (1965) noted that on the Guaitecas Islands off the coast of Chile the nests were placed well inside the nearly impenetrable forest, and so well hidden that they could not be found without the aid of a dog. When sitting, the female can often be caught and lifted off the nest bodily. The usual clutch is from 5 to 8 eggs, but the incubation period has not yet been established. Very probably it is between 30 and 40 days, as in the Falkland Island species. Foxes and caracaras are reportedly serious predators of the eggs, and large gulls and skuas almost certainly destroy significant numbers of young birds. Families with downy young have been seen in coastal caves on the small islands off Chiloé, where fresh water trickles down through the rocks (Murphy, 1936).

Status. This species is still quite common in southern Chile and Argentina, and neither its habitat nor use by humans places it in any foreseeable danger.

Relationships. Presumably this flightless form evolved from an ancestral type relatively similar to the present-day flying steamer duck.

Suggested readings. Murphy, 1936; Weller, 1976.

Falkland Flightless Steamer Duck

Tachyeres brachypterus (Latham) 1790

Other vernacular names. Logger, loggerhead; Falkland-Dampfschiffente (German); canard-vapeur des Iles Falkland (French); pato vapor malvinero (Spanish).

Subspecies and range. No subspecies recognized. Resident on the Falkland Islands. See map 47.

Measurements and weights. Folded wing: males, 272–82 mm; females, 251–72 mm. Culmen: males, 53–61 mm; females, 52–58 mm. Weights: males, 4,303–4,420 g; females av. 3,400 g. Eggs: av. 82 x 57 mm, buff, 147 g.

Identification and field marks. Length, females 24" (61 cm), males 29" (74 cm). *Adult males* have a head that is predominantly gray and white (nearly pure white in old males) and a body that is mostly a scaled pattern of slate-gray feathers edged with blackish and reddish brown, especially on the breast, scapulars, and flanks. The cheeks and throat of males are reddish brown, and the bill is bright orange, with a black nail. The tail and wings are gray, the latter with a large white patch formed by the secondaries and their greater coverts. *Females* are smaller than males and have a yellow-green bill and a dark brown head with a white eye-ring and eye-streak. *Juveniles* and first-year birds resemble females but usually lack the white streak behind the eyes, and second-year males gradually acquire a grayish head and orange bill.

In the field, this species can be separated from the flying steamer duck by its shorter wings and heavier bill, and by a golden-yellowish collar at the base of the neck. Males utter a loud alarm note, *cheroo,* and similar but softer conversational notes, and females produce low, creaking notes while throwing the head vertically upward.

NATURAL HISTORY

Habitat and foods. Like other steamer ducks, this species concentrates on mollusks for its food; there is one record of a single bird with over 450 mussel shells in its crop and stomach. Other bivalves, limpets, chitons, and gastropod mollusks are likewise consumed, as are crabs and shrimps. Foraging is done by either upending in shallow water or diving in deeper waters, with each dive about 30 seconds in length. The birds are thus essentially coastline birds, and are especially numerous in sheltered harbors and along creeks and where there are large beds of kelp near the shore. They are seen as far as three miles from shore, and at night sometimes move to ponds (Woods, 1975; Murphy, 1936). Ducklings begin to dive for food within a few days of hatching, gradually increase their diving times, and feed on amphipods, isopods, and snails. Adults have been observed feeding in water probably over 9 meters deep, but also dabble and probe for their food in shallow waters (Weller, 1972).

Social behavior. Adult paired birds remain relatively segregated and may stay on or near their territories throughout the year, in conjunction with a very protracted breeding season. However, newly fledged young and older but still sexually immature birds gather in flocks that sometimes number in the hundreds, foraging in the coastal waters and sometimes resting together on beaches. The birds also at times use ponds close to the tide line. The length of time to sexual maturity is not yet certain, but must be at least two years, judging from the number of nonbreeding birds and the extended period required for the male's assumption of adult plumage and bill coloration. Although this is the least shy of all the steamer ducks, the birds seem to show fear of sea lions, and gulls and skuas are serious predators of their ducklings (Woods, 1975; Murphy, 1936). Pair-forming behavior is not well known in this species, but Pettingill (1965) states that the aggressive patterns are apparently identical to those of the flying steamer duck. The male's threat consists of rising in the water, throwing the head back and lifting the tail, and uttering a series of wheezy sounds and clicking notes. Fighting during the period of territorial defense is common, prolonged, and often bloody. The death of one of the males by drowning is probably rare but has been observed.

Reproductive biology. The territories of these birds are stretches of shoreline, and areas defended include both the offshore water and the shoreline up to the edge of vegetation. Pettingill (1965) reported that territories were spaced about 300 yards apart on Kidney Island. Eggs have been found in every month of the year, but most nesting occurs between mid-September and December. The nests are placed in grass, dry kelp, "diddle-dee" (*Empetrum*), tussock grass (*Poa*), or even at times in an old penguin burrow. Nests are usually within 200 yards of a beach, but have been found as far as a quarter mile away. The nest is well lined with down, and the usual clutch is from 5 to 8 eggs, occasionally as many as 10 or more. Observations of a captive pair at the Zurich Zoo (Schmidt, 1969) suggest that the eggs are laid at approximately two-day intervals. In this study, eggs hatched over a variety of incubation periods that ranged from 28 to 40 days (estimated average 34 days), under various forms of incubation. The male closely guards the nest during the period of incubation, often patrolling offshore and calling to the female at the first sign of danger. After hatching, he joins the family, and upon disturbance will remain behind to ward off intruders as the female leads her brood to safety (Murphy, 1936). The fledging period is about 12 weeks (Weller, 1972).

Status. This species, which is still very widespread and common in the Falklands, seems to be in no danger.

Relationships. As Murphy (1936) has suggested, it seems apparent that the flying steamer duck is nearest the ancestral type which has given rise to near-flightless birds both on the Falkland Islands and on mainland South America. The steamer ducks are clearly related to the shelduck group, but Moynihan (1958) has suggested tribal separation, with which I (1965a) have agreed. Brush (1976), on the basis of his studies of feather proteins, judged, however, that the steamer ducks might be included in the shelduck tribe rather than being recognized as a distinct group of their own.

Suggested readings. Pettingill, 1965; Woods, 1975; Weller, 1976.

Tribe Cairinini (Perching Ducks)

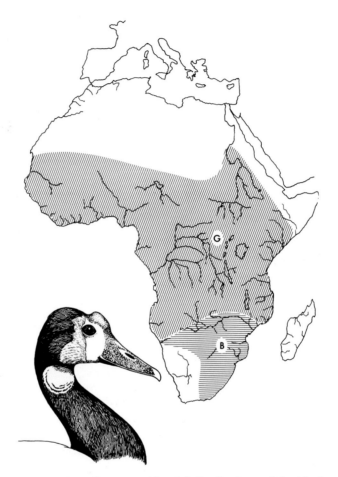

MAP 48. Breeding or residential distributions of the black
("B") and Gambian ("G") spur-winged geese.

Drawing on preceding page: Spur-winged Goose

Spur-winged Goose

Plectropterus gambensis (Linnaeus) 1766

Other vernacular names. None in general English use; Sporengans (German); oie armée (French); ganso africano de espolón (Spanish).

Subspecies and ranges. (See map 48.)

P. g. gambensis: Gambian spur-winged goose. Resident in Africa from Gambia, Kordofan, and the White Nile south to at least the Zambesi.

P. g. niger: Black spur-winged goose. Resident in Africa from the Cape of Good Hope northward perhaps through Botswana, but intergrading over a wide area with *gambensis.*

Measurements and weights. Folded wing: males, 530–50 mm; females, 422–40 mm. Culmen (from nostril): males, 59–63 mm; females, 57–59 mm. Weights: males, 5.4–6.8 kg (rarely to 10.0 kg); females, 4.0–5.4 kg. Eggs: av. 73 x 56 mm, white, 140 g.

Identification and field marks. Length 30–39" (75–100 cm). Plate 29. *Adults* are extensively iridescent bronze and green on the upperparts, including the neck and feathered parts of the head, which from the vicinity of the eyes forward is featherless and bluish or reddish. The underparts, including the under tail coverts, are white. The white extends up the breast in *gambensis* and also appears on the face in that race. The bill is red and variably extended into a knoblike caruncle on the crown; the legs and feet are pale pink. The wing is spurred at the bend and mostly iridescent bluish green, but with all the under wing coverts and variable amounts of the upper coverts white. The tail and upper tail coverts are blackish. *Females* are smaller than males and have duller facial coloration and smaller bill caruncles. *Immatures* of both sexes lack bare facial skin and an enlarged bill, and are browner on the face and neck. Most of the body feathers are fringed with brown, and the white areas on the wings and underparts are more restricted than in adults. The spur on the wing appears within a month of hatching and gradually enlarges.

In the field, the long-necked and long-legged appearance of this species and its black and white pattern make it unmistakable for any other African waterfowl (but the appearance is reminiscent of the Australian magpie goose). Males produce a high-pitched whistling note, and a similar repeated *chi* sound is uttered by females when agitated. A double-noted huffing sound is also made by males when threatened.

NATURAL HISTORY

Habitat and foods. This species is common and widespread over Africa, occurring in fields of crops or grasslands in the early morning, at dusk, or at night, and resting in swamps or open water during the rest of the day, sometimes perched in trees. In South Africa it is usually found along the larger rivers and the edges of lakes, vleis, dams, reservoirs, marshes, and swamps. During the dry winter months, flocks of nonmolting birds often fly into cultivated fields at daybreak to forage, returning a few hours later. They make a second flight out at about sunset, but remain in the fields at night only when there is moonlight to provide adequate light. During the rainy season, however, the birds remain in the fields all day, presumably because water is available there and thus they are not forced to return to their roosting areas for it. The adults feed primarily on the shoots and seed heads of grasses, the soft parts of aquatic plants such as water lilies, and a variety of grain and tuberous crops. A few observers have reported that they consume such animal materials as small fish (Clancey, 1967).

Social behavior. Flock sizes in this species are sometimes fairly large, especially in roosting areas. In favorable roosting or grazing areas, the birds are often seen in flocks of 50 or more, while on waters where molting birds congregate the numbers may go much higher. As many as 2,000 birds were reported in a molting area at Barbarspan, southwestern Transvaal, from June to August one year, while only half that many were seen the following one. The molting period is unusually long, lasting from six to seven weeks, and thus there is a need for safe and dependable water areas during this period. Although the birds are probably fairly sedentary over much of their range, there are marked seasonal changes in numbers in many areas. In the Luangwa Valley of Zambia, for example, the species is present mainly during the rainy season. In Zambia there are nesting records spanning nine months from September through May, but the majority are from January through March. In South Africa most breeding occurs between September and January, apparently corresponding to the rainy period of the spring and summer months (Clancey, 1967). Almost nothing is known of the pair-bonding behavior of this species; if it is like that of the other larger forms of perching ducks, pair bonds are virtually lacking. Likewise, no information is yet available on copulatory behavior.

Reproductive biology. Evidently a fairly wide variety of nest sites are accepted by females of this species, but most often the nests are built in grass or reed beds, and are substantial constructions of roots, twigs, herbaceous vegetation, and whatever other materials are immediately available. Nests have also been found among rocks in termite mound cavities, and on the nests of other large birds, such as the hammerhead (*Scopus umbretta*). The nest cup is relatively deep and well lined with down during the period of egg laying. Clutch sizes are quite variable, from 6 or 7 at the lower extreme to 14 or 15 at the upper extreme. In one case 8 eggs were known to be laid over a period of 9 days. Incubation is by the female alone, and requires from 30 to 32 days to be completed. Although it is said that the male remains until the clutch is completed, there seems to be no indication that the pair bond persists much longer or that the male participates in caring for the young. The fledging period is slightly more than two months (Clancey, 1967). However, since the molting period at Barberspan, South Africa, is from June to August, either nonbreeders must molt at a later time or the postnuptial molt of breeding adults must be considerably delayed.

Status. This species is still the commonest of the gooselike waterfowl in Africa, which is perhaps a testament to its wariness and adaptability to settlement conditions. Although it is regarded highly by sportsmen as a target, it is very hard to bring down and is often considered inedible except perhaps for the younger birds.

Relationships. This species seems to be one of the most primitive of the subfamily Anatinae, although its terrestrial adaptations may be secondary and its similarities to the magpie goose may be mostly superficial. In its general plumage sequences and its social behavior it seems fairly close to the genera *Cairina* and *Sarkidiornis*, and collectively these genera seem to be stem forms of the anatine stock. Curiously, Woolfenden (1961) noted a number of sheldrakelike osteological features in the spur-winged goose and considered it an aberrant member of this tribe. Assuming that the primitive Anatinae radiated from shelducklike birds, these similarities should not be too surprising.

Suggested readings. Clancey, 1967.

Muscovy Duck

Cairina moschata (Linnaeus) 1758

Other vernacular names. Musky or musk duck; Moschusente (German); canard de Barbarie (French); pato criollo or pato negro (Spanish).

Subspecies and range. No subspecies recognized. Resident from Mexico south to Central America and South America to the coast of Peru on the west and to Santa Fe, Argentina, on the east; accidentally to Buenos Aires. Also a rare resident on Trinidad. See map 49.

MAP 49. Breeding or residential distribution of the muscovy duck.

Measurements and weights. Folded wing: males, 350–400 mm; females, 300–315 mm. Culmen: males, 65–75 mm; females, 50–53 mm. Weights: males, 2.0–4.0 kg (domestic birds may exceed this maximum); females, 1.1–1.5 kg (Leopold, 1959). Eggs: av. 67 x 46 mm, white with greenish sheen, 74 g.

Identification and field marks. Length 26–33″ (66–84 cm). Plate 31. *Adults* are predominantly brownish black to black, with greenish to purplish iridescence on most of the upperparts. The under wing coverts, axillaries, and varying amounts of the upper wing coverts are white (varying with age); the legs and feet are black; and the bill is pinkish with black mottling. The skin around and in front of the eyes is bare and warty, and the base of the bill is enlarged. *Females* are smaller, have an entirely feathered head, and lack the bill enlargement. *Juveniles* are less iridescent dorsally, and have little or no white on their upper wing coverts.

In the field, muscovies are likely to be confused only with comb ducks but lack the white head,

breast, and underparts of that species. Muscovy ducks have few vocalizations; the male utters a weak hissing note, and the female has a very simple quacking call.

NATURAL HISTORY

Habitat and foods. The preferred habitats of this tropically adapted species are the rivers, lagoons, marshes, and similar areas of water at relatively low altitudes that are associated with tropical forests or heavy woodland. Slowly flowing rivers, or backwater swamps adjacent to such rivers, are apparently their favorite habitat. During dry seasons the birds may move into coastal swamps and lagoons, but fresh-water environments are their preferred breeding localities. In such areas the birds forage by dabbling or upending, and probably also do some grazing. Not much is known of specific foods taken, but small fish, insects, millepedes, small reptiles, and water plants have been reported. Evidently termites are a favorite food, and it is said that the birds tear nests apart to get at them. They will also chase crabs, and have been observed feeding on water lily seeds. Seeds of other aquatic plants, such as pickerelweeds (*Pontederiaceae*), and sedges (*Fimbristylis*), and the roots of *Mandioca* have also been reported among their foods.

Social behavior. In general, flocks of muscovies are not large, usually consisting of a half dozen or so. Sometimes larger flocks occur on common roosting sites such as trees, but these groups are not close-knit and upon disturbance usually disperse in all directions. There seems to be no indication that either pair bonds or family bonds are well developed in muscovy ducks. There is also no well-established migratory pattern in the species, which is generally restricted to areas without major seasonal temperature changes. Dry-season movements of a limited nature do occur, but there seems to be no flocking during the postbreeding molt period. The aggressive and sexual displays of muscovy ducks are simple and not differentiated from one another as they are in more typical ducks. The male's only obvious display is a crest-raising, tail-shaking, and back-and-forth head movement while making breathing or hissing sounds. Similarly, the female is relatively lacking in social displays, and apparently lacks even a well-defined inciting behavior. Copulation occurs on water, and usually takes the form of an apparent rape, with the male simply overpowering the much smaller female. However, at times the female may actively solicit copulation by extending herself flat on the water before the male. After treading, there appears to be no display behavior on the part of either sex before the female begins to bathe (Johnsgard, 1965a, 1975).

Reproductive biology. Very little is known of the nesting of this species under natural conditions. Eggs or young have been reported between November and June in Central and South America. Most of these dates seem to conform to the occurrence of the rainy season. They include Bolivia in November, Guayana from February to May, Peru in March, Surinam in March and April, and Panama in June. The usual nest site is a tree hollow, but the birds occasionally nest amid rushes. The clutch size is relatively large, as is typical of hole-nesting waterfowl, and usually ranges between 8 and 15 eggs. But up to as many as 20 eggs have been reported, probably laid by two females. The male plays no apparent role in nest-site selection, nor does he guard the nest or otherwise play any paternal role in assuring its success. The incubation period is 35 days, judging from data on cap-

tive birds. Virtually nothing is known of brood-rearing behavior, nor has the fledging period been established (Johnsgard, 1975).

Status. This species is still widespread and abundant over much of South and tropical Central America, where it is an important game species in some areas. It is readily domesticated, and provides natives with an important source of food under these conditions.

Relationships. The genus *Cairina* as recognized here includes the closely related muscovy and white-winged wood ducks; the genera *Sarkidiornis* and *Pteronetta* are obviously also closely related birds and might readily be considered members of this genus when broadly interpreted. These birds, together with the spur-winged goose, collectively make up the generalized "core" of the perching ducks, and the others in the tribe are variably more specialized in their foraging and social behavior patterns (Johnsgard, 1965a).

Suggested readings. Leopold, 1959; Johnsgard, 1975.

White-winged Wood Duck

Cairina scutulata (S. Müller) 1842

Other vernacular names. None in general English use. Malaienente (German); canard à ailes blanches (French); pato de alas blancas (Spanish).

Subspecies and range. No subspecies recognized. Resident in Assam, especially in the North East Frontier tracts (Arunachal Pradesh), Manipur, Bangladesh, and Burma. Locally south through the Malay Peninsula to Sumatra and Java. Rare throughout its range and probably endangered. See map 50.

Measurements and weights. Folded wing: males, 360–400 mm; females, 305–55 mm. Culmen: males, 58–66 mm; females, 51–61 mm. Weights: males, 2,945–3,855 g; females, 1,925–3,050 g. Eggs: 62 x 45 mm, greenish yellow, 89 g.

Identification and field marks. Length 26–32" (66–81 cm). *Adults* have a white or black and white spotted head, with the white usually terminating at the upper breast but at times extending to the abdomen. The rest of the body is chestnut brown and blackish with green iridescence. The tail and primaries are brownish black, while the axillaries, under wing coverts, and upper wing coverts other than the greater secondary coverts are white. The latter coverts are black and the secondaries are mostly bluish gray, while the first adjacent tertial is striped with black and white. The iris color of adult males is yellow. The legs, feet, and bill are yellow to orange, the bill reportedly being swollen at the base and reddish during the breeding season. *Females* are more heavily spotted with blackish coloration on the head, and are considerably smaller than males. They also have brownish eyes. *Immatures* are duller and browner, and lack iridescence on their dorsal feathers.

In the field, the white wing coverts and dark underparts distinguish it from the comb duck of the same general region. Loud, single-syllable calls are typical of both sexes; that of the female is a honking sound reminiscent of a rusty pump in need of oiling.

NATURAL HISTORY

Habitat and foods. Marshy swamps and lakes surrounded by extensive jungle, and smaller patches of

MAP 50. Known current (inked) and presumptive original (hatched) ranges of the white-winged wood duck.

jungle containing pools of water are the preferred habitat of this species. The birds evidently avoid moving waters, and also disappear when human activities encroach on the jungle environment. Water-logged depressions in tropical evergreen forest, especially with dead trees in the water, are often used, as are jheels in dense canebrakes and tall elephant grass jungle (Ali & Ripley, 1968). Foods taken by this species are little known, but apparently include snails, some small fishes, and rice. In captivity the birds eat a variety of foods, including grain, water plants, and various greens.

Social behavior. There is no indication of a strong flocking tendency in these birds. Usually they are to be found in pairs or small groups (families?) of 5 or 6 birds on small ponds. Larger numbers seen have included a flock of 11 and two parties totaling 30 birds. The birds spend most of the daylight hours perching in trees and toward dusk fly to rice fields, open waters, and marshes to forage at night, returning again at dawn. There are no known migrations.

Reproductive biology. Almost no nests have ever been located, but one was found in a tree standing beside a stream. The nest was in a decaying hollow at the first major branching some six meters up. The birds reportedly sometimes nest on the ground or on masses of branches in trees, presumably the deserted nest of another species. The probable clutch size is about 10 eggs, ranging from 6 to 13 in captivity. Records of captives indicate that the incubation period is 33 to 35 days. Probably the female performs all incubation and brood-rearing duties, judging from what is known of related species, but pair bonds are present and the male may remain with the family (Mackenzie & Kear, 1976).

Status. As of 1970, the species was very rare and decreasing in Assam, but was still present in Manipur, Bangladesh, and upper Burma. Its current status is uncertain, but the birds were once thought to be widely distributed in Thailand and Vietnam. It has recently been reported for the first time from central Laos (Ripley, in Hyde, 1974). Mackenzie and Kear (1976) discuss its past and present status, and indicate that although firm data are lacking, the species may still exist on Sumatra, Java, and Malaya but has not been seen for more than 17 years in Thailand. It has recently been observed in central and southern Sumatra, where it may still be fairly common, at least in Lampung Province (*Wildfowl* 28: 61–64).

Relationships. Anatomical similarities indicate that this species is a fairly close relative of the muscovy duck, and probably also of the Hartlaub duck, both of which are similar in their ecology and behavior.

Suggested readings. Ali & Ripley, 1968; Mackenzie & Kear, 1976.

Comb Duck

Sarkidiornis melanotos (Pennant) 1769

Other vernacular names. Knob-billed duck, knob-billed goose; nukhta (Indian); Höckerglanzente (German); sarcidiorne (French); pato arrocero (Spanish).

Subspecies and ranges. (See map 51.)
  S. m. melanotos: Old World comb duck. Resident in Africa from Gambia and the Sudan south to the Cape, Madagascar, India, Burma, Thailand, Laos, and southeastern China.
  S. m. sylvicola: South American comb duck. Resident in eastern Panama, Trinidad, and South America from Colombia and Venezuela southward east of the Andes to northern Argentina.

Measurements and weights. Folded wing: males, 350–80 mm; females, 305–55 mm. Culmen (from back of caruncle): males, 57–66 mm; females, 42–52 mm. Weights: males, 1,300–2,610 g; females, 1,230–2,325 g. Eggs: av. 62 x 43 mm, pale cream, 47 g.

Identification and field marks. Length 22–30" (56–76 cm). Plate 29. *Adults* have a black and white spotted head and neck, and an entirely white breast and abdomen. The flanks are pale gray (*melanotos*) or blackish (*sylvicola*), while the scapulars, back, tail, and inner secondary feathers (tertials) are iridescent green to purple. The secondaries are bronzy green, as are most of the upper wing coverts, while the primaries and under wing coverts are blackish. The under tail coverts are white, tinged during the breeding season with yellow, which also seasonally appears on the rear of the head and neck. The bill, legs, and feet are grayish black, and males have a fleshy caruncle at the base of the bill that is largest during the breeding season. *Females* are appreciably smaller and lack the enlarged bill and yellow on the head and under tail coverts. *Immatures* resemble

females but are less glossy above and are heavily barred with blackish on the back of the neck. *Juveniles* have a distinctive, an entirely brownish plumage that somewhat resembles that of a female Australian wood duck.

In the field, comb ducks are most likely to be confused with muscovy ducks or with white-winged wood ducks, but its white breast and abdomen separate the comb duck from those species, and the *Cairina* species also have conspicuous white wing coverts when in flight. Vocalizations, which are very weak, include a faint *churrr* sound in the males and a very weak quack by females.

NATURAL HISTORY

Habitat and foods. This is a bird typically found in waters associated with open woodlands rather than heavy forests, such as grassy ponds or lakes in savannah country and open woodlands along some of the larger rivers and occasionally very large lakes. In the breeding season it is sometimes found in grassland areas, but prefers areas where trees or other perching sites are readily available. The birds also spend time on sandbars, muddy flats, or similar areas during the middle of the day, and restrict their foraging largely to early morning and late afternoon. They are primarily vegetarians, and may be pests in areas where corn, oats, and similar grains are planted, as they often graze on seedlings of these grasses. Other grasses and grass seeds, the soft parts of aquatic plants, and some invertebrates such as insect larvae and adult locusts have been found in their stomachs (Clancey, 1967).

Social behavior. The social groupings of this bird include pairs, males with several females, small parties

of a single sex, and sometimes also fairly large flocks, especially outside of the breeding season. In some areas of Zambia the birds arrive by the thousands during the rainy period for breeding but are gone during the dry season, while in southernmost South Africa they are present during the nonbreeding season. The birds move considerable distances as the rainfall pattern dictates; one individual banded in Rhodesia was later taken 2,150 miles to the north in the Sudan. However, no clear-cut migratory patterns are yet evident. In many areas the breeding season is greatly prolonged, and in such cases the birds probably never move far from their breeding areas. Pair bonding is evidently quite weak or even absent in this species, and the period of time to reproductive maturity has apparently not been firmly established. During much of the year the females seem to avoid males, and even during the period of courtship the females may flee when approached by displaying males, which at that time are highly aggressive toward other males. The major aggressive display is head-pumping movement performed while swimming in a very high and erect posture and accompanied by a weak *churrr* sound. The typical display toward females usually occurs on land, and is begun by a slow wing-flapping display, followed by display-preening on the breast and behind the wing. There does not appear to be an inciting display by females, which seems understandable in view of the apparent absence of pair bonds (Johnsgard, 1965a). In several observations of copulation that I have recently made, the male approached the female with repeated head-pumping and drinking movements alternated with gaping and hissing with a diagonally outstretched neck. The female sometimes performed slight head pumping and assumed a prone posture, and at other times attempted to escape by diving, whereupon she would be chased and usually caught by the male. After treading, the male typically held the female's nape for several seconds as he held his wings slightly spread in the usual display posture. After releasing the female, the male shook its tail and swam in a circle around her while in this display posture, then finally bathed and flapped its wings.

Reproductive biology. The breeding season is generally associated with the rainy season. In India, it occurs during the monsoon period, mainly from July to September. Ceylon records of breeding are for February and March. In Zambia, records extend from November to March, but the great majority are for January and February. Most west African records

MAP 51. Breeding or residential distributions of the South American ("S") and Old World ("O") comb ducks.

are for August and September, but relatively little is known of its breeding times in the northern and western parts of its African range. Likewise, almost nothing is known of breeding periods in South America. A variety of nesting sites are reportedly used, and ground locations are said to be used more commonly than tree hollows. Tall grass cover near water, rocky hills, and hollows of trees or stumps have all been found as nesting sites. Tree hollows seem to be used more frequently in India and South Africa than in northern parts of Africa; mango and banyan trees are said to be especially favored. Some observers have reported seeing both sexes together on nest-site searches, which casts doubt on the idea that pair bonding is essentially absent in comb ducks. As in other tree-nesting ducks, clutch sizes seem to be fairly large and are often further inflated by dump-nesting tendencies. Records from captivity suggest a normal clutch range of 8 to 12 eggs, and an incubation period of 30 days (Johnstone, 1970). Incubation is the sole responsibility of the female. There is no indication that more than one brood is raised per year, even where breeding seasons are prolonged (Clancey, 1967).

Status. This species is apparently still common over most of its range and in general is not highly sought by sportsmen because of its poor palatability. It is apparently now extirpated in Ceylon, where at one time it was a resident (Ali & Ripley, 1968).

Relationships. Most of the morphological evidence suggests that this species is closely related to *Cairina*, but Woolfenden (1961) found a number of unique skeletal features that he felt deserve separation from all other perching ducks. Perhaps *Sarkidiornis* represents the best living link between the more primitive perching ducks previously considered and the more specialized forms such as *Chenonetta* and *Nettapus*.

Suggested readings. Clancey, 1967; Ali & Ripley, 1968.

150 ❖ ❖ ❖

Hartlaub Duck

Pteronetta hartlaubi (Cassin) 1859

Other vernacular names. Hartlaub's teal; Hartlaubsente (German); canard de Hartlaub (French); pato de Hartlaub (Spanish).

Subspecies and ranges. No subspecies recognized here (*albifrons* is sometimes recognized as an eastern race). Resident from Guinea west to Zaire and southwestern Sudan. See map 52.

Measurements and weights. Folded wing: males, 263–82 mm; females, 248–66 mm. Culmen: males, 46–50 mm; females, 44–48 mm. Weights: both sexes, 800–940 g. Eggs: av. 55 x 42 mm, cream, 51 g.

Identification and field marks. Length 22–23″ (56–58 cm). Plate 32. The *adult male* has a head that is black except for a variably large white area behind the bill (sometimes extending to the nape and chin) and a body that is almost uniformly rich chestnut brown. The tail and rump are olive brown, as is the upper wing surface except for the upper coverts, which are bluish gray. The bill is black, and is seasonally enlarged at the base, with pale pink to yellowish markings near the tip and below each nostril. The legs and feet are dusky brown. *Females* closely resemble males, but the bill is not enlarged basally and the pale markings are pinkish gray rather than yellow. *Juveniles* have straw-colored tips on the breast and abdominal feathers.

In the field, no other species of forest-inhabiting duck in Africa is predominantly chestnut-colored. Females have a loud quacking call that is uttered frequently, and males produce a quiet, high-pitched wheezing noise. The bluish upper wing coverts are no doubt evident in flight, but unlike those of *Cairina,* the under wing coverts are brown.

Natural History

Habitat and foods. Throughout the year the Hartlaub duck is found in the rain forests and adjacent gallery forests of western Africa. Small forest brooks that may be hidden beneath the adjoining trees, rather than large rivers, are the birds' favored habitats, and they spend a good deal of time perching in the trees. Not a great deal is known of their foods, but a small sample contained such food items as aquatic insect larvae (mainly dragonflies), freshwater snails, small bivalve mollusks, shrimps, a spider, and many small seeds. It has been reported that the ducks also eat the roots of cassava that are placed by natives in streams for soaking (Chapin, 1932).

Social behavior. Generally this species has been reported in pairs and very small groups, with 15 being about the largest number ever reported in a flock. It seems likely that the breeding season is greatly extended, and thus the breeding birds are likely to remain in or near their territories throughout the year. At least, no migrations or marked seasonal changes of abundance have yet been noted. What little flocking has been noted occurred on the open waters of some larger rivers and presumably was done by molting birds. The time of nesting is still poorly known. Eggs or broods have been reported in September in the Sudan and November in Cameroon, apparently coinciding with the wet season.

Reproductive biology. Evidently no nests have yet been found in the wild, but the birds presumably nest in tree hollows. The first attempted nesting in captivity occurred in 1958 at the Wildfowl Trust, and a second attempt in 1959 was successful. In 1958 three clutches totaling 24 eggs were laid, while in 1959 a clutch of 9 was produced. Only the female incubated, and the incubation period was established to be 32 days. Both sexes were very aggressive in the defense

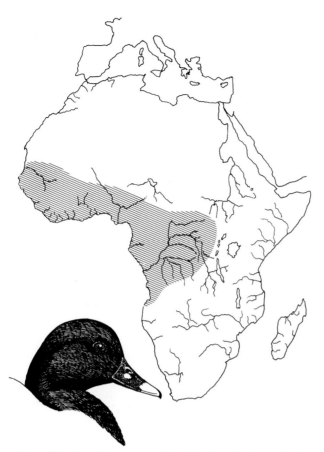

MAP 52. Breeding or residential distribution of the Hartlaub duck.

of their brood; in this respect the Hartlaub duck seems to differ strongly from *Cairina*. However, after the ducklings became fully feathered at eight weeks of age, the male began to threaten them, which is of interest considering the absence of a well-defined juvenile plumage in this species.

Status. The true status of this rain forest bird is hard to judge, owing to its inaccessible habitat. No doubt it will survive only as long as these rain forests remain relatively undisturbed.

Relationships. Although often included in the genus *Cairina*, the numerous plumage and behavior differences present in this species seem to warrant the retention of a monotypic genus for it (Johnsgard, 1965a). The skeleton of this species also shows a number of peculiarities that suggest generic distinction (Woolfenden, 1961).

Suggested readings. Chapin, 1932; Phillips, 1923–26.

Green Pygmy Goose

Nettapus pulchellus Gould 1842

Other vernacular names. Green goose; Grüne Zwergglanzente (German); sarcelle pygmée d'Australie (French); ganso pigmeo verde (Spanish).

Subspecies and range. No subspecies recognized. Resident in Ceram, Buru, southern New Guinea, and northern Australia from the Fitzroy River to Rockhampton, Queensland. See map 53.

Measurements and weights. Folded wing: males, 160–80 mm; females, 150–80 mm. Culmen: males, 23–28 mm; females, 21–29 mm. Weight: males, 300–430 g (av. 310 g); females, 245–340 g (av. 304 g). Eggs: 44 x 32 mm, creamy white, 30 g.

Identification and field marks. Length 13″ (33 cm). *Adults* have a brown and iridescent green crown, while the cheeks and throat are white (some birds have a gray mark extending below the eyes). The nape and neck are also green, and the green extends posteriorly over the mantle and scapulars to the tail coverts, which are finely barred greenish black and white. The upper breast and flanks are broadly barred with white and dark green or brown, and the lower breast and abdomen are white. The tail and wing are mostly blackish, except that the secondaries are mostly white, forming a conspicuous speculum. The bill is mostly grayish black except underneath and at the nail, where it is pinkish, and the legs and feet are greenish gray. *Females* are slightly smaller and duller than males, with less distinctive flank barring, and the green of the neck is not continuous around the front. *Immatures* initially resemble females, but are heavily spotted with brown on the face, chin, and neck.

In the field, this species is most likely to be confused with the cotton pygmy goose, but that species lacks barred flanks and has a white or mostly white nape and hind neck. Both sexes of pygmy geese have relatively high-pitched calls, which may be shrill or pure whistling sounds.

NATURAL HISTORY

Habitat and foods. In Australia, the favorite habitats of this species are tropical lagoons covered with water lilies and submerged aquatic vegetation. These are typically permanent and fresh-water areas in coastal or subcoastal regions. During the dry season the birds are more widespread, and sometimes gather in considerable numbers on some larger lakes where water lilies are scarce or absent. Streams, shallow pools, and brackish waters are avoided. In New

Guinea, where the swamps are often greatly overgrown with emergent vegetation, the birds are more often found in the open and deeper parts of such waters. The favored food wherever it is available is the water lily; budding flower heads and seeds of lilies are avidly eaten whenever they are present. When lilies are not present, the birds make shallow dives to obtain leafy materials of such aquatic plants as muskgrass and pondweeds, or they may swim near shore to strip the seed heads of shoreline grasses, especially wild millet (*Echinochloa*). Virtually no animal materials are consumed except those that might be ingested by accident (Frith, 1967).

Social behavior. Throughout the year these small ducks are found in pairs or in small groups that may represent family units. The pair bond is obviously strong, and Frith (1967) suggests that it may be per-

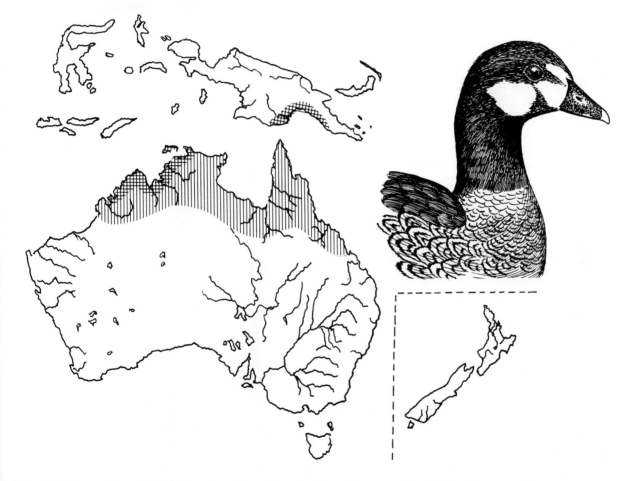

MAP 53. Primary (cross-hatched) and peripheral (hatched) breeding or residential distributions of the green pygmy goose.

manent, since males have been seen evicting other males from the vicinity of their mates at all times of the year. Males also strongly defend small feeding areas, and in spite of their size fight frequently. Seasonal movements are evidently on a very small scale and are associated with the availability of foraging areas during the dry season. Some dry-season refuges, such as the East Alligator and Adelaide rivers, may support several hundred birds seasonally. Breeding is evidently timed to coincide with the wet season, when the highest water levels are just past and the lagoons are either at their maximum size or are starting to decline again. Thus, sexual displays begin about October, at the onset of the rains, so that nesting can get underway by January. Courtship displays of the male are still only poorly known but appear to be surprisingly simple, consisting of neck stretching, preening the wing, and head turning. Copulatory behavior has not yet been described (Frith, 1967).

Reproductive biology. Near Darwin, the nesting season probably extends from January until March, and is no doubt preceded by territorial occupancy and defense by the male. The male also reportedly takes the initiative in searching for nest sites by flying up into trees and sitting near hollows. These are then examined by the female, who may visit several potential sites before settling on one of them. The usual clutch size is still uncertain, but probably ranges between 8 and 12. Likewise, the incubation period has not been established yet. It is known that the male remains on the territory through the incubation period and rejoins his mate and the brood at the time of hatching. They spend the daylight hours in fairly deep water, but forage in the shallows in early morning and late afternoon. In deep water, ducklings have been observed to dive well for food, and they also dive when disturbed, presumably coming up under water lily leaves. When swimming away from danger, the female takes the lead and the male follows behind the brood (Frith, 1967).

Status. According to Frith, this species is not numerous anywhere, but is common in favored habitats of its restricted range. Since the heart of its range is not now in any danger of development or destruction, the species appears to be relatively safe.

Relationships. The genus *Nettapus* is obviously fairly closely related to *Chenonetta*, judging from their downy plumage patterns, but the bill-shape similarities of these two genera are quite possibly the result of convergent evolution. A number of skeletal similarities also exist between *Nettapus* and *Chenonetta* that suggest close relationships (Woolfenden, 1961).

Suggested readings. Frith, 1967.

Cotton Pygmy Goose

Nettapus coromandelianus (Gmelin) 1798

Other vernacular names. Cotton teal, white pygmy goose, white-quilled pygmy goose; Zwergglanzente (German); sarcelle de Coromandel (French); ganso pigmeo de la India (Spanish).

Subspecies and ranges. (See map 54.)
 N. c. coromandelianus: Indian pygmy goose. Resident in India, Ceylon, and Burma east to southern China, south to the Malay Peninsula and the Indo-Chinese countries, northern Luzon, Borneo, Sumatra, Java, northern Celebes, and northern New Guinea.
 N. c. albipennis: Australian pygmy goose. Resident in eastern Australia, mostly between Cape Melville and Cape Townsend.

Measurements and weights. Folded wing: males, 172–88 mm; females, 161–86 mm. Culmen: males, 25–26 mm; females, 23–26 mm. Weights: males, 311–430 g (av. 403 g); females, 255–439 g (av. 380 g). Eggs: 47 x 35 mm, creamy white, 27 g.

Identification and field marks. Length 13″ (33 cm). *Adult males* in breeding plumage have a blackish crown and a narrow black eye-ring, while the rest of the head, neck, breast, and underparts are white, except for a narrow breast band of black. The back and scapulars are iridescent green, while the flanks and tail coverts are peppered with black and white and the tail is black. The upper and lower surfaces of the wings are dark except for a large white speculum formed by the tips of the secondaries and a subterminal band across the primaries. The iris is red, and the legs, feet, and bill are blackish. *Males in nonbreeding plumage* resemble females but have a more extensive white wing speculum. *Females* have a brown iris, lack a breast band, and exhibit less iridescent color dorsally. *Immatures* resemble females, but have no iridescence and have a more distinct dark stripe through the eye.

In the field, this tiny duck could be confused only with the green pygmy goose where these species overlap, and differs from it in having unbarred flanks, more extensive white on the wings (extending to the primaries), and a unicolored white or gray rather than bicolored hind neck and cheek pattern. When in flight, this species utters series of rapid staccato cackling notes sounding like "Fix bayonets, fix bayonets," while green pygmy geese produce a continuous chirrup sound. The female is said to also have a soft quacking note, but the vocalizations of this species are only poorly known.

NATURAL HISTORY

Habitat and foods. In Australia this species primarily inhabits deep lagoons that are associated with rivers and creeks in the tropical coastal regions, or similar fresh-water ponds that are well vegetated with water lilies and various other submerged and floating-leaf aquatic plants. In India it likewise favors lagoons, jheels, and ponds that are extensively vegetated. Sometimes lakes are also used, as are swamps and creeks, depending on the nature of their associated plant life. Swamps and creeks are largely wet-season habitats, while rivers and lakes are utilized primarily during the latter part of the dry season. In spite of the similarities in their bill shapes, the cotton pygmy goose and green pygmy goose seem to have marked differences in food preferences, with the water lilies being of little importance to the former, which prefers pondweeds, hydrilla (*Hydrilla*), and the seeds of aquatic and shoreline grasses. In neither species are insects or other animal foods of substantive significance (Frith, 1967).

Social behavior. In India this species is usually to be found in pairs during the breeding season but occurs in small flocks at other times, and on rare occasions has been seen in numbers of up to as many as 500 birds (Ali & Ripley, 1968). In Australia, dry-season concentrations of up to several hundred birds occur in a few areas such as near Townsville, but again, most observations are of pairs or small groups. There are some regular seasonal movements in Australia associated with seasonal variations in water, but the distances and number of birds involved are apparently small in comparison with some of the nomadic

and more mobile species. Definite seasonal migrations are more typical on the Asian continent, especially at the northern part of the range in China. Little has been recorded on the time of pair formation or the nature of the pair bond in this species, but it is assumed that pair bonds are well developed, as in the other pygmy geese. Likewise, courtship displays have not been adequately described. One early observer reported a male arching his neck and jerking open his wings to expose the white patch, but it seems probable that the displays are more complex and varied than this. The precopulatory behavior consists of mutual bill dipping while the birds face each other, and the male attempts to mount the female. The probable major postcopulatory display consists of the male arching his neck, partially spreading his folded wings, and turning to face the bathing female (Johnsgard, 1965a).

Reproductive biology. In northern India, the breeding season extends from June to September, coinciding with the wet season, but is concentrated in July and August, while in Ceylon it occurs from February to August (Ali & Ripley, 1968). In Australia the breeding period is mainly from January to March, or during the latter half of the wet season (Frith, 1967). Tree hollows are evidently the preferred nesting sites in most or all parts of the range, but a variety of other elevated locations have been used. These include holes in buildings or other man-made structures. As with the green pygmy geese, males typically help their mates search for suitable sites, remaining on nearby branches as the female inspects each prospective hollow. The nest is often from 2 to 5 meters above the ground, and may be as high as 20 meters. The clutch size is variable and generally estimated to range from about 6 or 8 to 14 or 15 eggs. As in the green pygmy goose, the incubation period is unknown; an early estimate of 15 or 16 days is obviously erroneous. Very probably only the female incubates, and the possible role of the male in brood care is still not clear. The raising of two broods

MAP 54. Breeding or residential ranges of the Australian ("A") and Indian ("I") cotton pygmy geese.

per year has been suggested for Ceylon and also for the lower Yangtze Valley but has not been confirmed (Phillips, 1923–26).

Status. Over much of its range this species is still very common, although very locally distributed, as is the case in India, Burma, and Ceylon. It is uncommon on the Malay Peninsula, and its status in Thailand and Vietnam is unknown. It seems to be quite local and fairly rare over the East Indian islands on which it occurs, and in Australia it is largely limited to a small part of Queensland. The Australian race is thus the most vulnerable to possible loss, and its status there should be closely monitored (Frith, 1967).

Relationships. The relationships within the genus *Nettapus* are not very obvious; all three species seem to be fairly distinct from one another. This is the most widespread species, and the only one that

shows marked seasonal differences in male plumage patterns. The female's plumage shows some similarities to that of the *Sarkidiornis* female, and a general similarity in proportions to *Chenonetta* is also evident among pygmy geese.

Suggested readings. Ali & Ripley, 1968; Frith, 1967; Phillips, 1922–26.

African Pygmy Goose

Nettapus auritus (Boddaert) 1783

Other vernacular names. Dwarf goose; Afrikanische Zwergglanzente (German); sarcelle pygmée d'Afrique (French); ganse pigmeo africano (Spanish).

Subspecies and range. No subspecies recognized. Resident in Africa from Senegal and Ethiopia south to the Cape, and on Madagascar. See map 55.

Measurements and weights. Folded wing: males, 150–65 mm; females, 142–58 mm. Culmen: males, 25–27 mm; females, 23–25 mm. Weights: males, 285 g (Kolbe, 1972); females, 260 g (Lack, 1968). Eggs: 43 x 33 mm, creamy white, 23 g.

Identification and field marks. Length 11–12" (28–30 cm). Plate 33. *Adult males* have a white face and foreneck bounded behind with iridescent greenish black on the crown and an oval sea-green patch on the side of the neck and head that is bordered in front by black. The breast and flanks are tawny brown, the abdomen is white, and the scapulars and back are greenish black, while the tail and tail coverts are blackish. The wing is mostly black to greenish black, except for the secondaries, which become progressively more tipped with white from the inside out, until the outermost ones are entirely white on the exposed vanes. The bill is yellow, with a black nail, and the iris is brown or reddish. *Females* resemble males but lack the distinctive green and white facial pattern, and instead are mostly grayish on the head, with a dark eye-stripe and darker crown and hind neck. The bill is also duller yellow. *Immatures* are like the adult female, but have a more distinct eye-stripe and are buffier on the breast and flanks.

In the field, this duck's tiny size and brownish breast and flank coloration set it apart from all other African waterfowl. It is relatively silent, but the male

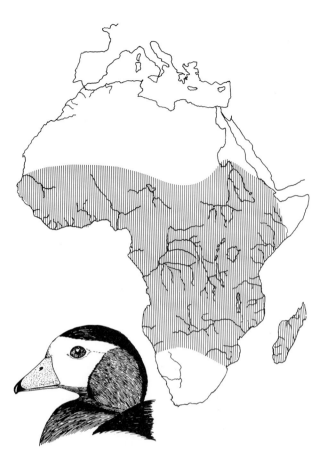

MAP 55. Breeding or residential range of the African pygmy goose.

produces several high-pitched whistling or twittering sounds, and the female utters a weak quack.

NATURAL HISTORY

Habitat and foods. In Africa this species is found on quiet backwaters of slowly flowing rivers, on shallow bays of lakes, and on marshes and swamps wherever there is an ample growth of water lilies (*Nymphaea*) and associated aquatic plants. Like that of the green pygmy goose, its distribution probably closely coincides with the distribution of water lilies. What little is known of its foods indicates that the seeds and other parts of water lilies are its primary items of diet. No doubt other aquatic plants are also consumed at times, and reportedly even some aquatic insects and their larvae are eaten, but these must be only secondary foods. The birds are known to forage during the daylight hours, and are able to dive very well during foraging or when escaping from danger.

❖ ❖ ❖ 157

They also perch well, and usually roost on partially submerged trees (Clancey, 1967).

Social behavior. Most accounts indicate that these birds are normally found in pairs or at most in small groups. Flocks numbering in the hundreds have been reported from near Freetown, in Sierra Leone. There are some movements associated with rainfall patterns, but in general the birds are fairly sedentary. Pair-bond patterns are not well understood, but in general appear to be well developed. Courtship displays are likewise almost wholly undescribed, and one of the few available descriptions suggests that the male swims "proudly" near the female, exposing his colorful head pattern to her view. Alder (1963) states that when displaying to its mate on land, the male holds its bill well down on its breast and utters a musical *chip, chip, chirrup, chiroo,* in a songlike manner. The female responds to this "song" with a sharper, twittering whistle while bobbing her head up and down. The same call and head movement are made when her mate returns from repelling an intruder.

Reproductive biology. The time of breeding in this species appears to be quite variable. In Madagascar it is during February and March, and on Pemba Island (off Tanzania) it is July and August. Records for Zambia range from September to March, but are concentrated in January and February, and Uganda's records are for February and March, as well as July to October. Breeding probably takes place in Nigeria in August and September, and in the southern Congo from July to October. Nests are usually in hollow trees, but have also been reported in cliff holes, termite hills, and even in the thatched roof of a hut. The birds are also reported sometimes to use the old nests of hammerhead storks (*Scopus umbretta*). Elevated nests may be as high as 20 meters above the ground, and on the other hand, the bird has been found nesting on a river bank in a thick clump of grass (Clancey, 1967). The clutch size ranges from about 6 to 12 eggs. So far as is known, the male plays no role in incubation, which has recently been determined to require 23½ to 24 days (Zaloumis, 1976). However, the fledging period and the possible role of the male in brood care are still unknown.

Status. Over much of its range this species is of local or sporadic occurrence; it is fairly common in Mozambique, northern Botswana, and Zambia. Although regarded as excellent table fare, its small size makes it unimportant as a game species.

Relationships. See the preceding two accounts for comments on evolutionary relationships.

Suggested readings. Clancey, 1967.

Ringed Teal

Callonetta leucophrys (Vieillot) 1816

Other vernacular names. Ring-necked teal, red-shouldered teal; Rotschulterente (German); sarcelle à collier (French); cerceta di collar (Spanish).

Subspecies and range. No subspecies recognized. Ranges from eastern Bolivia, Paraguay, and southwestern and southern Brazil to northwestern Argentina and Uruguay, but known to breed only in northwestern Argentina and Paraguay. See map 56.

Measurements and weights. Folded wing: males, 165–70 mm; females, 160–75 mm. Culmen: males, 36–37 mm; females, 34–36 mm. Weights: males, 190–360 g; females, 197–310 g (Weller, 1968a; Kolbe, 1972). Eggs: 45 x 36 mm, white, 32 g.

Identification and field marks. Length 14–15" (35–38 cm). Plate 34. *Males* have a finely streaked gray head, with a black stripe from the crown over the hind neck, and extending forward over the base of the neck to form a half ring. The breast is buffy pink with small black spots, the abdomen and sides are gray, the latter finely vermiculated with black, and the dorsal surface is olive brown centrally and chestnut brown laterally. The rump, tail, and tail coverts are black except for a white patch on the sides of the rump, and the wings are mostly black except for a white oval formed by the secondary coverts and iridescent green secondaries. The legs and feet are pink and the bill is gray. *Females* are quite different, with a mottled gray, buff, and brown pattern that lacks vermiculations. The head is brown, with a buffy white streak above the eyes and white cheeks and throat except for a brownish "fingerprint" in the ear region. The breast and flanks are mottled and barred with brown and gray, and the upperparts are generally olive brown. The wing is like that of the male. *Immature males* resemble females but lack a distinctive facial pattern and barred flanks. *Immature females* very closely resemble adult females.

In the field, the small size, perching tendencies, and distinctive green and white speculum should serve to identify this species. The Brazilian teal is much the same size and also perches, but its speculum has the white portion behind the green, rather than ahead of the green, as in this species. The usual call of the male is a soft, wheezy whistle, *wheee'-ooo*, and is apparently used only in courtship. The female's loudest call is a sharp, catlike *hou-iii* that rises sharply in pitch. Their wings produce a whistling noise in flight.

NATURAL HISTORY

Habitat and foods. Little is known of this species' ecology, but it inhabits tropical, mostly forested country, sometimes being found on isolated forest ponds or brooks. However, its preferred habitat seems to be periodically flooded areas that are covered with cutgrass (*Zizaniopsis*) and marshes surrounded by xerophytic forests (Olrog, 1968). It perches very well, and when foraging spends all of its time swimming in shallow water, picking objects from the surface or dipping its head below water. I have never observed it diving for food, and the bill shape suggests that small aquatic seeds are likely to be the major food.

Social behavior. Although social behavior has not yet been studied under natural conditions, observations on captive birds suggest that pair bonds in the ringed teal are strong and probably relatively permanent. The birds are fairly migratory, and move northward several hundred miles into southern Brazil during the nonbreeding season. Flock sizes at this time do not seem to have been noted. During flood years, the birds sometimes reach as far east as Cape

San Antonio, where they are seen as individual pairs (Phillips, 1922–26). The courtship display of the male is relatively simple, as is typical of many perching ducks, and consists of his bill-tossing courtship call, uttered either on water or on land. A rapid preening behind the wing, quickly exposing and again hiding the iridescent speculum, is another of the male's displays. Much of the time the male simply swims beside the female and turns to face her whenever the opportunity arises. Copulation occurs on water of swimming depth, and is preceded by slight head-bobbing or bill-dipping movements on the part of the male. The female usually becomes prone for some time prior to mounting by the male; and after treading is completed, the male utters a single call with his bill-tossing movement and turns and faces the female

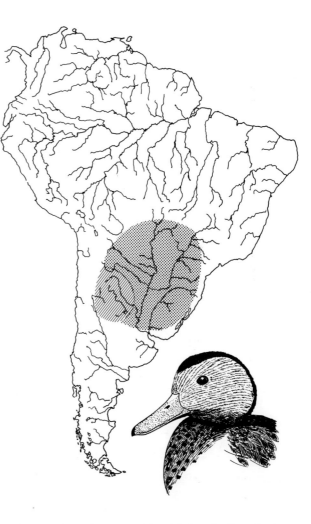

MAP 56. Breeding or residential (hatched) and wintering (stippling) distributions of the ringed teal.

in a motionless posture for several seconds (Johnsgard, 1965a).

Reproductive biology. Almost nothing is known of this species' breeding under natural conditions, but Olrog (1968) states that the birds usually nest in the large stick nests of the monk parakeet (*Myiopsitta monarcha*). The nesting season is not certain, but Weller (1968a) collected two males in May that had just completed their wing molts, and Wetmore (1926) indicated that birds in Paraguay were in breeding condition in September. The collection of young there in January and February also suggests late summer and fall breeding. In captivity, females prefer to nest in cavities, and the clutch size is from 6 to 12 eggs. The incubation period is 26 to 28 days (not 23 as sometimes reported), and incubation presumably is carried out entirely by the female. However, males remain close at hand and even enter the nesting hole either to help incubate or to protect the female. The male is more concerned with parental care of the brood than is the female (E. Dale Crider, pers. comm.).

Status. The status of this bird is poorly known, although there is no indication that it is unusually rare.

Relationships. I have reviewed (1960a) the anatomical and behavioral traits of this species, which has in general been placed in the genus *Anas*, and concluded that it should be considered a monotypic genus of perching duck. This suggestion was supported by Woolfenden (1961) on the basis of the species' skeletal anatomy. Both of us concluded that *Amazonetta* is probably the nearest living relative of *Callonetta*. However, the feather proteins of the ringed teal are definitely *Anas*-like, with lesser similarities to *Amazonetta* (Brush, 1976).

Suggested readings. Johnsgard, 1960a, 1965a.

North American Wood Duck

Aix sponsa (Linnaeus) 1758

Other vernacular names. Carolina duck, woodie; Brautente (German); canard carolin (French); pato del bosque de Carolina (Spanish).

Subspecies and range. No subspecies recognized. Breeds in North America from British Columbia to California, and from Lake Winnipeg eastward to southern Nova Scotia, and south to Texas, Florida, and Cuba. Winters in the southern half of its breeding range. See map 57.

Measurements and weights. Folded wing: males, 250–85 mm; females, 208–30 mm. Culmen: males, 30–35 mm; females, 550–680 mm. Weights: males, 539–879 g (av. 680 g); females, 482–879 g (av. 539 g). Eggs: av. 52 x 40 mm, whitish, 44 g.

Identification and field marks. Length 17–20" (43–51 cm). *Males* in breeding plumage are unmistakable, with a strongly crested head that has mostly iridescent colors on the crown and sides, but with a white throat, a white stripe leading from the eye to the end of the crest, and a second, narrower white stripe from the base of the bill to the tip of the crest. The upper wing surface and back are also mostly iridescent with a blue and green sheen; the breast is maroon with white spotting, separated from vermiculated yellow flanks by vertical black and white bars. Narrow black and white markings occur on the upper edge of the flanks, and the abdomen is white. The tail coverts are brown, blackish, and maroon, while the tail is iridescent greenish black. The upper wing surface is mostly iridescent blue or greenish blue, but the secondaries are narrowly tipped with white and the primaries have a silvery white sheen on the outer vanes. The eyes are red, the legs and feet are yellow, and the bill is brightly patterned with black, white, and red. *Males in eclipse* plumage resemble females but have a pinkish bill and retain their white cheek and throat pattern. *Females* are mostly olive brown above, with large whitish spotting on the flanks and breast, grading to white on the underparts. The head is generally grayish on the sides and greenish black on the crown, with a large white eye-ring that extends posteriorly as a pronounced eye-stripe, and a whitish throat. The upper wing surface generally resembles that of the male, but the iridescence is not so highly developed and the white spots at the tips of the secondaries are larger. The iris is dark brown, the bill is blackish, and the legs and feet are like the male's. *Immatures* resemble adult females, but have streaked and mottled brown bellies, and immature males soon begin to exhibit the white chin and throat pattern typical of adult birds.

In the field, the crested head and the long tail, which is carried high above the water by swimming birds, are distinctive field marks for both sexes.

Wood ducks in flight exhibit a generally grayish underwing surface, with a narrow white trailing edge on the secondaries. The birds move their heads almost constantly when in flight, producing a unique "rubber-necked" appearance. The female's flight call is a drawn-out and owllike *u-ih*, and her courtship note is a sharp-noted whistle; neither sex exhibits typical quacking notes.

NATURAL HISTORY

Habitat and foods. The breeding habitat of wood ducks is typically one of fresh-water areas such as sloughs, ponds, and slowly moving rivers, where large hardwoods such as oaks, cottonwoods, and willows are present. Trees large enough to have cavities with openings at least 3.5 inches wide and interiors at least 8 inches in diameter are needed for nesting sites. Preferred foods include nuts, particularly acorns, hickory nuts, and beechnuts, and the seeds of various floating-leaf aquatic plants, including water lilies (*Nuphar, Nymphaea*). The birds also consume large quantities of the vegetative parts and seeds of other aquatic plants, including pondweeds, wild rice, arrow arum, and duckweeds. Oak

species that produce small acorns are utilized more often than those producing large ones, and the acorns are either gleaned from the forest litter or plucked from the trees before they have fallen. When foraging on the water, wood ducks often tip up and can gather materials from water about as deep as 18 inches, but they only very rarely resort to diving for food (Johnsgard, 1975).

Social behavior. Wood ducks are highly social only on migration and in their wintering areas, where common roosting sites are used and the number of birds using them may range from fewer than 100 to several thousand, with a single instance of a roost being used by 5,400 birds (Hester & Quay, 1961). By the time they arrive at their breeding grounds the birds are already paired and usually in groups of no more than a dozen. Territorial boundaries are apparently lacking among wood ducks, and although males defend their mates from the attentions of other males, they freely share common ponds. Nesting concentrations are limited by available nest sites rather than territorial needs. Pair formation occurs in wintering areas, and is preceded by a period of intense courtship display, most of which occurs on water. Inciting by the female is an integral compo-

MAP 57. Breeding (hatched) and wintering (stippling) distributions of the North American wood duck.

nent of this process, and allows each female to "choose" a particular male, which usually responds to such inciting by a variety of display postures and calls. These include mock drinking, ritualized or display preening of the iridescent wing feathers, whistling with erected crest and stretched neck (burping), and various display shakes. Turning of the back of the head toward the inciting female is apparently a fundamental component of pair-forming behavior (Johnsgard, 1960b). Copulation is preceded by the female's assuming a prone posture on the water, without obvious prior display activity, and by the male's swimming about her, performing drinking, bill-dipping, or gentle pecking movements. After treading is completed, the male first swims rapidly away from the female while turning the back of the head toward her, and then turns and faces the female as she bathes (Johnsgard, 1965a).

Reproductive biology. Shortly after the pair arrives at its nesting area, they begin to seek out a nesting site. Males accompany their mates on these flights and may help locate potential nesting holes, but the female apparently is responsible for investigating and choosing the specific cavity to be used. Wood ducks prefer trees having natural cavities that are located fairly high and with entrances too small for raccoons to enter. Cavities in trees over water are preferred to those over land, and locations in groves or rows of trees are preferred over solitary tree sites. Clutch sizes tend to be large, and are often inflated further by dump nesting. Probably from 13 to 15 eggs represent a normal clutch, with the eggs laid on a daily basis. Incubation is performed only by the female and on the average requires 30 days. Generally, the male will remain with his mate until about the fourth week of incubation, at which time he deserts her to begin his postnuptial molt. When hatching is finished, the female usually keeps her brood in the nest through one night, then calls them out of the nest the following morning. After jumping to the ground, the female leads them to the nearest water. Little contact between broods occurs during the first two weeks of life, but thereafter brood mergers become increasingly frequent. Females probably normally leave their broods and begin their own molt between six and seven weeks after the young have hatched, and the young complete the rest of their 60-day fledging period on their own (Grice & Rogers, 1965; Johnsgard, 1975).

Status. The population status of the wood duck was extremely serious in the late 1930s, but an intensive program of nest-box erection and protection has brought the species back to a generally excellent status, and its range continues to expand into the Midwest. Bellrose (1976) summarized recent breeding population estimates that suggest a total population of more than 1.3 million birds, with Ontario having the largest single component. Probably the increase in beaver ponds and nesting cavities due to the activities of pileated woodpeckers has supplemented hunting restrictions to bring about Ontario's population increase, but in the southern part of the breeding range increased swamp drainage and forest clearing have caused population reverses in recent years.

Relationships. The strong plumage and behavioral similarities between the wood duck and mandarin duck argue for their close relationships, in spite of the fact that they only very rarely successfully hybridize (Johnsgard, 1968b).

Suggested readings. Grice & Rogers, 1965; Bellrose, 1976; Johnsgard, 1975.

Mandarin Duck

Aix galericulata (Linnaeus) 1758

Other vernacular names. None in general English use. Mandarinente (German); canard mandarin (French); pato mandarín (Spanish).

Subspecies and range. No subspecies recognized. Breeds in eastern Asia from the Amur and Ussuri rivers through Korea, eastern China, and Japan to the Ryukyu Islands (Okinawa). Feral in Great Britain and northern Europe. Winters in the southern part of its breeding range and south to southeastern China, rarely to Taiwan. See map 58.

Measurements and weights. Folded wing: males, 210–45 mm; females, 217–35 mm. Culmen: males, 26–32 mm; females, 26–30 mm. Weights: both sexes combined, 444–550 g (Tso-hsin, 1963). Eggs: av. 49 x 36 mm, whitish cream, 41 g.

Identification and field marks. Length 17–20" (43–51 cm). Plate 35. *Males* in breeding plumage have an iridescent crown that extends to a long crest, bounded below by a white to buff area extending from the bill to the crest tip, and downward to below the eyes, where it merges with chestnut cheeks and extended hackles. The breast is a rich maroon, separated from the flanks by three black and two white vertical stripes; the abdomen and under tail coverts are white; and the flanks are a finely vermiculated gold and black. The back and tail are olive brown, while the longer upper tail coverts are bluish green. The scapulars are mostly iridescent blue, as are the tertials, except for the outermost one (12th secondary), which is iridescent on its outer web but orange gold and greatly enlarged on the inner web into a saillike shape. The other secondaries are iridescent green, tipped with white, while the rest of the upper wing surface is mostly olive brown. The bill is red, with a whitish nail, and the feet and legs are yellow. *Eclipse-plumage males* resemble females, but have a reddish bill color. *Females* have gray heads, with a white eye-ring that extends narrowly backwards into an eye-stripe, a buff and brown mottled breast and flank pattern, white underparts, and a greenish brown upperpart coloration. Their wing pattern resembles that of the male, but lacks the specialized "sail" feather. The bill is grayish black, and the legs and feet reddish yellow. *Immature* males closely resemble females, but have a pinkish bill.

In the field, mandarins are unlikely to be confused with any other species when males are in full plumage. Females closely resemble female wood ducks, but have a lighter gray head color and a less distinctive white eye-ring and eye-stripe. The male produces a nasal whistling note in display, and the female a brief, high-pitched courtship call, *kett,* similar to the wood duck's, as well as other, softer notes. A specific flight call other than quacking has not been mentioned in the literature.

NATURAL HISTORY

Habitat and foods. The mandarin's typical breeding habitat is river valleys with wooded islands, forest lakelets with willow-lined banks, and small forest ponds. Swamps surrounded by broad-leaved trees, with reed-covered water surfaces, are also favored habitats. During the daytime mandarins spend much time in shady areas, such as river banks, and are most active in foraging at dawn and dusk. Their foods are quite variable with season and locality, but acorns are a favorite fall food, as well as cultivated grains such as buckwheat and rice. Spring foods are also variable, and include insects, snails, small fish, and vegetation. Seeds of aquatic plants, the roots and stems of such plants, shoots of horsetails (*Equisetum*), and the fruits or seeds of such terrestrial plants as grapes, roses, rhododendrons, and even pines have reportedly been found in mandarins (Dementiev & Gladkov, 1967; Tso-hsin, 1963).

Social behavior. Mandarins are highly social, and flock sizes may be quite large in winter, when groups of 100 or more are commonly seen. Drakes also col-

Map 58. Breeding (hatched) and wintering (stippling) distributions of the mandarin duck.

lect in groups during molting periods in late summer in some areas, but elsewhere reportedly remain solitary. Although seasonal, the pair bonds are very strong; and if both sexes remain alive through two breeding seasons, they re-form old pair bonds rather than establishing new ones. Courtship display begins in fall, and during the fall social display period incipient pair bonds may form slowly and subsequently be broken, whereas pairing during spring is much more direct and strong pair bonds may be formed in only a week or so (Bruggers, 1974). Displays are elaborate and complex, involving several ritualized shaking movements, display drinking, and display preening, the last display being restricted to the "sail" feather and oriented toward a specific female (Lorenz, 1951–53). Females take the initiative in choosing mates by orienting inciting behavior toward a preferred male. They also often initiate copulatory behavior, performing slight head-

pumping movements before extending themselves prone on the water. The male performs a series of bill-dipping movements before mounting, and after treading swims rapidly away from the female while orienting the back of his head toward her (Johnsgard, 1965a). Probably most pair formation occurs on the wintering grounds and during the spring migration back to the breeding areas.

Reproductive biology. In Manchuria and China the nesting season extends from late April to July; the birds have been reported to arrive already paired and ready to establish nesting territories. Females take the initiative in looking for suitable nests, but are always accompanied by the males as they examine tree hollows. Nearly all nests are in such locations, usually in trees near or overhanging water, and often in trees with trunks covered by grapevines. Sometimes tree stumps are used, and rarely the birds nest on the ground, under fallen trees or other dense vegetation. The clutch size ranges from 9 to 12 eggs, and in captivity averages 9.5 eggs. The incubation period is from 28 to 30 days, with the female incubating about 80 percent of the daylight hours and all through the night. She is joined by her mate during the short daytime breaks in incubation (Bruggers, 1974). Newly hatched young occur from late May until late July, and newly fledged birds have been seen as early as the first part of June. The fledging period has been reported as six weeks (Scott & Boyd, 1957). Males abandon their mates at some point before hatching and undergo their flightless molt at this time, while females molt at a somewhat later period (Dementiev & Gladkov, 1967).

Status. The mandarin is apparently still quite common over most of its native range, and is hunted in only a few areas, such as the Maritime territory of the U.S.S.R. The species is also now well established in parts of England (Beames, 1969), and feral birds also occur in Scotland and some parts of northern Europe.

Relationships. All evidence indicates a very close relationship between the wood duck and mandarin duck, both behaviorally and morphologically. There has been some controversy about whether the two species ever hybridize, but a recent review (Johnsgard, 1968b) indicates that this does occasionally occur. The genus *Aix* is probably very closely related to *Chenonetta* and also shows affinities with *Cairina* (Johnsgard, 1965a; Woolfenden, 1961).

Suggested readings. Dementiev & Gladkov, 1967; Savage, 1952; Bruggers, 1974.

Measurements and weights. Folded wing: males, 254–90 mm; females, 252–84 mm. Culmen: males, 24–31 mm; females, 22–31 mm. Weights: males, 700–955 g (av. 815 g); females, 662–984 g (av. 800 g). Eggs: av. 54 x 42 mm, creamy, 54 g.

Identification and field marks. Length 19–22″ (48–56 cm). *Males* have a brown head and neck, with a short black mane; a breast and upper back that is mottled gray, black, and buff; a black abdomen; finely vermiculated grayish sides and lower back; a black tail, tail coverts, and outer scapulars; and an upper wing with gray coverts, black primaries, and secondaries forming an iridescent green and white speculum, the green being bounded narrowly in front with white and with a white posterior border that increases in size outward. The bill, legs, and feet are olive brown. *Females* have a buff and dark brown head with distinct eye-stripes and cheek-stripes, and their breasts and flanks are also mottled with buff and brown; no vermiculations are present. *Immatures* closely resemble females but are lighter. Vermiculations on the flanks of males begin to appear at about three months of age, and provide initial outward sex criteria.

In the field, Australian wood ducks are unlikely to be confused with any other Australian waterfowl except perhaps the Australian shelduck, which, however, has white on the forewings rather than on the secondaries. The black hindquarters of male wood ducks also separate them from other, similar waterfowl. The courtship call of the male is a single-syllable, catlike *wee-ow* sound; and the most common of the female's calls is a loud, hoarse *whroo,*

Australian Wood Duck

Chenonetta jubata (Latham) 1801

Other vernacular names. Maned goose; Mähnengans (German); bernache à crinière (French); ganso de melena (Spanish).

Subspecies and range. No subspecies recognized. Resident throughout Australia, including Tasmania. See map 59.

lasting nearly a second, but she also produces repeated nasal *wonk* sounds when inciting.

NATURAL HISTORY

Habitat and foods. In Australia, wood ducks are likely to be found wherever lightly timbered country is adjacent to water and short grass or herbaceous cover is present under the trees. Although the birds favor the inland rivers of New South Wales, the water can be of almost any type so long as it is not saline or brackish. Likewise, the heaviest forests and densest swamps are avoided. The most important component is the presence of green herbage suitable for grazing, which is usually done at night. Almost no animal food is consumed, and most of the plant material is vegetative rather than seeds. A wide variety of grasses are consumed, and second to this group are the sedges. Broad-leaved herbs, especially smartweeds and legumes, make up the third component of the diet. Evidently only during the first month of life are insects and other invertebrate foods of significance to this species (Frith, 1967).

Social behavior. Wood ducks are quite gregarious, and flocks numbering as many as 2,000 or more individuals have been reported. Such flocking occurs after the breeding period; the flocks are evidently comprised of family groups. Such groups of birds are likely to be found in the same area year after year, forming "camps" that combine suitable roosting sites and available foraging areas within a few miles. The duration and size of such camps varies from year to

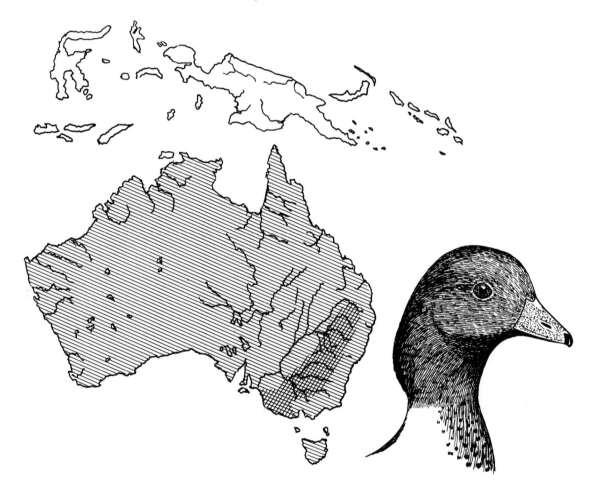

MAP 59. Residential distribution of the Australian wood duck, showing area of densest populations (cross-hatching) and total potential breeding range (hatching).

year, depending on local rainfall patterns, and the species may be better regarded as nomadic than migratory in the usual sense. Judging from banded birds, it appears that most seasonal movements are under 100 miles; less than 10 percent of the banded birds had moved over 200 miles when shot (Frith, 1967). Pair bonds apparently are strong in this species, and quite possibly are relatively permanent, as there is no seasonal variation in male plumages and breeding times are often erratic. The sexual displays are relatively simple, consisting of ritualized shaking movements by the male, performed either on land or in the water. The male also frequently utters his display call, or burp, while extending the neck and raising his short mane. Females incite their mates or potential mates to attack other birds, but very little fighting seems to be done. Display preening almost certainly occurs, judging from the presence of an elaborate speculum, but it has not yet been described. Copulation occurs on water, and is preceded by bill-dipping, head-dipping, or head-pumping movements by the male and perhaps also the female. The postcopulatory display consists of the male swimming around and away from the female with his hindquarters strongly raised and his head and neck extended forward (Johnsgard, 1965a).

Reproductive biology. To a very large extent, the breeding season of wood ducks in Australia is irregular and dependent upon the timing of rainfall, since it dictates when green grass will be available. In the southern highlands, where rainfall is fairly regular, breeding occurs in the spring, mainly September and October. However, farther north in New South Wales, the season is much more extended and may occur anytime during the year. However, it is concentrated between January and March, following the major rainy period. In the more arid inlands, the breeding is wholly controlled by rainfall and can occur anytime. In years when no rain of significance occurs, there is no breeding and the birds may leave the area to breed elsewhere. Mated pairs spend a good deal of time searching together for suitable nest sites. Females select hollows in trees, usually live ones at varying distances from water. At times they may be at least a mile from water, or they may directly overhang the edge of a lagoon. The hollow is lined with gray down, and a clutch ranging from 9 to 11 eggs is laid. The incubation period is 28 days, and only the female incubates, although the male remains close at hand. When there is no water near the nest, the male occupies the nearest available water area and keeps other males away from the vicinity. After hatching, the female and brood join the male, and as the young grow, the family gradually moves to other areas where they join similar groups (Frith, 1967).

Status. Wood ducks are shot extensively in Australia for sport, and farmers are also allowed locally to shoot them to avoid damage to rice crops. Damage to young rice stands can be substantial if flocks are allowed to feed on them repeatedly, but a small amount of foraging does little or no damage, and may even stimulate tillering. Damage to newly sown pastures has also been reported in some areas. The birds are unusually wary and difficult to hunt, and thus they are not especially sensitive to sport hunting. There is no evidence that the species is currently declining or otherwise in any danger (Frith, 1967).

Relationships. The genus *Chenonetta*, although distinctive, exhibits some strong similarities to several other perching duck genera, including *Nettapus*, *Aix*, and *Amazonetta*. The social displays indicate such affinities, and so too do several aspects of the skeletal anatomy (Woolfenden, 1961).

Suggested readings. Frith, 1967.

Brazilian Teal

Amazonetta brasiliensis (Gmelin) 1789

Other vernacular names. None in general English use. Amazonasente (German); sarcelle du Brésil (French); cerceta brazileña (Spanish).

Subspecies and ranges. (See map 60.)
 A. b. brasiliensis: Lesser Brazilian teal. Resident from Colombia and Venezuela south to the state of São Paulo and the southern part of the Matto Grosso.
 A. b. ipecutiri: Greater Brazilian teal. Resident in southern Brazil, eastern Bolivia, Uruguay, and northern Argentina south to Buenos Aires Province. Probably partially migratory.

Measurements and weights. (For *brasiliensis*.) Folded wing: males, 170–92 mm; females, 168–85 mm. Culmen: males, 34–39 mm; females, 32–36 mm. Weights: males, 380–480 g; females, 350–90 g (Kolbe, 1972). Weller (1968a) reports two males of *ipecutiri* averaging 600 g and a female at 580 g. Eggs: 49 x 35 mm, pale cream, 31 g.

Identification and field marks. Length 14–16″ (35–40 cm). *Males* have a head pattern of gray, brown, and black, with the crown and hind neck darkest and the ear region lightest, like those of the male ringed teal. The breast and anterior flanks are barred and spotted with rusty brown, buff, and black, and the posterior flanks, rump, and tail coverts are more uniformly brown, as is the back. The tail is black, and the entire upper wing surface is iridescent green to purplish, except for the secondaries, which are green anteriorly and white posteriorly, the two colors separated by a narrow black bar. The bill, legs, and feet are bright red. *Females* are similar to males but have a grayish bill and buffy white markings in front of the eyes and at the base of the bill, as well as a whitish throat. *Immatures* closely resemble adult females, but are duller in patterning.

In the field, this species is most likely to be confused with the ringed teal, but the bright red bill of the male should separate it easily. Females quite closely resemble female ringed teal, but lack white markings behind the eyes and on the sides of the rump. In flight, both sexes show a large white triangular mark on the trailing edge of the secondary feathers. Females have a typical ducklike quacking note quite different from that of the ringed teal, and males utter a strong, piercing whistle, *whee-whee-whee.*

NATURAL HISTORY

Habitat and foods. This species inhabits heavily vegetated lagoons, especially those surrounded by woodlands. They evidently avoid coastal and mangrove lagoons, and seem to prefer small water areas to larger ones. Almost nothing is known of their foods under natural conditions; three specimens that were examined contained marsh "fruits," bulbous roots, a planarian, and an insect that was probably a locust (Phillips, 1922–26).

Social behavior. In general, flock sizes of Brazilian teal appear to be quite small, and flocks probably consist of family units. Flocks of from 10 to 20 birds have been reported by some observers. The pair bond is seemingly quite strong and, considering the absence of a definite eclipse plumage in males, is quite probably permanent. Almost nothing is known of the time of nesting, but in northern Argentina, Weller collected females in mid-May that were ready to lay, indicating late-fall nesting in that region. However, males acquiring new wing feathers

MAP 60. Breeding or residential distributions of the greater ("G") and lesser ("L") Brazilian teals.

in mid-February have been collected in Paraguay, suggesting a breeding season several months earlier in that area (Weller, 1968a). The pair-bonding behavior of this species is relatively simple, and no complex displays have yet been described. Females incite their mates or potential mates with a series of rapidly alternated chin-lifting and lateral bill-pointing movements, making an associated *week* note, to which the male typically responds with a repeated whistling note while swimming rapidly ahead of her. A wing-preening display has been reported, but it seemingly is not nearly so common as the iridescent upper wing surface might suggest. However, wing flapping is common during social display and may display not only the upper wing surface but also the blackish underwing and contrasting white axillary feathers. Copulation is preceded by mutual bill-dipping or head-dipping movements, and is followed by the male's calling once and then swimming around the female in a tight circle as he holds his head and bill rigidly downward (Johnsgard, 1965a).

Reproductive biology. It appears that hole nesting is not the normal mode of behavior in Brazilian teal, in contrast to typical perching ducks. Very few nests have actually been found in the wild, but current evidence supports the idea that the birds normally nest on sedge hummocks surrounded by water. Some early observers suspected that they might nest on cliff sides, inasmuch as pairs have been seen in such locations. At least one nest has been found in a tree about eight feet above the ground, on the top of a previous year's blackbird (*Pseudoleistes*) nest. The clutch size is from 6 to 8 eggs, which are incubated entirely by the female. The incubation period is 25 days; and as soon as hatching occurs, the female and brood are joined by the male, who helps protect the young. Indeed, there is one case of a pair in captivity in which the male took over the care of the young completely, freeing the female to begin a second clutch (Phillips, 1922–26).

Status. Brazilian teal are said to be among the commonest ducks in Brazil (Delacour, 1954–64), and there is no reason to believe that they are not relatively plentiful throughout most of their original range. Hunting is an insignificant cause of mortality over this species' range, and its habitat requirements do not appear to be particularly specialized. Thus, there should be no concern for its status at present.

Relationships. Although frequently placed in the genus *Anas*, it has long been recognized that this species is not a typical teal (Phillips, 1924). Finally the genus *Amazonetta* was suggested for it by J. M. Derscheid, who recognized its perching duck affinities with the ringed teal. Woolfenden (1961) found skeletal similarities supporting the relationship between these two species, but suggested that they deserve generic separation, and my own (1965a) behavioral studies have likewise supported this interpretation.

Suggested readings. Phillips, 1922–26.

Tribe Merganettini (Torrent Duck)

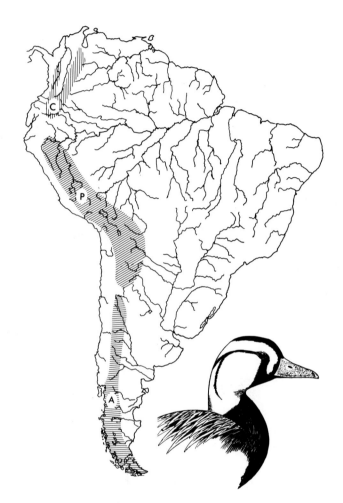

MAP 61. Breeding or residential distributions of the Colombian ("C") Peruvian ("P"), and Argentine ("A") torrent ducks.

Drawing on preceding page: Peruvian Torrent Duck

Torrent Duck

Merganetta armata Gould 1841

Other vernacular names. None in general English use. Sturzbachente (German); canard de torrents (French), pato corta-corrientes (Spanish).

Subspecies and ranges. (See map 61.)

M. a. colombiana: Colombian torrent duck. Resident in the Andes of Venezuela, Colombia, and northern Ecuador.

M. a. leucogenis: Peruvian torrent duck. Resident in the Andes of southern and central Ecuador, Peru, Bolivia, northwestern Argentina, and extreme northern Chile. (This encompasses the ranges of *turneri, garleppi,* and *berlepschi,* which do not appear to be sufficiently distinct from *leucogenis* to be recognizable.)

M. a. armata: Argentine torrent duck. Resident in the Andes of Chile and adjoining parts of Argentina from Mendoza to Tierra del Fuego.

Measurements and weights. Folded wing: males, 142–84 mm; females, 132–65 mm. Culmen: males, 29–31 mm; females, 25–27 mm. Weights: males, ca. 440 g; females, 315–40 g (Niethammer, 1952). Eggs: av. 61 x 41 mm, buff, ca. 65 g.

Identification and field marks. Length 17–18″ (43–46 cm). Plate 36. *Adult males* are slim-bodied ducks with long (over 120 mm) tails that are somewhat stiffened, and with bony spurs at the bend of the wings. In all races the head is mostly white, with a black crown stripe that extends down the back of the neck, where it merges with a black stripe that extends back from the eye and branches to form a second neck stripe just ahead of the nape stripe. The upperparts are mostly made up of gray to blackish feathers with pale gray to white margins, while the flanks and underparts vary from predominantly white with small black spotting (northern forms) to mostly blackish with brownish feather margins (southern forms). The tail is grayish brown and the upper coverts are finely vermiculated with gray and black, while the under coverts vary from white to blackish. The upper wing coverts are grayish blue, the primaries are brown, and the secondaries are iridescent green, with prominent white borders in front and behind. The bill is bright red, the legs and feet are reddish with darker markings, and the iris is brown. *Females* are mostly a finely vermiculated gray pattern on the upper half of the head and

body, and a contrasting rusty brown on the flanks and underside of the head, neck, and body. The tail and wing patterns are like those of the male, as are the soft-part colors. *Juveniles* are generally grayish above and white below, with distinctive gray barring on the flanks.

In the field, torrent ducks are the only waterfowl that inhabit the turbulent Andean streams, and are impossible to confuse with other species. The sharp, clear whistle of the males may be readily heard above the noise of the rushing water, and is directed toward other males (as a territorial call?) as well as toward their own mates or families. Torrent ducks fly quickly but fairly low over the water, rarely higher than 20 feet, methodically following the twists of the river. Their wingbeats are unusually shallow and rapid.

NATURAL HISTORY

Habitat and foods. All of the areas where I (1966a) have observed torrent ducks consisted of rivers with rapids and waterfalls interspersed with stretches of more placid water. The width of the river is seemingly unimportant, but the waters are always cold (12° C in one case), clear, and well oxygenated. In various parts of their range they occur from near sea level (in Chile) to at least 4,500 meters (in Bolivia). River gradients on which they have been found vary from as little as about 5 to 100 meters of descent per kilometer of flow. The birds typically forage by diving into the water from large rocks, disappearing from sight, and remaining submerged for periods of up to nearly 20 seconds as they probe rock crevices for aquatic insects, especially caddis fly larvae. At times the birds also swim on the surface, with only the head submerged, or upend in the manner of dabbling ducks. They also at times crawl behind waterfalls to probe among the rocky ledges. Besides caddis flies, the larvae of stone flies and May flies are consumed, and perhaps also some mollusks. It is likely too that in some areas small fish may be caught and consumed in limited quantities (Johnsgard, 1966a).

Social behavior. Throughout the year torrent ducks are evidently not gregarious, and rarely are seen in groups other than pairs or families, except perhaps during courtship display. In general the birds occupy exclusive foraging and breeding territories along river stretches that space the population out

into a density of about one pair per kilometer. There is little doubt that the birds move only limited distances from these areas, and that pair bonds are strong and presumably permanent. This is indicated by the absence of marked nonbreeding plumages, the extended breeding season, and the participation of males in brood care. Pair-forming displays are still but poorly understood, and the few published descriptions (Johnsgard, 1966a; Moffett, 1970) suggest that sexual displays are quite different from those of other dabbling or perching ducks. A repeated bowing display, or body bend, is evidently used both as a pair-maintaining display and as a territorial boundary display. Aggressive displays between birds holding adjacent territories are complex and well developed. They include calling with simultaneous body bends, and mule kicking, in which water is kicked backwards with both feet, accompanied by a high-intensity bowing movement. Actual fighting between rivals has not yet been reported, in spite of the fact that males possess well-developed wing spurs that almost certainly are used in aggressive encounters. Copulatory behavior is relatively inconspicuous, and according to unpublished observations by Jan Eldridge, is typically preceded by the female's assuming a prone posture, with only bill dipping or head shaking as associated behavior. The male then performs bill dipping, head dipping, barging, and a double-shake display before mounting. Postcopulatory behavior includes bathing by the female and a body bend by the male (Eldridge, 1977).

Reproductive biology. The breeding season of torrent ducks in South America is evidently very long, at least in the northern part of the species' range. In the central Andes, the sightings of young during July and August suggest that breeding occurs during the dry season. In Chile and Argentina, where seasonal temperature changes are considerable, the observations of broods indicate that nesting occurs at the end of the wet winter period (Johnsgard, 1966a). Only a few actual nests have been found. The locations have included a ledge site with overhanging rocks about 4 feet above a stream, a bank tunnel location along a power plant canal about 3 feet above water, and a cliff nest about 75 feet above water at the base of a bush. Johnson (1965) found a nest in a kingfisher cavity about 20 feet above a boulder-strewn river bank and a second probable nesting site in a vertical crevice in the face of a cliff. Most published information indicates that 3 or 4 eggs constitute the clutch. Moffett's (1970) field study in Argentina is the most complete to date, and he reported a nest in a coihue (*Nothofagus*) tree-root cavity about 15 feet above water, one in a streamside cavity about 9 feet above water, and a third in a cliff crevice about 60 feet above water. In one nest Moffett found that the 4 eggs were apparently laid at weekly intervals, and that the incubation period from the laying of the last egg was 43 to 44 days. This included an entire week between the pipping of the first egg and the emergence of the last chick. This amazingly long egg-laying and incubation period, if typical, would make the torrent duck's the longest of any known anatid. Only the female incubated, but the male always met her on her daily foraging trips of several hours each morning and afternoon. These long breaks in incubation and the cold temperatures at the time of incubation no doubt help to account for the extended incubation period Moffett observed. In one case, he found that the chicks left the nest two days after the last egg had hatched, and in a second nest he watched the ducklings drop from their cliff-side nest to the rocks 60 feet below at the call of their mother. After the young had reached the water, they were joined by the male, who thereafter remained with the family and guarded them as the young foraged in the shallows at the river's edge. The period of time to fledging has not yet been established.

Status. The torrent duck appears to be relatively rare and probably is declining in abundance over most of the northern parts of its range, but it is still quite common in Chile and Argentina. Its specialized habitat requirements of clear, cold, swiftly flowing waters are easily destroyed by impoundments or river pollution, and the introduction of insectivorous fishes could also prove harmful to the species.

Relationships. In an earlier review (Johnsgard, 1966a), I concluded that the torrent duck's closest affinities are probably with the perching ducks, and that it should either be included in that tribe or be placed in a separate tribe (Merganettini) of its own. Woolfenden (1961) likewise suggested a separate tribe for this species on the basis of its postcranial osteology. Quite possibly the bird is most closely related to the Salvadori duck, as Kear (1975) has suggested, but unless this or other strong dabbling duck affinities can be established, it seems most practical to keep it tribally separate from that group. Recently Brush (1976) concluded on the basis of its feather proteins that the torrent duck is probably not a member of the dabbling duck group and its affinities are more probably with the perching ducks or shelducks. He thus retained it in a separate tribe near the shelducks.

Suggested readings. Johnsgard, 1966a; Johnson, 1963; Moffett, 1970.

29. Comb duck, adult male

↑ 30. Gambian spur-winged goose, pair with male in foreground   ↓ 31. Muscovy duck, male

↑ 32. Hartlaub duck, family with male in foreground   ↓ 33. African pygmy goose, male and two females

↑ 34. Ringed teal, pair with brood   ↓ 35. Mandarin duck, pair with male in foreground

↑ 36. Chilean torrent duck, male   ↓ 37. American wigeon, female and two males

↑ 38. Chiloe wigeon, pair with male in foreground   ↓ 39. Falcated duck, pair with male in foreground

↑ 40. Baikal teal, two males   ↓ 41. Cape teal, pair with female in foreground

↑ 42. Crested duck, pair with male on right   ↓ 43. Bronze-winged duck, pair with male in foreground

↑ 44. White-cheeked pintail, pair with male in foreground   ↓ 45. Garganey, pair with female in foreground

↑ 46. Red-crested pochard, male　　　　　　　↓ 47. Eurasian pochard, pair with male in foreground

↑ 48. Redhead, male and two females ↓ 49. Greater scaup, male

↑ 50. American eider, pair with female in foreground   ↓ 51. King eider, pair with male in foreground

↑ 52. Spectacled eider, male preening

↓ 53. Steller eider, female and two males

↑ 54. Long-tailed duck, male in summer plumage   ↓ 55. Harlequin duck, male

↑ 56. Bufflehead, male   ↓ 57. Barrow goldeneye, pair with female in foreground

↑ 58. Hooded merganser, male

↓ 59. White-headed duck, male

Tribe Anatini
(Dabbling or
Surface-feeding Ducks)

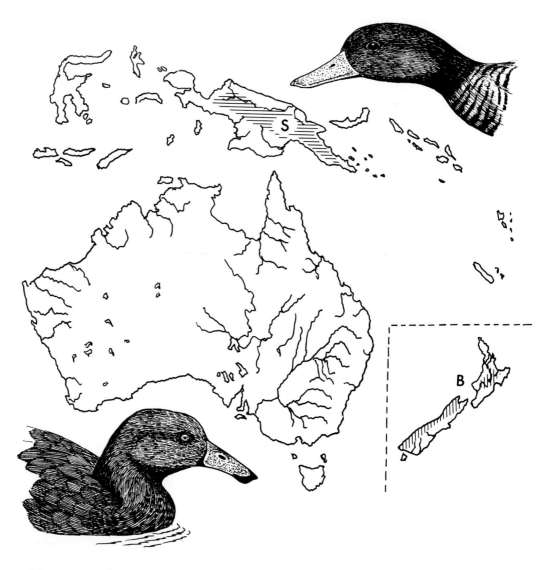

MAP 62. Breeding or residential distributions of the Salvadori duck ("S") and blue duck ("B").

Drawing on preceding page: American Wigeon

Blue Duck

Hymenolaimus malacorhynchos (Gmelin) 1789

Other vernacular names. Mountain duck; Saum-schnabelente (German); canard bleu (French); pato azul (Spanish).

Subspecies and range. No subspecies recognized. Limited to the highland streams of New Zealand, including North Island south of the Coromandel Peninsula and the western half of South Island. See map 62.

Measurements and weights. Folded wing: both sexes 235–49 mm (av. of 4 males 233 mm, of 6 females 217 mm). Culmen: males, 42–45 mm (av. 44 mm); females, 40–42 mm. Weights: males, 753–1,077 g (av. 887 g); females, 680–850 g (av. 750 g). Eggs: Average 65 x 45 mm, pale buff, 73 g. (Pengelly & Kear, 1970).

Identification and field marks. Length 21″ (54 cm). *Adults* of both sexes are an unusual dove gray color with a bluish sheen, except for a brownish gray crown, reddish brown spotting on the breast, and black edging on the edges of the inner secondaries. The six outer secondaries also have very narrow white tips, and the underwing is pale gray. The bill is pinkish white, with a black nail and pendant black lateral lobes near the tip; the eyes are yellow; and the legs and feet are dark brown. *Females* have less brownish spotting on the breast than males, and *immatures* lack breast spotting completely. The full plumage is evidently attained by six months, but birds raised in captivity retained a dark gray culmen band until they were more than a year old (Pengelly & Kear, 1970).

In the field, the unusual color and mountain stream habitat preclude confusion with any other species. The strong whistled call *whio* (which is the basis for the Maori name for the bird, whio) is uttered only by males; the female produces a rasping quack. The birds are quite tame and rarely fly, but they do fly strongly, with a fairly rapid wingbeat.

Natural History

Habitat and foods. Blue ducks at one time occurred on mountain streams of New Zealand from the tree line to sea level, but at present they are confined to mountainous regions supporting turbulent streams that are clean and well oxygenated. These streams, usually at least 10 feet wide in places, provide the invertebrate foods on which the birds depend almost entirely. Of these foods, caddis fly larvae are probably the most important single item (Kear & Burton, 1971), as is also true of the South American counterpart species, the torrent duck. May fly and stone fly larvae are also eaten, as is some algae. Blue ducks can dive readily and reportedly cling to boulders while probing under water for invertebrates, but most of the birds feed in water too shallow for diving. The soft-edged bill is used by adults for scraping and sucking food off rocks, and ducklings only a few days old are adept at probing their bill into crevices and moving it from side to side, or picking up solids by pulling the bill backward over the surface (Pengelly & Kear, 1970).

Social behavior. Except for family groupings, blue ducks are not gregarious, and at most will congregate at river mouths after flooding or on winter foraging areas when food shortages force them out of their breeding territories. For the rest of the year the birds occupy exclusive foraging and breeding territories of river habitat. These territories, proclaimed by the male and defended by him against all other blue ducks and even other species such as gray ducks, may occupy only a few hundred yards of river or may be several miles long, but Kear and Steel (1971) reported five territories in a stretch of about two and one-half miles of river. Pair formation certainly occurs after the birds are at least five months old, and probably before they are a year old, but the process is still essentially unknown. It is believed that the birds mate permanently, although this is still unproven. The role of the male in parental care and the strong attachment to a particular breeding area both favor the retention of pair bonds. Further, the extended egg-laying period

(August to at least January) also favors permanent pair bonds. Copulatory behavior is apparently very much like that of typical surface-feeding ducks, consisting of mutual head pumping, although the female has at times been observed to swim rapidly around and in front of the male in a manner resembling bathing. Following copulation the male assumes an erect posture, moves his head forward and back, but has not been heard to call at this time, as might be expected. The loud call of the male is largely aggressive in function, but he may also use it to attract unmated females (Kear & Steel, 1971).

Reproductive biology. Although the breeding season is very extended, with egg records extending over a six-month period, the majority of nest records are for October and November, and most hatching probably occurs in October. There is, however, a record of an adult with young as late as June. In spite of this extended breeding period, it is likely that the birds breed only once a year, as there is an extended incubation and fledging period in this species. Nest sites are usually natural burrows—under clumps of tussock grass, under logs or bushes, in rock clefts or cliff ledges, or in hollow logs or trees. Nests are often fairly close to water, and sites are probably used in successive years, especially if they were previously successful. The modal clutch size of the relatively few observed nests is 5 eggs, and the mean is 5.4, with a range of 4 to 9. The combination of relatively small clutch size and fairly large eggs is also characteristic of torrent ducks. Only the female incubates, while the male remains hidden nearby. He joins his mate on her daily foraging and bathing trips, which usually last less than an hour. The total incubation period is probably 31 to 32 days, judging from records of artificially incubated eggs. It is probable that the family remains at the nest for nearly two days before the parents lead the young to water, and both sexes closely attend the developing brood. The fledging period seems to be quite long, between ten and eleven weeks in ducklings raised in captivity; and even after fledging, the birds apparently remain within their parents' territory. How long the family remains intact is still uncertain, but it may be until the birds attain adult plumage at the age of about six months (Kear, 1972).

Status. This species is totally protected in New Zealand, but nevertheless has a range considerably contracted from its original one. On the North Island the bird is limited largely to the mountainous area around Urewera and Tongariro National Parks, while on the South Island it is most abundant in Otago and Southland. However, no estimates of its total population are available. Competing species may include various insectivorous birds and the introduced trout which now occupy many of the lowland streams. Introduced mammalian predators have doubtless caused considerable loss of eggs and young (Kear, 1972; Kear & Burton, 1971).

Relationships. A recent review (Kear, 1972) of this peculiar species' affinities has not provided any specific answers, other than to suggest that it seems to have its greatest affinities with the typical dabbling ducks of the genus *Anas*, although it also has some similarities to the perching ducks. Kear suggests that present evidence favors retaining it with the dabbling duck tribe Anatini. The analysis of feather proteins by Brush (1976) cast doubt on its affinities with the dabbling ducks, but failed to provide positive evidence of other relationships.

Suggested readings. Kear, 1972.

Salvadori Duck

Anas waigiuensis (Rothchild and Hartert) 1894

Other vernacular names. Salvadori's teal; Salvadoriente (German); canard de Salvador (French); pato de Salvadori (Spanish).

Subspecies and range. No subspecies recognized. Resident in mountain streams of New Guinea above 1,300 feet. Originally reported from, but of doubtful occurrence on, Waigeo Island. See map 62.

Measurements and weights. (From Kear, 1975.) Folded wing: males, 187–207 mm (av. 194 mm); females, 179–96 mm (av. 185 mm). Culmen: males, 35–39 mm (av. 37 mm); females, 34–38 mm (av. 36 mm). Weights: males, 400–525 g (av.

462 g); females, 420–520 g (av. 469 g). Eggs: av. 58 x 43 mm, creamy white, 58 g.

Identification and field marks. Length 15–17" (38–43 cm). *Adults* have a grayish black head and neck, with narrow pale edges on the feathers. The upperparts are black, with a slight greenish cast and strong white barring. The tail is dark brown, narrowly barred with white, and the central feathers are long and pointed. The lower breast and abdomen are white to buff, with small black spots that increase in size up the breast and on the flanks, where they merge with black barring. The upper wing surface is dark brown to blackish, with narrow white edging on the lesser coverts, while the greater secondary coverts are broadly edged with white. The secondaries are mostly iridescent green or purplish black, with broad white tips that form a white posterior speculum stripe. The iris is brown; the bill is bright yellow, sometimes tinged or spotted with black; and the legs and feet are yellow, with brown or blackish markings. *Females* are not obviously separable from males, and *immatures* differ from adults in being generally duller, with a dark gray to olive-colored bill and pinkish rather than yellow foot coloration.

In the field, the distinctive river habitat and limited New Guinea range separates this species from all others. The birds have slim bodies and a mallardlike upper wing pattern, but carry their tails in a distinctive cocked position much of the time, especially when alarmed. The male is known to produce a whistling note and the female a double-noted quacking call, but the birds apparently lack the shrill whistled notes of male blue ducks and torrent ducks. The whistled call is evidently not used as a territorial signal.

birds feed in the rapids while treading water to maintain their position on the surface as they probe in eddies between the rocks, or sometimes they dive below the surface. They dive readily and may remain submerged for periods averaging about 12 seconds, while the periods between dives average about 18 seconds (Kear, 1975). Most of the few food remains that have been found in Salvadori ducks are of aquatic insects, including water beetles. Caddis fly larvae, dragonfly nymphs, water beetles, and tadpoles are also known to occur in waters used by these birds, and all probably are consumed by them. They have also been observed foraging on water fleas (Bell, 1969).

Social behavior. Like blue ducks and torrent ducks, Salvadori ducks are normally found in pairs or in family units, and rarely if ever occur in larger groupings. The breeding season is known to be quite extended, and perhaps the birds never leave their territories after becoming mature and forming pairs. However, nothing is known of their period to reproductive maturity or their pair-bonding behavior, although they are presumed to pair permanently. Pairs occupy exclusive territories of river, with reported territories as close together as 160 meters and as widely separated as 1,500 meters (Bell, 1969; Kear, 1975). The female is known to possess an inciting display similar in form and context to that of other *Anas* species, but little is known of male displays other than their aggression toward other males. Like the torrent-dwelling and highly territorial blue ducks and torrent ducks, males have wing spurs that they probably use in such territorial fights rather than (as has been often speculated) for clinging to wet rocks when foraging. Copulatory behavior has not as yet been described.

NATURAL HISTORY

Habitat and foods. Salvadori ducks are found in three New Guinea habitats: rushing mountain streams, muddy and sluggish streams, and alpine lakes. Their vertical range is from about 500 to 4,000 meters, and they are known to be common around an elevation of 3,700 meters. The species' slim body suggests that, like torrent ducks, it is best adapted to mountain torrents, and studies of its foraging behavior indicate that it can feed in water much too swift for most ducks. In such torrents the

Reproductive biology. The breeding season is known to be prolonged; captive birds in New Guinea nest as early as May and breeding among wild birds continues at least into October and perhaps into January. It has been suggested that the birds might breed twice during this long period, but the evidence is still inadequate on this point. The few nests that have been found were in depressions near water, usually under clumps of grass or shrubs. The clutch size is remarkably small, apparently normally only 3 eggs, with a few instances of broods of 4 ducklings reported. However, the eggs are relatively large for the size of the female, as is also typical of torrent ducks and blue ducks. The incubation period is still unknown, but must be at least 28 days. Males remain with their females during the brood-raising period, and it has been reported that the young at times ride on their mother's back. The fledging period is as yet unreported, but by the age of two months the young are nearly full-grown and presumably nearly fledged (Kear, 1975). The young probably remain with their parents until they become sexually mature or until their parents' next breeding season.

Status. Kear (1975) believes that although this species is not yet rare, it may be endangered in the future by the introduction of trout or other insectivorous fish into New Guinea's rivers in attempts to provide additional protein for its native population. The species is now fully protected in Papua, New Guinea, but enforcement of such laws is difficult or impossible in that region. No natural predators are known to be a serious threat to the species.

Relationships. Kear (1975) has reviewed all of the available evidence regarding this species' probable evolutionary relationships, and supports the position that its similarities to the torrent duck, and possibly also the blue duck, are the result of actual phyletic ancestry rather than evolutionary convergence. Obviously, many of the similarities in foraging behavior are due to the latter, but others, such as the pattern of the wing speculum and the downy young, seem to be more basic. She believes that resurrection of the genus *Salvadorina* would be appropriate, at least until a more definite statement of the species' relationships can be achieved. However, Brush (1976) reported that the feather proteins of the species were essentially identical to those of 13 other *Anas* species he studied.

Suggested readings. Kear, 1975.

182 ❖ ❖ ❖

African Black Duck

Anas sparsa Eyton 1838

Other vernacular names. None in general English use. Schwarzente (German); canard noir d'Afrique (French); pato negro africano (Spanish).

Subspecies and ranges. (See map 63.)

A. s. sparsa: South African black duck. Resident in southern Africa, from South-West Africa through South Africa to Rhodesia and Mozambique, and perhaps intergrading with *leucostigma*.

A. s. maclatchyi: West African black duck. Resident in Gabon and probably Cameroon, the total range uncertain. Of dubious validity and possibly better considered a synonym of *leucostigma*.

A. s. leucostigma: Ethiopian black duck. Resident in central Africa, from Ethiopia and central Sudan south at least to Zaire and Tanzania, with the western and southern limits uncertain.

Measurements and weights. Folded wing: males, 245–72 mm; females, 232–48 mm. Culmen: males, 43–50 mm; females, 40–45 mm. Weights: males unreported; females, 952–1,077 g (Siegfried, 1968). Eggs: av. 59 x 45 mm, deep cream, 68 g.

Identification and field marks. Length 20–23″ (51–58 cm). *Adults* are mostly brownish black except for pale buff or white barring on the longer scapulars and buffy banding on the upper tail coverts. The head is mostly grayish brown with narrow buffy edging on the feathers, and sometimes with a whitish neck patch. The underparts are mostly dark olive brown, and the flanks are similar but some of the longer feathers are barred with buff. The axillaries are white and the underwing surface brown with white markings. The tail is dark brown, with two buffy bars, and the upper wing surface is brown

to blackish, except for the speculum, which is iridescent green to bluish, with black and white anterior and posterior stripes, as in mallards. The iris is dark brown, the bill grayish pink (northern forms) or lead blue (southern forms) with blackish markings, and the legs and feet orange with blackish markings. *Females* are nearly identical to males except for their smaller size, and may also differ in bill coloration (which is more pinkish, at least in *sparsa*), but this is not yet definite. *Juvenile* birds have buff rather than white barring on the upperparts and tail, and white abdomen coloration (Siegfried, 1968).

In the field, black ducks might be confused with African yellow-billed ducks, but lack the obvious yellow bill and the whitish edging on the flank and scapular feathers typical of that species. In flight, the white axillaries and white-tipped underwing feathers contrast with the dark body, and the white stripe in front of the speculum is conspicuous. Females quack loudly and almost continuously when in flight, and males utter a whistling "peep" sound, sometimes in flight, as well as on the water (Siegfried, 1968).

NATURAL HISTORY

Habitat and foods. In general, black ducks are thought to prefer fairly rapidly running rivers and streams, particularly in wooded and mountainous country. However, they have been seen (at least in South Africa) in open and arid habitats, and on waters that are either stagnant or at most are slowly flowing. Shallow waters with rocky substrates are nevertheless preferred habitats, probably because they provide the species' favored foods. To a degree, the species is a torrent-adapted form in spite of the absence of obvious morphological specialization for torrent feeding, and the birds often feed in rapidly running water by standing on protruding rocks and probing under stones with their bills. They also dive for food at the foot of rapids, and can readily negotiate small waterfalls and steep rapids, even when leading broods. Although no detailed studies have been made, Siegfried (1968) believed that the foods of wild black ducks might include waterweeds, other aquatic vegetation, and perhaps also grain, but he noted that he had never observed the birds feeding in fields where grain and other cereal crops are grown. Clancey (1967) mentions such animal foods as aquatic insects and their larvae, crustaceans, larval amphibians, and fish

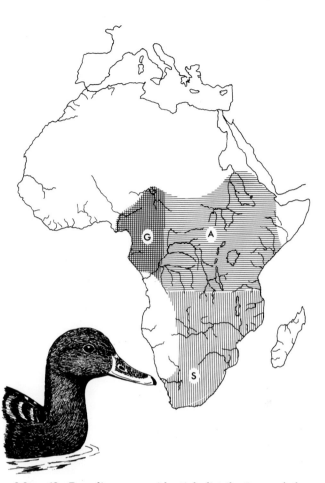

MAP 63. Breeding or residential distributions of the Gabon ("G"), Abyssinian ("A"), and South African ("S") black ducks.

spawn as probable or known foods, which would seem likely in view of their habitat and methods of foraging.

Social behavior. Like other river-adapted ducks, black ducks are not gregarious and are rarely seen in groupings larger than those of a family. Siegfried (1968) mentions that he never saw a group larger than four adult birds associating peacefully, and almost never are more than seven (presumably families) seen in a group. Spacing of pairs or families along rivers is the usual distributional pattern, with pairs separated by distances of as little as about 200 meters or as much as 700 meters. Pair-bonding patterns in black ducks are not yet well studied, but available data suggest that the male does not participate in brood care, and thus pairs may be renewed annually. The pair-forming displays are in general mallardlike, including inciting by the female, which in this species takes an unusual head-pumping form and to which the male responds in a similar manner (Johnsgard, 1965a). Contrary to my earlier observations, Siegfried (1968) reports that females do possess a typical *Anas* quacking call but agrees that they apparently lack a decrescendo call. Male displays include head pumping while calling in response to female inciting, a mock-feeding display toward females with the folded wings slightly raised, a mallardlike grunt-whistle display, and probably also a rudimentary form of the head-up-tail-up display of the mallardlike ducks (Johnsgard, 1965a). Siegfried (1968) describes various display flights, including the usual three-bird flights ("territorial defense flights" in some interpretations) and attempted rape chases. Copulatory behavior between paired birds is evidently not of the normal *Anas* type; Siegfried's observations and my own indicate a variety of precopulatory patterns, including mutual head pumping, mutual diving, and mutual bathing and preening with interspersed diving. I did not observe the anticipated mallardlike postcopulatory display (bridling), but Siegfried (1968) has since verified its occurrence.

Reproductive biology. The breeding season is relatively prolonged in African black ducks; in northern areas (Ethiopia to Tanzania and Zaire) breeding is generally reported for January to July, and as far south as Zambia the same is generally true, with most records for May through July. Thus in central Africa the species is mostly a dry-season breeder, in contrast to most of the other duck species. In South Africa most of the records occur between July and February, and in the southwestern Cape area there is a peak of egg laying in early September, or toward the end of the wet season (Siegfried, 1968). Nest sites are typically on grassy river banks or among driftwood, usually close to running water, or often on an island. Some elevated nesting sites such as trees have been reported, as have ground cavity nests. Close proximity to water and near invisibility from above are evidently major criteria for suitable nest sites. The female produces a relatively small clutch of from 4 to 8 eggs (averaging 5.9 for 42 clutches), which are laid on a daily basis. Only the female incubates, and she normally leaves the nest twice a day for foraging. The incubation period is 28 days; and if the first clutch or brood is unsuccessful, the female will normally renest. However, multiple brooding is not characteristic. The young are often brought back to the nest for brooding at night or during bad weather for the first week after

hatching; and as noted earlier, there is no strong evidence that the male participates in brood care or defense. The period to fledging is about nine weeks, and between one and two months after fledging the juveniles begin to disperse (Siegfried, 1968).

Status. In part because of its habitat requirements, this species is not often seen by casual observers, and almost no information on its numbers is available. It is still relatively plentiful in southern Africa at least, and no concern about its status there has been voiced. The more northerly forms are evidently considerably rarer, particularly the West African black duck, of which only a few specimens exist.

Relationships. Delacour (1954–64) believed that black ducks are not quite so closely related to the mallard group as they appear to be, and suggested a possible slight affinity with the shelducks. In my review of their behavior (1965a), I concluded that this species may be the most generalized living form of *Anas*, and possibly provides an evolutionary link with more primitive groups, such as the perching ducks. Siegfried (1968) agreed that this species may be the most generalized living *Anas*, but suggested that part of its behavioral simplicity is related to the fact that it is ecologically isolated from nearly all other *Anas* species.

Suggested readings. Siegfried, 1968; Clancey, 1967.

Eurasian Wigeon

Anas penelope Linnaeus 1758

Other vernacular names. European wigeon; Pfeifente (German); siffleur d'Europe (French); pato europeo (Spanish).

Subspecies and range. No subspecies recognized. Breeds in Iceland, the British Isles, and across northern Europe and Asia between about 60° and 70° north latitude, and more rarely farther south to Lake Baikal and northern Sakhalin. Winters from the British Isles south to northern Africa and east to India, Ceylon, and southeast Asia as well as Japan. See map 64.

Measurements and weights. Folded wing: males, 254–70 mm; females, 236–55 mm. Culmen: males, 33–36 mm; females, 31–34 mm. Weights: males (in winter), 465–970 g (av. 720 g); females, 415–800 g (av. 640 g) (Bauer & Glutz, 1968). Eggs: av. 55 x 39 mm, creamy, 44 g.

Identification and field marks. Length 17–21" (43–54 cm). *Males* in breeding plumage have a cinnamon red head and neck, with a buff crown and forehead, and sometimes a trace of iridescent green behind the eyes. The upper breast is purplish pink; and the lower breast, abdomen, flanks, and mantle are white, mostly finely vermiculated with dusky coloration. The rump is light gray and the tail coverts are black, except for the middle upper coverts, which are whitish. The tail is gray to brownish centrally and white to silvery gray outwardly. The upper wing coverts are mostly white, except for the secondary coverts, which are tipped with black. The secondaries are iridescent green, with blackish tips; the tertials are black, edged with white; and the primaries and their coverts are ashy brown. The axillars are mottled or flecked with brown or grayish coloration. The iris is brown, the bill pale bluish gray with a black tip, and the legs and feet gray. *Males in eclipse* closely resemble females, but have white upper wing coverts. *Females* have a cinnamon buff head and neck (grayish buff in some birds), flecked with dusky or greenish coloration. The upper breast and sides are buffy or reddish brown (grayish brown in some), marked with dusky coloration. The scapulars and rump are dusky brown, the longer scapulars being edged with buff or white. The upper wing coverts are mostly dusky gray or brown, with whitish tips, while the greater secondary coverts are tipped with black and white. The secondaries have a dull green to blackish speculum, with a narrow terminal bar. The axillars are gray, mottled with dusky coloration. The tail and the soft-part colors are as in the male. *Juveniles* resemble females, but males gradually assume nuptial plumage in their first fall. However, young males retain grayish brown upper wing coverts during their first winter.

In the field, the cinnamon-colored head and lilac-toned breast color of males is distinctive. Like those of all wigeons, the white upper wing coverts are conspicuous in flight. Females have an appreciably more brownish cast to their head and neck than do female American wigeons, and are generally more reddish throughout, although a grayish brown phase does occur. Males have a distinctive whistling call, *whew* or *wa'chew,* which is louder and has fewer syllables than the American species' call. The female has a very abbreviated decrescendo call of only one to three syllables, and her inciting call is a soft growling note.

Natural History

Habitat and foods. The habitat needs of this wigeon are those of most dabbling ducks, namely, shallow ponds with mud or silt bottoms, although it also favors those with gradually sloping and meadow-lined shorelines where grazing opportunities are present. In addition, it apparently is attracted to partially wooded shorelines, since it avoids both open tundra and small ponds enclosed by forests (Hilden, 1964). Wigeons are almost exclusively vegetarians, and their short and relatively stout bill makes them more efficient grazers than most of the other dabbling ducks. They feed on such items as grasses, eelgrass (*Zostera*), and algae (*Chara* and *Enteromorpha*) during the wintertime, utilizing both coastal areas where mud flats are abundant and inland waters with grazing opportunities. The vegetative parts of pondweeds and wigeon grass are also particularly important and consumed whenever they can readily be reached by these nondiving birds (Palmer, 1976); Ogilvie, 1975). Wigeons often feed in the vicinity of swans and diving ducks around pondweed beds, sometimes stealing the plants as they are brought to the surface by these birds.

Social behavior. The social behavior of this species is nearly identical to that of the American wigeon

MAP 64. Breeding (hatched) and wintering (stippling) distributions of the Eurasian Wigeon.

186 ❖ ❖ ❖

insofar as display posturing and calls are concerned, although the whistling notes of the males are somewhat louder and are only of one or two syllables. The birds begin to form pairs in the first winter of their life, and in Bavaria the incidence of mated females gradually rises from about 60 percent in October, presumably reflecting a high incidence of remating with previous mates, to 100 percent by May (Bezzel, 1959). The inciting behavior of females is marked by repeated chin-lifting movements and a growling *errr* note; and in addition, females fairly frequently perform a display preening behind the wing, momentarily exposing their grayish white coverts. Males respond to inciting by calling while chin lifting, preening behind the wing, and threatening other males with the head stretched forward and the folded wings raised over the back. Turning the back of the head is not very prevalent among wigeons, and instead the males tend to orient their bills in the direction of the female or toward other males. Copulation is preceded by mutual head pumping; and although postcopulatory behavior has not yet been specifically described as other than bathing and diving, it almost certainly is comparable to that of the American and Chiloé wigeons (Johnsgard, 1965a).

Reproductive biology. Female wigeons probably regularly nest at the end of their first year of life, and tend to have their nests well hidden and rather widely dispersed, with only a slight tendency to nest on islands when they are available, judging from studies in Iceland. There the favored nesting cover is low or high shrubs, with nonwoody cover being used substantially less often (Bengtson, 1970). In that area the clutch size usually ranges from 7 to 11 but averages 9 eggs, with very minor year-to-year variations that might be related to yearly variations in food supplies (Bengtson, 1971a). The eggs are laid on a daily basis, and the incubation period has been estimated at 22 to 25 days, with 24 days probably being close to the average. The fledging period is 40 to 45 days (Bauer & Glutz, 1968; Ogilvie, 1975). As in the other European species of *Anas*, males desert their mates early in incubation to begin their postnuptial molt, and at least in the U.S.S.R. these birds sometimes form large molting assemblages, as for example in the Anadyr Valley (Dementiev & Gladkov, 1967).

Status. Ogilvie (1975) reports that in northwestern Europe this species is very numerous during winter, with about half a million wintering birds, of which about half occur in the British Isles. Another half a million winter in the vicinity of the Mediterranean and Black Seas, particularly in Turkey and Tunisia. The species also winters commonly across all of southern Asia, and is especially common along the southeast coast of China and Japan, where it is the commonest of all wintering ducks.

Relationships. The American, Chiloé, and Eurasian wigeons are all ecological counterparts and close evolutionary relatives of one another. Although they form one extreme of the genus *Anas* in both behavior and bill morphology, they nevertheless are clearly a part of this large assemblage of birds (Johnsgard, 1965a; Woolfenden, 1961; Brush, 1976).

Suggested readings. Dementiev & Gladkov, 1967; Owen, 1977.

American Wigeon

Anas americana Gmelin 1789

Other vernacular names. Baldpate, widgeon; Amerikanische Pfeifente (German); siffleur d'Amérique (French); pato americano (Spanish).

Subspecies and range. No subspecies recognized. Breeds in northwestern North America from the Yukon and Mackenzie regions east to Hudson Bay and south to California, Arizona, Colorado, Nebraska, and the Dakotas, with infrequent breeding farther east. Winters along the Pacific coast from Alaska to Costa Rica, and along the Atlantic coast from southern New England to the Gulf of Campeche, as well as in the West Indies and extreme northern South America. See map 65.

Measurements and weights. Folded wing: males, 252–70 mm; females, 236–58 mm. Culmen: males, 45–48 mm; females, 33–37 mm. Weights: males and females shot in the fall average 770 and 680 g, respectively, with observed maxima of 1,133 and 861 g (Nelson & Martin, 1953). Eggs: av. 54 x 35 mm, creamy, 43 g.

Identification and field marks. Length 18–23" (46–58 cm). Plate 37. *Males* in breeding plumage have a white stripe from the forehead to the middle of the crown, and an iridescent green patch from the eye to the occipital region. The rest of the head and upper neck are white or buffy white heavily spotted with blackish coloration. The lower sides of the breast, the flanks, scapulars, back, and rump are all pinkish brown, finely vermiculated with black, and the breast is purplish pink. The underparts and sides of the rump behind the flanks are white. The tail coverts are mostly black, and the tail is dark gray to brown centrally, with the outer feathers silvery gray, edged with white. The upper wing coverts are all white, except for the lesser coverts, which are brownish along the edge of the wing, and the greater secondary coverts, which are tipped with black. The primaries and their coverts are ashy brown, and the secondaries are iridescent green shading to black toward the tips. The tertials are black, edged with white, and the underwing surface is uniformly ashy gray. The axillars are white or at most only faintly flecked with gray. The iris is brown, the bill grayish blue with a black tip, and the legs and feet are bluish gray. *Males in eclipse* resemble females, but have white upper wing coverts and may have brighter sides and flanks. *Females* have a brownish black crown, streaked with whitish coloration, and the rest of the head and upper neck are whitish with darker spotting. The back and inner scapulars are grayish brown, barred with buff and tipped with grayish, while the longer scapulars are browner and are tipped with white. The rump is grayish brown; the upper tail coverts are brown, edged with white; and the under tail coverts are white, barred with brown. The tail is grayish brown, with whitish edging, and the sides and flanks are pale reddish brown. The wings are as in the male, except that the lesser and middle coverts are grayish brown, edged and tipped with whitish coloration. The iris is brown, the bill bluish gray, and the legs and feet are ashy gray. *Juveniles* resemble adult females, but the back pattern is plainer and the lesser and middle coverts are grayish brown in both sexes. First-year males carry over their juvenile wing condition through their first breeding season.

In the field, the white forehead of males and pinkish breast and sides provide distinctive field marks, while in flight the white forewings of both sexes are highly conspicuous. Females on the water are best identified by association with males, but their short bills, rather uniformly colored heads, and slightly pinkish body tones are also good field marks. Males have a distinctive whistling call, usually a three-syllabled whistle with the middle note

the loudest, that is weaker and more wheezy than in the other wigeon species. It is used both as a courtship call and in other situations as well. The female decrescendo call consists of only one to three syllables, and the inciting notes are soft and growling.

NATURAL HISTORY

Habitat and foods. Breeding habitats preferred by the American wigeon seem to be characterized by lakes or marshy sloughs with fairly dry sedge-lined meadows around them, and with woody or brushy areas in the vicinity. Lakes that are fairly shallow but not covered by emergent vegetation, and with an abundance of submerged aquatic vegetation near the surface, are also clearly preferred over deeper or vegetation-choked water areas. The open parklands of western Canada provide the center of the species' breeding abundance, but the birds breed in limited numbers north to the coastal tundra. Their foods during this period consist largely of such aquatic plants as pondweeds, water milfoil (*Myriophyllum*) and naiad (*Najas*), while winter foods in more saline or brackish habitats are predominantly succulent aquatics such as muskgrass, wild celery (*Vallisneria*), wigeon grass, and eelgrass (*Zostera*). Grazing in wet meadows on wild greenery or in fields of lettuce and alfalfa is also more commonly done by wigeons than by other North American dabbling ducks. However, field feeding on grains such as sorghum and corn is rather rare (Johnsgard, 1975; Bellrose, 1976).

Social behavior. Wigeons are alert and excitable ducks that are usually found in fairly small, single-species flocks, or at times mingling with gadwalls. They are among the earliest of the waterfowl to arrive on their southern wintering grounds, and pair-forming behavior begins in Texas as early as November. Between then and March, by which time the majority of females have apparently formed pair bonds, the major period of courtship activity occurs. Evidently all females form pair bonds during their first winter, and probably all of them attempt nesting the following summer, although some birds raised in captivity do not attempt to nest in their first year. Pair bonds are held only for a single breeding season, and there is no indication that males ever remain to assist with brood rearing. There is also still no information on the incidence of remating with an earlier mate. Such remating might account for the early fall establishment of pair bonds seen in some females. Courtship is marked by intense competition among the males, with a repeated uttering of their shrill three-syllable whistling calls and a lifting of the folded wings nearly vertically above the back. Inciting by the females consists of repeated chin-lifting movements and soft growling calls, and often is carried on during aerial chases. Males also turn the back of the head toward inciting females and occasionally perform display preening behind the wing. Precopulatory behavior consists of the usual mutual head pumping; and although information is still minimal on postcopulatory behavior, it probably consists of the male's

turning and facing the female in a very erect and nearly motionless posture for several seconds (Johnsgard, 1965a). Late in the spring many of the aerial chases seen among wigeon are actually attempted rape chases, and their possible role in normal pair formation is dubious.

Reproductive biology. On arrival at their breeding grounds, pairs establish a home range that often centers on a pothole or small lake, frequently under an acre in size and often surrounded by hayfields or by ungrazed woodlands. Nests are generally placed rather far from water in upland meadows, at times several hundred yards from the nearest shoreline. Most often the nests are extremely well concealed in rushes (*Juncus*) or sedges, but at times the birds also nest at the bases of trees or under shrub clumps. The average clutch size as reported by various observers ranges from about 7 to 9 eggs, but no specific information is available as to the incidence of renesting by unsuccessful females. Crows and skunks were reportedly the most important nest predators in an Alberta study, but in general nesting success in this species appears to be rather high. Males typically desert their females during the first or at most the second week of incubation, and often migrate considerable distances before undergoing their flightless period. Estimates of the fledging period have been from 45 to 48 days in southern Canada and the northern plains states, while observers in Alaska have estimated fledging periods at 37 to 44 days. The flightless period of adults is probably about 35 days (Bellrose, 1976; Johnsgard, 1975).

Status. Surveys of wintering populations suggest that the American wigeon is one of the most abundant of North American game ducks, with about 1.5 million birds tallied in recent winter surveys and an estimated continental breeding population in excess of 3 million birds. Fall populations are likely to be twice as large as this, or more than 6 million (Bellrose, 1976).

Relationships. The three species of wigeons form a closely knit evolutionary group, within which the foraging behavior, general social behavior, and ecological adaptations are extremely similar. It is surprising that wigeons have not colonized southern Africa or Australia or that their ecological counterparts have not evolved there; none of the native species seem to exhibit quite the same niche characteristics.

Suggested readings. Munro, 1949b; Keith, 1961.

MAP 65. Breeding (hatched), including densest breeding concentrations (cross-hatched), and wintering (stippling) distributions of the American wigeon.

Chiloé Wigeon

Anas sibilatrix Poepping 1829

Other vernacular names. Chilean wigeon; Chilepfeifente (German); siffleur du Chili (French); pato overo (Spanish).

Subspecies and range. No subspecies recognized. Breeds in southern South America from Atacama to Tierra del Fuego in Chile, and from Northern Argentina around the vicinity of Buenos Aires south to the tip of the continent. Also resident in the Falkland Islands. In winter it migrates into Uruguay, Paraguay, and southern Brazil. See map 66.

Measurements and weights. Folded wing: males, 255–75 mm; females, 237–45 mm. Culmen: males, 33–35 mm; females, 34–36 mm. Weights: 5 males averaged 939 g; 3 females averaged 828 g (Weller, 1968a). Eggs: av. 58 x 40 mm, pale buff, 53 g.

Identification and field marks. Length: 17–21" (43–54 cm). Plate 38. *Adult males* have a head that is mostly iridescent green to blackish, except for a white to mottled white face patch in front of the eyes and a smaller whitish mark near the ears. The breast is a scalloped black and white; the underparts and tail coverts are white to yellowish white grading to rusty chestnut on the upper flanks; the tail is black; and the back and scapular feathers are mostly black, with broad white barring or margins. The upper wing coverts are mostly pure white, the primaries are brown, and the speculum is velvety black, with a greenish sheen, and is bordered anteriorly with a black bar on the greater coverts. The inner secondaries (tertials) are elongated and black, with white outer margins. The iris is brown, the bill bluish with a black tip, and the legs and feet are gray. There is no eclipse plumage. *Females* closely resemble males, but are somewhat duller throughout, especially on the head, and have mottling on their lesser wing coverts. *Immatures* are considerably duller than adults, with little iridescence on their heads, and have heavily mottled upper wing coverts.

In the field, the rusty-colored flanks and white hindquarters of both sexes serve as distinctive field marks, and in flight the white upper forewings are unique among South American dabbling ducks. The loud, three-noted whistling of males is heard almost constantly when the birds are agitated.

NATURAL HISTORY

Habitat and foods. According to Johnson (1965), these wigeons prefer the open waters of the center of

lakes and obtain nearly all of their food from near the surface by dabbling, diving only when necessary. The species is said to prefer the large lagoons and arroyos of western Patagonia (Phillips, 1922–26). On Isla Grande, Tierra del Fuego, it was found by Weller (1975a) to occur in both beech forest and steppe habitats, but was uncommon to rare in the former, and occurred on lakes with adjoining meadows and along slow-moving rivers. In the forest-steppe ecotone areas it was found on many lakes. Like the Northern Hemisphere species of wigeon, this bird is almost entirely vegetarian, and prefers to graze on green grasses whenever they are available. Other aquatic and terrestrial plants are also consumed in quantity; rush (*Juncus*) shoots made up a large part of a sample of eight specimens examined, while smaller quantities of water crow-foot (*Batrachium*), the seeds of water milfoil (*Myriophyllum*), and of bulrushes were present (Phillips, 1922–26).

Social behavior. Chiloé wigeons are strongly gregarious and highly sociable, at least during the nonbreeding season. Flocks often number in the dozens of birds and may reach as many as 200 on migration and in wintering areas (Phillips, 1922–26). Nevertheless, even in such flocks pair bonds probably persist; Weller (1968a) says that pairs are conspicuous in fall and winter flocks of this species. Pair formation evidently may occur immediately after the breeding season in this and other South American forms; Weller observed it among Chiloé wigeon in November, after a September to November breeding season. Pair formation may thus occur over a several-month period, and like that of the other wigeon species, is marked by inciting behavior on the part of females and by similar chin-lifting movements and whistled notes on the part of the males. This mutual display was compared by Lorenz (1951–53) to an anserine triumph ceremony, and related by him to the species' strong pair bond. Both sexes frequently perform display preening on each other, and males regularly turn the back of the head toward their mates as they swim ahead of them. Males also have an exaggerated general shake that clearly serves as a display, although like all wigeons, they lack a grunt-whistle display. Precopulatory behavior consists of mutual head pumping, and after one observed copulation the male called once and turned to face the female, who display-preened several times before starting to bathe (Johnsgard, 1965a).

MAP 66. Breeding or residential (hatched) and wintering (stippling) distributions of the Chiloé wigeon.

Reproductive biology. As noted earlier, nesting in central Argentina occurs from September to November, and in Chile Johnson (1965) found that it begins in August in the central provinces and a month or two later farther south. On the Falkland Islands nesting occurs between September and late December (Woods, 1975). In Tierra del Fuego females have been found incubating eggs as late as January, according to Johnson (1965). He states that the nest is located on dry ground, often in tall grass or among thistles, and frequently may be a considerable distance from water. The usual clutch ranges from 5 to 8 eggs, which are incubated only by the female. The normal incubation period is 26 days, a day or two longer than typical of the Northern Hemisphere wigeon, but the eggs are also slightly larger than in those species. With the hatching of the young, the male returns to help guard and rear the brood; Weller (1975a) reports that all of four broods that he

◆ ◆ ◆ 191

observed were tended by both parents, and that the broods ranged in size from 2 to 7 ducklings. The fledging period of this species has not yet been determined. It is possible that some birds undergo at least a limited molt-migration, since Weller noted a flock of about 5,000 molting birds on Lago Hantu, Isla Grande, in January or early February.

Status. This species is still relatively common; Johnson (1965) said that in Chile it is particularly abundant from Valdivia southward. Weller (1968a) reports that it was the sixth most common species of duck in a sample of hunters' kills from central Argentina, and constituted about 5 percent of the total. Johnson states that the birds become very wary and hard to approach after having been hunted, and thus are unlikely to be endangered by hunting.

Relationships. Except for the fact that the male has a strong pair bond and participates in brood care, this species differs little from the Northern Hemisphere wigeons and is clearly a close relative of them. It does have some unusual plumage features that distinguish it from them, such as the scalloped breast, but this pattern is shared with the falcated duck and the gadwall. Likewise, its unusually long and pointed tertials resemble those of the falcated duck, and indicate the close relationships between the wigeons and the more typical *Anas* species.

Suggested readings. Phillips, 1922–26; Johnson, 1965.

Falcated Duck

Anas falcata Georgi 1775

Other vernacular names. Bronze-capped teal, falcated teal; Sichelente (German); canard à faucilles (French); cerceta falcata (Spanish).

Subspecies and range. No subspecies recognized. Breeds in Asia south of the Arctic Circle from the Upper Yenisei River east to Kamchatka, south to Lake Baikal, eastern Mongolia, the Amur River, and Ussuriland. Winters in Japan, Korea, and eastern and southern China south to Burma. See map 67.

Measurements and weights. Folded wing: males, 230–42 mm; females, 225–35 mm. Culmen: males, 40–42 mm; females, 38–40 mm. Weights: males, 590–770 g (av. 713 g); females, 422–700 g (av. 585 g) (Tso-hsin, 1963). Eggs: av. 56 x 40 mm, creamish, 49 g.

Identification and field marks. Length 18–21″ (46–53 cm). Plate 39. *Males* in breeding plumage have a strongly crested head that is mostly iridescent bronzy green and chestnut purple, but with a white spot above the base of the upper mandible. The throat and foreneck are also white, with a narrow green collar. The body plumage is primarily gray, with fine black vermiculations, and with the black on the breast forming crescents. The under tail coverts are patterned with two buff triangles, separated medially and anteriorly with black, while the tail is gray, edged with white. The upper tail coverts are gray and black, the coverts very long and partially hiding the tail. The longer scapulars and tertials are gray and black, the latter being greatly extending and curved down over the other wing feathers. The speculum is iridescent green, bounded in front and behind with white lines, while the upper wing coverts are gray. The iris is brown, the bill blackish, and the legs and feet are bluish gray to yellowish. *Females* are mostly brown and gadwall-like, but a small crest is present and the speculum is iridescent green, as in males. The mandible is spotted with black, and yellow is extensive on the lower mandible. *Males in eclipse* resemble females, but have a more brilliant speculum and grayish rather than brownish upper wing coverts. *Juveniles* resemble adult females but lack the nape crest.

In the field, falcated ducks are best recognized by the distinctive shape and plumage of the male, since females are so gadwall-like. The females might also be mistaken for wigeon, but the longer bills and the rudimentary crests should help to separate them. The calls of the female are generally like those of a gadwall; the decrescendo call is from two to five syllables in length. Males utter a high-pitched whistle, *lililili*, and also produce a vibrating *rruh-urr* call during display.

Natural History

Habitat and foods. On spring migration, falcated ducks occupy lakes, rivers, and their floodwaters, but rarely coastal habitats. Breeding-season habitats include small lakes and the oxbows of rivers that have both open and wooded banks. On the wintering grounds the birds do occur on marine coastlines, on the open coves of lakes, and in particular forage in rice paddies and shallow fresh-water areas. There have been almost no quantitative studies of foods taken, but some samples from China indicate that the foods are almost entirely vegetable. Besides rice and grain, the seeds of knotweeds (*Polygonum*); leaves, stems, and roots of pondweeds; grass seeds of several species; and other hydrophytes have been found. Foods of animal origins include soft-shelled bivalve mollusks, gastropod mollusks, and aquatic insects (Tso-hsin, 1963; Dementiev & Gladkov, 1967).

Social behavior. Falcated ducks are relatively social, and on migration and wintering areas occur in flocks that range from small to fairly large. They reportedly also often associate with other species, including gadwalls and presumably other species of *Anas*, in wintering areas. Evidently pair formation occurs during fall migration, since by mid-December the birds are already in pairs. Pair bonding is obviously seasonal but nonetheless is strong, and courtship display among captive birds is neither conspicuous nor greatly prolonged. Females have an inciting call and display that is very gadwall-like, with strong lateral pointing alternated with chin lifting and uttering a soft *rrr* note. Females also often perform a malelike introductory shake display and preen behind the wing toward preferred males.

Lastly, they also often call at the precise moment that a male is displaying, uttering a hoarse *gak-gak* call. Males have the typical *Anas* repertoire of displays, including an exaggerated introductory shake, a neck-stretching burp call, a grunt-whistle, and a head-up-tail-up display. Males also have a strongly developed chin-lifting display, which is directed toward other males as well as females, and is accompanied by quickly repeated whistling notes. Like the chin lifting of wigeons, chin lifting in this species takes on an aspect similar to an anserine triumph ceremony. Turning of the back of the head is strongly developed and of course beautifully displays the crest, while display preening is not so fre-

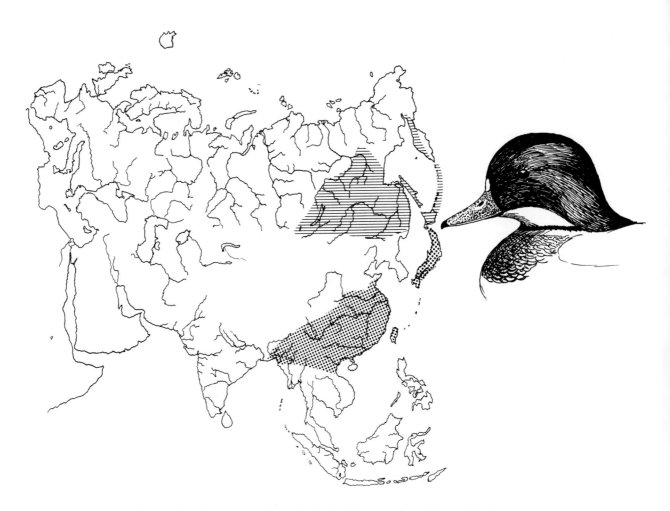

MAP 67. Breeding (hatched) and wintering (stippling) distributions of the falcated duck.

quently performed as in the wigeon group. Copulatory behavior has not yet been fully described, but precopulatory head pumping has been observed (Lorenz & von de Wall, 1960; Johnsgard, 1965a).

Reproductive biology. Nesting occurs in the U.S.S.R. in May, June, and perhaps early July, while in northern China it begins in May. The nest is placed in tall grass or small bush cover, usually near lake shores, but at times as far as 80 meters from the nearest water. The reported clutch range in wild birds is from 6 to 9 eggs, with 8 apparently being the most typical number. The male reportedly remains in the vicinity of the nest for the first half of the brooding period and joins the female during her foraging or rest periods. The incubation period is 24 to 25 days, with ducklings appearing in the U.S.S.R. as early as mid-June and as late as early August. The fledging period is as yet unreported. Flocking

by males before molting begins in June, and probably extends through July (Dementiev & Gladkov, 1967; Tso-hsin, 1963).

Status. The falcated duck is an important source of food in both the U.S.S.R. and China, and in China it is also harvested for its ornamental feathers. There seems to be no information on the population status of this species, but also no reason to believe that it is presently declining.

Relationships. It has generally been recognized that the falcated duck is a close relative to the gadwall. What is less apparent and equally interesting is that it has some strong behavioral similarities to the wigeons that suggest it is probably a link-form between these two groups and helps to unite the wigeons within the genus *Anas* (Johnsgard, 1965a).

Suggested readings. Dementiev & Gladkov, 1967.

Gadwall

Anas strepera Linnaeus 1758

Other vernacular names. Gray duck; Schnatterente (German); chipeau (French); pato ruidosa (Spanish).

Subspecies and ranges. (See map 68.)

A. s. strepera: Common gadwall. Breeds in Iceland, southeastern England, and across Europe and Asia to Kamchatka, south to Holland, Germany, central Russia, the Caspian Sea, and west to Transbaikalia. In North America breeds from the Prairie Provinces of Canada south to southern California, Colorado, Nebraska, and Wisconsin, with limited breeding in eastern Canada and on the Atlantic coast. Winters from southern Europe, Asia Minor, the Himalayas, China, and Japan south to northern Africa, the Nile Valley, India, Assam, and southern China, and in North America from British Columbia to Baja California, and from the Atlantic coastal states to the Gulf Coast, Yucatan, and the West Indies.

A. s. couesi: Coues gadwall. Extinct; originally limited to Washington and New York Islands of the Fanning Group, central Pacific Ocean.

Measurements and weights. (Exclusive of *couesi.*) Folded wing: males, 260–82 mm; females, 235–60 mm. Culmen: males, 38–45 mm; females, 36–42 mm. Weights: Adult males shot in the fall average about 990 g and females about 850 g, with maxima of about 1,180 and 1,050 g, respectively (Johnsgard, 1975). Eggs: av. 55 x 39 mm, pale pink, 44 g.

Identification and field marks. Length 19–23″ (48–58 cm). *Males* in breeding plumage have the forehead,

sides of the head, and upper neck buff, spotted with blackish brown, while the crown and area above the eyes is chestnut, barred with black. A short, bushy, paler crest is present, and is also barred with black. The lower neck, back, and shorter scapulars are all brownish black, with some vermiculations. The longer scapulars are dull brown, narrow, and pointed, with cinnamon margins. The rump is brownish black, and the tail coverts are glossy black. The tail is brownish gray; the breast grayish black, scaled with black scallops; and the flanks are finely vermiculated with dark gray and white. The upper wing coverts are chestnut, except for the lesser coverts, which are brownish gray. The outer secondaries are black or gray, the inner ones are white, and the tertials are pale grayish buff. The abdomen, underwing surface, and axillars are white. The iris is brown, the bill slate gray, and the legs and feet are grayish yellow. *Males in eclipse* resemble the females even to the color of the bill, but the upper wing coverts have more chestnut-colored feathers, the feathers of the upper breast are more finely barred, and the crown is plain blackish. *Females* are brownish black on the top of the head,

with pale buff or whitish streaking, and the sides of the head are blackish brown, with lighter markings. The chin and throat are whitish, and the lower neck and breast are blackish brown, with pale buff edges and central markings. The lower breast and abdomen are white, and the sides, flanks, and under tail coverts are grayish brown with pale buff edges and central markings. The back, scapulars, rump, tail, and upper tail coverts are brown, margined with buff. The lesser and middle wing coverts are mostly brownish gray, with some chestnut present; the middle greater coverts are black and the others gray. The outer secondaries are gray, the middle ones gray or black, and the inner ones gray with white edges and tips. The iris is brown, the legs and feet straw yellow to orange brown, and the bill is dusky, with yellow or dull orange sides and variable dark spotting that is strongest in late summer. *Juveniles* resemble adult females, but are darker and more heavily streaked below. *First-winter males* have less chestnut in the wing coverts than do adults, and in *young females* the black of the greater coverts and secondaries may be replaced with dusky coloration.

MAP 68. Breeding (hatched) and wintering (stippling) distributions of the common gadwall.

In the field, gadwalls are often rather nondescript, but males in breeding plumage appear rather uniformly grayish except for the black hindquarters. In flight, the white secondaries are evident in both sexes and provide the best field mark. Males have a low-pitched, reedy call which during display is interspersed with whistling notes, often in a distinctive *raeb-zee-zee-raeb-raeb* cadence. Females have the usual decrescendo call, which is mallardlike but somewhat more rapid and higher in pitch, and an alarm call of paced *quack* notes. The inciting call is also mallardlike.

NATURAL HISTORY

Habitat and foods. The common gadwall has a very large breeding range, but is most abundant where marshes or small lakes occur in grasslands, particularly where the marshes are somewhat alkaline and shallow but permanent, and where grass-covered islands provide nesting cover. Summer and fall foods are those that can be obtained on such water areas by tipping-up, and include the seeds and vegetative parts of a variety of aquatic plants, especially submerged species (Gates, 1957). A very low incidence of animal materials occurs in the diet, except in juvenile birds, and even on brackish or fresh-water estuaries of their wintering grounds the birds remain essentially vegetarians, concentrating on the vegetative parts of such plants as wigeon grass, muskgrass, eelgrass (*Zostera*), and pondweeds (Stewart, 1962).

Social behavior. Gadwalls become sexually mature in their first winter of life and, except for juvenile birds, social display begins remarkably early in the fall, well before the males have attained full nuptial plumage. In Austria this activity begins in August, and over half of the females appear to become paired within a month. However, there is also a spring peak of social display, after the birds are in full plumage, during which strong pair bonds are presumably established. The inciting behavior of female gadwalls serves as the initiating mechanism for social display, and is much like that of most *Anas* species in sound and posturing. Males perform a relatively large number of complex displays, including the grunt-whistle, head-up-tail-up, down-up (these last two linked in sequence), uttering a burp whistle, and turning the back of the head toward inciting females. A good deal of aerial chasing occurs

during social display, and burping by the males during such chases is regular. Copulation is preceded by the usual mutual head pumping, and is followed by the male's uttering his burp whistle, then turning to face the bathing female (Johnsgard, 1965a).

Reproductive biology. Gadwalls usually arrive on their nesting areas already paired, but nonetheless there is usually a delay of several weeks between the time of arrival and the establishment of a nesting site. During this period the pair occupies a fairly large home range of up to several hundred acres, especially where the birds nest on islands. Little intraspecific aggression is evident among gadwalls, which often nest colonially where islands provide favored nesting locations. Dense and dry herbaceous cover is preferred for nesting over wet and sparse cover, and usually the nest is well hidden from above and all sides. Eggs are laid at a daily rate until a full clutch of about 10 eggs is present in initial nesting efforts, but renesting and parasitic nesting often influence observed clutch sizes in this species. Males typically desert their females before the midpoint of the incubation period, which is normally 26 days. After hatching, females usually move their broods to deeper marshes or the edges of large impoundments, at times more than a mile from their nesting sites. Evidently little or no brood merging occurs in this species, but females sometimes adopt stray young, and remain with their offspring for most of the 50- to 60-day fledging period. About six weeks after hatching the broods, females begin their flightless period, which lasts for approximately a month. In southern Canada they regain their powers of flight by early September, or shortly before the onset of the fall migration (Gates, 1962; Oring, 1969; Duebbert, 1966).

Status. Bellrose (1976) reported that between 1955 and 1974 an average breeding population of 1,432,000 gadwalls occurred in North America, with most of these birds in the Dakotas and the Prairie Provinces of Canada. For still unknown reasons the species has recently become more abundant in the interior of the country, but wintering numbers in the Atlantic flyway have declined. Ogilvie (1975) indicated that in the western U.S.S.R. there are an estimated 163,000 pairs of gadwall, as well as much smaller numbers in Iceland, southern Sweden. Britain (100 to 200 pairs), Spain, and a few hundred pairs in Czechoslovakia, East Germany, West Germany, and France. There are no numerical estimates available for the breeding population of eastern Asia.

Relationships. Evidence from plumage and behavior rather clearly suggests that the gadwall helps to link the falcated duck and wigeons with the most typical *Anas* forms (Johnsgard, 1965a), and there seems to be no reason for retaining this species in a separate genus, as was the practice at one time. This position has also support by Woolfenden's (1961) analysis of the postcranial skeleton.

Suggested readings. Gates, 1962; Duebbert, 1966; Oring, 1969.

Baikal Teal

Anas formosa Georgi 1775

Other vernacular names. Clucking teal, Formosa teal, spectacled teal; Gluckente (German); sarcelle formose (French); cerceta del Baikal (Spanish).

Subspecies and range. No subspecies recognized. Breeds in eastern Asia to about 70° north latitude, east from the Yenisei River to the Kolima Delta and Anadyr, south to Lake Baikal, northern Sakhalin, and northern Kamchatka. Winters in Korea, Japan, and China, wintering as far south as Swatow in southeastern China. See map 69.

Measurements and weights. Folded wing: males, 200–220 mm; females, 180–210 mm. Culmen: males, 35–38 mm; females, 33–35 mm. Weights: males, 360–520 g (av. 437 g); females, 402–505 g (av. 431 g) (Tso-hsin, 1963). Eggs: av. 48 x 35 mm, greenish, 31 g.

Identification and field marks. Length: 16" (40 cm). Plate 40. *Males* in breeding plumage have a unique head pattern that includes a black crown, hind neck, and upper throat, with the face, sides of the upper foreneck, and lower throat buff, narrowly margined with white. A narrow black line extends from the eye to the upper throat, dividing the buff area, and an iridescent green patch that is narrowly bordered with black extends from behind the eye to the nape and downward along the sides of the neck. The breast is pink to vinaceous, with small black spotting, and the flanks and mantle are gray, with black and brownish vermiculations. The inner scapulars and tertials are elongated and pointed, with black, white, and cinnamon striping. Vertical white bands occur between the breast and flanks and on the sides of the rump in front of the black under tail coverts. The upper wing surface is brown, except for an iridescent green and black speculum that is formed by the secondaries and is bounded anteriorly with

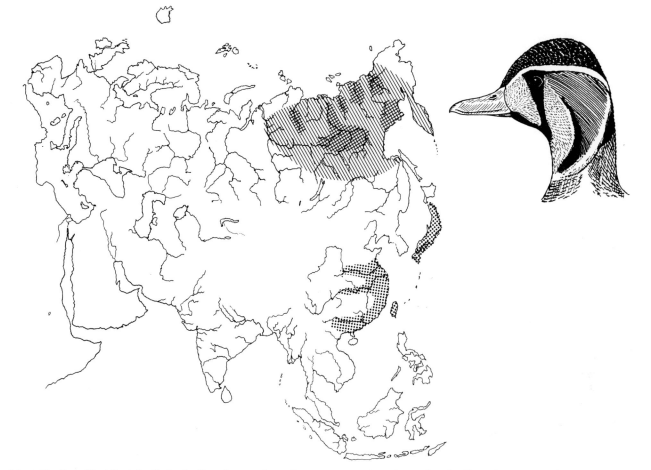

MAP 69. Breeding (hatched), including densest breeding concentrations (cross-hatched), and wintering (stippling) distributions of the Baikal teal.

buff and posteriorly with white. The iris is brown, the bill is bluish black, and the legs and feet are bluish gray or yellowish gray, with darker webs. *Females* closely resemble female green-winged teal, but have darker upperparts and a white or nearly white rounded spot on the cheeks just behind the bill. There is an interrupted pale buff line above the eye, and an extension of the nearly white throat up the side of the face behind the eye. Blackish centers on the back and scapular feathers are conspicuous; the scapulars are not pointed as in the male. The wing speculum pattern and soft-part colors are like those of the male. *Males in eclipse* closely resemble females, but lack the extension of the white throat up the sides of the head, and the white spot behind the bill is less conspicuous. *Juveniles* resemble adult females, but their undersides are spotted or streaked with brown.

In the field, the male in nuptial plumage is unmistakable if its facial markings can be seen clearly;

the white bars in front of and behind the flanks are also quite conspicuous. Females can be separated from female green-winged teal only with difficulty, and also resemble female garganeys. The calls of the female are typically *Anas*-like; the decrescendo call consists of one long and about five shorter and softer notes. During courtship, males utter a highly distinctive *ruk* or *ruk-ruk* call which is the basis for their name "clucking teal."

NATURAL HISTORY

Habitat and foods. The Baikal teal evidently occupies a variety of habitats in its natural range, but in the summer it is said to prefer small wooded bodies of water and islands in river deltas. On migration and wintering areas it occurs on rice fields, floodplain lakes, and river floodwaters, and probably consumes a wide variety of foods in these diverse areas. A few samples from China indicate that such foods as rice, knotweeds (*Polygonum*), pondweeds (*Blyxa* and *Hydrilla*), grass seeds, and aquatic invertebrates such as snails and insects are consumed (Tso-hsin, 1963). The bill shape of this species is much like that of the green-winged teal, and presumably the birds forage in much the same manner. They have not been observed diving while foraging, and instead seem to dabble at the water's surface. Captive birds spend a good deal of time on shore, and probably eat a variety of shoreline or terrestrial plants in addition to aquatic ones. Records from the U.S.S.R. indicate that the birds even feed on highways at night, consuming soybeans lost in transport, or seek out acorns in woods, often at great distances from water. The presence of seeds of such terrestrial plants as poppies, foxtail, and millet in the stomachs of collected birds also suggests a relatively terrestrial diet (Dementiev & Gladkov, 1967).

Social behavior. Wintering flocks of Baikal teal in Japan are at times incredibly large; Austin (1948) describes flocks of 10,000 or more as not uncommon in the Kansai area, and groups of up to 100,000 have been seen near Osaka. Males are among the latest of the Northern Hemisphere dabbling ducks to acquire their full breeding plumage, and sexual display does not begin until that time. Pair formation probably occurs between January and March, since the birds are reportedly already paired on their arrival at the breeding grounds. Displays of wild birds have not been described, but captive Baikal teal exhibit an unusual display repertoire. Females have a gadwall-like inciting display, with strong chin lifting, and a repeated series of *geg* notes. Females also utter a louder *geg-geg* call as the males display to them, but unlike green-winged teal, they do not perform nod-swimming. The male displays are very few, consisting mostly of aggressive bill tilting, which exhibits their black throats, and a distinctive burp display, in which the head is suddenly thrown up from a withdrawn posture, the crest is raised, and a single- or double-noted *ruk* call is uttered. This display evidently replaces all of the other elaborate displays (grunt-whistle, head-up-tail-up, down-up, etc.) found in related species, and is repeated endlessly by courting males. Males also turn the back of the head in a somewhat exaggerated manner as they swim ahead of inciting females and sometimes perform display preening behind the wing, but have virtually no other displays except drinking and the introductory shake. Other than the usual precopulatory head pumping, no copulatory behavior has yet been described for this species (Johnsgard, 1965a).

Reproductive biology. Nesting reportedly occurs in northern China during May, June, and July, and in the U.S.S.R. eggs have been found during these same three months. Clutches of incubated eggs in the latter area contained from 6 to 9 eggs, and nests have been found under willows or juniper bushes, beside very small forest ponds, and on open mossy tundra. In China the locations have included grassy thickets along river banks or near the shores of lakes, as well as small holes. Nests have also been reported under large branches and in piles of driftwood. Records of captive birds indicate an incubation period of 25 days. Ducklings in the U.S.S.R. have been seen from early July until August, but the fledging period is still undetermined (Tso-hsin, 1963; Dementiev & Gladkov, 1967).

Status. In the U.S.S.R. and China the Baikal teal is one of the most widely harvested of waterfowl, and in China it is hunted for meat as well as for the ornamental feathers of the male. In Japan it is captured in great quantities with throw nets; Austin (1948) reports one catch of 10,000 birds in a single day by only three men. There is no information on the total population of this species, but it apparently is still one of the most common surface-feeding ducks of eastern Asia.

Relationships. Delacour (1954–64) believed that this teal occupies a rather isolated position in the genus *Anas*, with no close relatives, but I (1965a) have concluded that it is probably a fairly close relative of the green-winged teal.

Suggested readings. Dementiev & Gladkov, 1967; Phillips, 1922–26.

Green-winged Teal

Anas crecca Linnaeus 1758

Other vernacular names. Common teal, greenwing; Krickente (German); sarcelle d'hiver (French); cerceta de alas verdes (Spanish).

Subspecies and ranges. (See map 70.)

  A. c. crecca: Eurasian green-winged teal. Breeds in Iceland, Great Britain, and Europe and Asia north to latitude 70° and south to the Mediterranean Sea from the Atlantic to the Pacific coast, and on Sakhalin and in the northern half of Japan. Winters from the southern parts of its breeding range to northern and central Africa, across Asia Minor, India, and in southeastern Asia to Malaya, eastern China, and southern Japan.

  A. c. nimia: Aleutian green-winged teal. Resident on the Aleutian Islands, east to Akutan.

  A. c. carolinensis: North American green-winged teal. Breeds from northern Alaska, northern Mackenzie, Great Slave Lake, and Fort Churchill south to California, northern New Mexico, and east to Michigan, Quebec, the Maritime Provinces, and Newfoundland. Winters in the southern United States, Mexico, and occasionally reaches the West Indies and Central America.

Measurements and weights. Folded wing: males, 257–440 mm; females, 250–374 mm. Culmen: males, 34–38 mm; females, 34–36 mm. Weights: males (of *carolinensis*) shot in the fall average about 360 g, with a maximum of 453 g, while females average 340 g, with a probable maximum of 510 g (Johnsgard, 1975). Eggs: av. 45 x 33 mm, yellowish white, 29 g.

Identification and field marks. Length 13–16" (38–43 cm). *Males* in breeding plumage have an iridescent

Map 70. Breeding or residential distributions of the Aleutian ("A"), North American ("Am"), and Eurasian ("E") green-winged teals. The wintering distributions of the two latter races is shown by stippling.

green to purple patch on the side of the head extending from the eyes to the nape, and narrowly outlined in white (less well developed in *carolinensis*). The rest of the head and neck are chestnut except for a purplish black occipital crest. The neck, scapulars, front of the back, and sides of the breast and flanks are vermiculated with grayish brown and white. Several outer scapulars are black with a narrow white shaft line (*carolinensis*) or are black and white (other races); the American race also has a conspicuous vertical white bar in front of the wing. The hind back, rump, and shorter upper tail coverts are grayish brown, and the longer ones dusky to black, with wide margins. The lower foreneck and breast are pinkish buff, spotted with dusky coloration, and the abdomen is white. The central under tail coverts are black, the longer ones margined with white, and the lateral ones forming a buffy yellow triangular patch. The tail and primaries are grayish brown, as are the upper wing coverts, except for the greater secondary coverts, which are tipped with brownish buff. The secondaries are black outwardly and iridescent green inwardly; the tertials are brownish gray. The iris is brown, the bill black, and the legs and feet are bluish gray. *Males in eclipse* resemble females but have plainer upperparts, less

distinct spotting on the underparts, and usually show some vermiculations on the scapulars. *Females* have a blackish crown, with buffy brown markings, and are whitish on the sides of the head and neck, with grayish brown spotting, except for the chin and throat, which are white. The hind neck and most of the upperparts are dull grayish brown, barred and margined with dull whitish coloration; the chest is deep buff, with blackish spots; and the abdomen is white. The sides and flanks are grayish brown, with dull whitish edges and central markings, and the under tail coverts are white, streaked with grayish brown. The wings and tail are as in the male. The iris is brown, the feet and legs brownish, and the bill is mostly blackish or purplish gray spotted with black. *Juveniles* resemble adult females but have spotting on the belly and lack buff markings on the upper tail coverts.

In the field, its tiny size separates this species from most other dabbling ducks, and the yellow triangular patches under the tail of males is also a useful field mark. The vertical white bar in front of the wings (American race) or the horizontal line above the wings (other races) also aids in identification. Females resemble female garganeys and female blue-winged or cinnamon teal, but have well-developed dark eye-stripes and crowns, are generally grayer in feather tone, and have shorter and weaker bills than any of these. The courtship call of the male is a distinctive "cricket" whistle, *krick'et*, which can be heard great distances, while the female has a high-pitched decrescendo call of about four notes, and various other weak quacking calls.

NATURAL HISTORY

Habitat and foods. Throughout its broad breeding range the green-winged teal occupies rather diverse habitats, but in general seems to prefer small and shallow but permanent ponds in the vicinity of woodlands, and with fairly dense herbaceous nesting cover available nearby. Ponds with an abundance of emergent vegetation are preferred to open water, and those shallow enough to allow for foraging by tipping-up rather than diving are preferentially utilized. The birds feed on a variety of fairly small plant seeds, mainly of grasses, bulrushes, and some forbs such as smartweeds (*Polygonum*). In addition the birds seem to relish the stems and leafy parts of pondweeds and the reproductive bodies of muskgrass. Small invertebrates such as tiny mollusks that can be found along tidal mud flats are also sometimes eaten, as well as some snails and amphipods. Tidal creeks and marshes associated with estuaries are preferred to more saline or open-water habitats, especially where mud flats for foraging can be found (Olney, 1963b; Stewart, 1962).

Social behavior. Perhaps because the species seems to prefer small ponds with limited food resources, it is rarely found in large flocks even in fall, when the sexes migrate relatively separately. Adult males tend to move southward earlier but do not travel as far as females and young birds. The birds mature their first winter, and adult birds may begin some display as early as September, although display activity does not peak until about mid-March, when 90 percent of the females have become paired. Male displays of this species are numerous, and include most typical *Anas* displays such as the grunt-whistle, head-up-tail-up, down-up, and bridling. This latter display, which occurs among the mallardlike ducks only as a postcopulatory display, is present in the green-winged teal as an independent courtship display as well, and furthermore is frequently performed by birds standing on land as well as when they are swimming. Aerial activity seems to be of

little significance during display, and may serve only to change the location of the displaying group. However, the *krick-et* whistle, or burp, is often uttered by males during such flights. Copulation is preceded by mutual head pumping; and after treading is completed, the male utters this whistle while in the bridling posture (Johnsgard, 1965a; McKinney, 1965a).

Reproductive biology. Green-winged teal arrive in their northern breeding areas surprisingly early and begin nesting activities shortly thereafter. They become well dispersed over the terrain, with resultant low population densities. In addition, the females become quite secretive and hide their nests extremely well. They use sites that may be mainly shrubs or herbaceous vegetation, but in either case the vegetation used is very dense, with little penetration of light. The usual clutch size is from 8 to 10 eggs, presumably laid at the rate of one per day. Males desert their mates at about the time incubation begins, and sometimes migrate some distance to special molting areas. The incubation period is from 21 to 23 days, and the female is a surprisingly strong defender of her nest or brood, in spite of her tiny size. It has been suggested that this very effective defense keeps the mortality of the tiny ducklings at a minimum. They grow very fast and fledge at the remarkably early age of about 35 days. It is probable that the female remains with her brood until this time, before undergoing her flightless period, but there have been a few reported cases of females undertaking fairly long flights to molting areas hundreds of miles from the nearest nesting grounds (Munro, 1949a; Keith, 1961; Bengtson, 1970).

Status. Wintering ground survey data are not very reliable for this species because of its small size and dispersal characteristics. Using indirect calculations based on banding and harvest data, Moisan et al. (1967) estimated a North American spring population of more than 3 million birds. There are no available estimates for the population of the Aleutian green-winged teal, but it is obviously quite small. Likewise, figures for the Eurasian race are incomplete, but Ogilvie (1975) indicated that about 250,000 winter in northwestern Europe and another 750,000 around the Mediterranean and Black seas, particularly in Turkey. It is impossible to guess the size of the more easterly segment, but in the U.S.S.R. it is one of the most abundant duck species, and the same is probably true in China.

Relationships. The green-winged teal has a close relative and ecological counterpart in South America, the speckled teal, and a somewhat less close relative in southern Africa, the Cape teal. It is likewise obviously rather closely related to the gray and chestnut teals of Australia and the South Pacific, but has a larger distribution than any of these forms.

Suggested readings. Munro, 1949a; McKinney, 1965a.

Speckled Teal

Anas flavirostris Vieillot 1816

Other vernacular names. Chilean teal, sharp-winged teal, South American teal, yellow-billed teal; Chile-Krickente (German); sarcelle du Chili (French); pato jergon (Spanish).

Subspecies and ranges. (See map 71.)
   A. f. flavirostris: Chilean speckled teal. Breeds from northern Argentina and central Chile south to Tierra del Fuego, and resident on the Falkland Islands and South Georgia. Migrates into Uruguay and Brazil in winter.
   A. f. oxyptera: Sharp-winged speckled teal. Resident in the Andean region of central and southern Peru, western Bolivia, northern Chile, and northern Argentina, mainly in the *puna* zone, but also reaching coastal valleys in Chile.
   A. f. andium: Andean speckled teal. Resident in the *páramo* zone of the central and eastern Andes of Colombia and northern Ecuador.
   A. f. altipetans: Merida speckled teal. Resident in the *páramo* zone of the Andes of Venezuela and eastern Colombia, south to Bogota.

Measurements and weights. Folded wing: males, 190–240 mm; females, 185–215 mm. Culmen: males, 33–41 mm; females, 30–40 mm. Weights: both sexes of *oxyptera*, 390–420 g (Koepke & Koepke, 1965); males of *flavirostris* average 429 g, females 394 g (Weller, 1968a). Eggs: av. 53 x 37 mm, creamy white, 39 g.

Identification and field marks. Length 15–17" (38–43 cm). *Adults* have a head and neck that is gray, finely barred with black, and upper parts grayish except for the scapulars, which are black with lighter edges. The breast is silvery gray, heavily speckled with small black spots, while the abdomen, flanks, and tail coverts are unspotted gray to brownish gray, varying with the subspecies. The tail is grayish brown, and the upper wing coverts are gray, except for the greater secondary coverts, which form a cinnamon band in front of the speculum. The inner secondaries are iridescent green and the outer ones are black, and all have buff to pale cinnamon tips. The iris is brown, the legs and feet are gray, and the bill is mostly yellow (*flavirostris* and *oxyptera*) or lead blue (*andium* and *altipetans*) with a black tip and culmen stripe. *Females* resemble males but are smaller and generally duller, with less colorful bills and slightly darker heads. *Juveniles* exhibit spotting on the underparts and have duller bill coloration.

In the field, speckled teals appear to be generally dark brown above and almost uniformly gray to grayish brown on the head, breast, and flanks. The species' small size and teallike shape separate it from the larger brown pintail, while the similar-sized silver teal exhibits a strongly bicolored head pattern. Males produce a repeated musical whistling note nearly identical to that of the green-winged teal, while females utter a high-pitched quacking call like those of other small teals.

NATURAL HISTORY

Habitat and foods. This is a relatively adaptable species and is found in a variety of fresh-water, brackish, and even marine habitats. It is especially common on small fresh-water ponds near rivers or on coastal lagoons, and during the breeding season is often found in forested areas. On Isla Grande, Tierra del Fuego, it is called the mud teal, and is usually found at the edges of lakes or marshes, or on the mud flats of streams and estuaries (Weller, 1975a). On the Falkland Islands it is widespread, but is especially common where there are small ponds with aquatic weeds and an abundance of small invertebrate life occurs. Weller (1972) reported on the foods of adult and young birds from that region, and found that small crustaceans, amphipods, the larvae of midges, and similar small-sized foods were prevalent. In the winter, the teal sometimes move to the seashore to feed on rotting kelp, and they also at times eat the seeds of pig vine (*Gunnera*) in winter (Woods, 1975).

Social behavior. The speckled teal is a gregarious bird and, except when breeding, is found in flocks that usually number from 10 to 20 birds, or at times

MAP 71. The breeding or residential distributions of the Merida ("M"), Andean ("A"), sharp-winged ("S"), and Chilean ("C") speckled teals. The wintering distributions of the Chilean speckled teal is indicated by stippling.

may approach 200. Pairs remain intact while in winter flocks (Weller, 1968a), and thus much of the flocking at this time is presumably for reasons of safety rather than to facilitate social courtship. Nevertheless, a good deal of courtship, presumably mostly by first-year birds, does occur in winter flocks, and Weller observed it within a month of the termination of breeding in Argentina. Although males do not invariably remain with their mates to help raise the young (Weller [1975a] reported seeing both members of the pair with three out of five broods), it seems likely that the majority of females remate with their previous mates. Displays of the various subspecies appear to be nearly identical, and closely resemble those of the Northern Hemisphere green-winged teal. Females perform a nearly identical inciting display, utter a decrescendo call of from 5 to 12 syllables, and sometimes nod-swim during social display. Both sexes perform display preening behind the wing, and the female rarely performs the "gesture of greeting," a gaping with open bill and neck extended forward. Males perform nearly the entire *Anas* repertoire of displays, including the burp, the grunt-whistle, the head-up-tail-up, nod-swimming, turning-the-back-of-the-head, and, in particular, bridling. This display, a rearing of the head back on the scapulars, is performed either on land or on water, and is perhaps more common in this species than in any other *Anas*. As a precopulatory display both sexes perform mutual head pumping, and after treading, the male makes a single bridling movement while the female bathes (Johnsgard, 1965a).

Reproductive biology. In Chile, nesting by the sharp-winged teal occurs in November and December, while the yellow-billed form of more southern areas nests from September or late August onward. Perhaps two broods are raised in central Chile, but in more southerly regions, where nesting does not begin until late October, only a single brood is raised (Johnson, 1965). Nest sites evidently vary greatly with locality; Johnson states that the sharp-winged race often nests in holes of banks or escarpments, while *flavirostris* typically nests on the ground among vegetation near water, or in forks of trees, especially the eucalyptus. In eastern Argentina, the yellow-billed form is referred to as the tree teal, since it nests in the tree nests of monk parakeets (*Myopsitta monarcha*), in eucalyptus trees, or in tala (*Celtis*) trees (Weller, 1967a), as well as on the ground or in clay banks. In the Falkland Islands it usually nests among grass, often a mile or so from water. From 5 to 8 eggs constitute a normal clutch, and the incubation is performed entirely by the female. The normal incubation period is 24 days, and after hatching, most males evidently rejoin their mates to help care for the brood. The fledging period is approximately six or seven weeks (Weller, 1972). During the molting period adults often gather in relatively large flocks; Weller (1975a) noted a flock of 175 that contained a few flightless adults.

Status. Johnson (1965) considered the speckled teal to be the commonest small duck in Chile, particularly in the lake district and southward. The species is widespread in Patagonia, and on the Falkland Islands Weller (1972) found it to be the fifth most common waterfowl species. He noted (1968a) that it ranked second in total number of birds examined among samples killed by hunters in eastern Argentina, constituting about 10 percent of the total sample. The Andean populations are doubtless far less abundant and the ones that should be watched more closely for possible declines.

Relationships. The close relationship between *flavirostris* and *crecca* is very evident in terms of their morphology, ecology, and behavioral characteristics; they were considered by me (1965a) to constitute a superspecies.

Suggested readings. Johnson, 1965; Woods, 1975.

Measurements and weights. Folded wing: both sexes 168–206 mm. Culmen: both sexes 34–44 mm. Weights: males, 352–502 g (av. 419 g); females, 316–451 g (av. 380 g). Eggs: av. 49 x 36 mm, pale to deep cream, 30 g.

Identification and field marks. Length 18″ (46 cm). Plate 41. *Adults* have buffy white heads, flecked with blackish brown on the sides and crown. The back and scapulars are dark brown, edged and mottled with pale reddish buff, while the breast, abdomen, flanks, and tail coverts are mostly buffy white, variably spotted with brown. The tail is gray, edged with whitish coloration, while the upper wing coverts are dark gray, except for the greater secondary coverts, which are tipped with white. The secondaries form an iridescent green speculum, bordered inwardly and outwardly with black, and with trailing black and white bars. The feet and legs are yellow, the iris pink, and the bill is pink, with narrow black edges and a black base. *Females* have slightly darker heads than males, and their bill coloration is somewhat duller; they lack the rudimentary occipital crest present in males. *Juveniles* closely resemble adults, but probably have less colorful iris and bill pigmentation.

In the field, the predominantly pale gray color of this small duck, and its preference for alkaline habitats, sets it apart from other African waterfowl. The only other species approximating its size and shape is the red-billed pintail, which has a strongly bicolored head pattern and lacks an iridescent speculum pattern. The species does not usually make much noise on the water or while in flight, but the female utters a definite decrescendo call of four to eight syllables, with the second the loudest, while

Cape Teal

Anas capensis Gmelin 1789

Other vernacular names. Cape wigeon, pink-billed teal; Kapente (German); sarcelle du Cap (French); cerceta del Cabo (Spanish).

Subspecies and range. No subspecies recognized. The overall African range is from the eastern Sudan and Ethiopia south to the Cape, but the major areas of occurrence are in South Africa (including the coast) and on the Ethiopian, Kenyan, and Tanzanian alkaline lakes. See map 72.

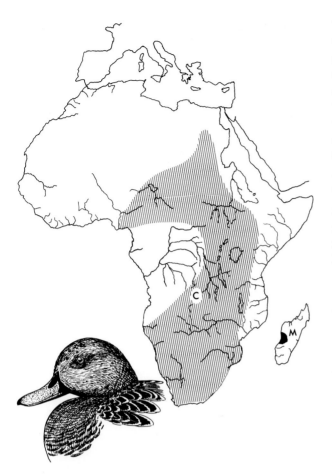

MAP 72. The breeding or residential distributions of the Madagascan teal ("M") and Cape teal ("C").

the male produces a clear and fairly loud whistle, *oo-whee'-oo*, during aquatic display.

NATURAL HISTORY

Habitat and foods. General agreement exists that standing open waters, particularly shallow and saline ones, are favored habitats. Less often the birds occur on rivers, coastal shorelines, and other areas, but saline pools, evaporation pools of sewage and mine works, and temporary or permanent vleis are strongly favored. Foraging on such waters is done both from the surface and by diving; the Cape teal is among the most accomplished of all *Anas* species in its diving abilities, at least judging from captive birds. The birds frequently remain submerged for five seconds or more when thus foraging, and when swimming under water keep their wings tightly closed. An analysis of foods of wild birds indicated that less than 20 percent is from vegetable sources, primarily the leafy parts and seeds of pondweeds. Most of the remainder consisted of insect remains, with smaller quantities of crustaceans and *Xenopus* tadpoles (Winterbottom, 1974). It would seem that the Cape teal's strong attraction to saline waters must be associated with abundant aquatic invertebrate life, presumably insects and crustaceans.

Social behavior. Flock sizes in Cape teal are said to be fairly small during nonbreeding periods, often of from 3 to 7 birds, presumably representing family groups. But records of larger parties have been noted, and molting aggregations of as many as 2,300 birds have been reported at De Hoop Vlei, Bredasdorp, South Africa. It has been suggested that adults and young of the year tend to remain segregated from each other in flocks, which would facilitate pair formation in young birds and tend to keep mated pairs together. C. J. Skead (pers. comm.) reports that there are May and November peaks in courtship activity at Barberspan, South Africa, but records from throughout South Africa confirm the fact that breeding can occur at any month of the year, probably most often after rainy periods. Pair-forming displays consist of the usual female inciting display, and females also regularly perform an active nod-swimming display that is exactly like that of males. Male displays include a burp whistle, uttered with erected nape feathers and a forward extension of the neck. The most typical displays are a rapid nod-swimming over the water surface in a scudding fashion, and a striking head-up-tail-up posture that exposes the speculum pattern. Males also turn the back of the head toward inciting females, but display preening is confined largely to the precopulatory situation. In this situation both sexes perform display preening in the early stages, and later have the usual mutual head-pumping display. After copulation, the male performs what appears to be a rudimentary bridling posture, then turns and faces the bathing female (Johnsgard, 1965a).

Reproductive biology. Although nearly 300 nesting records for South Africa encompass all 12 months, the largest numbers are concentrated between August and November. Fewer records are available for elsewhere in Africa, but a few records for eastern Africa are generally from March to May. Females apparently prefer to locate their nests on islands whenever possible, and they are often placed

❖ ❖ ❖ 207

among small trees or in thorny bushes. The nest is frequently some distance from water; Brand's (1966) study indicates an average distance of more than 20 meters from the nearest water. Clutch sizes range from 5 to 11 eggs and average about 7 or 8. The eggs are laid on a daily basis, usually in the morning. Only the female incubates, and although there is some disagreement on the length of the incubation period, 25 to 26 days is probably normal. Both sexes assist in brood care, and the fairly large average brood sizes (five to six) for well-grown young suggest relatively low brood mortality rates. Fledging occurs in the seventh week of life, at which time the young closely resemble their parents (Winterbottom, 1974).

Status. Although the species has a very broad range, it is abundant only locally where the water conditions are especially favorable. Sewage plant and mining activities apparently favor Cape teal, and its opportunistic breeding adaptations favor the bird's survival. It is a favored sporting species, both as a table bird and as a difficult target, but it seems doubtful that any special attention is needed at present to assure its survival.

Relationships. Although Delacour (1954–64) regarded this species as a member of the "spotted teal" group, and a probable relative of the marbled teal, virtually all the evidence is counter to this idea and points to a relationship with the green-winged teals (Johnsgard, 1965a).

Suggested readings. Brand, 1966; Clancey, 1967; Winterbottom, 1974.

Madagascan Teal

Anas bernieri (Hartlaub) 1860

Other vernacular names. Bernier's teal; Bernier-Ente (German); sarcelle de Madagascar (French); cerceta de Madagascar (Spanish).

Subspecies and range. No subspecies recognized. Extremely rare, and limited to western Madagascar (Malagasy); recently recorded only from Lake Bemamba and Lake Masama. See map 72.

Measurements and weights. Folded wing: males, 203–13 mm; females, 192–98 mm. Culmen: males, 38–39 mm; females, 37–38 mm. Weights: no records. Eggs: no records.

Identification and field marks. Length: 16″ (40 cm). This species closely resembles the gray teal, but *adults* have a velvety black speculum. Their chin and throat are buff instead of white, the feather edges of the upper surface are darker, and the spotting of the undersides is more obscure. The speculum is lined anteriorly by a broad white wing bar, and posteriorly by a narrow one. The iris is brown, and the bill, legs, and feet are reportedly light red, although *females* are said to have a browner red color on the bill and legs.

In the field, the rusty brownish color of the birds should separate them from red-billed pintails, with which they sometimes associate. The call of the male is similar to that of the white-faced whistling duck, but has two rather than three syllables.

NATURAL HISTORY

Habitat and foods. Scott and Lubbock (1974) observed that on the western shore of Lake Bemamba, the birds were present in moderate numbers in July 1973. The lake is shallow and saline, with many clumps of emergent reeds, and at times becomes dry in September and October, just before the rainy season. The birds fed mostly in morning and early evening, and spent the rest of the day sleeping on the mud banks. They fed in shallow water, by dabbling in the mud, and were not observed to upend or dive for food. The foods taken are unknown, and it is likewise uncertain if the birds need fresh water or can utilize saline water for drinking.

Social behavior. Scott and Lubbock (1974) found that the birds on Lake Bemamba were in isolated pairs and in small groups of three or four pairs, totaling as many as 30 or 40 birds. A good deal of aggres-

sion and courtship was observed by them, in which chasing flights were common. Females incited much as the gray teal, and in both sexes a good deal of head bobbing in a circular motion was evident. The most common male courtship display was an upright posture with neck extended and the wings and tail slightly raised, perhaps corresponding to the burp posture of gray teal. Precopulatory behavior consisted of the usual mutual head pumping, and in three cases, immediately after treading the male performed a bridling display followed by a down-up. He then chased the female a short distance. This behavior seemingly differs somewhat from the typical chestnut teal's behavior, in which bridling also occurs, but which then either nod-swims or turns to face the bathing female. Nevertheless, it seems likely that such minor variations are not significant and may be subject to individual variation.

Reproductive biology. No nests or eggs have yet been described, but Scott and Lubbock reported that local observers stated that the birds nested in November or in April. Perhaps they nest at both times, before as well as after the rainy season. Likewise, local informants differed as to the normal clutch size, with reports of as few as 2 to 4 eggs and as many as 8 to 10 given.

Status. Scott and Lubbock saw a maximum of 61 teal along the shore of Lake Bemamba and believed that no more than 120 birds occupied the lake. The only other large lake in the area is Lake Masama, where Salvan (1970) reported this species in 1970. No other records for it exist, and thus the population must be extremely low.

Relationships. Delacour (1954–64) considered this to be an erythristic form of the gray teal, and available evidence would support this view.

Suggested readings. Scott & Lubbock, 1974.

Gray Teal

Anas gibberifrons Müller 1842

Other vernacular names. Andaman teal, slender teal; Weisskehlente (German); sarcelle grise (French); cerceta gris (Spanish).

Subspecies and ranges. (See map 73.)

A. g. *gibberifrons:* East Indian gray teal. Resident in Java, Celebes, Sula Islands, Selajar, Sumba, Flores, Timor, and Wetar.

A. g. *remissa:* Rennell Island gray teal. Known only from Rennell Island.

A. g. *gracilis:* Australian gray teal. Resident in Australia, Tasmania, New Zealand, New Guinea, Aru and Kai islands, and occasionally elsewhere (New Caledonia, Lord Howe and Macquarie islands).

A. g. *albogularis:* Andaman gray teal. Resident on Andaman Islands, also Landfall and Great Coco islands.

Measurements and weights. Folded wing: males, 186–210 mm; females, 179–205 mm. Culmen: males, 33–41 mm; females, 30–40 mm. Weights (of *gracilis*): males, 395–670 g (av. 507 g); females, 350–602 g (av. 474 g). Eggs (of *gracilis*): 49 x 36 mm, creamy white, 35 g.

Identification and field marks. Length 16" (40 cm). Adult *males* have a brown head that is blackish above and nearly white below. (In *albogularis* much of the head and neck is white.) The mantle and scapular feathers are dark brown with a pale buff margin. The breast, flanks, and tail coverts are patterned likewise, while the abdomen is spotted with light and dark brown. The upper wing surface is dark brown, with a black and iridescent green speculum bounded posteriorly with a narrow white bar and anteriorly with a broader one. The iris is red, the legs and feet blackish, and the bill black. *Females* are nearly identical to males in color, but are slightly paler, and their iris coloration is not so bright. *Juveniles* are even paler than females, especially on the head and neck.

MAP 73. The breeding or residential distribution of the Andaman ("A"), Australian ("Au"), East Indian ("E"), and Rennell Island ("R") gray teals.

In the field, gray teal are most likely to be confused with chestnut teal, as they are of the same size and proportions. They lack the green head color of male chestnut teal, and female chestnut teal are generally a darker brown, with no gray tones predominating. Vocalizations in the two species are nearly identical, but the female gray teal's decrescendo call usually has at least a dozen or more separate notes, while the male has a soft whistled note. In flight, the double white wing bar is conspicuous, but unlike that of the larger gray duck, the underside of the wing is not white.

NATURAL HISTORY

Habitat and foods. In Australia the gray teal is widespread and occupies fresh-water, brackish water, and salt-water areas, but prefers to utilize billabongs, lagoons, and floodwaters of inland rivers. The birds are highly nomadic and are opportunistic colonizers of newly flooded areas. They are highly mobile and have colonized New Zealand in recent times, where they occur in habitats similar to those in Australia. The birds forage by probing in mud at the edge of water, upending in shallow water, filter-feeding along the surface, or stripping seeds from terrestrial vegetation near the water. They also dive well, and at least in captivity sometimes forage in this way. A large sample of stomachs from New South Wales indicates a vegetable diet high in the seeds of smartweeds (*Polygonum*), sedges, grasses, and broad-leaved plants, with about a third of the food remains of invertebrates, nearly all insects. A smaller sample from Queensland also had a predominance of plant foods, mainly sedges, pondweeds, and water lilies, but fewer grasses and insect remains (Frith, 1967).

Social behavior. The social groupings of gray teal may range from pairs to flocks that may number in the thousands of birds in Australia; and because of the irregular breeding seasons, there is no specific time of the year when flocking occurs. In addition, there are large year-to-year variations in numbers as breeding conditions vary. However, the timing of pair formation and breeding is most regular in the southern tablelands, where breeding occurs after the winter rains, and in the northern areas, where breeding occurs in the wet period of late summer and autumn. Sexual display is apparently initiated in the interior by rainfall and associated increases in local water levels, with the first eggs being laid within ten days of such water-level changes. In the southern tablelands, however, courtship occurs during the autumn and winter months, and presumably allows for the gradual establishment of pair bonds (Frith, 1967). Females incite unpaired males in the usual *Anas* manner, and also have a nod-swimming display that evokes strong display on the part of males. The males perform a rather large number of social displays, including burping, an introductory shake, grunt-whistle, and a head-up-tail-up display that is usually followed by facing the female and bridling, then nod-swimming, and finally turning the back of the head toward the courted female. Bridling also occurs independently of these other displays, and a final display, the down-up, is likewise independently performed. Preening behind the wing is also a fairly frequent display by males but has not been seen in females (Johnsgard, 1965a). Copulatory behavior has not yet been described, but is almost certainly like that of the chestnut teal.

Reproductive biology. In Australia, where the gray teal has been best studied, nesting can occur during any month of the year, while in New Zealand it is restricted to the period from September to January. In northern Australia and the tropical islands to the north nesting is apparently associated with the summer monsoon rains, which begin in June on the Andaman Islands. Nests are placed in a wide variety of locations, but are usually on the ground. In inland Australia the birds often nest in the hollow limbs of trees, sometimes at considerable elevations, but also use rabbit burrows or rocky crevices. Clutch sizes range from 4 to 14 eggs and in Australia average about 7 or 8. The clutch sizes of the other races are less well known, but, surprisingly, have been reported at about 10 in the Andaman form. Only the female incubates and the incubation period is normally 24 to 25 days. There is evidence (Janet Kear, pers. comm.) that in the Australian race the male participates in brood care, at least under captive conditions, and one early account of a captive male associating with a female and brood of the Andaman race suggests that in this form such paternal participation is also normal. The fledging period is probably very close to the 56-day period established for the chestnut teal (Frith, 1967).

Status. This is probably Australia's most common and certainly its most widespread species of duck, and additionally it is relatively common through most of its native range. It is evidently increasing in numbers in New Zealand, which it colonized about a century ago. Probably only the Andaman teal is in any potential danger of extinction; this belief is based on its intrinsically small population and apparent inbreeding characteristics. No estimates of its population seem to be available.

Relationships. The gray teal and chestnut teal are obviously very closely related, but at the same time are clearly distinct species. These two forms are apparently less closely related to New Zealand's brown teal, and the affinities of the gray teal with the Madagascan teal appear to be very close (Johnsgard, 1965a).

Suggested readings. Frith, 1967; Lavery, 1972.

Chestnut Teal

Anas castanea (Eyton) 1838

Other vernacular names. Chestnut-breasted teal, mountain teal; Kastanienente (German); sarcelle d'Australie (French); cerceta de pecho castaño (Spanish).

Subspecies and range. No subspecies recognized. Resident in Australia except for the north coast and interior, also on Tasmania. See map 74.

Measurements and weights. Folded wing: males, 204–31 mm; females, 197–210 mm. Culmen:

males, 40–43 mm; females, 37–42 mm. Weights: males, 340–708 g (av. 595 g); females, 368–737 g (av. 539 g) (Max Downes, pers. comm.). Eggs: 51 x 37 mm, deep cream, 40 g.

Identification and field marks. Length: 16″ (40 cm). *Adult males* in breeding condition have an iridescent green head and neck, dark brown mantle and scapular feathers that have lighter edges, a black rump and tail coverts, with a white patch on the lateral rump region, and chestnut brown breast and flanks, with darker spotting, especially on the flanks. The upper wing surface is generally brown, except for the secondaries, which form a black and green iridescent speculum, bounded both in front (broadly) and behind (narrowly) with white bars. The iris is brilliant red, the legs and feet are dark gray, and the bill is bluish gray. *Nonbreeding males* are slightly duller in plumage, and the iris coloration is less brilliant, but the pattern remains much the same. *Females* in general are a dark brown, with the crown of the head blackish and the chin and throat grayish brown, and with most of the body feathers also dark brownish with paler brown edging. The wing pattern is as in the male, but the iris is less brilliant red. *Juveniles* resemble adult females but have less well developed spotting.

In the field, chestnut teal are most likely to be confused with gray teal, but females are appreciably darker and browner in color, and males have the distinctive green head, which is lacking in the gray teal. The white marking on the side of the male's rump is also helpful at a considerable distance when the head color is not apparent. The calls of the two species are very similar, but the chestnut teal female has a decrescendo call that tends to consist of fewer than

ten syllables. The courtship notes of the males appear to be identical.

NATURAL HISTORY

Habitat and foods. The chestnut teal has a much more restricted range than the gray teal, and in Australia is especially characteristic of brackish coastal lagoons and estuaries, as well as of coastal creeks. In Tasmania and the southern highlands of Australia it is also found on deeper lakes, especially where islands occur. It is often associated with mangrove habitats in the southern and western portions of its range, and does not penetrate far into interior habitats in Western Australia. It forages in the same ways as the gray teal, and their bill shapes as well as their body conformations would suggest considerable similarity in foods taken. However, no detailed information on the kinds of foods that are taken is yet available (Frith, 1967).

Social behavior. In their general social tendencies, chestnut teal appear to be identical to gray teal; but because they occur in much smaller numbers, the enormous flocks typical of the gray teal are never encountered here. Further, since they are limited to the southern parts of Australia, which has fairly predictable periods of rainfall and cold weather, they do not exhibit the opportunistic breeding tendencies or the nomadic flock movements so typical of that species. Thus, the period of social display is also fairly predictable, and in the Canberra area it begins in March and continues until the start of breeding in September. Although no quantitative studies have been undertaken, the pair-forming displays of both sexes appear to be identical in form with those of the gray teal, in spite of the diverse array and complex sequences of displays typical of these species. This unexpected similarity, together with the apparent rarity of known wild hybridization, suggests that other factors must be operating to keep these two species as separate genetic entities. Perhaps the differences in male plumage are of special importance here, since effective ecological separation seems to be lacking. The precopulatory behavior of the chestnut teal consists of mutual head pumping; and after treading is completed, the male performs a bridling display, then usually nod-swims away from the female. Sometimes, however, he follows bridling with simply turning to face the female as she bathes (Johnsgard, 1965a).

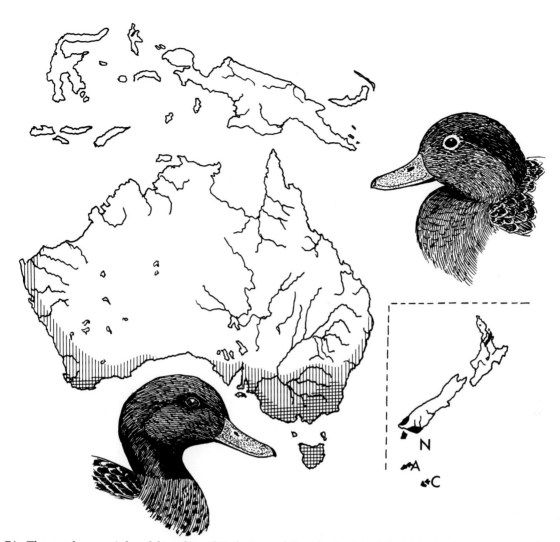

MAP 74. The total or peripheral breeding distribution of the chestnut teal (hatching), showing areas of major concentrations (cross-hatching), and the residential distributions of the New Zealand ("N"), Auckland Island ("A"), and Campbell Island ("C") brown teals.

Reproductive biology. In Victoria, where this species is relatively abundant, breeding occurs over a surprisingly long period extending from June to December, with a breeding peak in October. Evidently the peak of breeding in South Australia is also in October. Evidence from banded birds is strongly indicative of multiple brooding in this species, since of 80 marked females, 27 were known to have two clutches and 12 were found to have three. Usually the second clutch was begun within two to six weeks of the female's leaving her brood. Nests are usually placed on the ground, but they may also be in rock crevices or even in tree cavities. Island nesting is common in the mountain lakes of the interior. Clutch sizes are greatly variable, from 5 to 15 eggs, but the modal number is 9 or 10. Incubation requires 28

days; and although it is performed entirely by the female, the male evidently returns to the brood shortly after it is hatched and more often than not remains with the ducklings until they are fledged. This male participation in brood care, which also occurs in gray teal, doubtless facilitates the production of multiple broods by a single female. Fledging by the ducklings may occur as soon as 56 days after hatching, and sexual maturity is attained within the first year (Frith, 1967).

Status. Although still reasonably common in much of southern Australia, the chestnut teal has declined considerably in numbers in historic times, and close attention to the species' needs may be needed in the future. Less is known of its ecology and food requirements than of most Australian waterfowl; yet like most *Anas* species, it seems to respond well to nest-box erection and other management techniques.

Relationships. At times the chestnut teal was considered conspecific with the gray teal and also with the brown teal, but the present evidence strongly indicates that these three forms are all specifically distinct. Chestnut teal are probably more closely related to gray teal than to brown teal, even though they resemble the latter more closely. Additionally, this entire group has some plumage and behavioral similarities to the mallard group, and in this way help to link the "green-winged" teal assemblage to the mallards (Johnsgard, 1965a).

Suggested readings. Frith, 1967.

Brown Teal

Anas aucklandica (G. R. Gray) 1844

Other vernacular names. Brown duck, Auckland Island flightless duck, flightless teal; Aucklandente or Grünohrente (German); sarcelle de la Nouvelle Zálande (French); cerceta de la Isla Auckland y Castaña (Spanish).

Subspecies and ranges. (See map 74.)

A. a. chlorotis: New Zealand brown teal. Resident in New Zealand, where now restricted to the northern peninsula, Great Barrier, Fiordland, Stewart, and possibly Ruapuke and Codfish islands; formerly also on Chatham Islands and Mayor Island.

A. a. aucklandica: Auckland Island teal. Resident on the Auckland Islands, where now confined to Adams, Disappointment, Rose, Ocean, Ewing, and Enderby islands.

A. a. nesiotis: Campbell Island teal. Resident on the Campbell Islands, where extremely rare.

Measurements and weights: Folded wing: both sexes of *chlorotis*, 185–203 mm; both sexes of *aucklandica*, 125–44 mm. Culmen: both sexes of *chlorotis*, 39–45 mm; both sexes of *aucklandica*, 30–41 mm. Weights: males (*chlorotis*), 615–730 g (av. 665 g); females, 530–700 g (av. 600 g); females of *aucklandica* estimated 450 g (Reid & Roderick, 1973; Lack, 1968). Eggs: av. of *chlorotis*, 58 x 43 mm, dark cream, 62 g; av. of *aucklandica*, 65 x 44 mm, light tan, ca. 75 g.

Identification and field marks. Length 17–19" (43–48 cm). *Adult males* in breeding plumage closely resemble those of the chestnut teal, but have a dark gray head, with green limited to the ear and occipital regions, and with a whitish eye-ring. There is usually a white collar at the base of the neck, while the back and scapulars are dark grayish brown, with black and rust markings. The breast is chestnut, variably spotted with black, while the abdomen is gray and also spotted. The flanks are mostly brown, with indistinct vertical black barring, while the sides of the rump are gray (white in *chlorotis*) and the under tail coverts are black. The tail and upper tail coverts are dark brown, and the upper wing surface is generally brown, except for the secondaries, which are variable iridescent green with a broad anterior and narrow trailing edge of buff. The iris is brown, the legs and feet gray, and the bill is bluish gray, sometimes grading to yellowish near the base. *Males in eclipse,* as well as some dull-phase males in breeding plumage, closely resemble females. *Females* are very much like female chestnut teal, but have a whitish ring around the eyes and a dark brown iris. *Juveniles* closely resemble adult females.

In the field, brown teal are inconspicuous, dark brown ducks that are relatively nocturnal and are rarely seen in the open during daylight hours. The white markings on the sides of the rump in males, and the whitish eye-rings of both sexes, are the most conspicuous markings, but they are reduced or nearly absent in the Campbell and Auckland Island forms. Males utter a fairly loud piping or whistling note, and females have a low-pitched quacking call. The New Zealand form flys infrequently, and the other two forms are essentially flightless.

NATURAL HISTORY

Habitat and foods. The New Zealand brown teal is found primarily in swampy streams and on ponds and tidal creeks shaded by trees, as well as in sheltered portions of the coastline where kelp beds are present. Weller (1974) analyzed the habitat of the brown teal on Great Barrier Island and (1975b) of the Auckland Island teal. In the former area, the birds foraged almost entirely in tidal estuaries, and were also seen on fresh-water streams and brackish ponds. On the estuaries, foraging was regulated by the tide, with the birds dabbling or probing for invertebrates

in very shallow waters. In deeper waters the birds would upend or dive, apparently also feeding on invertebrates. Additionally, at night they would often move into grassy fields, apparently to probe for feed and to search for insect larvae on the plants there. The Auckland Island birds are largely associated with shorelines having easy access to higher ground, and especially to sites combining stream and marine habitats. During the day the birds often rested in petrel burrows or under rocky ledges, apparently to avoid attack by skuas (*Catharacta skua*), and they foraged in the uplands both at night and at times during daylight hours. Most foraging, however, was done by probing in windrows of kelp, and a few collected specimens contained mainly amphipods, isopods, and copepods, but no plant materials.

Social behavior. Current evidence indicates that the brown teal is monogamous, with pair bonds persisting through incubation and probably permanently. At least this is the pattern Weller (1975b) established for the Auckland Island form, and it is unlikely to differ in the other two races. Males vigorously defend foraging and nesting territories during the breeding season, with small numbers of nonbreeding birds remaining in groups. Territories of five males that were spaced along the coastline were apparently separated by distances of from about 30 to 50 meters, at which points aggressive encounters between the males were frequent. During such encounters the males would utter trilling whistles, with chin lifting, and at times breast-to-breast fighting. The males also utter a *pee-dit* call, which serves as an alarm note. No other elaborate courtship displays were seen by Weller, and his observations of two copulations also lacked the usual precopulatory head pumping of most *Anas* species. After treading, the male uttered its trilling whistle and also the *pee-dit* note, following which the two birds wing-flapped and bathed. In a study of captive birds, M. Williams

(unpublished honors project report, 1967) reported that the courtship displays of the New Zealand form are generally much like those of the mallard, but in addition there are independent burping and bridling displays, which are also characteristic of chestnut and gray teal.

Reproductive biology. In New Zealand the brown teal reportedly nests between July and December (Falla et al., 1967), while the Auckland Island race is said to nest from October to December, although Weller (1975b) found a nest in January. The nest he found was situated in fern (*Blechnum*) cover, near a soggy watercourse, and in general the nests appear to be well concealed under vegetation. Nests of the New Zealand form are said to be placed in dry and secluded places, but sometimes are close to creeks or other water areas. The clutch size of this race is evidently significantly larger than that of the Auckland Island form, ranging from 5 to 7 (averaging 5.9) in the former, as compared to 3 or 4 in the latter. However, the eggs are substantially larger in the Auckland Island form, although the adult bird is lighter in weight. The incubation period is 27 to 30 days for *chlorotis* (Reid & Roderick, 1973), and Weller's (1975b) observations indicate that during incubation the male of the Auckland Island race remains very close to the nest, often as little as four meters away. Of two or three broods he observed, none was definitely tended by both parents, although he did see a single duckling near a lone adult male. However, Janet Kear (pers. comm.) has seen males defending young in the New Zealand race. The fledging period is about 55 days (Reid & Roderick, 1973).

Status. The status of the brown teal has recently been summarized by Fisher et al. (1969) for the species in general, by Scott (1971) for the Auckland Island form, and by McKenzie (1971) for the Auckland Province. The Campbell Island form has at times been thought to be extinct, but the most recent information is that four of these teal were captured and released on Dent Island, a small islet off the west coast of Campbell Island, in 1975. The Auckland Island population is regarded by Weller (1975b) as secure, since it occurs on virtually all of the archipelago islands except the main island, which has both cats and pigs present. Its total population is almost impossible to judge, but Weller estimated a minimum of 24 breeding pairs on Ewing Island. In New Zealand, the species still occurs on both of the major islands, but was extirpated from Kapiti Island (where birds have recently been released again), Mayor Island, and Rotorua district of North Island. On the South Island it has been exterminated from the northern provinces of Marlborough and Canterbury, and is now known only from Southland and the Stewart Island archipelago, and perhaps also occurs on Ruapuke Island. It has not been seen on Codfish Island since 1948.

Relationships. The relationships of the brown and Auckland Island teal to one another, and their affinities with the chestnut teal of Australia, have been the subject of considerable controversy. There seems to be general agreement that the brown and Auckland Island teals must have evolved from a chestnut teal-like ancestor, but the species limits are still unsettled. There is also some doubt about the validity of the Campbell Island race. However, behavioral and plumage differences between the brown teal and the chestnut teal seem to preclude their consideration as subspecies, and it is impractical simply to consider the brown teal as an island derivative of the chestnut teal (Johnsgard, 1965a).

Suggested readings. Weller, 1974, 1975b; Scott, 1971.

Mallard

Anas platyrhynchos Linnaeus 1758

Other vernacular names. Greenhead, green-headed mallard, Hawaiian duck, Laysan teal, Mexican duck, northern mallard; Stockente (German); canard col-vert (French); pato comun (Spanish).

Subspecies and ranges. (See map 75.)
A. p. platyrhynchos: Common mallard. Breeds in Europe and Asia from the Arctic Circle south to the Mediterranean Sea, Asia Minor, northern Iran, Turkestan, Tibet, central China, and northern Japan, also in Iceland and the Azores. In North America breeds from northern Alaska to the west shore of Hudson Bay, south to lower California, New Mexico, Kansas, Missouri, Ohio, and Virginia. Winters from its breeding range southward to northern Africa, the Nile Valley, the Persian Gulf, India, southern China, Japan, and northern Mexico.

A. p. conboschas: Greenland mallard. Resident in coastal Greenland.
A. p. fulvigula: Florida mallard. Resident in peninsular Florida.
A. p. maculosa: Mottled mallard. Resident on the Gulf Coast of the United States and Mexico from Mississippi to central Tamaulipas, wintering south to Veracruz.
A. p. diazi: Mexican mallard. Breeds locally in the Rio Grande Valley of southern New Mexico and western Texas, southeastern Arizona, and the highlands of Mexico south to central Mexico. Winters over most of its breeding range.
A. p. wyvilliana: Hawaiian mallard. Resident in the Hawaiian Islands.
A. p. laysanensis: Laysan mallard. Resident on Laysan Island.

Measurements and weights. Both sexes of *laysanensis*, 190–210 mm; of *wyvilliana*, 210–28 mm; of *diazi, fulvigula,* and *maculosa*, 223–85 mm; and of *platyrhynchos* and *conboschas*, 240–85 mm. Culmen: both sexes of these respective groups have ranges of 38–41 mm, 41–48 mm, 48–59 mm, and 43–57 mm. Weights: Lack (1968) reports female weights as follows: *laysanensis*, 450 g; *wyvilliana*, 505 g; *fulvigula*, 970 g; and *platyrhynchos*, 1,000 g. Males and females of *wyvilliana* average 670 and 573 g, respectively (Berger, 1972); *fulvigula* adults average 1,030 and 968 g, respectively (Beckwith & Hosford, 1957) and common mallard shot in the fall average 1,261 and 1,084 g, respectively (Bellrose & Hawkins, 1947). Eggs: Range in average size from 58 x 33 mm to 58 x 40 mm, vary in color from white or buff to greenish, and range in weight from 43 to 56 g.

Identification and field marks. Length 16–26" (40–66 cm). *Males* (of *platyrhynchos* and *conboschas*) in breeding plumage have an iridescent green head and neck, a narrow white ring around the neck, and a grayish brown to brown hind neck and back, merging with black on the rump and upper tail coverts. The longest upper tail coverts are recurved. The scapulars are mostly grayish white, the outer ones pale chestnut. The lower neck and breast are dark chestnut, gradually merging with whitish underparts and with vermiculated gray and white sides and flanks. The tail is white to grayish brown centrally, the under tail coverts are black, with a white border anteriorly and laterally. The upper wing coverts except for the greater secondary coverts are ashy brown, and the latter are white, tipped with black.

The secondaries are iridescent bluish to purplish, barred with black subterminally and with white tips. The tertials are blackish to reddish brown. The bill is greenish yellow to bright yellow, with a black nail; the legs and feet are orange red; and the iris is brown. Males of these two races have eclipse plumages that are very similar to the female's plumage, but the bill lacks spotting and is dull olive and some vermiculations are often present on the back feathers. Males of the other races are either entirely femalelike, have a variable amount of green iridescence on the head (as in *wyvilliana*), or may show variable degrees of albinism on the head (*laysanensis* and some *wyvilliana*). *Females* of most races have a buff-colored head, streaked with brownish coloration on the cheeks and blackish on the crown and through the eyes. The upperparts are generally blackish brown, the feathers with buff edges and central markings, while the lower neck and breast feathers are brownish black with paler margins. These paler areas increase in size posteriorly, forming whitish buff underparts streaked or mottled with brown, and flanks strongly marked with brown and buff crescents. The tail is grayish brown, paler and strongly edged with white or buff on the outer feathers, and the wings are very similar to those of the male. The iris is brown, and the bill is orange yellow to greenish yellow or olive green, with a darker nail. Females of *laysanensis* and *wyvilliana* differ from this description in being generally darker throughout except for the head, which in older birds often is whitish, and in having a speculum of reduced iridescence. Females of *fulvigula*, *maculosa*, and *diazi* tend to be darker and, in the first two, more tawny throughout, and in these two forms the white portions of the greater secon-

dary coverts are virtually obscured by this tawny color. Females of *diazi* are sometimes scarcely separable from common mallards, but have browner outer tail feathers, darker tail coverts, and a more olive-colored bill. *Juveniles* resemble adult females, but are more heavily streaked on the underparts and are generally darker throughout.

In the field, the familiar green-headed plumage of the male common mallard needs little experience to distinguish, but females may be readily confused with female pintails, gadwalls, or other *Anas* species unless the bill coloration or the speculum pattern can be seen. Separation of the various North American races is even more difficult, and is further confused by frequent hybridization among these forms and sometimes also by hybrids between them and black ducks. The problems of such identification have been discussed elsewhere (Johnsgard, 1975) and cannot be dealt with here. The typical quacking calls of female mallards are familiar to most people, and are similar in all of the races, although the smaller races have noticeably higher-pitched calls. Males utter reedy quacking sounds and during display produce sharp single- or double-noted whistles.

NATURAL HISTORY

Habitat and foods. The breeding habitats of the common mallard are extremely diverse, ranging in climate from subtropical marshes to arctic tundra. They evidently require only the combination of shallow-water foraging areas and suitable nesting cover, which ranges from grassy or herbaceous vegetation to shrubs or even trees. In the last-named situation the birds often nest on stumps or in tree holes or crotches. Generally, however, open country is preferred to forested areas, and in North America the densest breeding occurs on the northern prairies of

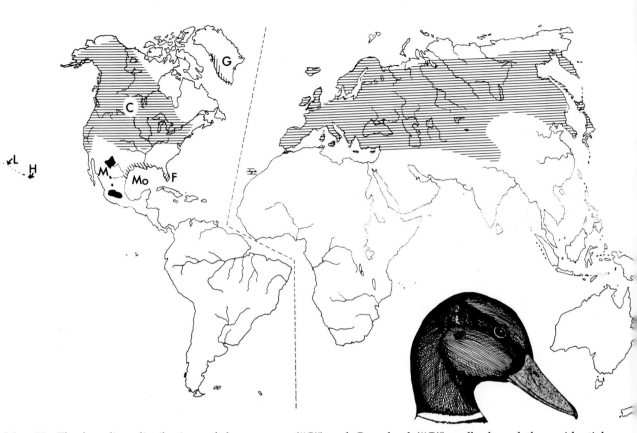

MAP 75. The breeding distributions of the common ("C") and Greenland ("G") mallard, and the residential distributions of the Laysan ("L"), Hawaiian ("H"), Mexican ("M"), mottled ("Mo"), and Florida ("F") mallards.

the Great Plains. Mallards of the more southerly populations (Florida, mottled, and Mexican mallards) seem to prefer coastal prairies and marshes for breeding, while the Greenland mallard breeds around fresh-water lakes and streams and on islands of fiords. The Hawaiian mallard is now confined to the coastal wetlands and hillside streams of Kauai, from sea level to 3,500 feet. Pen-raised birds have also been released on Oahu and Hawaii, where they were once native (Berger, 1972). The small Laysan form is limited to the fresh and brackish ponds of that tiny mid-Pacific island. The Laysan race has greatly modified its ecology to become largely nocturnal and insectivorous (Warner, 1963), but the wider-ranging North America populations are basically daytime foragers feeding on materials predominantly of vegetable origin (Johnsgard, 1975). Little is known of the summer foods of the Greenland mallard, but in the winter it evidently feeds largely on mollusks and amphipods (Palmer, 1976).

Social behavior. The social behavior of the common mallard has been extensively studied, but less is known of the insular and sedentary southerly popu-

lations. The available information indicates that the latter are scarcely different from the common mallard in their social adaptations, while the Hawaiian and Laysan Island forms remain virtually unstudied. All forms are thought to have a yearly renewal of pair bonds, and in all except the Greenland race the birds probably mature their first year; a two-year period to sexual maturity has been suggested for both sexes of the Greenland mallard. In most forms there is a several-month period of social display, which in the common mallard begins in the late fall and continues at a fairly high level through winter and early spring. Most pair-forming activity occurs on water, but several kinds of aerial chases also occur, including not only true courtship flights but also attempted rape chases, the latter occurring mainly late in the season when few or no females remain unmated. The major aquatic displays of common mallard males include the grunt-whistle, head-up-tail-up, down-up, and nod-swimming, as well as display preening behind the wing and turning the back of the head toward inciting females. Females stimulate courtship with inciting calls and movements, and in addition perform nod-swimming as an additional strong stimulus to male display. These same displays are known to occur without obvious modification in the Gulf Coast and Mexican populations of mallard, while in the Hawaiian and Laysan Island forms they show increasing degeneration of form, judging from limited observations of captive birds. In all of them, copulation is preceded by mutual head pumping and is typically followed by the male's performing a single bridling movement and nod-swimming away from the female (Johnsgard, 1965a).

Reproductive biology. Little can be said of the reproductive biology of the Laysan race, other than that it probably breeds between February and July and nests in various ground-level sites, such as under cyperus (*Cyperus*) and shrubs (*Chenopodium* and *Scaevola*) (Warner, 1963). Its clutch size in captivity ranges from 4 to 8 eggs. The breeding season of the Hawaiian form apparently extends mainly from December through May, and the birds also nest on the ground, often under honohono grass (*Commelina*). The reported average clutch size is 8.3 eggs, the incubation period 28 days, and the fledging period nine weeks (Berger, 1972). In the case of the Gulf Coast and Mexican populations the breeding season is also fairly extended, allowing for repeated renesting efforts, and clutch sizes average about 9 or 10 eggs. Nests are placed in a variety of locations, but often are located under grasses or sedges, and a 25- to 27-day incubation period is apparently normal. A fledging period of 60 to 70 days has been reported for the mottled mallard (Bellrose, 1976). Finally, in the common mallard the nesting period tends to expand or contract with the length of the growing season, and renesting efforts by unsuccessful females are typical in most parts of the breeding range. As noted earlier, nests are placed in diverse situations, including elevated ones in wooded habitats, and may be found some distance from water. Clutch sizes of early or initial nesting efforts average about 9 or 10 eggs, with later or renesting clutches of from 6 to 8 eggs. Eggs are laid on a daily basis, and incubation typically requires about 28 days. Males desert their mates early in the incubation period, and often migrate some distance to favored molting areas. The fledging period for common mallards is from 50 to 60 days, but there seems to be considerable individual and probably regional variation in this (Johnsgard, 1975; Bellrose, 1976; Palmer, 1976).

Status. The status of the common mallard is so secure that it needn't be discussed here, but the Laysan mallard's survival is, by contrast, extremely precarious.

Warner (1963) and Fisher et al. (1969) have reviewed its history; the most recent estimate in 1977 was that 69 birds existed in the wild (*Auk* 92, Suppl. 7B). Likewise, the Hawaiian mallard, which now is entirely limited to Kauai, was estimated in the mid-1960s to be approximately 3,000 birds (Swedburg, 1967). The rarest of the North American subspecies is the Mexican mallard, which in 1975 probably numbered 30,000–40,000 birds (Williams, 1978). In contrast, the Florida mallard's fall population averaged about 50,000 birds in the 1960s, and the mottled mallard's fall population is perhaps between 200,000 and 250,000 (Bellrose, 1976). The population status of the Greenland mallard is not known.

Relationships. The relationships of these populations to each other and to other mallardlike ducks is complex, and has been dealt with earlier (Johnsgard, 1961d). Several problems of establishing realistic species limits and the acceptance of certain subspecies still exist, reflecting the dynamic nature of the speciation problem.

Suggested readings. Warner, 1963; Berger, 1972; Johnsgard, 1975; Bellrose, 1976; Swedburg, 1967.

North American Black Duck

Anas rubripes Brewster 1902

Other vernacular names. Black mallard; Dunkelente (German); canard obscur (French); pato negro (Spanish).

Subspecies and range. No subspecies recognized. Breeds from Manitoba southeast to Minnesota, east through Wisconsin, Illinois, Ohio, Pennsylvania, Maryland, West Virginia, and Virginia, and in the forested portions of eastern Canada to northern Quebec and northern Labrador. Winters in the southern parts of its breeding range and south to the Gulf Coast, Florida, and Bermuda. See map 76.

Measurements and weights. Folded wing: males, 265–92 mm; females, 245–75 mm. Culmen: males, 52–58 mm; females, 45–53 mm. Weights: adult males average about 1,330 g in fall, while adult

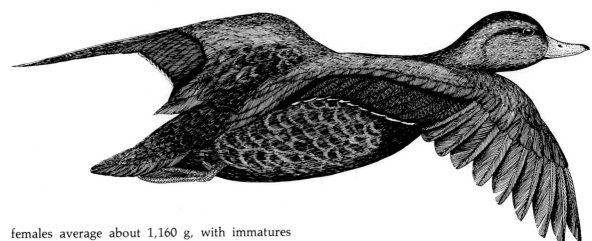

females average about 1,160 g, with immatures averaging from 50 to 100 g lighter. Eggs: av. 58 x 40 mm, greenish, 63 g.

Identification and field marks. Length 21–26″ (54–66 cm). *Males* in breeding plumage have a buffy head that is heavily streaked with black, especially through the eye and on the tip of the head; and the upperparts, including the tail and wing coverts, are blackish brown, the feathers with narrow lighter edges. The underpart feathers are dark sooty brown, with pale reddish and buff margins and with buff U-shaped markings. The secondaries are iridescent bluish purple, with a black subterminal border and a narrow white tip (sometimes lacking). The tertials are glossy black next to the speculum, but otherwise are gray to blackish brown; the underwing surface is silvery white. The iris is brown, the bill is greenish yellow to bright yellow, with a black nail, and the feet and legs are orange red, with darker webs. *Males in eclipse* lack the buff U-shaped markings on the breast and side feathers, and *females* tend to have more V-shaped markings. Females also have a greenish to olive-colored bill, with small black spotting, and dusky to olive-colored legs and feet. *Juveniles* resemble adults, but are more heavily streaked on the breast and underparts, since these feathers have broader buff margins but dark tips.

In the field, the mallardlike shape of this species may cause it to be confused with female mallards, which are appreciably lighter and more buff in overall coloration, or with the Florida and Gulf Coast races of mallards, with which black ducks might associate during winter. These races tend to be more tawny-colored than black ducks, and thus appear generally reddish brown rather than blackish brown in general tone. In flight, black ducks appear to be nearly black, and their white underwing colora-

tion contrasts strongly with the rest of the plumage. Their vocalizations are identical to those of mallards.

NATURAL HISTORY

Habitat and foods. Breeding habitat preferences of this essentially forest-adapted species are rather broad, and include alkaline marshes, acid bogs and muskegs, lakes, ponds, and stream margins, as well as tidewater habitats such as fresh, brackish, and salt marshes and the margins of bays and estuaries (Stewart, 1958). Wintering habitats are likewise rather diverse, but the most favored areas are brackish estuarine bays with extensive adjacent agricultural lands. As a reflection of its distinctly coastal wintering preferences, the black duck's foods tend to show a higher proportion of animal materials than do those of mallards. However, like mallards, black ducks regularly feed in grain fields, and at seasons other than winter their diet is predominantly of plant material relatively similar to that of mallards.

Social behavior. During the fall and winter, black ducks are highly gregarious and may occur in flocks numbering in the thousands of birds. However, as pair formation begins about the end of September, paired birds begin to break away from the unpaired segment of the population and flock sizes gradually diminish. Pair-forming behavior occurs over a several-month period that probably peaks in midwinter, and by April nearly all the females will have formed pair bonds. Social display is virtually identical to that of mallards, and, indeed, mixed courting groups may often be seen wherever the two species come into wintering contact (Johnsgard, 1961d). Mixed pairings are, however, not yet as common as might be expected on the basis of the great incursions of mallards into black duck wintering habitats during the past 75 years, although the incidence of hybridization appears to be slowly increasing (Johnsgard, 1967b; Johnsgard & Di Silvestro, 1976). Like the pairforming postures and vocalizations, behavior associated with copulation in black ducks is also identical to that of mallards, and thus, once mixed pairs are formed, there is probably little pressure for their disruption other than the somewhat different habitat preferences of the two species.

Reproductive biology. So far as is known, nearly all first-year females attempt to nest; probably less than 10 percent of them do not breed at that age. Older females usually return to their nesting areas of previous years and very frequently use an old nest site, or at least nest within 100 yards of an old site (Coulter & Miller, 1968). Eggs are deposited in the

MAP 76. The breeding distribution of the North American black duck (hatching), including areas of densest concentration (cross-hatching). Wintering distribution indicated by stippling.

nest at the approximate rate of one per day, and clutch sizes generally average between 9 and 10 eggs, with smaller clutches typical of renests, of first-year females, and of clutches initiated relatively late in the nesting season. The time at which pair bonds are disrupted varies somewhat, with males typically remaining with their females about two weeks into the incubation period, but with some staying only a week and others about three weeks. In a few instances apparently paired birds have been seen together through an entire nesting season, but participation by the male in brood rearing has not been reported. Female black ducks are relatively persistent renesters, and estimates of the incidence of renesting by unsuccessful females have ranged from 16 to 31 percent. The incubation period is the same as that of mallards, averaging about 27 days, and a fairly high rate of nest destruction by predators such as crows and raccoons appears to be typical. After hatching, the female typically remains with her offspring during their entire period to fledging, which requires approximately 60 days. At about the time they fledge, or shortly before that time, she begins her postnuptial molt and enters her flightless period of about four weeks. Some flightless females may be seen in southern New Brunswick even after mid-September, or shortly before the fall migration begins (Wright, 1954).

Status. Calculations by Geis et al. (1971) indicated that during the 1950s the early fall population of black ducks averaged nearly 4 million birds, but annual winter inventories fail to confirm such high numbers and during the late 1960s averaged less than half a million ducks. Winter inventory data for the period from the mid-1950s to the early 1970s show a fairly consistent decline in numbers of about 40 percent, and Audubon Christmas Count data for the past several decades have indicated a tremendous decline in wintering populations of black ducks relative to mallards throughout virtually all of the former's range. To what extent this may be an actual rather than a relative population change is uncertain, but it is clear that the black duck is effectively losing out to the mallard as the typical breeding duck of eastern North America.

Relationships. A study of the evolutionary affinities of all the North American mallardlike ducks (Johnsgard, 1961d) indicated that the black duck is certainly closely related to and has only quite recently been derived from earlier mallard stock, and in some respects is only a poorly characterized species that has become adapted to more forested breeding and maritime wintering habitats than is the mallard. The recent expansion of the mallard's range and associated increases in hybridization rates are threatening the genetic integrity of the black duck's gene pool, and will further tend to blur the taxonomic lines between these two forms in the foreseeable future.

Suggested readings. Wright, 1954; Coulter & Miller, 1968.

Meller Duck

Anas melleri Sclater 1864

Other vernacular names. None in general English use. Madagaskarente (German); sarcelle de Madagascar (French); pato de Meller (Spanish).

Subspecies and range. No subspecies recognized. Resident in eastern Madagascar (Malagasy), introduced on Mauritius. See map 77.

Measurements and weights. Folded wing: males, 245–60 mm; females, 241–53 mm. Culmen: males, 58–62 mm; females, 52–58 mm. Weights: no record. Eggs: 59 x 42 mm, yellowish white, 45 g.

Identification and field marks. Length 25–27″ (63–68 cm). *Adults* of this species are generally mallardlike but have brown feathers with reddish edges. The head and neck feathers have dusky streaks and are edged with reddish coloration, and the speculum is iridescent green, bordered broadly behind with black and narrowly tipped with white, while the greater secondary coverts produce a broad black band and a narrow dull reddish terminal band. The iris is brown, the bill olive green with a black nail and base, and the legs and feet are orange red. *Females* are nearly identical to males, and *juveniles* are more reddish than adults, especially on the underparts.

In the field, the mallardlike shape and appearance separates this bird from other Madagascan waterfowl. Vocalizations of both sexes are also almost exactly like those of mallards.

NATURAL HISTORY

Habitat and foods. These birds occupy the swamps and morasses of interior Madagascar, from sea level to about 6,000 feet, and are said to occur on such habitats as open ponds, bays, and sluggish streams, especially in humid country close to forests. They are also found in rice fields, and no doubt consume rice where it is available, but no specific studies of their foods have been undertaken. They are evidently rare in coastal regions, and several have been collected from the rushy swamps in the vicinity of Lake Aloatra, where they are perhaps as common as anywhere on the island.

Social behavior. These ducks are reportedly most often found in pairs or in only small groups, and in captivity are said to be quite aggressive toward other duck species. Lorenz (1951–53) states that drakes defend their mates more strongly than do mallards, and have to fight other males much more frequently. Although they are as large as mallards, the females have unusually high-pitched inciting and decrescendo calls, and the males tend to utter their *raeb* notes in series of three rather than in groups of two. Additionally, the males differ from male mallards in that they perform a femalelike independent nod-swimming display. Otherwise the species evidently has a typical mallard repertoire of social displays.

Reproductive biology. No nests have yet been described from the wild, but the breeding time is said to extend from July until September. This species has been kept in captivity at various times, and has been bred regularly. The nest and eggs are mallardlike, and the incubation period is likewise reported by Delacour (1954–64) to be 28 days, although Kolbe (1972) reports it as 29 days for the Berlin zoo. The average clutch is 10 eggs, laid one per day, and fledging occurs in 9 weeks (McKelvey, 1977).

Relationships. This species is obviously part of the mallard complex, but no detailed studies of it are

available that might help to provide a more accurate assessment of its probable relationships.

Status. No specific information is available on this species' status. Since the mallardlike ducks seem to be highly adaptable to human-altered environments, it seems likely that this species will be able to survive indefinitely.

Suggested readings. Lorenz, 1951–53; McKelvey, 1977.

Yellow-billed Duck

Anas undulata DuBois 1837

Other vernacular names. African yellow-billed duck, yellowbill; Gelbschnabelente (German); canard à bec jaune (French); pato africana de pico amarillo (Spanish).

Subspecies and ranges. (See map 77.)
A. u. ruppelli: Abyssinian yellow-billed duck. Resident in the upper Blue Nile and the Ethiopian lake region of Africa.
A. u. undulata: South African yellow-billed duck. Resident in Africa from Angola, Zaire, Uganda, and Kenya south to the Cape.

Measurements and weights. Folded wing: males, 245–65 mm; females, 225–43 mm. Culmen: males, 48–51 mm; females, 44–49 mm. Weights: males, 678–1,208 g (av. 954 g); females, 630–1,114 g (av. 817 g). Eggs: av. 56 x 41 mm, buff, 55 g.

Identification and field marks. Length 20–23" (51–58 cm). *Adults* have a blackish gray head and neck, the lighter edging of the feathers producing a streaked effect. The feathers of the upper parts of the body are mostly brownish black with buff to whitish edging, producing a scaled appearance, while the underparts are mostly dark brown, with narrower buffy edging, and the flanks are both edged and marked centrally with buff. The tail is blackish brown, with buffy edging or barring; and the upper wing surface is mostly olive brown, except for the mallardlike speculum, which is iridescent green, bordered in front and behind with black and white bars. The underside of the wing is almost entirely white. The iris is brown; the legs and feet are brown to brownish black; and the bill is bright lemon yellow, with a black patch on the culmen down to the nostrils, black edging on the upper mandible, and a black nail. *Females* are nearly identical in appearance to males, but are slightly smaller, tend to be slightly lighter in color, and the bill coloration is slightly less bright. *Juveniles* resemble adults, but have coarser streaking on the head, which thus tends to be darker, while the pale fringing of the body feathers is more buff and the underparts are more heavily spotted.

NATURAL HISTORY

Habitat and foods. Yellow-billed ducks occur over a broad geographic area in southern Africa, but are most common in areas of temperate climate, and inhabit waters that are not extremely acidic and do not have high concentrations of sodium chloride. Rapidly flowing rivers are used relatively little, but slow rivers, their adjacent flooded fields, or coastal lagoons and estuaries are often utilized. Artificial reservoirs associated with mining, and sewage disposal impoundments are also highly favored habitats, indicating a mallardlike flexibility and adaptability in habitat use by yellow-billed ducks. In spite of the species' abundance, almost nothing has been reported on its foods, but wild harvested birds have been found to contain such items as grain, greenstuffs (possibly alfalfa), and seeds. The birds typically forage at dusk and after dark, often flying some distance from the areas where they loaf during the daylight hours. They probably feed on a very large variety of wild and agricultural plants, and to some extent on invertebrate foods (Rowan, 1963; Clancey, 1967).

Social behavior. It is still uncertain whether the pair bond persists in this species beyond the female's initiation of incubation, since some apparent "pairs" of adults tending broods may simply represent two females. In any case, a seasonal monogamy probably prevails, and it is also clear from South African observations that social displays occur essentially throughout the year, but are concentrated in the period between June and August, or about five to seven months before the peak of the breeding season (Rowan, 1963). However, since the species is an opportunistic breeder, some birds are probably paired at all times of the year, and breeding cycles are not

nearly so predictable as they are in mallards. The social displays of yellow-billed ducks and common mallards are nearly identical, judging from my own observations (1965a) and other published reports, with the exception of the fact that independent nod-swimming by males is a regular part of this species' repertoire. Aerial pursuit flights have frequently been observed among wild birds, and D. M. Skead (pers. comm.) reports that territoriality and territorial defense flights are present in this species. Copulatory behavior is essentially identical to that of the other mallards, and is usually followed by a rapid nod-swimming on the part of the male, which is terminated by a variable number of nodding movements.

Reproductive biology. Although even in temperate South Africa nesting records extend throughout the entire year, there are clear indications that variations exist. In areas of winter rains such as the southwest Cape region, nesting occurs shortly after the peak of the rainy season (June to September), while in areas of summer rains, such as the Transvaal, nesting has a peak in the wet season, although breeding does extend throughout the entire year. Evidently the species is physiologically adapted to breed when water conditions are suitable, regardless of the time of year. The nesting sites are likewise highly adaptive, but usually consist of areas where thick, grassy vegetation occurs near water. Relatively few nests are actually constructed above water, and elevated sites are rarely if ever used. The clutch size is quite variable, but usually ranges between 4 and 10 eggs, with 8 being the most commonly encountered number. Eggs are evidently laid at daily intervals, and the female may begin to incubate before the clutch is complete. The incubation period is normally 27 days, and it is probable that during this period the male abandons his mate. Brood counts suggest that relatively few

brood mergers occur, and approximately 9 to 10 weeks are required for the young to fledge. Even after fledging, however, the brood may remain more or less intact and in their brood-rearing area, not dispersing until the subsequent winter. Wing molt in adults occurs about three or four months after the peak of the breeding season, and a flightless period averaging 36 days follows. There seems to be no evidence yet that multiple brooding occurs, but renesting is regular if the first clutch should be destroyed (Clancey, 1967; Rowan, 1963).

Status. At least in the southern part of its range, the yellow-billed duck is perhaps Africa's most common member of the genus *Anas*, and there is no indication that its range or numbers are significantly diminishing at present.

Relationships. Anatomical and behavioral evidence indicates that this is a typical member of the mallard group, but it is not yet clear if it has been derived from the more northerly common mallard stock or from the gray duck stock of southern Asia. Murton and Kear (1975) suggest that the yellow-billed duck is an earlier derivative of the ancestral mallard stock than either the gray duck or the Philippine duck, but probably was derived from temperate-zone ancestors.

Suggested readings. Rowan, 1963; Clancey, 1967; Skead, 1976.

Gray Duck

Anas poecilorhyncha Forster 1781

Other vernacular names. Black duck, spot-billed duck; Augenbrauenente, Fleckschnabelente (German); canard à bec tacheté, canard à sourcils (French); pato pico manchado, pato australiano (Spanish).

Subspecies and ranges. (See map 77.)
  A. p. poecilorhyncha: Spot-billed gray duck. Resident throughout the Indian subcontinent, from east of the lower Indus River and Kashmir to western Assam, and south to Mysore and occasionally Ceylon.

A. p. haringtoni: Burmese gray duck. Resident in eastern Assam, Burma, and Yunan.

A. p. zonorhyncha: Chinese gray duck. Breeds in eastern Siberia, Manchuria, northern China, southern Sakhalin, the Kurile Islands, and Japan. Winters south to southern China and Formosa (Taiwan).

A. p. superciliosa: New Zealand gray duck. Resident in New Zealand and adjoining islands, including the Macquarie, Auckland, and Campbell islands.

A. p. pelewensis: Lesser gray duck. Resident of Melanesia on Fiji, Society, Cook, Tonga, Samoa, New Caledonia, Loyalty, New Hebrides, Solomon, Bismarck, Palau, and Caroline islands, and in northern New Guinea.

A. p. rogersi: Australian gray duck. Resident in Australia, Tasmania, East Indies, Celebes, Moluccas, southern New Guinea, and the Louisiade Archipelago.

Measurements and weights. Folded wing: males, 224–80 mm; females, 221–60 mm. Culmen: males, 46–65 mm; females, 49–55 mm. Weights: males (of *zonorhyncha*), 1,156–1,340 g; females, 750–980 g (Tso-hsin, 1963); males of *poecilorhyncha* 1,230–1,500 g; females, 790–1,360 g (Ali & Ripley, 1968); males of *rogersi*, 870–1,400 g; females, 806–1,280 g (Frith, 1967); males of *superciliosa*, 765–1,275 g; females, 623–1,275 g (Balham, 1952). Eggs: av. of *zonorhyncha*, 58 x 40 mm, creamish white; av. of *rogersi*, 58 x 41 mm, creamish white to greenish, about 60 g.

Identification and field marks. Length 21–24″ (54–61 cm). Both sexes of *adults* are nearly identical, but racial variation is considerable. The head has a black

crown stripe and a second stripe through the eyes; except in *poecilorhyncha* and *haringtoni,* a cheek stripe is present. The mantle, breast, and flanks are spotted with dark brown and buff to grayish edges. The tail and tail coverts are blackish or dark brown, the coverts usually with lighter edging, and the tertials are either entirely dark brown or have white on the outer vanes (*poecilorhyncha, haringtoni,* and *zonorhyncha*). The upper wing surface is brown, with an iridescent green to purple speculum line in front and behind with black and white bars. The iris is brown, the legs and feet yellow green to orange red, and the bill is lead gray, with a pale gray (*rogersi, pelewensis,* and *superciliosa*) or yellowish (*poecilorhyncha, haringtoni,* and *zonorhyncha*) tip. Swollen reddish spots occur at the base of the bill in *poecilorhyncha. Females* are nearly identical to males, but are slightly smaller and sometimes slightly lighter in color. *Juveniles* resemble adults, but tend to be paler and have spotted underparts.

In the field, the mallardlike shape and behavior of these birds will separate them from all species except the common mallard, which sometimes occurs in the same area. In such cases, the white tertials (in *poecilorhyncha, haringtoni,* and *zonorhyncha*) or the strong cheek stripe (in the other three races) will serve to separate the species. The vocalizations of both sexes are virtually identical to those of the common mallard.

NATURAL HISTORY

Habitat and foods. This widespread and adaptable species occupies a wide range of tropical and temperate habitats, and like the common mallard is difficult to pinpoint as to ecological needs. In Australia it is found in all fresh-water, brackish, and sometimes saline habitats; and in New Zealand the birds are found in mountainous areas, estuaries, sheltered inlets, and all inland waters. The continental races are likewise widely distributed, but they are most often associated with shallow fresh-water areas with emergent or floating vegetation, and tend to avoid mountainous or heavily wooded situations. Foods have been reported on by Frith (1967) for the Australian race, by Balham (1952) for the New Zealand form, and by Tso-hsin (1963) for the Chinese race. The largest samples are from Australia, where the most important foods were found to be the larger-seeded aquatic and swamp plants, such as

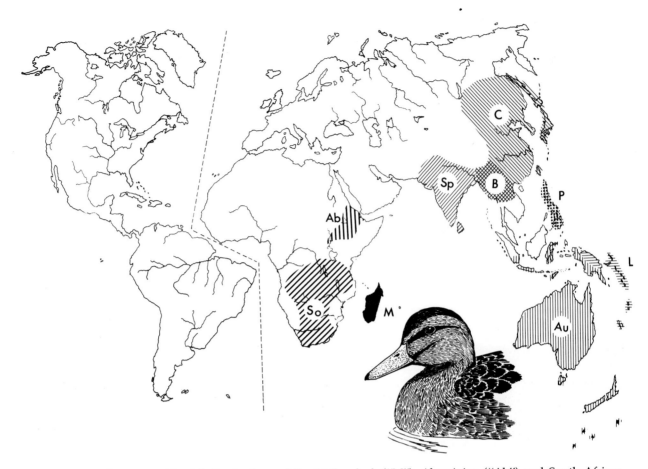

MAP 77. Breeding or residential distributions of the Meller duck ("M"), Abyssinian ("Ab") and South African ("So") yellow-billed ducks, Philippine duck ("P"), and Australian ("Au"), Burmese ("B"), Chinese ("C"), and spot-billed ("Sp") gray ducks.

sedges, smartweeds, grasses, and, where available, water lilies. Except for young birds, animal materials comprised a minority of the volume in all the samples. At least in Australia, gray ducks very rarely feed in grain stubble fields like common mallards, although they do take terrestrial foods along shorelines and banks.

Social behavior. As in the common mallard, pair-forming behavior occurs over a prolonged period between fall and spring, beginning in May in southern Australia and continuing through the end of the year. In northern Australia, where the breeding seasons are nearly the reverse of those in the south, courtship begins with the first rains of October or November and persists until the start of the breeding season in March or April. Thus, the species' flexibility to adapt to variable climates is no doubt a reason for its general abundance and wide distribution throughout

southeastern Asia and the Australian regions. The pair-forming displays are virtually identical to those of the common mallard, the only major exception being the presence of an independent nod-swimming display by males (Johnsgard, 1965a). No doubt because of these similarities, hybridization with mallards is fairly common in areas where mallards have been introduced in historical times, such as in New Zealand (Sage, 1958) and also in Australia (Braithwaite & Miller, 1975).

Reproductive biology. The timing of breeding in gray ducks is highly flexible, as indicated for the Australian race, but in temperate-zone areas it is timed to occur in spring and summer. It occurs between September and December, during the austral spring, in New Zealand, while in China, Japan, and the U.S.S.R. it occurs between April and June. In northern and peninsular India it extends from July to about

October, while in south India it occurs during November and December, as water conditions dictate. Clutch sizes are moderately large in most of the species' range; in Australia they average about 9 eggs (Frith, 1967), in New Zealand the average is the same or slightly less (Balham & Miers, 1959), and in India the usual number is from 7 to 9 (Ali & Ripley, 1968). Nest locations are quite flexible, as in common mallards, and may be on dry ground, in elevated situations such as tree holes, or on the deserted nests of other birds (Frith, 1967). The incubation period is from 26 to 28 days, and the incubating is entirely performed by the female. There is some doubt about the role of the male in brood care; Ali and Ripley (1968) state that the male does participate in this activity in the Indian race, but in New Zealand males begin flocking when the females begin incubation (Balham, 1952), and presumably play no further role in reproduction. The period to fledging in the Australian race is about 50 to 60 days. It has been suggested that two broods per year may be raised in some areas, such as in India and China, but this remains unproven.

Status. In Australia the gray duck is probably the most numerous species of duck (Frith, 1967); and in New Zealand, although it has suffered from competition with the introduced mallard as well as from ecological changes, it is still common in most areas. Less is known of the East Indies populations and those of the Indian and Southeast Asian areas, but they are seemingly less common in general.

Relationships. The close similarities of gray ducks and mallards in behavior and plumage, as well as their tendency to hybridize whenever they are placed in proximity, indicates that these species are very closely related. It is very difficult to draw lines separating species in this group because of these intermingling characteristics, and several authorities regard the continental ("spot-billed") forms as specifically distinct from the gray (and "black") ducks of Australia, New Zealand, and Melanesia. On the basis of their photoperiod responses, Murton and Kear (1975) suggest that the dull-colored Australian, New Zealand, and Melanesian populations were derived from the brighter northern "spot-billed" forms, in a manner somewhat comparable to the dull-colored races of the common mallard and the North American black duck.

Suggested readings. Frith, 1967; Dementiev & Gladkov, 1967; Ali & Ripley, 1968.

Philippine Duck

Anas luzonica Fraser 1839

Other vernacular names. Philippine mallard; Philippenente (German); canard des Philippines (French); pato de Filipinas (Spanish).

Subspecies and range. No subspecies recognized. Resident on the Philippine Islands. See map 77.

Measurements and weights. Folded wing: males, 240–50 mm; females, 234–40 mm. Culmen: males, 48–52 mm; females, 46–50 mm. Weights: males, 803–977 g (av. 906 g); females, 725–818 g (av. 779 g) (Rand & Raber, 1960). Eggs: av. 54 x 41 mm, pale green, 51 g.

Identification and field marks. Length: 19–23" (48–58 cm). *Adults* have a rusty cinnamon head and neck color, except for a darker crown and nape, and a similar dark stripe extending from the bill back through the eyes. The rest of the feathers are mostly gray and relatively uniform, although the breast is slightly fulvous and the back and rump are darker than the rest of the body plumage. The upper wing surface is grayish brown to brownish, except for the iridescent green speculum on the secondaries, which is bounded in front and behind with a black bar, and trailing and leading white bars. The iris is brown, the bill lead blue, and the legs and feet brownish black. *Females* differ from males only in being slightly smaller. *Juveniles* are paler on the head and throat, having little indication of reddish coloration, and the speculum is less well developed.

In the field, the mallardlike shape and behavior provide the best field marks for this endemic species, since no other mallardlike ducks occur on the Philip-

pine Islands. The vocalizations appear to be virtually identical to those of the common mallard.

Natural History

Habitat and foods. This species apparently is of widespread occurrence, and has been reported on rivers, mountain lakes, tidal creeks, and small ponds. Its foods and foraging behavior have not been investigated in the wild, but captive Philippine ducks appear to be normal mallards in these respects.

Social behavior. Early observers (cited in Phillips, 1922–26) indicate that these birds are often seen in pairs or small flocks, with a few cases of flock sizes numbering up to two hundred individuals. There is no information on the possibility of permanent pair bonding in Philippine ducks, although it is not known to occur in any of the other mallard group. The social displays have been seen only in captivity and take the normal mallard form. The female has a decrescendo call of about seven syllables which seems to be squeakier than that of the common mallard or spot-billed duck. Inciting is performed in exactly the same way as in those two species, but the associated call is somewhat weaker. Although it may be expected to occur, display preening behind the wing by females has not yet been specifically noted. The male displays are notable for the frequency of an independent nod-swimming display, which is even more frequent than in spot-bills and has an exaggerated nodding component much like that of the chestnut teal. Nod-swimming not only occurs as a separate display, but also almost invariably follows the head-up-tail-up display. The two other mallard displays, the grunt-whistle and the down-up, are both performed in typical mallardlike fashion and with the usual vocalizations. Males very frequently turn the back of the head to inciting females while swimming ahead of them, and on a few occasions display preening has been seen. The copulatory behavior is of the usual form, consisting of precopulatory mutual head pumping; and after treading, an extremely pronounced nod-swimming occurs. Lastly, the male turns the back of the head to the bathing female (Johnsgard, 1965a).

Reproductive biology. Apparently no nests or eggs have been found under wild conditions, so all available information comes from captive birds. The relatively bulky nest is constructed on dry land, and is placed in rank grasses, bushes, or similar vegetation.

Eggs are laid on consecutive days, and the clutch normally consists of about 10, with a range of 8 to 14. The incubation period is from 25 to 26 days, and the incubating is performed by the female alone. The period to fledging has not been reported, but probably is in the vicinity of eight weeks.

Status. There seems to be no specific recent information on the status of this species, and the few earlier observers who have commented on it reported that it was anywhere between "rare" and "very abundant." At least in early times it was abundant in northern Luzon and on the west end of Masbate (Phillips, 1922–26).

Relationships. While clearly a member of the mallard group, this species has a more distinctive plumage pattern than most, and its displays are also somewhat divergent from those of most other mallards. It would seem most likely that the Philippine duck is derived from an ancestor similar to the present-day gray duck, whose current range comes within a few hundred miles (Celebes Islands) of the Philippine duck's present distribution.

Suggested readings. Delacour, 1954–64.

Bronze-winged Duck

Anas specularis King 1828

Other vernacular names. Spectacled duck; Kupferspiegelente (German); canard aux ailes bronzies (French); pato anteojillo or pato perro (Spanish).

Subspecies and range. No subspecies recognized. Breeds in forested areas of Chile south from Cautín, and perhaps Curicó, to Tierra del Fuego, and in western Argentina at comparable latitudes, ranging in the nonbreeding season northward to Santiago in Chile and Mendoza in Argentina. See map 78.

Measurements and weights. Folded wing: males, 260–80 mm; females, 252–77 mm. Culmen: males, 49–52 mm; females, 43–44 mm. Weights: adult females ca. 960 g (Lack, 1968). Eggs: av. 64 x 44 mm, deep cream, 75 g.

Identification and field marks. Length: 21″ (54 cm). Plate 43. *Adults* have a dark brown head and neck, with a contrasting white oval facial patch in front of the eyes and a second white patch on the throat and extending up the side of the neck. The mantle and scapular feathers are dark brown to blackish, with lighter edging, while the breast and underparts are barred or mottled with tan and dark brown, with colors on the flanks becoming large dark brown or blackish crescents on a tan background, while on the tail coverts dark brown tones predominate. The tail is likewise dark brown, and the upper wing surface is dark brown to purplish black, except for the secondaries, which form a brilliant bronzy speculum, bordered behind by narrow black and white bands. The iris is brown, the bill bluish gray, and the legs and feet are orange, with darker markings. *Females* closely resemble males, but may be slightly duller. *Juveniles* are similar to adults, but may have little or no white on the face and are heavily streaked on the breast.

In the field, few other ducks occupy the forested river habitat of this species, except possibly torrent ducks. The doglike "bark" of the female is distinctive and the basis for the Spanish name pato perro, while the male utters a loud trilled whistle during display. The conspicuous white facial marking is the best visual field mark, and when in flight the birds tend to follow the course of the river rather than to fly over land.

NATURAL HISTORY

Habitat and foods. Most observers agree that heavily forested rivers that are relatively swift-flowing are the preferred habitat of this species, although they also occur on slow-moving rivers and on pools or ponds of the adjoining forest areas. They are said to consume both vegetable and animal materials, and have been observed eating small snails that abound on stony shingle beaches. Stomach remains from two birds that were examined contained the seeds of water crowfoot (*Batrachium*), water milfoil (*Myriophyllum*), and a bulrush, leaves of water crowfoot, foliage and seeds of a pondweed, and caddis fly larvae as well as a few other aquatic insect remains (Phillips, 1922–26). In captivity at least the birds seem to spend a good deal of time on land and have not been observed diving for food.

Social behavior. In general, bronze-winged ducks are to be found in pairs or family groups, with an observation of 18 the largest flock size that has been recorded. It seems probable that the birds, like other river ducks, are relatively sedentary and may space themselves out along such habitats by establishing exclusive foraging and breeding territories. It is presumed that pair bonds are relatively strong in this species, but it does not seem to be established whether the male regularly participates in brood care. The decrescendo call of the females is a repeated, raucous quacking sound of five or six syllables, while inciting is a loud and gutteral note that is accompanied by strong chin-lifting movements, exposing the white throat markings. Male displays have not been adequately seen, but I have recently observed a captive male responding to female's inciting—loud, trilled whistle—by facing the female and performing vigorous diagonal neck stretching and chin lifting. The chin lifting, which also strongly exposed the white throat, was so strong that the head of the nape sometimes even touched the back. Once the male gaped toward the female with outstretched neck, in the manner of crested ducks, and then preened behind the wing toward her. Although an early observer said that the male's display was pintaillike, this is certainly not the case, and instead it appears to be relatively unique. Unfortunately, copulatory behavior has not yet been described.

Reproductive biology. Almost no nests of the bronze-winged duck have been found, but Johnson (1965) provides an account of finding several in the Andes of Nuble province in November. Examining some small islets in the middle of a river, he found nests hidden in tall grass, half covered with bent-over grasses and abundantly lined with down. Subsequent nests that he found were also always on river islets, which he concluded were the safest location against foxes or other terrestrial predators. November is evidently the usual time for nesting, with hatching occurring in late December or early January. The clutch size is usually from 4 to 6 eggs, and the incubation period has been established to be 30 days in captivity (Johnstone, 1970). Johnstone (1965) noted that the downy young are quite different from the only published illustration in Delacour (1954–64), and are somewhat wigeonlike but have distinctive white areas on the cheeks and throat much like the adult plumage. It is believed that the male normally participates in brood care (Murton & Kear, 1975).

Status. Johnson (1965) states that the bird is present in "fair" numbers in southern Chile, and in Argentina it is probably equally common, but no real basis for estimating its population is yet available.

Relationships. Phillips (1922–26) clearly regarded the crested duck and bronze-winged duck as close rela-

tives, and commented on the similarity in the structure of the tracheal bulla. However, Delacour (1954–64) evidently regarded the species as being nearest the mallard group, although he said that its behavior suggests some slight relationship to the shelducks (in which group he regarded the crested duck to belong). My own (1965a) review of the situation led me to conclude that both the crested duck and the bronze-winged duck provide an evolutionary link between the mallardlike ducks and the pintail group. The absence of pintaillike displays and the surprisingly distinctive downy plumage of the bronze-winged duck have forced some reconsideration of this position, and I now believe that both of these species, although clearly related to each other, are fairly isolated from the rest of the genus *Anas.*

Suggested readings. Johnson, 1965.

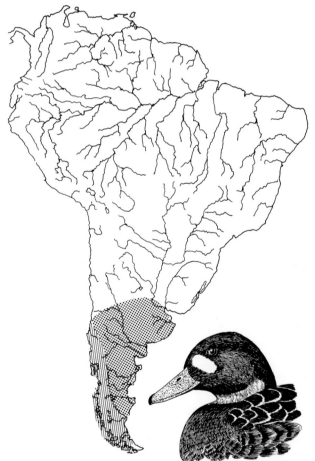

MAP 78. Breeding or residential (hatching) and wintering (stippling) distributions of the bronze-winged duck.

Crested Duck

Anas specularioides King 1828

Other vernacular names. None in general English use. Schopfente (German); canard huppé (French); pato crestón (Spanish).

Subspecies and ranges. (See map 79.)

A. s. alticola: Andean crested duck. Resident in the Chilean Andes from Arica to Talca, and the corresponding portions of northwestern Argentina northward through Bolivia and Peru to Lake Junín.

A. s. specularioides: Patagonian crested duck. Breeds from Talca in Chile and Mendoza in Argentina south to Tierra del Fuego and the Falkland Islands, and moves northward along both coasts during the nonbreeding periods.

Measurements and weights. Folded wing: males, 268–310 mm; females, 250–90 mm. Culmen: males, 43–51 mm; females, 40–48 mm. Weights: males, 1,070–1,180 g; females, ca. 900 g. Eggs: av. 63 x 46 mm, deep cream, 56 g.

Identification and field marks. Length 20–24" (51–61 cm). Plate 42. *Adults* have a head and neck that is pale grayish brown, with a darker brown stripe behind the eyes that fades to gray as it approaches a long but ragged crest. The breast is fulvous-toned, with subdued spotting, while the abdomen, flanks, and tail coverts are mostly light gray, with darker spotting and barring. The scapular feathers are brown, with grayish centers and darker tips. The tail is distinctly elongated and black, as are the under tail coverts. The upper wing surface is gray brown to olive brown, except for the secondaries, which are iridescent coppery to greenish, with a broad black band behind, and edged narrowly with white. The legs and feet are dark gray, the bill is also gray, and the iris color varies from yellow (*alticola*) to reddish (*specularioides*). *Females* are nearly identical to males, but have a less definite crest and are appreciably smaller. *Juveniles* lack crests, have light brown faces, and have considerably paler abdomens than do adults.

In the field, the slim-bodied appearance, the long tail, and the crested head are all distinctive features. In flight, the dark speculum, with a narrow white trailing line, is distinctive. The birds are relatively noisy, and the male frequently utters a buzzy whistling *sheeooo*, while the female's voice is low, grating, and somewhat like the barking of the dog (as is also true of bronze-winged ducks).

NATURAL HISTORY

Habitat and foods. Patagonian crested ducks are reported by Johnson (1965) to occur on mountain lakes, on the plains, or along coastlines, where in winter they gather in small groups to forage among the rocks and kelp beds. In the Falkland Islands they are most often encountered on salt water, but also are found on ponds near the sea and in creeks (Woods, 1975). In creeks they apparently forage on small crustaceans that are found by dabbling and upending. Brackish inland lakes on Tierra del Fuego are utilized, as is the coastline itself, especially where quiet bays and estuaries occur (Weller, 1975a). Weller's (1972) studies on the Falkland Islands indicate that the adults and young forage on marine isopods and amphipods, invertebrate larvae, and tiny clams; and although they often feed in growths of filamentous algae, they consume very little plant materials. They are thus almost entirely carnivorous, and generally feed in water shallow enough to stand in, or at least in areas where diving is not necessary.

Social behavior. Perhaps in conjunction with a basically animal diet, crested ducks are notably nongregarious and tend to be highly territorial. Rarely do large flocks develop, except perhaps in areas of unusually rich food supplies; and in general, pairs will expel others of their own as well as alien species from their foraging areas. Weller (1975a) reported seeing some large flocks on Tierra del Fuego, which he judged to be immature (first-year) birds, although the period to reproductive maturity is still uncertain in this species. Social display has not yet been adequately studied, but inciting by the female plays an important role in mate selection and pair formation;

indeed Delacour's belief that the species is an aberrant shelduck is largely based on this characteristic. Females also perform an unusual nod-swimming display, which differs from that of other *Anas* forms in that inciting movements sometimes are interspersed with its performance. Females also perform an exaggerated general body shake as an apparent display, like the bronze-winged duck. I have recently observed that males utter a wheezy whistled note during courtship as they stretch their necks vertically, and also perform a ritualized body shake that is frequently followed by a head-up-tail-up posture. An *Anas*-like grunt-whistle display is also produced, and males frequently follow their introductory shake display with a preening display toward a particular female. Finally, males sometimes gape with horizontally extended neck toward the female exactly like bronze-winged ducks, and this display presumably also corresponds to the "gesture of greeting" used by the speckled teal. Copulation is preceded by the usual mutual head pumping by the pair, and in the few instances I have observed, the male swam away from the female afterwards in a manner resembling nod-swimming (Johnsgard, 1965a).

Reproductive biology. Little is known of the nesting biology of *alticola*, but Johnson (1965) reports that in Chile the birds evidently nest over a prolonged or irregular period, and usually on small islets in areas of open water. However, they also place nests in such other locations as exposed salt flats, in holes in banks, or on abandoned coot nests. The Patagonian race often nests in boggy areas on clumps of marsh grass around mountain lakes, or almost anywhere in tall grass if it is fairly close to a lagoon. The nesting period in the Falkland Islands is very extended, and Woods (1975) suggested that double-brooding is frequent there. The major nesting period there is from September to November, while Weller (1975a) noted that most of the broods he observed on Tierra del Fuego had evidently hatched in early or mid-January. Clutch sizes are normally from 5 to 8 eggs in both races, and the incubation period is 30 days. Only the female incubates, but participation of the male in brood care is normal. Weller (1975a) reported seeing both parents in attendance with about 90 percent of the broods he observed. The fledging period is approximately 10 or 11 weeks (Weller, 1972).

Status. Johnson (1965) reports that the Andean crested duck is the commonest duck of the *puna* zone of the Chilean Andes, and that the Patagonian race is relatively common in southern Chile from Aysén

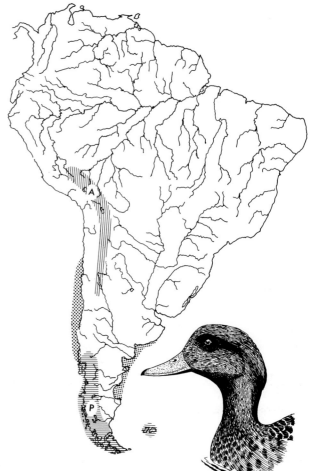

Map 79. Breeding or residential distribution of Patagonian ("P") and Andean ("A") crested ducks. Wintering distribution indicated by stippling.

southward. In Patagonia the species is apparently common only near the mountains, and especially to the south, where it is perhaps the commonest duck on the Magellan Strait (Scott, 1954). In the Falkland Islands it is generally common, especially around West Falkland (Woods, 1975). Thus, the species seems to be in no real danger at present.

Relationships. Although the crested duck has traditionally been placed in the genus *Anas*, Delacour (1954–64) urged recognition of a separate genus (*Lophonetta*) because of its unusual shape, proportions, and its shelducklike behavior. He placed the genus in the shelduck tribe, but regarded it as representing a transitional form between the shelducks and dabbling ducks. However, strong similarities in specific display traits between crested ducks and bronze-winged ducks, as well as similarities in their tracheal structure, indicate that both belong in the genus *Anas*. The shelducklike aggressiveness of crested ducks is almost certainly correlated with their feeding ecology, and is not indicative of phyletic affinities.

Suggested readings. Johnson, 1965; Woods, 1975.

Pintail

Anas acuta Linnaeus 1758

Other vernacular names. Common pintail, sprig; Spiessente (German); pilet ordinaire (French); pato cola puntiaguda común (Spanish).

Subspecies and ranges. (See map 80.)

A. a. acuta: Northern pintail. Breeds in Iceland and in northern Europe and Asia west to Kamchatka and the Commander Islands. In North America breeding occurs from the Aleutian Islands and St. Lawrence across arctic Alaska and Canada to Victoria and Banks islands, and south to the southern United States and occasionally to northern Baja California. Winters to northern Africa, the Persian Gulf, India, southeastern Asia, southern China, Taiwan, the Hawaiian Islands, and in North and Central America to Panama and the West Indies.

A. a. eatoni: Kerguelen pintail. Resident on Kerguelen Island.

A. a. drygalskii: Crozet pintail. Resident on the Crozet Islands.

Measurements and weights. Folded wing: males of *acuta*, 254–87 mm; females, 242–66 mm. *Eatoni* males, 205–30 mm; females, 190–210 mm. Culmen: males of *acuta*, 48–59 mm; females, 45–50 mm. Both sexes of *eatoni* measure 30–33 mm. Weights: males of *acuta*, 710–1,250 g, averaging about 850 g. Eggs: av. 55 x 39 mm (in *acuta*), cream-colored, 45 g. Eggs of *eatoni* average 52 x 30 mm.

Identification and field marks. Length 20–29" (51–74 cm) for males, females 17–25" (43–63 cm). *Males* in breeding plumage have a dark brown head and upper foreneck, with a darker crown and a black nape patch, blending with a gray back, and a white stripe on each side of the neck that is continuous with the white breast and underparts. The sides of the lower neck, breast, and sides are all finely vermiculated with black and white, and the longer scapulars are pointed and purplish black, with white margins. The outer scapulars and tertials are similar, as are the longest upper tail coverts, while the outer ones are gray. There is a white patch on each side of the rump, and the under tail coverts are black. The tail feathers are mostly gray with white margins, and the central ones are black and elongated. The underwing surface is dusky and white, while most of the upper coverts are ashy gray and the primaries are light grayish brown. The greater secondary coverts are tipped with pale cinnamon, and the secondaries form an iridescent green speculum with a narrow black and white border. The iris is brown, the legs and feet are bluish slate, and the bill is black, with a grayish blue stripe along the sides of the upper mandible. *Females* are generally grayish brown, with darker brown, buff, and white streaking on the head. The upperparts and flanks are brownish black, with buff and white marginal and central V-shaped markings; the breast is buff to reddish brown, with lighter mottling; and the underparts are whitish, with sparse spotting. The tail is dark brown, with buff margins and central markings, the underwing surface is dusky and white, and the upper wing coverts are grayish brown, margined with white. The secondaries are brownish, often with a slight greenish gloss, and are tipped with white. The iris is brown, the bill grayish black, and the legs and feet bluish gray. *Males in eclipse* resemble females but have a bright wing speculum, and often show vermiculated feathers on the upperparts. *Juveniles* closely resemble females but are more heavily spotted underneath, and have plainer and less buffy upperparts.

In the field, the greatly elongated body profile and the long, white neck and breast of males make pintails among the most easily identified of waterfowl. The contrast between the black under tail coverts and white on the sides of the rump is also conspicuous in males, and females lack the blackish eye-stripe and crown markings typical of many other large *Anas* females, and also lack the yellow bill coloration of gadwall and mallard females. The quacking notes of female pintails are softer than those of mallards, and the decrescendo call is usually fairly abbreviated. The most conspicuous male call is a fluty whistle, uttered both on water and in the air, and accompanied by marked neck stretching. In flight, the long,

pointed wings and elongated tail are conspicuous, as is the rapid wingbeat and slim body profile.

NATURAL HISTORY

Habitat and foods. The tremendously broad breeding range of the northern pintail encompasses a wide variety of habitats, but the species is most abundant on prairie and tundra habitats that provide open vistas that contain shallow marshes, quiet rivers, or shallow lakes. Ponds with dense but low vegetational growth around the shore are favored, as are areas with interspersed brushy thickets or copses, and the birds prefer to forage in waters they can reach by upending. Summer and fall foods are predominantly vegetable in origin, and often include the seeds and vegetative parts of pondweeds and wigeon grass, and the seeds of bulrushes and smartweeds (*Polygonum*). However, during the laying and prelaying period, nesting females consume a high proportion of in-vertebrate foods, mainly obtained from shallow and seasonally wet habitats (Krapu, 1974). During the winter the birds occur widely, but in coastal areas they concentrate on shallow fresh or brackish estuaries that have adjacent agricultural areas with scattered impoundments. In areas where corn or, preferably, smaller grains are available they often feed in fields in the company of mallards, but they seem to be very opportunistic in their feeding and to take whatever is available in the local habitats (Stewart, 1962).

Social behavior. The social behavior of the Southern Hemisphere insular populations of pintails has yet to be studied, and thus only the northern pintail can be discussed here. Pintails are known to mature their first winter, and probably all females mate and attempt to nest as yearlings. By December, the largely sex-segregated flocks of the fall migration period begin to merge, and display begins. Evidently pair formation then proceeds fairly rapidly, and most

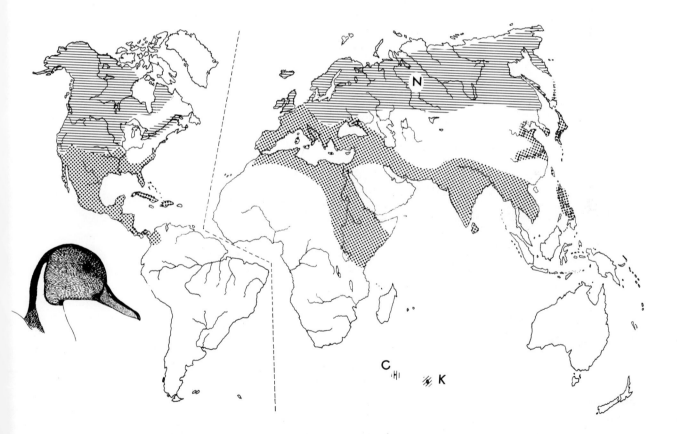

MAP 80. Breeding (hatched) and wintering (stippling) distributions of the northern pintail ("N"), and residential distributions of the Crozet Island ("C") and Kerguelen Island ("K") pintails.

females are paired before the end of winter, although some courtship continues through the spring migration period. Social display is marked by numerous male displays on the water, and courtship flights are easily recognized by their erratic nature and the repeated whistling burps of the drakes. Females may also incite during such flights, and this is the major female display on the water. Besides the burp whistle, uttered with stretched neck and slightly down-tilted bill, males also perform the grunt-whistle and head-up-tail-up displays repeatedly, and turn the back of the head to inciting females whenever the opportunity is presented them. Copulation is preceded by mutual head pumping, and afterward the male usually performs a single bridling display (Johnsgard, 1965a; Smith, 1968).

Reproductive biology. Nest-site selection begins shortly after the paired birds return to their breeding grounds; female pintails are more prone to nest in areas well away from their place of hatching than are most ducks, and thus they are prone to "pioneer" new habitats. Further, because of their tendency to nest in relatively dry locations as far as a mile from the nearest water, pintails tend to be well dispersed and show no territorial defense of limited water areas. Nests are often placed in dead herbaceous cover of the past year's growth, and frequently show almost no concealment; their trait of initiating nests before the new season's growth is underway no doubt in part accounts for this behavior, which in turn often produces a high rate of nest loss to predators that hunt visually. Eggs are laid at the rate of one per day, usually early in the morning, and the average size of early clutches is around 9 eggs. Where the length of the breeding season permits, renesting by unsuccessful females is fairly frequent. The incubation period is a rather short 21 days, and is followed by an equally short fledging period of 40 to 45 days in southern Canada and an even shorter 35- to 42-day period in Alaska. Males usually leave their mates at about the time that incubation begins, and often migrate to favored molting areas some distance away. About a month is required for the birds to complete their flightless period, and in the northern parts of their range the fall migration begins immediately, with the females regaining flight nearly a month later than the males (Sowls, 1955; Keith, 1961).

Status. The northern pintail is certainly one of the most abundant of North American ducks, with breeding-ground surveys between 1955 and 1973

averaging over 6 million birds (Bellrose, 1976). Likewise, in Eurasia it is extremely abundant, but continent-wide figures are not available. Ogilvie (1976a) provides some estimates for western Europe. On the other hand, the populations of the Kerguelen and Crozet pintails are inherently small, although no population estimates seem to be available. Watson (1975) states that the birds are protected and "now abundant" on Kerguelen Island, but are apparently uncommon on Crozet. They have also been introduced on the small islets of Amsterdam and Saint Paul.

Relationships. This extremely broadly ranging form might be thought of as nearest to the ancestral pintail, with Southern Hemisphere relatives in South America (brown pintail, white-cheeked pintail) and southern Africa (red-billed pintail). Of these, the brown pintail seems to be closest in ecology and behavior to the Northern Hemisphere species, and probably represents its nearest relative.

Suggested readings. Sowls, 1955; Smith, 1968.

Brown Pintail

Anas georgica Gmelin 1789

Other vernacular names. Chilean pintail, South Georgian teal, yellow-billed pintail; Spitzschwanzente (German); pilet à bec jaune (French); pato jergon (Spanish).

Subspecies and ranges. (See map 81.)

A. g. georgica: South Georgian brown pintail. Resident on South Georgia.

A. g. spinicauda: Chilean brown pintail. Breeds from the highlands of Peru and Bolivia south along the Andes through most of Chile and Argentina to Tierra del Fuego, and on the Falkland Islands. Winters north to southern Brazil.

A. g. niceforoi: Niceforo brown pintail. Once bred in the eastern Andes of central Colombia, now apparently extinct.

Measurements and weights. Folded wing: males, 210–60 mm; females, 195–240 mm. Culmen: males, 32–43 mm; females, 31–41 mm. Weights:

males of *spinicauda*, 740–827 g (av. 776 g); females, 663–769 g (av. 705 g) (Weller, 1968a); males of *georgica*, 610–60 g; females, 460–610 g (Weller, 1975c). Eggs of *spinicauda*: av. 56 x 40 mm, creamy, 42 g.

Identification and field marks. Length: 19–21″ (48–54 cm). *Adults* are generally fulvous brown, with chestnut crowns; buff throats, forenecks, and abdomens; and with the larger body feathers having darker brown centers. The brownish tail is long (over 75 mm) and pointed. The upper wing coverts are grayish brown, except for the greater secondary coverts, which are buff-tipped. The secondaries form a glossy black speculum and are tipped with buff. The iris is brown, the legs and feet gray, and the bill bright yellow, with a black nail and culmen mark, and with bluish coloration around the nail. *Females* are somewhat whiter on the underparts and have a dark brown speculum with most buffy coloration behind. *Juveniles* resemble adults, but lack the chestnut color on the top of the head, and they are streaked rather than spotted on the underparts and breast.

In the field, the generally fulvous brownish coloration and slim-bodied profile, with an elongated tail, separate this species from other South American ducks except perhaps the crested duck and white-cheeked pintail. The former is more grayish and is distinctly crested, and the latter has conspicuous white cheek and foreneck markings. The yellow bill of the brown pintail also separates it from other South American species except for the speckled teal, which is much smaller. Vocalizations are not loud or prevalent in brown pintails, and consist of the typical quacking notes of the females and a clear whistling note by males.

Natural History

Habitat and foods. This extremely widespread species occurs in a great variety of habitats in both fresh-water and marine situations. On South Georgia it shows a strong preference for ponds rimmed with tussock grasses, especially during the breeding season (Weller, 1975c). On Isla Grande, Tierra del Fuego, the birds were observed by Weller (1975a) to occupy a variety of steppe lakes, and they also occurred in estuaries. In both pond and seashore situations, pintails prefer to forage by dipping their heads and upending, but at times they will also dive for favored foods such as fairy shrimp. They also at times feed on

land or in soggy areas, rooting for food or reaching for overhead grasses. In seashore situations they feed on a variety of invertebrates and also some algae (Weller, 1975c). In Chile they have been reported to favor *Anacharis*, and at times also visit rice fields (Johnson, 1965). The long necks of pintails enable them to reach foods that most other dabbling ducks are unable to harvest effectively, and they thus probably compete relatively little with other *Anas* species in the same area, such as speckled teal (Weller, 1975c).

Social behavior. Flock sizes and sociality may well differ between the abundant mainland form and the relatively rare South Georgia race, but the latter is much better studied, owing to Weller's (1975c) observations there. During pair formation he noted that birds gathered in groups of up to 25 on sheltered ponds, and even during mid-November many of the birds seemed already to be in pairs. Since it is known that at least occasionally the male assists with brood rearing (Weller reported it in four of five cases on South Georgia), it seems likely that most pair bonds never are disrupted during the breeding season. Nonetheless, Weller did observe courtship frequently in November (first eggs being found in December), and most display was apparently initiated by excess drakes approaching already paired females. The most commonly performed male display is the burp whistle, a neck stretching accompanied by a simultaneous wheezy *geeeegeeee* sound and a multiple-noted whistle. Less frequently the grunt-whistle is performed, but other than turning of the back of the head and an apparently very rare display preening, no other elaborate postural displays are present. Evidently the South Georgia race differs scarcely at all from the mainland form in its postures and vocalizations, which may not be surprising considering the already simplified repertoire of the latter. Behavior associated with copulation is apparently identical to that of northern pintails (Johnsgard, 1965a).

MAP 81. Residential distributions of the South Georgian ("S") and Niceforo ("N") brown pintails, and breeding (hatched) and wintering distributions (stippling) of the Chilian brown pintail ("C").

Reproductive biology. Breeding seasons undoubtedly vary widely across the enormous range of this species, and even in temperate regions such as Chile they appear to be quite prolonged. Johnson (1965) suspected that two broods are raised in the central provinces there, the first nesting starting in August and the second brood appearing in January or February. In the mountains and to the south the nesting is more restricted, and occurs between October and December. On the Falkland Islands nesting is also between September and December, and double-brooding is suspected (Woods, 1975). On South Georgia, eggs evidently are laid between late October and February (Weller, 1975c). Nesting sites chosen by brown pintails are simple, merely scrapes on dry ground reasonably close to water (Johnson, 1965) or hidden in coarse grass, rushes, or tussock grasses (Woods, 1975). The clutch size of the Chilean race is normally between 4 and 10 eggs, probably averaging (in cap-

tivity) about 7. However, on South Georgia the brown pintail apparently never lays more than 5 eggs, which probably reflects its smaller body size as well as local ecological factors. Only the female incubates, and, at least for the Chilean race, the incubation period is 26 days. The male apparently normally joins the female after hatching to help rear the brood, but the fledging period is still unestablished.

Status. Although one Andean subspecies of this pintail is apparently extinct, the insular population on South Georgia is apparently not endangered. Weller (1975c) estimated it at several thousand birds in 1971, and it is currently protected. The widespread Chilean race is generally abundant in most areas; Johnson (1965) regards it as the most abundant and widely distributed of all Chilean waterfowl. It also constituted about half of the total number of ducks examined among hunters' kills by Weller (1968a) in eastern Argentina, although he regarded this as a biased sample. In any case, the species is in no present apparent difficulty.

Relationships. All the morphological and behavioral evidence indicates that the brown pintail and common pintail are very closely related; I (1965a) considered them to form a superspecies. The other pintails are slightly less closely related, but both the white-cheeked pintail and the red-billed pintail exhibit some behavioral and structural similarities to the brown pintail that are indicative of phyletic affinities.

Suggested readings. Weller, 1975c; Murphy, 1916, 1936.

White-cheeked Pintail

Anas bahamensis Linnaeus 1758

Other vernacular names. Bahama duck, Galapagos Island duck; Bahamaente (German); canard de Bahama (French); pato gargantillo (Spanish).

Subspecies and ranges. (See map 82.)

    A. b. bahamensis: Lesser white-cheeked pintail. Resident in the Bahama Islands, the Greater and northern Lesser Antilles, northern Venezuela, the Guianas, and northern Brazil.

A. b. rubrirostris: Greater white-cheeked pintail. Occurs in southern Brazil, Paraguay, Uruguay, and Bolivia, but apparently now breeds only in Argentina; previously also nested in Chile but now an occasional migrant in the north.

A. b. galapagensis: Galapagos white-cheeked pintail. Resident in the Galapagos Islands.

Measurements and weights. Folded wing: males, 190–235 mm; females, 180–221 mm. Culmen: males, 40–45 mm; females, 37–43 mm. Weights: males (*bahamensis*) 474–533 g; females, 505–633 g (Haverschmidt, 1968); males of *rubrirostris* averaged 710 g; 4 females averaged 670 g (Weller, 1968a). Eggs (*bahamensis*): av. 57 x 38 mm, creamy, 34 g.

Identification and field marks. Length 18–20″ (46–51 cm). Plate 44. *Adults* have the top of the head dark mottled brown, while the cheeks, chin, and foreneck are white to grayish (mottled with brown in *galapagensis*). The back and scapular feathers are dark brown, with fulvous edging, while the breast and underparts are also rich fulvous, with extensive blackish spots that are larger on the flanks. The tail and tail coverts are reddish buff, and the upper wing olive gray, except for the secondary coverts, which are tipped with fawn. The secondaries form a speculum pattern that is iridescent green anteriorly, followed by a narrow black bar and a broad trailing bar of buff. The legs and feet are dark gray, the iris is bright reddish brown, and the bill is reddish behind the nostrils and blue beyond, with a black nail. *Females* have slightly less colorful bills, shorter tails, and are less pure white on the face and throat. *Juveniles* closely resemble females, and have reduced iridescence on their speculums.

In the field, these pintails generally resemble brown pintails, except for their bill coloration and conspicuous two-toned head pattern. No other South American duck has a white face and throat, and only the distinctive torrent duck has a red bill. Females sometimes utter a rather weak decrescendo call, of only a few syllables, and the calls of the male (a soft

MAP 82. Residential distributions of the Galapagos ("G") and northern ("N") white-cheeked pintails, and breeding distribution of the southern white-cheeked pintail ("S"). Stippling indicates areas of wintering or uncertain breeding status.

geeee and a double-noted whistle) are also not especially loud or conspicuous.

NATURAL HISTORY

Habitat and foods. This species seems to show a definite preference for brackish water or salt-water habitats, and is infrequent to rare in fresh-water locations. Various early naturalists described its affinity for pools in mangrove areas or salt-water ponds. The Galapagos Islands race also is often found on salt lagoons and pools, as well as in the fairly few freshwater habitats of these islands. Only a few specimens have been examined as to their foods, and a sample of eight adults had been feeding exclusively on vegetable matter, including the seeds of wigeon grass (*Ruppia*), and particularly the foliage of algae (*Chara*) (Phillips, 1922–26). In captivity they have been observed to dive for food, but like the other pintails they almost certainly obtain most of their food by upending and reaching down with their long necks.

Social behavior. Small flocks appear to be the usual rule in this species, with a few observations of flock sizes numbering more than 100 birds. The birds usually are found alone, but at times also associate with brown pintails, whose foraging behavior and preferred foods must doubtless be similar. There is virtually no information on the nature of the pair bond in this species, but in the absence of contrary information it may be presumed that it is relatively permanent. In captivity, at least, courtship displays are performed fairly often nevertheless, and take a form that is much like that of the other pintails. Observations on the greater and lesser white-cheeked pintails show no apparent differences in displays, but the isolated Galapagos form has not yet been carefully studied in this regard. Scott (1960), however, said he was unable to detect any display differences in this form. Females produce a stiff-necked inciting display marked by rather creaking notes, and such birds are closely followed by males, which often attempt to turn the back of the head to inciting birds. Females also preen behind the wing as a display toward preferred males, and sometimes call at the moment of a male's display. Males precede their major displays with a repeated general body shake ("introductory shake"), and with their neck-stretching call (burping). The major display is a down-up that is immediately followed by a head-up-tail-up posture and a burp call as the male points his bill toward the courted female. Only very rarely is the head-up-tail-up display given independently of the preceding down-up, although in the other pintails this is the usual situation. Further, this species lacks a grunt-whistle display, which is also present in the typical pintails. Copulation is preceded by the usual head pumping by both birds, and is followed by the male's performance of a single bridling display and then uttering the burp call several times as the female bathes (Johnsgard, 1965a).

Reproductive biology. Few nests of these pintails have been found in the wild, but in the Bahamas the birds have been reported to nest in thickets of mangroves, with the eggs being placed in a simple nest among their roots. Breeding in Trinidad and Tobago occurs from August to November, and in Surinam breeding records extend from May to October (Haverschmidt, 1968). Apparently the greater subspecies nests in the austral spring in the southern parts of its range, specifically October and November, while the Galapagos population is said to have a more extensive nesting period, probably lasting from October until July (Phillips, 1922–26; Delacour, 1954–64). In captivity, the usual clutch size is from 6 to 10 eggs, and the incubation period is 25 days. There is no certainty about male participation in brood care, but in the case of a few broods of the Galapagos race the male was not in evidence.

Status. There is no specific information on the status of any of the three subspecies of this pintail, but probably only the Galapagos population warrants attention. The relatively small amount of suitable waterfowl habitat on those islands and the introduction of various mammalian scavengers and predators places all ground-nesting birds in danger there. Close attention to that population's status would certainly seem warranted.

Relationships. Although clearly a member of the pintail group, this species exhibits some strong affinities with the African red-billed "teal," which is perhaps most accurately regarded a member of the pintail assemblage as well (Johnsgard, 1965a).

Suggested readings. Phillips, 1922–26; Lorenz, 1951–53.

Red-billed Pintail

Anas erythrorhyncha Gmelin 1789

Other vernacular names. African red-billed teal; Rotschnabelente (German); canard à bec rouge (French); pato pico rojo africana (Spanish).

Subspecies and range. No subspecies recognized. Resident in Africa from Benguela (western Angola), the upper Katanga in southern Congo, Zambia, Malawi, northern Mozambique, Tanzania, Uganda, Kenya, southern Sudan, and southern Ethiopia, south to the Cape. Also resident in Madagascar. See map 83.

Measurements and weights. Folded wing: males, 218–28 mm; females, 207–16 mm. Culmen: males, 42–46 mm; females, 41–47 mm (Clancey, 1967). Weights: males, 503–755 g (av. 617 g); females, 434–786 g (av. 566 g). Eggs: av. 56 x 43 mm, greenish white, 39 g.

Identification and field marks. Length 17–19" (43–48 cm). *Adults* of both sexes are nearly identical in appearance, with a dark brown crown and nape, sharply demarcated from a buff face, chin, and throat. The buff coloration extends down the neck to the breast, where brown to blackish speckling and spotting increases in size, so that the underparts are heavily barred or spotted with light brown, and the flank feathers are brown, with buff borders. The upper back and scapular feathers are also olive brown, with narrow buff borders, producing a scaly appearance, while the lower back and upper tail coverts are blackish. The tail is olive brown and relatively pointed (75–90 mm long). The upper wing surface is mostly brown to olive brown, with the secondaries forming a speculum that is mostly buffy pink, lined

in front with a black bar, and the secondary coverts forming a second narrow buff bar. The iris is brown, the legs and feet slate gray, and the bill is bright pink, with a darker culmen and black nail. *Females* differ from males only in being slightly smaller and having a slightly duller bill coloration. *Juveniles* are more grayish and less buffy than adults, and are more streaked on the underparts.

In the field, the slim body and the two-toned head pattern separate this species from all other African waterfowl except possibly the Hottentot teal, which is smaller, has a dark ear-patch, and lacks red on the bill. The birds are not very vocal. The female's decrescendo call is given infrequently and consists of about four notes of equal length and pitch, with a weaker introductory note sometimes added. The most typical call of the male is a very soft *geeee* sound uttered only during display.

brood rearing, it may be assumed that the pair bond is renewed each year. The social display of this species is extremely inconspicuous, and among captive birds (in England) occurs over several winter months. The inciting behavior of females is practically identical to that of the other pintails, and the soft *rrrr* call associated with it closely resembles that of the white-cheeked pintail. The male only infrequently performs the introductory shake, but the burp display is correspondingly increased, and indeed is the only major display of the species. It consists of a rather slow vertical neck stretching, with a soft *geeee* call being uttered as the neck is stretched and the bill opened. Males also occasionally perform display preening behind the wing, and regularly turn the back of the head toward inciting females. None of

Natural History

Habitat and foods. This species reportedly prefers shallow fresh-water areas containing large amounts of submerged, floating, and peripheral vegetation, the areas varying from lakes to marshes, vleis, farm impoundments, and temporarily flooded areas. The birds evidently rarely if ever dive for food, and prefer to dabble in shallow water or even forage on mud flats, although upending is also a common foraging method. They probably spend more time foraging on land than many of the *Anas* species, and in some areas are considered pests in fields of rice or other crops. There is little specific information on the species' other foods, which are said to include aquatic plants and such invertebrates as insects, worms, crustaceans, tadpoles, and even fish, but these latter items must be extremely unusual in its diet (Clancey, 1967).

Social behavior. Red-billed pintails are highly social and gregarious birds, with flock sizes often of hundreds of birds, and one estimate of half a million birds on Lake Ngami in October of 1954. They are also tolerant of other ducks and frequently mingle with Cape teal and yellow-billed ducks. The birds are evidently fairly nomadic and mobile, and in Zambia typically arrive at the end of the dry season or shortly after the rainy season begins. The length and strength of the pair bond has not been established with certainty; but since the male does not assist with

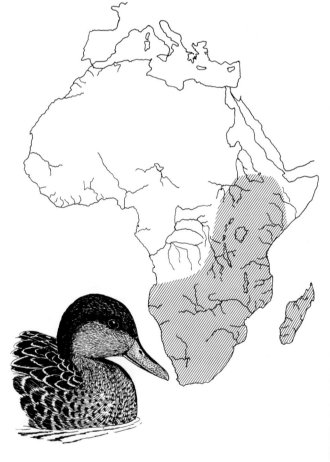

MAP 83. Breeding or residential distribution of the red-billed pintail.

the other complex *Anas* displays appear to be present. A mutual head pumping is known to be the precopulatory behavior, and D. M. Skead (pers. comm.) reports that bridling is the male's postcopulatory display.

Reproductive biology. As might be expected with a rather mobile and nomadic species, the breeding season is relatively long and somewhat irregular in the red-billed pintail. South African records of eggs or newly hatched young occur mostly between August and May; Rhodesian records are concentrated in the period of February through April; and Zambian records extend from December through May, with a peak in February. Farther north, as in Kenya, April through June appear to be major breeding months. The nest is usually located near water, and may be in grass, rushes, weeds, or stunted shrubs, suggesting some adaptability in nest-site selection. Females typically lay a clutch of about 10 eggs, but the range in clutch size is from 5 to 12 (Siegfried, 1962a; Clancey, 1967). Eggs are laid at daily intervals, during early morning hours. Incubation requires from 25 to 27 days, and is undertaken by the female alone, although the male may rejoin her during breaks in her incubation behavior. However, the pair bond is evidently terminated before hatching, since the female rears her brood without the male's assistance. Siegfried (1962b) reports that the young are fully grown within three months of hatching. During the posthatching period the adult birds molt, undergoing a flightless period that lasts only about 24 to 28 days, since the birds become able to fly again before the primaries are completely regrown (Clancey, 1967).

Status. This species is widespread over a large part of Africa, and at least locally is very common, especially in eastern Africa and from Zambia to South Africa. It is extensively hunted and regarded as a favorite sporting species, but is evidently nowhere considered to be threatened in any way.

Relationships. Although the red-billed pintail is regarded by Delacour (1954–64) as one of the "spotted teal," I have provided evidence (1965a) to favor the view that it is a member of the pintail group, with its closest affinities to the white-cheeked pintail of South America. In the simplicity of its social display behavior it does, however, have some resemblances to the silver teal and the Hottentot teal, and these two groups of *Anas* are obviously fairly closely related.

Suggested readings. Clancey, 1967; Skead, 1976.

Silver Teal

Anas versicolor Vieillot 1816

Other vernacular names. Gray teal, pampa teal, puna teal, versicolor teal; Versicolorente (German); sarcelle versicolore (French); pato capuchino or cerceta versicolor (Spanish).

Subspecies and ranges. (See map 84.)

A. v. puna: Puna silver teal. Resident in the *puna* zone of the Andes of Peru, Bolivia, and Chile south to Antofagasta.

A. v. versicolor: Northern silver teal. Breeds from southern Brazil, Uruguay, Paraguay, and Bolivia south to south-central Argentina (38° south latitude), and in central Chile (33° to 42° south latitude). Migrates into Bolivia, Paraguay, and Brazil during winter.

A. v. fretensis: Southern silver teal. Breeds from southern Chile (42° south latitude) and southern Argentina (40° south latitude) south to Tierra del Fuego, migrating an uncertain distance north during winter. Resident on the Falkland Islands.

Measurements and weights. Folded wing: males, 180–231 mm; females, 175–215 mm. Culmen: males, 38–54 mm; females, 36–41 mm. Weights: 10 males of *versicolor* averaged 442 g; 3 females averaged 373 g (Weller, 1968a); males of *puna* 546–60 g (Koepcke & Koepcke, 1965). Eggs: av. (*versicolor*) 45 x 35 mm to (*puna*) 58 x 44 mm, cream, 34 g (*versicolor*).

Identification and field marks. Length 16–20″ (40–51 cm). *Adults* have heads that are blackish brown above the eyes and pale buff below, the two colors sharply demarcated. The lower neck, breast, and anterior flanks are generally buff, with increasingly larger black spots, which on the flanks become vertical bars or stripes, and on the rump and tail coverts form fine vermiculations of black and white. The upper wing coverts are dull slate blue, except for the greater secondary coverts, which are tipped with white. The secondaries form an iridescent green speculum, which is limited posteriorly with a narrow black and finally a white bar. The iris is brown, the legs and feet grayish, and the bill is mostly pale blue, with a black nail and black culmen stripe, and (except in *puna*) a yellowish spot below and behind the nostrils. *Females* have a less brilliant speculum pattern and slightly less colorful bills. *Juveniles* are generally duller in patterning and have less contrasting head patterns, together with less iridescence on the speculum.

In the field, the teal-sized body and generally bluish bill provide useful field marks, as does the vertical barring pattern of the flanks. In flight, the green speculum, lined in front and behind with white, is conspicuous, and helps distinguish the species from the similar-sized speckled teal. Silver teal are notably quiet ducks, although females do sometimes utter a decrescendo call of four or five (*puna*) to ten or more (*versicolor*) syllables. The only calls reported for males are a fairly weak whistle and a mechanical rattling note.

NATURAL HISTORY

Habitat and foods. The various races differ considerably in habitats and no doubt also in foods consumed, although little evidence on the latter exists. The *puna* race is found on the high lakes and rivers of the central Andes, in country dominated by grasses and arid vegetation. The lowland forms are also open-country birds ("pampa teal" being the name applied in the Falklands), favoring fresh-water ponds, especially shallow areas well lined with marshy vegetation. The silver teal has only rarely been observed diving for food, and its short neck doubtless limits the depth to which it can effectively forage. Yet, virtually nothing is known of its food; Weller (1972) found clams and amphipods in one bird collected on the Falkland Islands, and a sample of seven stomachs from Argentina contained a predominance of vegetable materials, particularly the seeds of water milfoil (*Myriophyllum*), sedges, bulrushes, dock (*Rumex*), and a few other plants, while the animal remains were of amphipods, caddis fly larvae, midge larvae, and various aquatic insects (Phillips, 1922–26).

Social behavior. Nearly all observers have commented on this species' tendency to occur in small groups, presumably pairs and family units, but sometimes it also associates with speckled teal and Chiloé wigeon. The pair bond is not yet studied, but Weller (1967a) states that males accompanied some of the broods he observed, and that paired birds were evident in early winter, suggesting a relatively permanent attachment between the sexes. He also observed courtship in November in Argentina, following a September to November breeding season, and thus indicative of a very rapid establishment or reestablishment of pair bonds after the breeding season. In its courtship behavior the silver teal is remarkably undemonstrative, and the casual observer is very likely to overlook it altogether. The most frequent display, the burp, is a slow raising of the head and tail, accompanied by an extremely soft rattling note. Males also produce a rather exaggerated version of the general body shake during display, but little or no sound is associated with it. Males do perform display preening, usually after drinking, and also turn the back of the head toward inciting females at every opportunity. Copulation is preceded by the usual mutual head pumping, and although only a few observations have been made, the postcopulatory display appears to vary according to the subspecies (Johnsgard, 1965a).

Reproductive biology. The period of breeding in silver teal is probably long and irregular, at least in the *puna*-nesting populations, although Johnson (1965) found that race nesting only between November and January, with some nests in coarse vegetation around a lake at nearly 16,000 feet elevation. The lowland forms also nest on the ground in long grassy or rush vegetation, according to Johnson. Evidently the *puna* silver teal has a smaller clutch than do the lowland populations as well as appreciably larger eggs; reported clutches of the former average about 6 eggs and rarely exceed 7, while the latter range from 7 to 10 eggs and probably average at least 8. In all cases the incubation is performed by the female alone, and the incubation period is 25 or 26 days, with the longer period apparently being typical of *puna*. No information on fledging periods is yet available, but it would be of interest to know if population differences also exist in this area.

Status. Johnson (1965) does not consider the northern silver teal to be common in Chile, but states that the southern form is abundant in the southern provinces, and that the *puna* race is relatively common, especially to the north. It is also probably the commonest duck in the *puna* zone of the Peruvian Andes. In Argentina, the silver teal is evidently a relatively common nester in northwestern Patagonia (Phillips, 1922–26). In eastern and southern parts of Argentina it is evidently not quite so prevalent, being considered by Weller (1967a) to be the third most common nesting duck in the vicinity of Cape San Antonio. It is considered by Woods (1975) to be locally numerous on the Falkland Islands.

Relationships. Considered by Delacour (1954–64) to be a member of the *Anas* group he called "spotted teal," the silver teal is probably most closely related to the Hottentot teal of Africa (Johnsgard, 1965a).

Suggested readings. Johnson, 1965.

MAP 84. Breeding distributions of the *puna* ("P"), northern ("N"), and southern ("S") silver teals. Wintering distribution indicated by stippling.

Hottentot Teal

Anas hottentota Eyton 1838 (replaces *punctata*, which has been suppressed [*Bulletin of Zoological Nomenclature* 34: 14–15])

Other vernacular names. None in general English use. Hottentottenente (German); sarcelle hottentote (French); cerceta hotentote (Spanish).

Subspecies and range. No subspecies recognized. Resident in Africa from Angola, Zambia, eastern Congo, Malawi, northern Mozambique, Tanzania, Kenya, Uganda, southern Ethiopia, and the Sudan south nearly to the Cape, also resident in Madagascar. See map 85.

Measurements and weights. Folded wing: males, 145–50 mm; no data for females. Culmen: males, 33–40 mm; no data for females. Weights: both sexes 224–53 g (Kolbe, 1972). Eggs: av. 43 x 33 mm, creamy, 25 g.

Identification and field marks. Length 12–14" (30–35 cm). *Adult males* have a black crown contrasting with a face and throat that is buffy white except for a blackish "thumbprint" mark in the ear region. The back of the neck is also spotted and tipped with black. This spotting extends down the sides of the neck and expands on the breast, forming a heavily spotted lower breast, the spots becoming larger and less well marked on the tawny buff flanks and abdomen, and the posterior underparts and under tail coverts becoming vermiculated with black. The back, rump, and tail are dark brown to black, and the upper tail coverts are vermiculated with black. The upper wing surface is mostly dark brown to blackish, with the coverts having a greenish gloss. The secondaries form an iridescent green speculum, bounded posteriorly by narrow black and terminal white bars. The iris is brown, the legs and feet bluish gray, and the bill is light bluish gray, with the culmen and nail blackish. *Females* have browner crowns, less contrasting facial markings, more rounded scapulars, under tail coverts that are not vermiculated, and less glossy wing iridescence. *Juveniles* resemble adult females but are duller throughout.

In the field, the very small size, two-toned head with the blackish ear-patch, and bluish bill coloration are relatively distinctive marks. The birds exhibit white on the under wing surface and on the trailing edge of the secondaries when in flight, but otherwise appear nearly uniformly brown. The female reportedly sometimes utters a quack when taking off, but few other female vocalizations have been heard, and the only sound attributed to males is an extremely soft series of *took* notes during courtship display.

NATURAL HISTORY

Habitat and foods. Evidently the preferred habitats of this tiny duck are shallow fresh-water marshes and ponds with fringed edges of reeds or papyrus, and with an abundance of floating-leaf plants. It prefers to spend the twilight and nocturnal hours dabbling in very shallow waters, and moves to deeper and safer parts of the marsh to spend the daylight hours. Not many specimens have been examined as to their foods, but a sample of 15 from Malawi indicated that the birds have a preference for grass (mainly *Sacciolepis*) seeds, but also consume adult water insects, ostracods, and insect larvae (Schulten, 1974). No diving while foraging has ever been reported, and instead the birds seem to forage by dabbling at the surface, upending, or foraging in the mud or sand near the water's edge. The tiny bill, like that of the green-winged teal, no doubt limits the size of seeds or other materials that can effectively be obtained.

Social behavior. So far as is known, the pair bond of the Hottentot teal does not extend beyond the female's incubation period, and thus is presumably reestablished yearly. Although breeding is largely associated with the summer months, it also occurs during winter, and thus courtship activities can be observed throughout the entire year. No territorial defense has been described, and the social behavior patterns consist of the usual array of *Anas* postures

and calls. Inciting by the female is apparently performed silently or nearly silently, but the lateral movements are like those of its near relatives. Males often respond to inciting by swimming ahead and turning the back of the head. This combination of inciting and turning the back of the head seems to be the most common display of this species. However, the male may respond with drinking, a vertical neck raising accompanied by a soft mechanical series of call notes (burping), or the combination of these two displays in a burp-drink sequence. When the head is maximally stretched it may be silently and rather quickly turned from side to side through an angle of about 90°, and this display is also sometimes performed by females. Additionally, during social display the birds frequently perform a wing-flapping and both-wings-stretch sequence of behavior that seems to have developed display significance. Preening behind the wing by males has been seen a few times, but apparently either is not utilized as a display or is used only rarely. Precopulatory behavior consists of mutual head pumping, and on the basis of ten observations by Clark (1971), postcopulatory display by the male may vary from no perceptible activity to a swimming shake, wing flapping, or burping. The female most often only bathes after copulation.

Breeding biology. Most information on breeding in this species comes from South Africa, where in the Witwatersrand area breeding is concentrated in the summer months of December through April. Rhodesian records are for May and June, and Zambian records extend from December through April. Farther north, from Kenya to Malawi, the records extend from May to August. Of 14 nests described by Clark (1969a), 10 were in clumps of cattails, 2 were in reed grass (*Phragmites*), and the others were in sedges or grass cover. Clutches usually range from 6 to 8 eggs, with 7 the most frequently encountered number, and apparently the eggs are laid on a daily basis. Incubation period estimates include 24 days for eggs hatched in incubators and 25 to 27 days for naturally incubated clutches. The male may remain nearby as the female incubates, but there is no indication of further male participation in brood rearing (Clark, 1969a). So far as is known, the species is not multiple-brooded, although renesting following nest failure is probable (Clancey, 1967).

Status. This species has a wide but very local distribution pattern, and apparently is not extremely common anywhere in its range. It is too small to be a significant sporting bird, and is relatively tame in most areas. So far, there is no indication that the species is seriously declining or in danger anywhere in its range.

Relationships. Behavioral evidence suggests that the Hottentot teal and the silver teal are very closely related, and their plumages also exhibit a number of strong similarities. To a lesser extent, the garganey must also be regarded as a relative of this group, in spite of its more obvious connections with the "blue-winged" ducks. It is of some interest that, like those of the blue-winged ducks, the sexual displays of the "spotted teal" are relatively simple, are surprisingly quiet, and are remarkably inconspicuous.

Suggested readings. Clark, 1969a, 1971; Clancey, 1967.

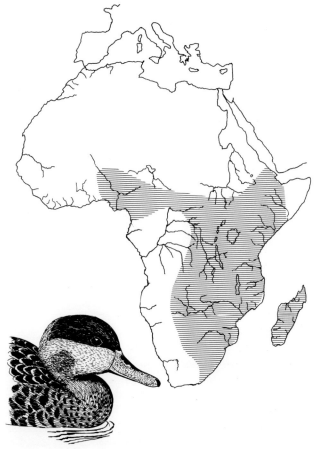

MAP 85. Breeding or residential distribution of the Hottentot teal.

Garganey

Anas querquedula Linnaeus 1758

Other vernacular names. None in general English use. Knäkente (German); sarcelle d'eté (French); cerceta (Spanish).

Subspecies and range. No subspecies recognized. Breeds in southeastern England, southern Sweden, Finland, and northern Russia east across Asia to Kamchatka, south to France, Italy, the Black and Caspian seas, Turkestan, and Manchuria. Winters from the Mediterranean to West Africa, upper Egypt, and Kenya, southern Arabia, India, Indochina, the Philippines, and sometimes the East Indies. See map 86.

Measurements and weights. Folded wing: males, 187–98 mm; females, 175–94 mm. Culmen: males, 35–40 mm; females, 34–39 mm. Weights: males, 240–542 g; females, 220–445 g (Bauer & Glutz, 1968). Eggs: av. 46 x 33 mm, light straw, 27 g.

Identification and field marks. Length 15″ (38 cm). Plate 45. *Males* in breeding plumage have a blackish brown tip of the head and occipital region, which are separated from the streaked brownish red head by a white band extending from the eye to the back of the neck. The chin is black, and the region from the breast to the middle of the back is brown, with crescent-shaped black bars. The back, rump, and upper tail coverts are blackish with lighter barring, while the abdomen is white, becoming vermiculated with black on the flanks. The under tail coverts and sides of the rump are buff, with blackish spots, and the tail is blackish. The outer scapulars and upper wing coverts are bluish gray, the scapulars having black and white stripes, and the greater secondary coverts are tipped with white. The secondaries form an iridescent green speculum, with a narrow white trailing border, and the primaries are dark brown. The iris is brown, the bill is dark gray, and the legs and feet bluish gray. *Males in eclipse* resemble females, but have a more colorful upper wing pattern. *Females* are generally dark brown above and on the flanks, with lighter feather edges. The head is brownish, with a broad whitish supraocular stripe, a brownish stripe passing through the eye posteriorly (and extending slightly above the eye to nearly obscure the supraocular stripe), and a pale buff spot near the base of the bill. The chin and throat are white; otherwise the head is buff, streaked with darker brown. The breast is buff, heavily streaked or spotted with brown, shading to white on the abdomen. The upper wing surface is similar to the male's, but the speculum is less brilliant and the coverts are more brownish. The soft-part colors are like those of the male. *Juveniles* resemble adult females, but are more finely streaked and spotted ventrally.

In the field, the male's white eye-stripe is highly distinctive and contrasts with the brownish red head. The brownish breast also contrasts sharply with the white sides, and the undertail area lacks the black or black and yellow markings of the other European teal-sized ducks. Females somewhat resemble female Baikal teal, but are generally more buffy on the lower face where the Baikal teal is white or nearly white. In flight, the pale gray upper wing surface of males, and to a lesser extent of females, is obvious and similar to that of the blue-winged teal. Males produce a highly distinctive wooden rattling call that sounds like a mechanical rattle, while the female has a typical teal vocabulary, including a decrescendo call of only a few notes, the last one or two being "swallowed."

NATURAL HISTORY

Habitat and foods. The breeding habitat of this species is apparently much like that of the blue-winged teal, consisting of small and shallow ponds in an environment that is mostly grass-dominated. It also prefers ponds with an abundance of emergent vegetation, and with its small and dainty bill tends to forage on items it can obtain at the water's surface or just below it, rarely if ever resorting to diving for its food. Its foods consist of both plant and animal materials, with mollusks apparently one of the important spring foods, and with smaller quantities of insects and crustaceans (ostracods and phyllopods especially) also being taken then, but probably more so later in the summer. The vegetative parts of aquatic plants such as hornwort (*Ceratophyllum*) and naiad (*Najas*) are also important summer foods. With the coming of fall, the birds gradually shift over to a plant-dominated diet, including seeds of such plants as pondweeds, smartweeds (*Polygonum*), sedges, and dock (*Rumex*). In the winter, the birds are usually on larger water areas, occasionally including coastal marshes, but in general they prefer inland habitats (Dementiev & Gladkov, 1967; Bauer & Glutz, 1969).

Social behavior. Garganeys mature in their first winter of life and begin pair formation on their wintering grounds. However, considerable numbers of females are already paired by September or October, suggesting a substantial amount of renewal of pair bonds among older individuals. In its social behavior the garganey exhibits several surprising similarities to the shoveler group, a point which Konrad Lorenz was perhaps the first to establish. This is well exhibited by the female's inciting behavior, which has a strong chin-lifting component, and in her abbreviated decrescendo call, which is very infrequently uttered. In common with males, females frequently perform display preening behind the wing (but rarely if ever in front of the wing, as Lorenz (1951–53) believed to be typical of this species), and Lorenz also reported that females sometimes perform the laying-of-the-head-back display, which seems to be an exaggerated burp display. Males produce a typical burp, involving a sudden and nearly vertical neck stretching, and the more extreme laying-of-the-head-back posture, in which the crown touches the back. In either form the display is followed by a drinking movement. In both cases a mechanical, wooden rattling sound is produced at the moment of maximum neck extension; the sound and display movements are much like those of the Hottentot teal and silver teal. Further, like the Hottentot teal and some shovelers, male garganeys have apparently incorporated wing flapping into their display repertoire, and additionally they often swim ahead of inciting females while turning the back of the head toward them in the usual *Anas* manner. Copulatory behavior is still incompletely described in this species, although mutual head pumping is known to be the preliminary display (Johnsgard, 1965a; Bauer & Glutz, 1969).

Reproductive biology. Garganeys are among the last of the waterfowl migrants to arrive on their breeding

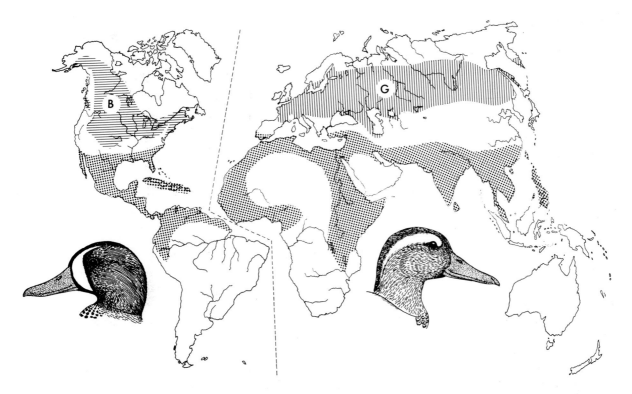

MAP 86. Breeding distributions of the blue-winged teal ("B") and garganey ("G"), and their wintering distributions (stippling).

grounds, and typically arrive in small flocks of paired individuals, which soon begin to establish nesting territories. Nests are typically situated in tall grasses or dense herbage, but generally are not placed under shrubs. Meadows are favored nesting areas, with nests being located up to 150 meters from water and frequently being hidden under rushes (*Eleocharis*) or tall grasses such as manna grass (*Glyceria*). The clutch usually consists of 8 to 11 eggs, with 9 perhaps most typically present, and exceptionally as many as 14. The incubation period is usually 22 to 23 days, with males remaining in the vicinity until incubation is underway, then assembling with other drakes and broodless females for their molt. Observations in the U.S.S.R. indicate that some males molt close to their nesting areas, while others undertake considerable molt migrations to certain lakes and bay areas. The young require from five to six weeks to attain fledging, and females reportedly provide excellent brood defense for much of this period. By late summer many of the nesting ponds have become dry, and long overland movements to new water areas are frequently required (Dementiev & Gladkov, 1967; Bauer & Glutz, 1969).

Status. Very few surveys of this species' breeding or wintering numbers are available, even for western Europe, but Ogilvie (1975) reported that the population that breeds in Europe and Russia and winters in Africa may number anywhere between half a million and 2 million birds. The Niger Delta in Mali and the Senegal Delta in Senegal and Mauretania are two of the major wintering areas for this population. In addition, large numbers winter in India, Pakistan, and southeastern Asia, but no numerical estimates are available.

Relationships. Behavioral evidence (Johnsgard, 1965a) places the garganey quite neatly between the "spotted teal" group such as the silver teal and Hottentot teal, and the "blue-winged ducks" that comprise the remainder of the genus *Anas* as here designated. The nearest living relative of the garganey is perhaps the blue-winged teal, although their similarities are at least in part a reflection of their common niche adaptations.

Suggested readings. Owen, 1977; Dementiev & Gladkov, 1967.

Blue-winged Teal

Anas discors Linnaeus 1766

Other vernacular names. Bluewing; Blauflügelente (German); sarcelle soucrourou (French); cerceta de alas azules (Spanish).

Subspecies and range. No subspecies recognized here (the Atlantic coast breeding population has at times been separated as *orphna*). Breeds from British Columbia east to southern Ontario and Quebec, south to California and the Gulf Coast, and along the Atlantic coast from New Brunswick to North Carolina. Winters from the Gulf Coast south through Mexico, Central America, and into South America, sometimes as far south as central Chile and central Argentina. See map 86.

Measurements and weights. Folded wing: males, 180–96 mm; females, 175–92 mm. Culmen: males, 38–44 mm; females, 38–40 mm. Weights: males, 273–410 g (av. 360 g); females, 266–375 g (av. 332 g) (Kolbe, 1972). Data summarized in Johnsgard (1975) suggest slightly higher average weights. Eggs: av. 46 x 33 mm, creamy, 29 g.

Identification and field marks. Length 14–16" (35–40 cm). *Males* in breeding plumage have a slate gray head and neck, with a black-bordered white crescent in front of the eyes and a blackish crown. The foreback and shorter scapulars are blackish brown, with buff markings, and the hind back, rump, and upper tail coverts are olive brown, with pale buff shaft-streaks and some with blue outer webs. There is a white patch on the side of the rump, and the under tail coverts are black. The breast, abdomen, sides, and flanks are pale reddish with brownish black spots that become stripes on the rear flanks; the tail is brown. Most of the upper wing coverts are light blue, while the greater secondary coverts are tipped with white, especially outwardly, and the primaries and their coverts are blackish brown. The secondaries form an iridescent green speculum narrowly tipped with white and bordered inwardly and outwardly with black, while the tertials are greenish black, with buff shaft-streaks. The underwing surface is mostly white, the iris is dark brown, the legs and feet are orange yellow with dusky webs, and the bill is bluish black. *Females* have a streaked grayish brown and whitish head, with a darker crown and eye-stripe and an unmarked buffy white patch below the eye at the base of the bill. The chin and throat are white, and the upperparts are dull olive brown, with U-shaped central markings and buffy to white feather edges. The breast, sides, and flanks are brown, with buff margins, the underparts are whitish with some spotting or streaking, and the tail is brown. The wings are like those of the male, but the blue of the coverts is paler, the greater coverts are tipped with gray, the iridescence of the secondaries is less well developed, and the tertials are brownish, with pale edges. The soft-part colors are like the male's. *Males in eclipse* closely resemble females, but have brighter wing markings. *Juveniles* also resemble adult females, but are more heavily spotted on the underparts and have plainer upperparts, lacking the U-shaped feather markings.

In the field, the small size and blue wing coverts (often invisible in swimming birds) separate this species from all others except the cinnamon teal. Males of the two species can be readily separated when in breeding plumage, but females and eclipse-plumage males are best distinguished by the shorter and less spatulate bill of this species and the fairly obvious whitish mark behind the bill of females. In flight the bluish upper wing surface is highly conspicuous, and separation of flying blue-winged teal from cinnamon teal is nearly impossible except in the case of males in breeding plumage. Like the other teal, females utter a high-pitched quacking call when alarmed, and have a poorly developed decrescendo call of only three or four notes, usually muffled at the end. The typical male call is a high-pitched whistled *tsee* note, most often associated with courtship and uttered on the water as well as when in flight.

NATURAL HISTORY

Habitat and foods. The preferred breeding habitats of this species are the shallow marshes and sloughs of the native prairie grasslands of central North America, with more limited breeding occurring in wetlands of the drier mixed prairies and short-grass plains to the west and in the Pacific Northwest grasslands. Additionally, some coastal nesting occurs in wet meadows adjoining tidal ponds or creeks. In the prairie areas that the species favors it uses a variety of small and often temporary water areas, including flooded ditches and shallow, temporary ponds that often have little open water. Wintering habitats are quite diverse, and include a large number of subtropical to tropical wetlands, such as mangrove swamps, fresh-water or brackish estuarine

lagoons, and other shallow wetlands. Prominent among the species' foods are the seeds and vegetative parts of aquatics such as pondweeds, naiad (*Najas*), wigeon grass, and duckweeds. Most of birds' food is taken near the water surface by tipping-up or skimming surface waters, with diving done only very rarely. Their small bill size probably in part restricts the size of foods taken, and field-feeding is not done by this species (Johnsgard, 1975; Bellrose, 1976).

Social behavior. Blue-winged teal are relatively social birds, remaining in flocks from early fall through their spring migration, even though by that time most of the birds are already paired. The early fall departure to winter quarters is not matched by an early assumption of the nuptial plumage, and pair-forming behavior is probably not begun until about mid-January or later. It continues well into the spring migration, although by mid-March most females are firmly mated. Courtship is performed largely on the water, with the males periodically making short jump-flights toward the female, followed by such displays as ritualized forms of foraging (mock feeding, head-up and upend) and of comfort movements such as shaking, bathing, preening, and wing flapping. The female's primary display consists of inciting, and the usual response of the male is to swim ahead of her while turning the back of the head toward her. Males also regularly perform aggressive chin lifting toward one another, a display sometimes confused with precopulatory head pumping. Following copulation, the male utters a single call, then assumes a rather rigid body posture, with his bill pointing sharply downward (Johnsgard, 1965a; McKinney, 1970). There is no information on the incidence of females remating with earlier mates, but after pair bonds are initially established they appear to be quite strong. Females have a relatively low tendency to return to their natal areas for nesting, and as a result the species is relatively capable of pioneering into new habitats (Sowls, 1955).

Reproductive biology. Current evidence suggests that virtually all females, including yearlings, attempt to nest, and shortly after arriving on their breeding grounds the pair establishes a relatively small home range, usually of less than 100 acres. Both members of the pair actively search for suitable nest sites, which most often are in grassy cover or in fields of alfalfa (*Medicago*). Grasses of medium height, averaging about one foot, are preferred over taller cover, and often the nests are placed about midway between water and the highest adjacent

point of land. Egg laying is done at the rate of one egg per day until the clutch is completed. Clutch sizes of early nests average about 10 to 11 eggs, with the number declining appreciably in later nests and renesting efforts (Dane, 1966). The pair bond tends to be broken early in incubation, and the male remains near his mate anywhere from a few days until about three weeks into the incubation period. Incubation may last about 21 to 23 days (Glover, 1956; Bennett, 1938), or sometimes as long as 27 days under wild conditions. There is a moderately high loss of nests to predators and accidental destruction, but as many as half of the females other than yearlings may attempt to renest following such losses. There are also substantial posthatching losses of young during the approximately 40-day fledging period, and during this period the males undergo their flightless period of four to five weeks. Females molt somewhat later, at about the time their brood has nearly fledged. Shortly after regaining their flight abilities, males begin to leave the nesting grounds, followed later by females and young (Johnsgard, 1975; Bellrose, 1976).

Status. Along with the mallard and pintail, the blue-winged teal is one of the most abundant of the North American dabbling ducks, and these three species probably all have populations numbering in the millions. Bellrose (1976) reports that in the 20-year period from the mid-1950s to 1974 the average breeding population of the blue-winged teal was more than 5 million. He states that there was about a 20 percent decline in estimated numbers of breeding birds during that period, although the significance of this change is still uncertain.

Relationships. The blue-winged teal and cinnamon teal provide an interesting pair of species in terms of evolutionary relationships, with the cinnamon teal presumably evolving in South America and the blue-winged teal in North America, and the two generally having retained complementary rather than overlapping breeding ranges. Ecologically they have few obvious differences; in addition, the blue-winged teal has a close ecological counterpart in Europe, the garganey. The extent to which the blue-winged teal and cinnamon teal are serious competitors in areas of overlap is still unknown but deserves future attention, as does the nature of their isolating mechanisms.

Suggested readings. Bennett, 1938; Glover, 1956; Dane, 1966.

Cinnamon Teal

Anas cyanoptera Vieillot 1816

Other vernacular names. None in general English use. Zimpente (German); sarcelle cannelle (French); pato colorado (Spanish).

Subspecies and ranges. (See map 87.)

A. c. cyanoptera: Southern cinnamon teal. Breeds from the Amazon Basin of southern Peru, Bolivia, and southern Brazil south to Tierra del Fuego, ranging farther north into Brazil during winter. Resident on the Falkland Islands.

A. c. orinomus: Andean cinnamon teal. Resident on the Andean plateau of Peru and Bolivia.

A. c. borreroi: Borrero cinnamon teal. Resident in the Andes of western Colombia between 7,500 and 11,500 feet, in the savannahs and moist highlands.

A. c. tropica: Tropical cinnamon teal. Resident in the upper Cauca Valley of Colombia below 3,000 feet, and in the lower parts of the Cauca and Magdalena drainages.

A. c. septentrionalium: Northern cinnamon teal. Breeds in North America from British Columbia and Alberta southward as far east as Montana, Wyoming, Nebraska, Texas, and northwestern Mexico. Winters in the southwestern states and southward through Mexico, Central America, and northwestern South America.

Measurements and weights. Folded wing: males, 168–200 mm; females, 167–208 mm. Culmen: males, 37–49 mm; females, 33–45 mm. Weights (of *septentrionalium*): 26 males shot in the fall averaged 408 g (max. 543 g), and 19 females averaged 362 g (max. 498 g). Eggs: av. 48 x 35 mm, pale cream, 32 g.

Identification and field marks. Length 15–19" (38–48 cm). *Males* in breeding plumage have a brownish black crown, and the rest of the head, neck, breast, underparts, sides, and flanks are cinnamon brown, with blackish brown on the back, rump, and upper tail coverts. The scapulars are dull greenish black, the longer ones having pale brownish buff shaft-streaks. The tail is blackish brown, with rusty margins, and the under tail coverts are brownish black. The wings are as in the blue-winged teal, except that the outer tertials are blackish brown, streaked with buff, rather than greenish black with buff streaks. The bill is black, the iris is orange red,

and the legs and feet are orange yellow, with dusky webs. *Females* are much like female blue-winged teal, but are more heavily streaked on the sides of the head and chin, and the body feathers are more rusty in tone, especially on the underparts. Although the female blue-wing has a clear, whitish oval area as large as or larger than the eye behind the base of the upper mandible, in the cinnamon teal this is tinged with yellow, and because of the increased cheek streaking is either much smaller or completely absent. Soft-part colors are identical in the two species. *Males in eclipse* resemble females but retain brighter wing coloration and are a richer cinnamon buff in the head and neck areas. Even when in eclipse, males retain a distinctly pale yellow to reddish iris coloration. *Juveniles* resemble adult females, but are generally more heavily streaked, especially on the underparts.

In the field, cinnamon teal are most likely to be confused with blue-winged teal, or possibly with red shovelers in southern South America. The slightly shovelerlike bill, with a nearly straight culmen profile, can be easily detected when the two teal species are side by side, and the female's more heavily streaked cheeks are usually also evident on close inspection. In flight, the blue upper wing coverts are highly visible, and the generally cinnamon coloration of the male is also evident in spring. Vocalizations of the female appear to be identical to those of female blue-winged teal, but the male lacks a whistling note and instead utters only a series of *chuk* notes much like those of shovelers.

NATURAL HISTORY

Habitat and foods. Although no detailed analysis of the cinnamon teal's breeding habitats has been performed, it is clear that the birds prefer small and shallow water areas that are often distinctly alkaline and that are surrounded by grasses and other low herbaceous cover no more than about two feet tall. Their habitat differences from blue-winged teal are evidently few, and shallow ponds that have both emergent vegetation and zones of open water in addition to submerged aquatic plants are very attractive to these birds. Likewise, the dietary intake of the two species is apparently very similar, with a high proportion of seeds of aquatic plants. The cinnamon teal's bill is slightly more shovelerlike than that of the blue-winged teal, but only a detailed study would be likely to show significant differences in plankton-

sized materials that may be consumed (Johnsgard, 1975; Spencer, 1953).

Social behavior. Cinnamon teal are much like blue-winged teal in their general sociality, moving south from their breeding grounds early in the fall, migrating considerable distances into Mexico and Central America for the winter months, and apparently beginning their annual pair-forming activities while in their wintering areas. Captive birds have been observed to start courtship about the end of February, and observations of spring migrants indicate that pair-forming behavior is still occurring at a fairly high level at the time of northward movement in March and April. Male displays are very much like those of the blue-winged teal and also the shovelers, consisting primarily of ritualized forms of foraging behavior, comfort movements, and turning the back of the head toward inciting females (Johnsgard, 1965a; McKinney, 1970). Short jump flights of males toward females are also frequent, and aerial pursuits are likewise common. Strong chin-lifting behavior occurs in females during inciting and also between competing males; as in the shovelers, this may readily be confused with precopulatory head pumping. Copulation also takes a form like that in shovelers, being terminated by the male calling once and assuming a posture lateral to the female, with his bill pointed downward and his hindquarters and wings slightly raised (McKinney, 1970).

Reproductive biology. Cinnamon teal are fairly late arrivals on their breeding grounds, and shortly after arriving the pair establishes a home range that often overlaps with those of other pairs. Little hostility occurs between such pairs, and most activities of the pair are carried out within an area as small as 30 square yards, with the nest site either located within this limit or no more than 100 yards away. Nesting densities in favorable habitats often exceed 100 pairs per square mile, or are appreciably higher than those typical of blue-winged teal. Nests are very commonly placed in grassy cover, such as salt grass (*Distichlis*), while sedges, bulrushes, and broad-leafed weeds are used less frequently. Island nesting is very frequent, especially where the islands provide herbaceous cover a foot or so high and excellent concealment characteristics (Spencer, 1953). Clutches are laid at the rate of about one egg per day, and for initial attempts average about 9 or 10 eggs. Cinnamon teal are persistent renesters, and in one case a female was found to attempt three nestings. Various predatory mammals, including skunks, and bird predators such as California gulls seem to be the major causes of nest losses. The incubation period is usually about 24 or 25 days, but may be as short as 21 days. A few males abandon their mates almost as soon as incubation has begun, while most are present well into incubation and a small proportion remain until pipping occurs (Oring, 1964). During the approximately seven-week fledging period the female moves her brood into ditches, canals, or ponds where both submerged and emergent aquatic plants are abundant. In contrast to blue-winged teal, there is no indication that males undergo major movements or form large concentrations before molting; instead, they molt on or quite near their breeding areas. They usually begin their fall flight out of Utah by mid-September, followed within a month by females and immatures (Spencer, 1953).

Status. Virtually nothing can be said of the relative abundance of the southern races of cinnamon teal, but Bellrose (1976) has estimated that the North American population consists of about 260,000 to 300,000 breeding birds and an annual fall population about twice that size. Probably nearly half of this entire population nests in Utah, which therefore represents a critical part of this subspecies' basis for survival. Most birds have left before hunting seasons are underway, and therefore hunting is probably not significant in population regulation. Drainage of desirable wintering habitats in central Mexico may be much more significant; in addition, cinnamon teal breed in an area where botulism has traditionally been a significant source of waterfowl mortality.

Relationships. In morphology and behavior, the cinnamon teal stands clearly between the blue-winged teal and the more shovelerlike species of *Anas,* and in particular bears some similarities to the South American red shoveler. It seems likely that the cinnamon teal evolved in the temperate parts of South America and only later expanded its range into North America, bringing it into contact with the blue-winged teal, which perhaps effectively prevented further range expansion to the east through competition.

Suggested readings. McKinney, 1970; Spencer, 1953.

MAP 87. Residential distributions of the tropical ("T"), Borrero ("B"), and Andean ("A") cinnamon teals, and breeding distributions of the northern ("N") and southern ("S") cinnamon teals. Wintering distribution of the two latter races is indicated by stippling.

Red Shoveler

Anas platalea Vieillot 1816

Other vernacular names. Argentine shoveler, South American shoveler; Südamerikanische Löffelente (German); souchet roux (French); pato cuchara sud americano (Spanish).

Subspecies and range. No subspecies recognized. Breeds in central Chile (Aconcagua to Chiloé, infrequently farther south) and in Argentina from about Sante Fe south to the Straits of Magellan and perhaps Tierra del Fuego. Occurs at least during winter in Bolivia, Paraguay, Uruguay, and southern Brazil; breeding there uncertain. See map 88.

Measurements and weights. Folded wing: males, 213–22 mm; females, 202–10 mm. Culmen: males, 63–67 mm; females, 56–60 mm. Weights: 10 males averaged 608 g, 7 females 523 g (Weller, 1968a). Eggs: 52 x 36 mm, creamy, 46 g.

Identification and field marks. Length: 21″ (54 cm). *Adult males* have pinkish buff heads, with darker spotting, which is heavier on the crown and absent on the throat. The back and scapular feathers are reddish, with oval black markings, and the longer scapulars are black with light shaft-streaks. The breast, underparts, and flanks are reddish chestnut, extensively marked with rounded black spots. The upper and lower tail coverts are black, bordered anteriorly on the sides of the rump with a white patch, while the tail feathers are relatively pointed and brownish gray, with lighter edging. The upper wing coverts are light blue, except for the greater secondary coverts, which are tipped with white. The secondaries form an iridescent green speculum, narrowly edged behind with blackish coloration. The bill is dark grayish black, the legs and feet are yellow,

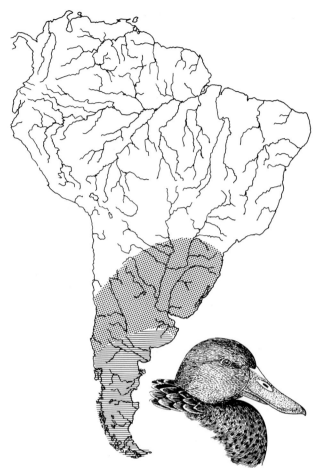

MAP 88. Breeding (hatching) and wintering (stippling) distributions of the red shoveler.

and the iris is white to straw-colored. *Females* resemble the female northern shoveler, but are more grayish, have more grayish-colored legs and feet, and a more pointed tail. *Juveniles* closely resemble adult females, but young males may have a more brilliant speculum.

In the field, the elongated and spatulate bill provides the best field mark. In flight, the birds closely resemble cinnamon teal, which occur in the same area but are somewhat larger and have distinctly longer and heavier bills. Neither sex is very vocal; and although the female probably has a decrescendo call, it evidently has not been described. The male utters a wooden series of notes *ttoka-tooka-tuk-tuk* in courtship and apparently also as an alarm call.

NATURAL HISTORY

Habitat and foods. Brackish lagoons and shallow inland waters are said to be the preferred habitats of

this little-studied species (Phillips, 1922–26). Johnson (1965) reported that it prefers estuaries and coastal lagoons to inland lakes. The birds filter-feed by immersing the bill, or sometimes by upending, but rarely if ever dive to obtain their food, which is presumably partly planktonic materials. It seems probable that this species' ecological requirements and foraging behavior must be nearly identical to those of the northern shoveler and the other Southern Hemisphere shovelers, which are better studied.

Social behavior. Red shovelers are virtually unstudied in their natural habitats, and nothing can be said with certainty about their pair-bonding patterns or timing of courtship. Weller (1968a) reported seeing courtship in November in Argentina, following a September to November breeding season, suggesting a very rapid restructuring of pair bonds. He also noted occasional pairs among fall and winter flocks of shovelers there. The birds are fairly strongly migratory, and they are said to arrive in Uruguay in great numbers to spend the winter there. Probably the birds nesting in Patagonia move into central and northern Argentina during this time; Weller (1968a) reported that shoveler numbers in east-central Argentina increased during the winter months. Their pair-forming behavior, based on observations of captive birds, is much like that of the other shovelers. The female has a strong inciting display, with extreme neck-stretching and chin-lifting movements and an associated soft *rrrr* call. Males most often respond to such inciting by performing a mock-feeding display, in which the bill is dipped in the water and the bird remains almost motionless beside the inciting female. Males also utter their typical *tooka* call frequently, often as an introduction to the mock-feeding display. Apparently preening behind the wing is only poorly developed as a display, and turning of the back of the head has not yet been specifically noted. Probably aerial displays, which are well developed in northern shovelers, are also present, but display by full-winged birds has not yet been studied. Copulation is preceded by the usual mutual head pumping, which differs appreciably from the head pumping associated with inciting; and after treading, the male utters his call, then turns and faces the female as she bathes (Johnsgard, 1965a).

Reproductive biology. According to Johnson (1965), red shovelers prefer to nest in dry areas surrounding coastal lagoons. Birds in breeding condition have been collected in Argentina between September and November, in the austral spring. In the wild, the clutch size is reported to be from 5 to 8 eggs (Johnson, 1965), and records from captivity indicate that the incubation period is 25 days. Undoubtedly only the female incubates, and although it may be presumed that the male participates in brood care, this has not been specifically noted in the wild.

Status. No quantitative information on this species' populations are available, but at least in earlier times it was reported as the most common wintering duck in Uruguay (Phillips, 1922–26). Weller (1968a) listed it as the sixth most common duck in the Cape San Antonio area of Argentina during the breeding season, and Johnson (1965) says that it occurs in "moderate" numbers in central Chile during the breeding season. No doubt the bird's specialized foraging requirements limit the habitats that it can effectively utilize, and thus limit its population. It is shot by hunters in moderate numbers in eastern Argentina (Weller, 1968a).

Relationships. As noted earlier (Johnsgard, 1965a), the red shoveler seems to link the typical shovelers with the cinnamon teal and other "blue-winged" teals, in both behavior and morphology. It is of interest that each major continental area has a single species of shoveler, all of which are nearly identical in size and foraging adaptations.

Suggested readings. Phillips, 1922–26.

Cape Shoveler

Anas smithii (Hartert) 1891

Other vernacular names. African shoveler, South African shoveler; Südafrikanische Löffelente (German); souchet du Cap (French); pato cuchara del Cabo (Spanish).

Subspecies and range. No subspecies recognized. Resident in Africa, primarily limited to South Africa, but sometimes ranging north to South-West Africa (rarely Angola), Botswana, and Rhodesia. Subject to seasonal movements, and records from farther north are apparently all of nonbreeders. See map 89.

Measurements and weights. Folded wing: males, 222–53 mm; females, 208–35 mm. Culmen: males, 56–65 mm; females, 52–60 mm. Weights: males, 550–830 g (av. 688 g); females, 480–690 g (av. 597 g) (Siegfried, 1965). Eggs: av. 54 x 36 mm, creamy, 45 g.

Identification and field marks. Length 20″ (51 cm). *Adult males* have a grayish brown head and neck that are streaked with black, especially on the crown and nape. The shorter scapulars and mantle are blackish brown, edged and barred with buff, while the longer scapulars, rump, and tail coverts are blackish, with green to bluish iridescence. The breast and ventral surface are mottled with umber brown and cinnamon, while the flanks have crescent-shaped umber markings that are edged with cinnamon. The tail is blackish brown, with each feather edged with white or buff. The upper wing coverts are pale bluish gray, except for the secondary coverts, which are tipped with white, forming an anterior border to the iridescent green speculum. The iris is pale yellow, the legs and feet yellow to orange, and the bill is black. *Females* are distinctly paler than males, are less cinnamon and more buff in general coloration, and have no iridescence on the upperparts. The upper wing coloration is duller on the coverts, and the speculum iridescence is reduced. The iris is brown, the legs and feet are brownish olive, and the bill grayish black. *Juveniles* resemble adult females, but are paler and duller, with the buff edging of the feathers more pronounced.

In the field, this species is most likely to be confused with the northern shoveler, which is a winter visitor to Africa and has probably been responsible for many of the apparent extreme northern records of this species. Winter-plumage males of the northern shoveler are sufficiently colorful to avoid confusion, but female northern shovelers may be very readily confused with Cape shovelers, tending only to be rather more buff-colored and having a yellowish to olive-colored bill rather than a black or grayish one. The calls of the females are nearly identical, while the male Cape shoveler utters a very characteristic wooden note, often in series of four.

NATURAL HISTORY

Habitat and foods. Cape shovelers are by nature of their bill specializations dependent on fairly specific habitats and foods, and in southern Africa are largely confined to fresh-water environments, with some occurrence on tidal estuaries, saline lagoons, and salt pans. Its preferred areas are temporary water areas that are open and shallow, such as marshes and flooded areas. It often occupies alkaline waters, but

MAP 89. Residential or breeding distribution of the Cape shoveler.

is generally absent from acidic areas, and particularly favors fertile waters such as sewage disposal ponds that are rich in planktonic organisms. The birds are rarely found on rivers except for the most sluggishly flowing waters or those with swampy borders. Foods include about 70 percent animal materials, on the basis of a study of 48 birds analyzed by Brand (1961), and these are mostly insects, snails, and crustaceans. Leaves and stems of aquatic plants, especially pond-weeds, were the major plant materials reported by Brand. Like the other shovelers, this species feeds primarily by dabbling, and often forages in a co-operative "circle-swimming" fashion, with several birds foraging in one another's wake (Siegfried, 1965).

Social behavior. Evidence available for the Cape shoveler indicates that the pair-bond pattern, like that in the northern shoveler, is one of seasonally reestablished monogamy, with total disruption of the pair bond near the end of the breeding season, but a slight tendency for previously paired birds to re-form pair bonds in the subsequent pair-forming period. The most typical aquatic displays associated with pair formation consist of inciting by the female, which takes the form of strong head-pumping movements and the associated call notes. Males respond to inciting in various manners, but the usual response of a favored male is to swim ahead of the inciting female and turn the back of the head toward her. Males also utter a burp sounding much like a snort or belch, with extended neck and the head either held motionless or raised in the prominent chin lifting that is so characteristic of all of the shoveler species. Mock feeding, or display feeding, is especially well developed in this species, and may be accompanied by the calls just mentioned. Very frequently mock feeding is terminated by upending, or tipping-up. Preening behind the wing has also been observed in this species. A number of aerial displays associated with courtship have been described by Siegfried (1965), including three-bird flights (or territorial defense flights), pursuit flights involving a single female and several unpaired males, display or jump flights by males that are accompanied by noisy wing sounds and are used by males to approach females more closely or to induce them to fly away with them. Copulation is preceded by the usual mutual head pumping, and postcopulatory display seemingly consists of the male's calling one or more times and then turning the back of the head or performing a short jump flight (Johnsgard, 1965a; Siegfried, 1965).

Breeding biology. Breeding records from South Africa indicate that breeding in this species occurs throughout the entire year, but at least in the southwestern Cape region and the Witwatersrand area the largest number of clutch records are for August and September, with diminishing numbers through the end of the year. In general they seem to be correlated with increasing photoperiods and temperatures, but only indirectly with precipitation patterns. Nests are usually located near highly fertile shallow-water areas that have abundant invertebrate life, generally within ten meters or less of the edge of such ponds. Most nests that have been found have been in cover that did not exceed 15 inches in height, and there is no evidence for nest spacing through territorial behavior. Indeed, there is a report of 31 nests on a single 200-square-yard island; no nests have been found constructed over water. Only the female constructs the nest, although males have been seen inspecting potential nest sites in the company of females, and they also remain nearby during the actual egg-laying process. Eggs are deposited at the usual rate of one per day, in the morning hours, and a large sample of nests indicate a normal range of from 5 to 12 eggs, with a mean of between 9 and 10 eggs for completed clutches. Records from incubator-hatched eggs as well as those hatched by females indicate an incubation period of from 27 to 28 days, or somewhat shorter than that typical of the northern shoveler. Only the female tends the young, and pen-reared birds have been found to fledge in about eight weeks after hatching. During the postbreeding molt period, the adults are flightless for at least thirty days (Siegfried, 1965).

Status. Although its total range is small as compared with the other shovelers, the Cape shoveler is fairly common, and in some areas such as the Cape Peninsula it is second only to the yellow-billed duck in abundance. It is most abundant on the southern and southwestern Cape, the Transvaal Highveld, and in the northern parts of the Orange Free State. Outside of these areas the populations are sparse and local (Siegfried, 1965).

Relationships. In general, the Cape shoveler and northern shoveler have been regarded as extremely closely related or even conspecific (Siegfried, 1965). However, in its behavioral characteristics the species comes closer to the red shoveler and also has some similarities with the Australasian shoveler (Johnsgard, 1965a). Further, in contrast to the other three shovelers, the Cape shoveler has a "Type A" breeding cycle (associated with breeding around the time of the summer solstice), while the others stop breeding while day lengths are still increasing (Murton & Kear, 1975), suggesting that it may be a relatively more primitive species than the other shovelers.

Suggested readings. Siegfried, 1965; Clancey, 1967; Skead, 1977.

Australasian Shoveler

Anas rhynchotis Latham 1801

Other vernacular names. Blue-winged shoveler, spoonbill; Australische Löffelente (German); souchet d'Australie (French); pato cuchara de Australia y Zelandia (Spanish).

Subspecies and ranges. (See map 90.)
  A. r. rhynchotis: Australian shoveler. Resident in Australia, primarily in the eastern, southeastern, and southwestern portions, and in Tasmania.
  A. r. variegata: New Zealand shoveler. Resident in New Zealand, including both islands.

Measurements and weights. Folded wing: males, 210–61 mm; females, 210–97 mm. Culmen: males, 57–67 mm; females, 57–62 mm. Weights (of *rhynchotis*): males, 570–852 g (av. 667 g); females, 545–745 g (av. 665 g). Eggs: 55 x 38 mm, greenish, 43 g.

Identification and field marks. Length 18–22" (46–56 cm). *Males* in breeding condition have a grayish black face and crown, with variable greenish iridescence, and with a white or mottled crescent in front of the eyes, and black and white mottling or spotting also on the lower neck and breast, where it gradually becomes suffused with chestnut, especially on the flanks and underparts. The back and scapulars are dark brown to blackish with lighter borders, and some of the longer scapulars have white shaft-streaks. The rump and tail coverts are black, with white or barred black and white patches on the sides of the rump, while the tail is dark brown, with whitish edges. The lesser wing coverts are light blue, except for the greater secondary coverts, which are tipped with white, and the secondaries form an iridescent green speculum. The legs and feet are orange, the bill black, and the iris bright yellow. *Nonbreeding males* approach the female in pattern, but retain some chestnut on the sides and may also retain a yellowish iris. *Females* are generally various shades of brown on the head and body, and closely resemble females of the northern shoveler, but tend to be darker, more heavily spotted, and perhaps have a less well developed iridescence on the speculum. The iris is brown, the bill is blackish, and the legs and feet are greenish gray. *Juveniles* resemble females, but are paler and less distinctly marked.

In the field, the typical shovelerlike bill separates this species from all other Australian waterfowl except the pink-eared duck. Otherwise, the bird looks very much like any other *Anas* species, and the bluish upper wing coverts should be looked for in flying birds. The female's calls include the typical *Anas* sounds, presumably including a decrescendo call, although it has not yet been specifically noted. The male utters a repeated *chuck* note during courtship and when disturbed.

NATURAL HISTORY

Habitat and foods. In Australia, this shoveler has been reported to show a preference for the relatively permanent swamps, including cattail swamps in the interior and tea tree swamps of the coastal areas. The birds rarely breed in lagoons and billabongs of the interior except during times of major flooding (Frith, 1967). In New Zealand the birds occupy nearly all of the inland waters and coastal lagoons, but tend to avoid streams in forested areas (Falla et al., 1967). Like other shovelers, the birds feed almost entirely by filtering materials from the surface of the water or just below it. They sometimes also upend, or probe in the muddy shorelines, but diving is apparently done rarely if ever. Often a shoveler will feed in the wake of another bird, usually another shoveler, and sometimes circles of foraging birds are thus formed. A sample of nearly 50 birds collected in New South Wales indicated that about three-fourths of the food remains were of animal materials, especially water beetles and other aquatic insects. Some snails and mussels were also present. The plant materials that were found included the seeds of a wide diversity of species, including sedges, grasses, and a great number of broad-leaved herbs (Frith, 1967). In another sample of 161 gizzards from a permanent swamp in New South Wales, a larger proportion of plant foods was found, particularly of the small floating forms such as duckweed and azolla (*Azolla*). Pink-eared ducks from the area did less mud dabbling at the edge of the swamp, used smaller food items, and fed to a larger degree on zooplankton (Frith et al., 1969).

Social behavior. In Australia, shovelers are usually found in small flocks that are widely dispersed, al-though during times of drought they may concentrate in limited areas in numbers of up to several hundred. To some extent the birds are nomadic and breed opportunistically in flooded lowlands, but the majority of the population apparently breeds in the same permanent marshes and swamps from one year to the next (Frith, 1967). The timing of the breeding, and perhaps also the length and strength of the pair bonds, may thus vary in different areas. There is no indication that the male ever participates in brood care, however, and thus pair bonds are probably renewed or initially established each year. Detailed studies of this species' displays are still lacking, but in general they appear to be very similar to those of the northern shoveler, and the same is true of the associated vocalizations (Johnsgard, 1965a).

Reproductive biology. In New Zealand, the nesting season is said to extend from October to January, while in those parts of Australia with a regular breeding season it reportedly occurs between August

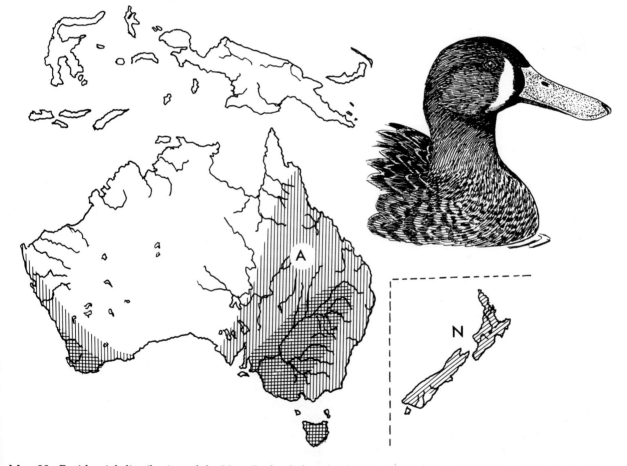

MAP 90. Residential distribution of the New Zealand shoveler ("N"), and breeding distribution of the Australian shoveler ("A"), showing total (hatched) and primary (cross-hatched) ranges of the latter.

and December. However, during years when conditions favor waterfowl breeding in the interior, nests can be found at any time of the year. They are placed on the ground, usually in the cover of fairly open grasses, and often are some distance from the nearest water. They have also been found in weeds or short grasses in pasture land, in low cavities of dead stumps standing in the water, and in other sites. The clutch size in both subspecies apparently ranges from about 9 to 12 eggs, with 10 eggs probably representing the normal clutch. The incubation period is probably 26 days, although a 24-day period has also been reported. The fledging period has not been established, but is likely to be between six and eight weeks. There is so far no indication of multiple brooding in this species.

Status. In New Zealand, the shoveler is widespread and probably is increasing in numbers (Falla et al., 1967). In Australia it is not extremely common anywhere, but is legally hunted in some states, even though it is not a good table bird nor unusually wary. Preservation of the permanent swamps in the interior probably provides the best insurance of the species' continued existence (Frith, 1967).

Relationships. Nearly all the evidence currently available favors the view that this species is very closely related to the northern shoveler, and the two might even be regarded as a superspecies. It is of interest that a blue-winged teal-like white crescent mark should be present in this form, but such ancestral markings often occur in shoveler hybrids that have no blue-winged teal influence (Johnsgard, 1965a).

Suggested readings. Frith, 1967.

Northern Shoveler

Anas clypeata Linnaeus 1758

Other vernacular names. Spoonbill; Löffelente (German); souchet ordinaire (French); pato cuchara común (Spanish).

Subspecies and range. No subspecies recognized. Breeds through much of the Northern Hemisphere, including the British Isles, Europe except for northern Scandinavia, most of Asia except for the high arctic, and in North America from western and interior Alaska south to California and east to the Great Lakes, with some breeding along the middle Atlantic coast. Winters from the southern parts of its breeding range south to northern and eastern Africa, the Persian Gulf, India, Burma, and southern China, and in North America from Puget Sound and Chesapeake Bay south to Honduras and sometimes beyond. See map 91.

Measurements and weights. Folded wing: males, 225–45 mm; females, 220–25 mm. Culmen: males, 62–64 mm; females, 60–62 mm. Weights: males, 410–1,100 g; females, 420–763 g. Eggs: av. 55 x 37 mm, greenish, 40 g.

Identification and field marks. Length 17–22" (43–56 cm). *Males* in breeding plumage have an iridescent green head except for a brownish black crown, and a brownish black back, with light feather edging. The rump and tail coverts are blackish with a green gloss, and there is a large white patch anterior to the tail coverts. The tail is mostly white, except for the central feathers, which are dusky brown. The breast is white, the abdomen, sides, and flanks are chestnut with dusky mottling, and the flanks are finely vermiculated with black. The scapular feathers are mostly white, but the longer ones have light blue on their outer vanes, and the upper wing coverts are also blue, except for the greater secondary coverts, which are broadly white-tipped. The primaries and their coverts are grayish brown, and the secondaries form a green to greenish black speculum, edged posteriorly with white. The tertials are black, glossed with green and narrowly edged with white; the underwing surface is mostly white. The iris is orange yellow, the bill is black, and the legs and feet are orange red. *Females* have a blackish crown, buff sides of the head and neck that are streaked with brown, and a white chin and throat. The upperparts are dark brown, the feathers barred or marked with buff, and the breast and underparts are brownish buff, with small brown spotting. The sides and flanks have larger brown spots and light buff edges, and the tail is brown, with variably broad buffy white markings. The upper wing coverts are mostly grayish blue, except for the greater secondary coverts, which are grayish and tipped with white. The secondaries are slate brown, with a slight greenish sheen, especially inwardly, and the tertials, primaries, and primary coverts are brown. The iris is brown, the legs and feet orange, and the bill is dull greenish brown, with the lower mandible and base of the upper mandible orange and with small black spots scattered on the upper mandible. *Males in eclipse* resemble females, but have more colorful upper wing patterning, and the iris coloration is pale yellow. The bill color approaches that of the female during this period. *Juveniles* resemble adult females, but are darker above, have little or no blue on the upper wing coverts, and show little or no iridescence on the secondaries.

In the field, the shovelerlike bill should separate this species in nearly all areas where it occurs naturally, except perhaps in southern Africa, where confusion with the Cape shoveler is conceivably possible. Males in breeding plumage exhibit a greater amount of white on the breast and in front of the tail than any other shovelerlike duck, and females are much like female cinnamon or blue-winged teal except for their

larger bill. In flight, both sexes exhibit blue upper wing and white underwing coloration, and the large bills make them appear "front-heavy." Females have a typical *Anas*-like quacking call and a decrescendo call of about five notes, with the initial note loudest and the last one or two notes muffled. The male's courtship call is a series of low *took'-a* notes; additionally a noisy wing rattling accompanies takeoff during courtship flights.

Natural History

Habitat and foods. The breeding habitat of this widely distributed species consists of shallow marshes that are usually mud-bottomed and rich in invertebrate life. They are typically in open country such as grasslands, but coastal shorelines and even subtundra habitats are sometimes utilized. Submerged aquatic plant life that shelters an abundance of small planktonic invertebrates such as waterweeds (*Anacharis*) are valuable, for the birds can utilize the

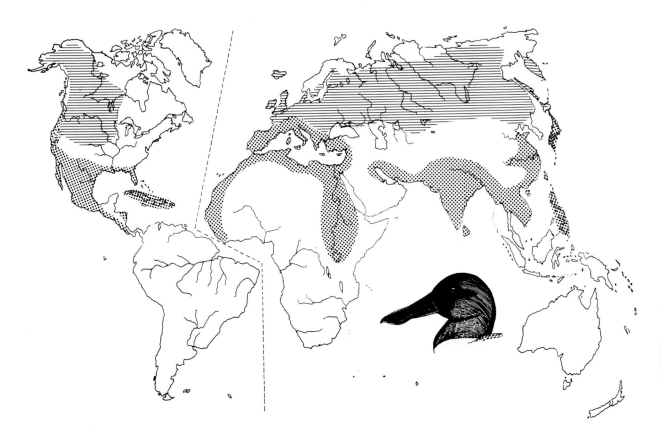

MAP 91. Breeding (hatched) and wintering (stippling) distributions of the northern shoveler.

plants as well as the associated invertebrate life. The birds feed on a variety of larger invertebrate foods such as the larvae of caddis flies, the nymphs of damselflies and dragonflies, and adults of various beetles and bugs. Duckweeds (*Lemmnaceae*) are also highly preferred, and these small floating plants can readily be gathered by the birds during their typical surface foraging. During the winter shovelers occur in both interior and coastal areas, and in the latter they inhabit fresh and brackish estuarine bay marshes, and are most common on still-water ponds with little tidal fluctuation. The birds generally avoid salt-water habitats whenever possible (Stewart, 1962; Johnsgard, 1975).

Social behavior. In general, shovelers seem to be found in relatively small flocks or pairs during most of the year; their specialized bill shape restricts their foraging capabilities and probably serves as a dispersal device. This might also explain in part the high level of hostility evident in social displays and territorial defense exhibited in shovelers, which are an extreme in the genus *Anas* in this regard. Pair forma-

tion begins in the wintering areas when the birds molt into breeding plumage, usually before the end of December, and continues at a fairly high level during the spring migration. Although hostile chin lifting by both sexes is one of the most conspicuous displays, and is the motor pattern associated with inciting by females, several other displays are performed by males, most of which are clearly derived from foraging movements such as dabbling, head dipping, and tipping-up. Turning the back of the head to inciting females is a very common male display, and males can continue to perform it toward their mates for some time after pair bonds seem to have become established. Copulation is preceded by mutual head pumping, which superficially resembles hostile chin lifting but involves a lower angle of the head and bill; and after treading is completed, the male utters a single nasal note followed by a series of wooden sounds as he remains in a stationary posture beside the female (McKinney, 1970; Johnsgard, 1965a).

Reproductive biology. Shovelers are among the late species of dabbling ducks to arrive on their breeding

264 ❖ ❖ ❖

grounds, and in addition the female may spend several weeks searching for a nest site while her mate establishes a territory. Home ranges averaging about 50 acres have been estimated by Poston (1969) in wild shovelers, but observers have reached different conclusions as to the degree to which territorial defense (as distinct from defense of the female) is important as a spacing device. Males certainly exhibit strong pair bonds, which may even persist until shortly after hatching, but there is no evidence that males either participate in brood care or that polyandry ever exists in this species. The nest site is frequently in grasses under a foot in height, less often in taller grasses, and in addition tends to be situated at a relatively great distance from water. The eggs are laid at a daily rate, and clutch sizes of initial or early nests tend to average about 10 eggs. Renesting attempts by unsuccessful females apparently occur, but at a rather low incidence, judging from available information. The incubation period under natural conditions has been estimated from as little as 21 to as much as 28 days, and probably normally falls somewhere near the midpoint of these figures. Some observations indicate that female shovelers may be quite aggressive toward ducklings other than their own, even killing them, but other studies indicate that brood mergers may occur that cause terminal brood sizes to average larger than those at hatching. The fledging period may be as little as 36 days in the northern parts of the species' range to as much as 50 to 60 days in the northern plains states (Girard, 1939; Johnsgard, 1975).

Status. According to Bellrose (1976), the breeding-season population of shovelers in North America has averaged about 1.9 million birds in recent years, and has shown a fairly consistent increase in the period 1955 to 1974, especially in the Central Flyway. There is no complete estimate available for the Eurasian population, but Ogilvie (1975) reported that about 65,000 birds winter in northwestern Europe and an additional 60,000 in the Mediterranean and Black sea region.

Relationships. The northern shoveler has the largest range of any of the four species of shovelers and is the most strongly dimorphic in plumage of the four. It is appealing to think that this is the nearest to the ancestral type, with the South American, southern African, and Australasian forms representing three independent offshoots, but plumage and bill structure differences tend to favor the view that the South American and African forms are older and less spe-

cialized types and are closely related to each other, while the northern shoveler and Australasian shoveler represent another superspecies of later derivation.

Suggested readings. Girard, 1939; McKinney, 1967, 1970.

Pink-eared Duck

Malacorhynchus membranaceous (Latham) 1801

Other vernacular names. Pink-ear, whistler, zebra duck; Spatschnabelente (German); canard à oriellons roses (French); pato orejas rosadas (Spanish).

Subspecies and range. No subspecies recognized. Resident throughout Australia and northern Tasmania, particularly in the Murray and Darling river basins. See map 92.

Measurements and weights. Folded wing: males, 172–213 mm; females, 152–200 mm. Culmen: males, 44–74 mm; females, 53–67 mm. Weights: males, 290–480 g (av. 404 g); females, 272–423 g (av. 344 g). Eggs: av. 49 x 36 mm, white or creamy white, 31 g.

Identification and field marks. Length 15–16" (38–40 cm). *Adults* of both sexes have a white head, finely barred with gray toward the rear, a dark brown patch around the eye that extends to the back of the neck, and a narrow white eye-ring. A small pink area is present in the region above the ears. The lower neck, breast, underparts, and flanks are strongly barred with brown and white markings, while the mantle and scapulars are finely barred with gray and black. The upper tail coverts are white, the under tail coverts are buff, and the tail is dark gray, with white edging. The upper wing surface is mostly dark grayish brown, with the secondaries forming a gray speculum that is bordered in front and behind with white bars. The spatulate bill is gray, with an elongated upper mandible that has soft flaps near the tip which hang below the lower mandible. The iris is brown, and the legs and feet are lead gray. *Females* cannot be externally distinguished from males except

perhaps by size, and *juveniles* resemble adults but are paler, with a less distinct pink ear patch.

In the field, the enormous bill and the zebralike plumage pattern are unmistakable marks at any distance. The birds fly much like teal, but make a loud chirping noise that is highly distinctive. Females lack quacking notes altogether, and instead both sexes have rather high-pitched trilling or chirping sounds that are uttered under many circumstances.

Natural History

Habitat and foods. The pink-eared duck is adapted to the shallow and often temporary saline ponds of Australia's dry interior, and is especially fond of residual flood-water areas or recently filled claypans. Only when such shallow waters are not available will the birds use deeper waters such as permanent swamps, and coastal estuaries or swamps are also avoided. Since its favored habitats are transient ones, the birds are highly nomadic, but throughout their entire range they forage on essentially the same foods, planktonic organisms they strain from the water with their highly specialized bills. Foraging is done in the same manner as the shovelers do it, but the pink-eared duck's bill is even more modified for filter-feeding than are the bills of these species. Like shovelers, they often forage socially, with each bird closely following in the wake of the one ahead of it, with its bill partially submerged and its tongue pumping water through the finely spaced lamellae. In a large sample of food remains reported by Frith (1967), animal materials made up over 90 percent of the food volume, and the majority of this consisted of insects, followed by planktonic crustaceans. The relatively few plant materials were mostly seed and algae. In a later study (Frith et al., 1969) of birds from a permanent swamp the proportion of animal food remains found was lower, but small aquatic invertebrates still provided the largest single source of foods.

Social behavior. Like the gray teal, the pink-eared duck is an opportunistic breeder that has no set breeding season, although in areas where precipitation patterns are predictable, the breeding is likewise regular. In general, however, the birds are flocked during nonbreeding periods and probably little if any sexual behavior occurs. Where favorable habitats do develop as a result of inland flooding, enormous flocks may appear to take advantage of them, and likewise during drought periods the birds may be greatly concentrated on remaining suitable areas. The pair-forming displays of this unusual species have yet to be well observed and described. On the basis of a few observations of wild birds (Johnsgard, 1965b), it appears that aggressive chin lifting and associated calling is perhaps the most common social display, and that females have no other specific display that might be regarded as inciting. Males showed no tendency to turn the back of the head toward females, and instead tended to face them in an erect, stiff-necked posture, from which they would at times utter a bill-tossing call. Pink-eared ducks reportedly also perform a form of ritualized

feeding in a manner similar to shovelers, except that both sexes participate and the birds revolve bill-to-bill in a tight circle only about six inches in diameter. Only a few copulations have been seen, and in one case observed and filmed by Thomas Lowe (in litt.) the treading was preceded by preening behavior on the part of the male, in a manner apparently similar to that of Cape teal or marbled teal. After copulation, both birds typically perform vigorous movements similar to those of normal bathing and foraging.

Reproductive biology. According to Frith (1967), pink-eared ducks can breed at any time of the year, according to local conditions. However, in the southern parts of their range, where winter precipitation is the rule, they usually nest between August and October, while in the northern and summer-rainfall areas they typically nest between March and May. Nests are always placed in elevated situations, anywhere from a few inches to 30 feet or more above the water

MAP 92. Total (hatched) and principal (cross-hatched) breeding distributions of the pink-eared duck.

level. They are often built on the tops of flooded shrubs, on fence posts or logs, or on the forks, limbs, or in holes of flooded trees. Because of the occasional crowding of birds into limited breeding areas, multiple use of available nest sites is not uncommon, and may result in as many as 60 eggs being placed in a single tree hollow. However, the normal clutch ranges from 3 to 11 eggs, and averages about 7. Incubation is performed by the female and requires 27 days (Frith, 1967). Warham (1958) found that males remain in the nest vicinity and not only defend the nest site but also help call the young down from their nesting site after hatching. The young pink-ears already have specialized bills and begin to dabble like the adults immediately after hatching. Although the male evidently remains with the brood, it is the female that often defends it against intruders such as coots, in Warham's observations.

Status. Frith (1967) reports that although pink-eared ducks are protected in New South Wales, where they are relatively common, they can be legally hunted in adjoining states. He states that such a policy for a highly nomadic species makes no sense, and further notes that since they are such poor game birds many of those that are shot are simply discarded by hunters. With increasing control of inland rivers and less frequent extensive flooding, the future of species like the pink-ear and gray teal becomes increasingly uncertain, in Frith's view.

Relationships. Most taxonomists have followed the tradition of considering the pink-eared duck as an "aberrant" dabbling duck; this procedure was followed by Delacour (1954–64) and adopted by me (1965a) in the absence of contrary evidence. Woolfenden's (1961) osteological studies indicated that the genus fell within the range of variations he encountered in the dabbling ducks and perching ducks. The absence of a female decrescendo call might suggest perching duck affinities, and the male's trachea is also said to be much like that of the pygmy geese. Similarities between the downy young of this species and such perching ducks as the ringed teal are also suggestive of such relationships. Yet, unlike nearly all perching ducks, this form totally lacks iridescent coloration. Brush (1976), studying feather proteins by electrophoresis, concluded that this genus is quite distinct from *Anas* and may be more closely related to the perching ducks and shelducks.

Suggested readings. Warham, 1958; Frith, 1967.

268 ❖ ❖ ❖

Marbled Teal

Marmaronetta angustirostris (Menetries) 1832

Other vernacular names. Marbled duck; Marmelente (German); sarcelle marbrée (French); cerceta marmorata (Spanish).

Subspecies and range. No subspecies recognized. Breeds in southern Spain, northern Africa, Cyprus, Syria, and Iraq, north to the Caspian and Aral seas, and south to West Pakistan. Winters in northern Africa and eastward to northern India. See map 93.

Measurements and weights. Folded wing: males, 205–15 mm; females, 198–205 mm. Culmen: both sexes, 40–49 mm. Weights: males, 240–600 g; females, 250–550 g (Bauer & Glutz, 1968). Eggs: av. 46 x 32 mm, creamy, 31 g.

Identification and field marks. Length 15–19″ (38–48 cm). *Adults* have a pale gray head, except for a brownish area through the eyes that extends back to a short crest. Brown speckling also occurs on the crown, cheeks, and sides of the neck, and the breast is grayish white, spotted with brownish coloration. The underparts including the under tail coverts are nearly white, while the flanks, rump, back, and scapular feathers are generally light brown, with buffy white tips. The upper wing surface is pale grayish brown, the secondaries being slightly lighter than the rest of the wing. The iris is brown, the legs and feet are olive brown, and the bill is grayish black with a subterminal pale gray line (males) or a pale triangular area at the base (females). *Females* also have a slightly shorter crest and less strongly patterned eyestripe than do males. *Juveniles* have more uniform

MAP 93. Breeding (hatched) and wintering (stippling) distributions of the marbled teal.

underparts and less distinctly mottled sides and upperparts than do adults.

In the field, marbled teal are among the palest of the surface-feeding ducks, and the white mottling on the sides and scapulars are distinctive, as is the poorly defined brown eye-stripe on the otherwise gray head. In flight the teallike shape and flight behavior, combined with the absence of a distinctive speculum, provides for identification. Females lack the decrescendo call of female *Anas* species, and instead apparently have only a squeaky note uttered during display. Even an inciting call is either very weak or lacking. Males utter a distinctive nasal *eeeep*

note during display, but have few if any other loud calls.

NATURAL HISTORY

Habitat and foods. During the breeding season marbled teal are generally found in fairly dry, steppe-like areas, where they breed on fresh-water or alkaline ponds with well-vegetated shorelines. They also breed on delta marshes, and in southern Spain occur on the Guadalquivir Delta, where as winter floods recede large areas of shallow water with abundant sedges and bulrushes gradually develop and provide excellent brood-rearing habitats (Hawkes, 1970). A study of foods has not been undertaken, but the birds are known to be able to dive readily while foraging, much like pochards, although the bill shape is unusually long and narrow and is clearly not highly adapted for sieving food materials from silt or mud.

Social behavior. So far as can be learned from observations of captive birds, marbled teal are monogamous, with the pair bonds being re-formed each fall and winter. Dementiev and Gladkov (1967) noted that spring migrant flocks consist of already paired birds. Wintering birds often occur in very large flocks; groups of up to 2,000 individuals were seen in southern Turkey during 1967–68 (Savage, in Hawkes, 1970). At least among captives there is little overt aggression evident in groups of adults, and inciting behavior by females consists of direct threatening rushes toward another bird, followed by a rapid swim back toward the mate or potential mate. There appears to be no decrescendo call as found in most *Anas* species. Social display occurs over a rather long period between late fall and spring among captive birds, and among wild birds has been seen in January and March. The male displays also diverge considerably from those of *Anas,* and include a general body shake, a bill shake, and a sequence of neck-stretching, pausing, and a quick withdrawal of the head downward and backward into the scapulars as a weak *eeep* sound is uttered. Females perform this same display, but with less vigor than males. Males also often extend their head and neck forward over the water toward another bird, in a posture very much like the sneak of pochards, and turn the back of the head toward inciting females as they swim ahead of them. Both sexes perform display drinking, but in general it seems to serve more as a prelude to copulation than as part of the basic pair-forming patterns. Copulation is preceded by both sexes performing a series of bill-dipping, bill-shaking, and dorsal preening displays, often in synchrony or near-synchrony. After treading, the male calls once with his head fully extended diagonally, then he draws his head back, points the bill downward, and swims in a partial circle around the female (Johnsgard, 1965a).

Reproductive biology. In the U.S.S.R. this species reportedly locates its nests on the ground, under the cover of low bushes or coarse, dry grass, with a lin-ing of dry grasses and a small amount of down. The birds have also been reported to nest in old raven nests, and on the Guadalquivir Delta have recently been described as nesting in the thatched roofs of peasant huts (Hawkes, 1970). Here the clutches were reported as quite large, of from 16 to 20 eggs, but these almost certainly were the result of multiple nesting efforts; in captivity clutches of 10 or 11 are the general rule. The breeding period appears to be rather protracted, not only in southern Spain and northern Africa, but also in southern Russia, which probably reflects the general southerly breeding distribution of this species. Incubation is performed by the female alone, and probably lasts 25 days on the average. Evidently males abandon their mates fairly early and soon gather into small flocks, but no specific molt-migration has been reported for this species. Instead they apparently molt on the breeding grounds and remain there until the generally transitory ponds dry out, after which the fall dispersal begins. The fledging period has not yet been reported, nor is there any specific information on brood behavior or mortality factors associated with nesting (Dementiev & Gladkov, 1967).

Status. The marbled teal is one of the rarest of the European ducks, with southern Spain now representing the only region outside of Russia where it still regularly breeds. There it occurs in very small numbers (under 100 pairs in 1971), although it was once quite abundant (Hudson, 1975). It likewise breeds in very small numbers in Turkey, and in unknown numbers in Morocco and Algeria. A substantial part of the Russian breeding population must winter in Iran, since a winter survey there in 1973 indicated more than 21,000 birds. Regardless of this, it is perhaps the rarest of the European dabbling ducks, and its status should be closely watched.

Relationships. I concluded (1961e) that the marbled teal should be removed from the genus *Anas* and placed in a monotypic genus near the pochards, to emphasize the many aberrant features it exhibits. Woolfenden (1961) was unable to study the post-cranial osteology of this species, and Brush (1976) found that its feather proteins were *Anas*-like when analyzed electrophoretically, so there is still no confirmatory evidence supporting my position.

Suggested readings. Dementiev & Gladkov, 1967; Cramp & Simmons, 1977.

Tribe Aythyini (Pochards)

Drawing on preceding page: Canvasback

Pink-headed Duck

Rhodonessa caryophyllacea (Latham) 1790

Other vernacular names. None in general English use. Rosenkopfente (German); canard à tête rose (French); pato de cabeza rosada (Spanish).

Subspecies and range. No subspecies recognized. Extinct; previously resident in northern India, probably including Assam, Manipur, Bengal, Bihar, and Orissa.

Measurements and weights. Folded wing: males, 250–82 mm; females, 250–60 mm. Culmen: males, 50–56 mm; no record for females. Weights: males, 793–990 g; av. (of 5) 935 g; females, 840 g (Schönwetter, 1960) to 1,360 g (Ali & Ripley, 1968). Eggs: 44 x 41 mm, white, 45 g.

Identification and field marks. Length 24″ (60 cm). *Adult* males have a bright pink head, which is slightly tufted behind, the color extending down the hind neck, while the foreneck, breast, underparts, and upperparts are brownish black, except for some pale pinkish markings on the mantle, scapulars, and breast. The tail and upper wing surface are brownish, except for the secondaries, which are pale fawn to salmon, with white tips. The underwing surface is also shell pink. The bill is bright pink, the iris brownish orange, and the legs and feet brown. *Females* have a less bright pinkish head, while *juveniles* have whitish rose heads and hind necks, and their

bodies are lighter brown, with dull brown underparts and the edges of the feathers whitish.

In the field, the bright pink head and bill of males are unique, although many recent reports of this species are the obvious result of confusion with male red-crested pochards, in which the head is chestnut rather than pink.

Natural History

Habitat and foods. These birds apparently occupied swampy lowland grass jungles, and even in very early times the birds were shy and not easily observed in this heavy growth. They evidently fed largely on the surface, but were known to be capable of diving readily. Apparently only a single specimen was ever examined as to its food contents, and it contained water weeds and various small shells (Ali & Ripley, 1968).

Social behavior. During the winter months this species apparently gathered in flocks of from 6 to 30 or even 40 birds, in lagoons adjoining large rivers. Their displays were observed by Delacour (1954-64) in captives, and he believed that they showed a rudimentary form of dabbling duck postures. The call of the male was a "whizzy whistle," uttered with the neck extended vertically, while the female was said to produce a low quack. Apparently during display the males would swim about with the neck shortened and head resting on the back, then suddenly extend it and call. None ever attempted to nest in captivity.

Reproductive biology. Reports summarized by Ali (1960) indicate that pairing occurred by April, nesting was begun in May, and eggs could be found in June and July. Nests were usually placed within 500 yards of water, in clumps of tall grass, and well concealed. The clutches were of from 5 to 10 eggs, which were remarkable in being nearly round. The incubation period was never reported, but the young were normally flying by September, at which time the birds returned to jungle lagoons.

Status. According to Ali (1960), the last fairly certain record of this species in the wild was in 1923 or 1924, while the last surviving bird in captivity lived no later than 1936. Prestwich (1974) states that 1935 and 1939 represent the probable last dates for birds being reported from the wild and captivity, respectively. A few sight records have been made in recent years (Singh, 1966; Mehta, 1960), but no convincing evidence for the species' existence has been forthcoming.

Relationships. Until fairly recently, the pink-headed duck has taxonomically been placed with the dabbling ducks, usually adjacent to *Anas* as an "aberrant" member of that group. However, I (1961b) located a trachea from a male of this species, and recognized its close affinities with the pochard group (*Aythyini*). At about the same time, Woolfenden (1961) reached the same conclusions on the basis of osteological evidence. These conclusions were confirmed by Humphrey and Ripley (1962), who reexamined the tracheal anatomy, the humerus, and the skeletal elements of the foot. They believed that the dabbling duck-like locomotor adaptations of the genus were related to environmental changes and modifications from a more typically pochardlike condition during Tertiary times. The studies by Brush (1976) on feather proteins suggest a close relationship between this genus and *Netta*, the more generalized pochards, and a distinction between these groups and the more typical pochards of the genus *Aythya*.

Suggested readings. Ali, 1960; Humphrey & Ripley, 1962; Prestwich, 1974.

Red-crested Pochard

Netta rufina Pallas 1773

Other vernacular names. None in general English use. Kolbenente (German); brante roussâtre (French); zambullidor de cresta roja (Spanish).

Subspecies and range. No subspecies recognized. Breeds chiefly from the lower Danube through southern Russia east across the Kirghiz steppes to western Siberia, south to northern Syria, Iran, and western China. Winters largely in India and Burma, but also in northern Africa and the eastern end of the Mediterranean. See map 94.

Measurements and weights. Folded wing: males, 256-78 mm; females, 249-58 mm. Culmen: males, 48-52 mm; females, 44-50 mm. Weights: males, 900-1,170 g (av. 1,135 g); females, 830-1,320 g (av. 967 g) (Dementiev & Gladkov, 1967). Eggs: av. 58 x 42 mm, greenish or light stone, 56 g.

Identification and field marks. Length 22″ (58 cm). Plate 46. *Males* in breeding plumage have a uniformly pale chestnut head, the feathers forming a bushy crest. The back of the head, mantle, and lower back are black, with the lower back and upper tail coverts grading to a greenish sheen. The scapulars are brown, and are separated from the breast by a white patch. The breast and lower parts are entirely black, with the flanks a sharply contrasting white, with some brown mixed in. The tail is gray; the upper wing coverts are generally gray to brown, except for a white area at the carpal joint; and the secondaries and most of the exposed vanes of the primaries are white to cream, with only the tips of the primaries conspicuously darker. The iris is red, the bill is vermilion, with a slightly lighter tip, and the legs and feet are orange to yellowish orange. *Males in eclipse* are nearly identical to females except for the iris and bill coloration, which remains reddish. *Females* have a two-toned head of dark brown from the eyes upward and backward along the hind neck, and a uniform grayish brown on the sides of the head, cheeks, and throat. The rest of the body consists of brown to brownish gray tones without strong patterning, although the wing pattern is nearly identical to that of the male. The iris is brown, the bill is dull grayish blue, becoming slightly reddish toward the tip, and the legs and feet are dull yellow. *Juveniles* closely resemble adult females, but young males soon begin to acquire a crest and exhibit reddish coloration on the bill and iris.

In the field, the reddish shaving-brush crest of the male, contrasting with a black breast and nearly white flanks, is a distinctive field mark; only the Eurasian pochard is at all similar among the species with which it might be found under natural conditions. Females have a strong resemblance to female black scoters in their general patterning, but have

longer and narrower bills. In flight both sexes exhibit white underwings and a white stripe on the upper flight feathers that extends the entire length of the wing. The male utters a very distinctive sneeze call during display, and females utter both soft purring sounds and harsher *gock* calls similar to those of other pochards.

NATURAL HISTORY

Habitat and foods. The breeding habitats of this rather sporadically distributed species seem to be predominantly still, brackish or salt-water areas; only secondarily or locally does it use fresh-water or slowly flowing water areas with extensive submerged and emergent shoreline vegetation. The species is structurally not so well adapted for diving as are the true pochards of the genus *Aythya*, and additionally the bill is more like that of a typical dabbling duck. The birds apparently do not dive very deeply, and usually remain under water for only short periods, often less than ten seconds. The birds are primarily vegetarians, and feed on such submerged plants as musk grass, hornwort (*Ceratophyllum*), pondweeds, mare's-tail (*Hippurus*), and milfoil (*Myriophyllum*). Small mollusks are sometimes also eaten, probably mainly through accidental ingestion along with the leafy plant materials, and seeds have been found in small quantities. Red-crested pochards are the only waterfowl species known to perform ritualized or ceremonial feeding, in which the male typically dives, surfaces with vegetation or even inedible materials in his bill, and then swims to his mate and passes it to her. This activity seems to be confined to actually paired birds, rather than to birds that are still in the mate-selection phase of courtship (Johnsgard, 1965a).

Social behavior. Red-crested pochards breed in the first year of their life, with a seasonally established monogamous pair bond situation prevailing. Sexual display begins as early as late fall. Female inciting behavior develops from the greeting ceremonial behavior of ducklings after the birds are two to three months old, and the major courtship call of the male develops at the same time, and from the same origins. This posture and call, the sneeze, serves for both aggressive and sexual purposes. Another major display, head raising (or neck stretching) develops after the birds are six months old, and is accompanied by *brumm* calls, which are used both in social

MAP 94. Known European breeding areas (inked) and presumptive Asian breeding distribution (hatched) of the red-crested pochard. Wintering range indicated by stippling.

display and as an alert signal (Platz, 1974). A posture in which the male stretches his head and neck forward over the water and utters a nasal call is also sometimes seen, and seems to correspond to the sneaking posture of the typical pochards. In addition, males fairly frequently preen behind the wing and perform display drinking, but both of these displays are probably more closely associated with precopulatory behavior than typical courtship. Males initiate copulation by interspersing bill-dipping and dorsal preening movements with head shaking, preening behind the wing and elsewhere, and with rudimentary head-pumping movements that are less conspicuous than but distinctly similar to those of *Anas*. Females respond with the same

kind of head-pumping movements, and gradually become prone in the water. After treading, the male utters a single sneeze call, and then swims away in the usual pochard bill-down posture (Johnsgard, 1965a).

Reproductive biology. In the U.S.S.R., birds arrive on their nesting areas already paired, although some display and fighting occurs there as a result of the activities of unpaired or "bachelor" drakes. Pursuit flights, involving either a male chasing out of his home range any females that intrude into it, or the chasing of breeding females by one or several males in attempted rape flights, are fairly common early in the nesting season. The nest site is often on dry land

276 ◈ ◈ ◈

or at the edge of water, in dense vegetation, but also may be in floating mats of vegetation or amid reed beds. Normal clutch sizes vary considerably, from about 6 to 12, but there is a strong tendency in this species for dump-nesting, with as many as 39 eggs reported in one case, and involving up to three females. Males abandon their mates at about the onset of incubation, which requires from 26 to 28 days. Another ten or eleven weeks are needed for the young to attain fledging (Platz, 1974; Dementiev & Gladkov, 1967; Bauer & Glutz, 1969).

Status. The generally southerly distribution of this species is indicated by the fact that about 10,000 birds winter in northwestern Europe, compared to about 50,000 in the Mediterranean and Black sea region. Probably many of the birds of western Europe winter farther south into Africa, since western Russia is believed to support at least 90,000 pairs. From 3,000 to 6,000 pairs breed in Spain, and smaller numbers are present in southern France, the Netherlands, Austria, East Germany, Czechoslovakia, and Hungary (Ogilvie, 1975). The birds also breed east to Central Asia, but in unknown numbers.

Relationships. Behavioral information (Johnsgard, 1965a) as well as data on the postcranial osteology (Woolfenden, 1961) indicates that *Netta* provides a transitional evolutionary stage between the dabbling duck group and the typical pochards of the genus *Aythya*. Woolfenden believed that the southern pochard should be kept generically separate from the red-crested pochard, but Brush (1976) reported that in their feather proteins these two species and the pink-headed duck appeared to be virtually identical and yet distinct from *Aythya*.

Suggested readings. Platz, 1974; Owen, 1977.

Southern Pochard

Netta erythropthalma (Eyton) 1838

Other vernacular names. African pochard; Rotaugenente (German); canard plongeur austral (French); zambullidor austral (Spanish).

Subspecies and ranges. (See map 95.)

  N. e. erythropthalma: South American southern pochard. Resident in western South America, breeding primarily from northwestern Vene-

zuela to southern Peru, and extending in winter along the coasts of Peru, Colombia, Venezuela, and Brazil. Breeding in these latter areas is uncertain.

  N. e. brunnea: African southern pochard. Resident in Africa, primarily in Ethiopia, the southern Sudan, eastern Zaire, Angola, Uganda, Kenya, Tanzania, Malawi, Zambia, Rhodesia, and South Africa.

Measurements and weights. Folded wing: males, 202–28 mm; females, 201–21 mm. Culmen: males, 40–49 mm; females, 38–49 mm. Weights: males, 600–977 g; females, 533–1,000 g (Middlemiss, 1958). Eggs: av. 54 x 44 mm, creamy white, 59 g.

Identification and field marks. Length 20" (51 cm). *Adult males* have a purplish black head and neck, merging with a glossy black breast, while the rest of the body coloration is a nearly uniformly dark chestnut brown, the mantle tending toward olive brown and the abdomen and under tail coverts grading to fulvous. The tail is dark brown, and the upper wing surface is olive brown to dark brown, except for a white speculum formed by the bases of the secondaries, which have broad brown tips. The iris is bright red, the legs and feet are bluish gray and blackish, and the bill is bluish gray with a black nail. *Females* have a brownish black forehead, crown, and nape, grading to umber on the cheeks, with off-white markings extending from the chin to the base of the upper mandible, as a streak behind the eye, and as an ill-defined crescent extending from the throat up the sides of the cheeks to the ear region. The upper parts of the body are generally uniformly olive brown, vermiculated with buffy brown, while the flank feathers are more fulvous, fringed with rusty coloration. The underparts are dull brown, except for the under tail coverts, which are mottled with white. The wing coloration is very similar to that of the male. The iris is brown, the bill dark slate gray, and the legs and feet are as in the male. *Juveniles* resemble adult females, but the tip of the head is more brownish, the whitish eye-stripe is less pronounced, and the body coloration is generally a lighter brown.

In the field, this species is likely to be confused only with the female rosybill in South America, which also has a bluish bill and a brown body, but lacks the white facial markings. In Africa no other pochard occurs within this species' range, and confusion with other diving ducks such as the Maccoa seems unlikely. Both sexes exhibit a conspicuous white speculum in flight, which extends into the primaries along their bases. The calls of both sexes are relatively quiet, and no decrescendo call has yet been described for the female. Males are said to utter a repeated *prerr* note in flight; their usual courtship call is a soft and mechanical *eeroow.*

NATURAL HISTORY

Habitat and foods. In southern Africa, the southern pochard occurs from coastal areas to about 8,000 feet of elevation, using primarily permanent water areas that may be vegetation-free or have emergent aquatic plants. Generally, rivers are avoided, as are turbid waters, shallow temporary ponds, and flooded areas. The birds are usually found in open water or near the edges of emergent shoreline vegetation. They feed primarily by diving, but also upend in shallow water and occasionally go to shore to forage on vegetation at the water's edge. The available data indicate that the birds are primarily vegetarians, eating the vegetative parts or seeds of such plants as water lilies (*Nymphaea*), bladderwort, duckweeds, bulrushes, cattails, and other similar aquatic or shoreline plants (Schulten, 1974; Middlemiss, 1958). Animal materials that have been found in samples from adults or young include snails, aquatic beetles and hemipterans, ants, and crustaceans.

Social behavior. Southern pochards are relatively gregarious, with flock sizes of several hundred birds recorded at times, and one concentration of 5,000 noted. These, however, are nonbreeding season

Map 95. Breeding or residential distribution of the African ("A") southern pochard, and major breeding (hatched) and nonbreeding or peripheral (stippling) distributions of the South American southern pochard ("S").

278

counts, and usually occur during the dry season, when the reduction of water areas causes concentrations to develop. The timing of pair formation is not yet clear, but presumably temporary monogamous pair-bonding behavior is the rule. Display behavior has only infrequently been observed in wild birds (Middlemiss, 1958), but observations on captive individuals indicate a nearly typical pochard display repertoire. Inciting is the primary female display, and is accompanied by a harsh *rrrr-rrrr* call; a threatening *quarrk* note has also been reported. The most common display of the male is a courtship call, *eerooow*, sounding like the rapid unwinding of a spring, and accompanied by withdrawal of the head into the shoulders, with the bill held horizontally. The same or a very similar call is uttered during a rapid head-throw display, and a soft three- or four-noted call is uttered during the sneak display, as the male slightly extends his head and neck toward a female. Very frequently a male will respond to inciting by swimming ahead of the female and turning the back of the head toward her, while depressing his crown feathers. In addition, a display preening behind the wing is often performed, usually in response to the same display by females. Copulation is preceded by slight head-pumping movements on the part of the female and by bill dipping and display preening (usually on the back, but also elsewhere) by the male. After treading, the male calls once and then swims rapidly away from the female in a bill-down posture (Johnsgard, 1965a).

Reproductive biology. Records of eggs and broods, when plotted over the entire African range of the species, encompass all months of the year. However, in the more northern parts of the range (Mozambique, Zambia, Malawi, Kenya, Uganda) most records occur between March and August, while in Cape Province, Natal, and Transvaal they are concentrated between August and December. They thus would seem to be associated more with the timing of the rainy season than with the calendar year. The nest is usually located near water and hidden by rank grass, weeds, or low shrubs, or it may be built over water in emergent vegetation, as in sedges, reeds, or in papyrus beds. The clutch size ranges rather widely, from 6 to 15 eggs, but averages 9. Only the female incubates; and the probable normal incubation period is 26 days, although shorter estimated periods of 20 to 21 days appear in the literature. There is some doubt about whether the male ever participates in the rearing of the brood; one report

(Middlemiss, 1958) indicates that the male has been seen helping to tend young ducklings. This is not known to be a normal condition in any of the other species of pochards. The fledging period has not yet been established, but the flightless condition of adults during the postnuptial molt is known to last about 31 days.

Status. This species is apparently much more common and widespread as a breeding species in Africa than in South America, but even there is reported as a common resident species in only a few areas such as Kenya, Uganda, Malawi, and Rhodesia.

Relationships. The recommendation by Delacour (1954–64) that this species be included in the genus *Netta* rather than among the typical pochards (*Aythya*), as it traditionally has been, seems to have some merit, at least on behavioral grounds (Johnsgard, 1965a). However, Woolfenden (1961) has argued that the other two species of *Netta* as here constituted do not seem to be closely related, and thus the relationships of these "primitive" pochards has not yet been resolved to everyone's satisfaction.

Suggested readings. Clancey, 1967; Middlemiss, 1958; Clark, 1966.

Rosybill

Netta peposaca (Vieillot) 1816

Other vernacular names. Rosy-billed pochard: Peposakaente (German); canard peposaca (French); pato picazo or pato negro (Spanish).

Subspecies and range. No subspecies recognized. Breeds in central Chile from Atacama to Valdivia, and on the eastern side of the Andes from central Argentina (about 40° south latitude) northward to southeastern Brazil, Uruguay, and Paraguay, and occasionally south to Tierra del Fuego. Winters farther north, to Bolivia and south-central Brazil. See map 96.

Measurements and weights. Folded wing: males, 228–45 mm; females, 220–40 mm. Culmen: males, 61–66 mm; females, 54–60 mm. Weights: 6 males averaged 1,181 g; 5 females averaged 1,004 g (Weller, 1968a). Eggs: 56 x 42 mm, green or creamy, 60 g.

Identification and field marks. Length 22″ (56 cm). *Adult males* are purplish black on the head and neck, while the back and scapulars are greenish black or blackish, finely speckled with white or gray. The breast is black, and the abdomen and flanks are finely vermiculated with black and white. The rump, upper tail coverts, and tail are black, while the under tail coverts are white. The upper wing coverts are greenish brown, with a white patch at the carpal joint, while the secondaries are white, tipped with black. The inner primaries are also white, with black tips, while the outer four are cream-colored on the inner vane and black on the outer vane. The bill is bright red, with a bulbous enlargement at the base and a black nail, the iris is yellow to orange, varying seasonally, and the legs and feet are yellow to orange, with darker webs. *Females* are nearly uniformly brown, with a darker back and white on the chin, throat, and under tail coverts. The wing pattern is like that of the male. The iris is dark brown, the bill bluish slate with a black nail, and the legs and feet dull orange yellow to grayish. *Juveniles* resemble females, but have brown rather than silvery white underparts.

In the field, the bright red bill, contrasting with the blackish head and upperpart coloration, readily identifies the male. Females lack the white head markings of the otherwise similar southern pochard, and appear to be a uniform and nondescript brown except for their white under tail coverts. Females utter both a rather harsh quacking note and a *krrr* sound during

inciting, while males produce a similar growling sound and a faint *whee-ow* during aquatic display.

NATURAL HISTORY

Habitat and foods. This marsh-dwelling pochard is almost like the dabbling ducks in its preference for shallow waters and its terrestrial inclinations. Weller (1967a) never observed it diving, and noted that it was highly terrestrial, spending as much time on land as on water. It preferred to occupy duckweed-covered ponds, and was only rarely found on open lakes or rough water. Its foods have yet to be studied, but appear to be predominantly vegetable in nature. A single specimen collected in Argentina had a large number of seeds of water milfoil (*Myriophyllum*), as well as the seeds and rootstalk fragments of a grass or sedge (Phillips, 1922–26).

MAP 96. Breeding or residential (hatched) and wintering (stippling) distributions of the rosybill.

Social behavior. This is a highly social species, and even during the breeding season some flocking occurs, presumably by nonbreeding birds. Weller (1967a) doubted that permanent pairing occurs in this species, and he neither observed males participating in brood care nor saw pairs in the early postnesting period. He did observe a possible territorial defense flight, but surprisingly little courtship in view of the presumed annual renewal of pair bonds. In captivity, however, pair-forming displays may be readily observed, and in general are closely similar to those of the other pochards. Females incite with strong neck stretching alternated with bill pointing toward a threatened drake, and utter a harsh *krrr* with each movement. They also at times produce a multinoted call rather like the decrescendo call of *Anas* females, and frequently perform a conspicuous display preening behind the wing. They likewise perform the highly stereotyped and exaggerated drinking movements used by males as a greeting ceremony, and extend their heads and necks forward in a sneak posture in the manner of males. These two displays, ritualized drinking and the sneak posture, are the two most conspicuous and frequent of the male postures, but in addition the males sometimes also perform a head-throw display, with an associated *wheee-ow* call. Males utter two other distinctive calls without strong head or neck movements, and quite frequently preen behind the wing as a display. Preferred males also swim rapidly ahead of inciting females while turning the back of the head toward them. Before copulation, the male alternates bill-dipping and dorsal preening movements, and perhaps also makes rudimentary head-pumping movements, while the female has been observed only to perform head pumping. After copulation the male probably calls, and definitely assumes the usual pochard bill-down posture (Johnsgard, 1965a).

Reproductive biology. Nesting in central Argentina occurs mainly between October and December, with a probable peak in early November according to Weller (1967a). Contrary to earlier descriptions, he never found a nest on land; instead, they were located in water from 10 to 22 inches deep, amid dense vegetation adjacent to open pools. Evidently there is a strong tendency for dump-nesting in available nests of other species; Weller found eggs in an old coot nest, and a nest containing not only 24 rosybill eggs but also 6 eggs of the black-headed duck. The average clutch in nests used by a single female is probably about 10 eggs. He also observed a seemingly low hatching success among rosybill nests, with only one of six successful. Only the female incubates, and Weller found no evidence of male participation in brooding. The incubation period is 28 days, which is relatively long for the pochard group. The fledging period seems not to have been established, but apparently brood mergers are not uncommon, with one early report of a female tending a group of 52 ducklings (Phillips, 1922–26).

Status. Weller (1967a) said that the rosybill was the most common of the marsh-nesting Anatidae in the Cape San Antonio Province of eastern Argentina, which is near the center of its breeding range. Among a survey of ducks killed by hunters in the same general area, Weller (1968a) found that rosybills constituted 16 out of 263 total birds examined, but judged that their fall migration pattern, as well as the type of hunting being done there, affected this relatively low harvest. In general, it would appear that the rosybill is among the most abundant of the South American diving ducks.

Relationships. Although at one time the rosybill was generically separated from the other pochards, Delacour (1954–64) has argued that such separation obscures the close relationships between this bird and the other species he placed in the genus *Netta*, particularly the southern pochard. Woolfenden (1961) rejected this position, while I (1965a) advanced the view that this species of *Netta* is the one most closely approaching *Aythya* in its affinities. Weller (1967a) suggested that perhaps the rosybill should actually be removed from *Netta* and merged with the typical pochards, in *Aythya*, but in most skeletal features the red-crested pochard is closer to *Aythya* than is the rosybill (Woolfenden, 1961).

Suggested readings. Phillips, 1922–26; Weller, 1967a.

Canvasback

Aythya valisineria (Wilson) 1814

Other vernacular names. Can; Riesentafelente (German); milouin aux yeux rouges (French); pato lomo cruzado (Spanish).

Subspecies and range. No subspecies recognized. Breeds in North America from central Alaska south to northern California and east to Nebraska and Minnesota. Winters from southern Canada south along the Atlantic and Pacific coasts to central and southern Mexico. See map 97.

Measurements and weights. Folded wing: males, 225–42 mm; females, 220–30 mm. Culmen: males, 55–63 mm; females, 54–60 mm. Weights: adult males, 850–1,600 g (av. 1,252 g); adult females, 900–1,530 g (av. 1,154 g) (Ryan, 1972). Eggs: av. 63 x 45 mm, bright olive, 68 g.

Identification and field marks. Length 19–24" (48–61 cm). *Adult males* in breeding plumage have a dark reddish head and neck, grading to blackish on the face and crown. The midback and scapulars are white, vermiculated with blackish coloration, and grading to brownish black on the rump and upper tail coverts. The under tail coverts are blackish, but the rest of the underparts are white with blackish vermiculations, especially on the sides and flanks. The upper wing coverts are white with gray vermiculations, and the secondaries are pearl gray, the inner ones with narrow black margins. The primaries and their coverts are slate brown, while the underwing surface is white to pale gray. The iris is bright red in spring and duller in winter, the bill is blackish, and the feet and legs are grayish blue. *Adult females* have a reddish brown head and neck, with a darker crown and more buff around the eyes and on the cheeks, chin, and throat, while the lower neck has a reddish cast. Most of the upperparts are dark brown (more grayish in winter), the feathers having lighter edges

and often vermiculated toward their tips. The wing coverts and tertials are mostly grayish brown with vermiculations, the primaries and their coverts are brown, and the secondaries are gray, with paler tips. The sides and flanks are dull brown, the feathers somewhat vermiculated, while the rest of the underparts are mostly grayish white, and the tail is sooty brown. The bill is blackish, the iris brown, and the legs and feet are grayish blue. *Males in eclipse* resemble females, but the iris remains reddish and the head darker. *Juveniles* resemble adult females, but have darker backs and more mottled and browner underparts. The head color of young males is darker than that of females, and a yellowish iris color is acquired during the juvenile plumage.

In the field, canvasbacks are most likely to be confused with redheads, which are appreciably darker in both sexes and have a shorter and less sloping bill and forehead profile. Canvasbacks also closely resemble Eurasian pochards, but both sexes lack the conspicuous pale band that occurs near the tip of the bill in that species. In flight, canvasbacks appear to be unusually long-necked and fly swiftly, with rapid and strong wingbeats. Calling by the female seems to be largely limited to the inciting display, when a soft *krrr-krrr* note is uttered, while males produce a dovelike cooing note during courtship display and a softer breathing sound.

NATURAL HISTORY

Habitat and foods. Breeding habitats of the canvasback typically consist of shallow prairie marshes surrounded by cattails, bulrushes, and similar emergent vegetation, and which are both permanent and large enough to have sufficient open water for easy landings and takeoffs. Such marshes also usually have an abundance of submerged aquatic vegetation such as pondweeds, which are the single most important group of food plants for this species. On migration and in wintering areas they concentrate on lakes or marshes where wild celery (*Vallisneria*), arrowhead (*Sagittaria*), water lily (*Nymphaea*) and similar succulent aquatics are to be found, and in bays and estuaries where eelgrass (*Zostera*) and wigeon grass beds are abundant. Brackish estuarine bays, rather than salt-water or fresh-water ones, provide the preferred wintering habitat; in such areas not only these submerged plants but also clams, crabs, and

MAP 97. Breeding (hatched) distribution of the canvasback, including areas of major concentrations (crosshatching). Wintering distribution indicated by stippling.

other small invertebrates are abundant (Stewart, 1962).

Social behavior. Probably at least in part because of their preference for such localized food sources as wild celery and certain pondweed species, canvasbacks typically occupy specific and traditional rivers, lakes, and marshes on their migratory and wintering areas, and in such locations often concentrate in the thousands or even tens of thousands. Chesapeake Bay, San Francisco Bay, and several Gulf Coast areas represent the major areas where the birds concentrate in winter and where pair formation begins in late winter. Courtship is actively performed during the spring migrations, with some females still remaining unpaired at the time of arrival on their Manitoba nesting grounds (Hochbaum, 1944). Most displays are performed on water, but aerial chases are also frequent, especially late in the pair-forming period. Males perform all of the typical pochard displays, in-

cluding a courtship note uttered in a kinked-neck posture, the same call uttered during a head-throw, a lowered head posture called the sneak, and aggressive chin lifting usually given in response to the same posture in females, who incite in this manner. Wing preening is rare or even absent as a male display, but preening of the dorsal region is the most conspicuous male precopulatory display, being alternated with bill-dipping movements. Females sometimes respond with the same displays, but treading may often occur without such mutual behavior. As the male releases his hold of the female's nape he utters a single courtship call and swims away from her in a rigidly held posture with the bill pointed sharply downward (Johnsgard, 1965a). During aerial chases the males may at times also try to bite the tail of the female they are chasing, but such behavior may simply be a reflection of attempted rape rather than a specific display.

Reproductive biology. Shortly after their arrival on the breeding grounds nearly all females will have formed pair bonds, although it is believed that yearling females are less inclined to attempt nesting than are older birds, and some may breed later or possibly not at all if conditions are not ideal. Paired birds establish rather large home ranges that may include several ponds and at least in males may exceed 1,000 acres; home ranges of adjacent pairs often overlap and little if any aggressive activity occurs among such mated pairs. Females typically construct their nests in the midst of emergent beds of bulrushes, cattails, or reeds (*Phragmites*), usually in vegetation between 14 and 48 inches high, fairly close to and often within 40 feet of areas of open water that measure at least 50 by 50 feet in size. One or more nests may be started before a complete clutch is produced, or eggs may be randomly dumped in the nests of other ducks. Eggs are laid at the approximate rate of one per day until a complete clutch, averaging 9 to 10 eggs for initial nesting efforts, has been produced. The addition of "parasitic" eggs of redheads and other canvasbacks often increases the actual clutch to a dozen or more eggs. Incubation periods under wild conditions seem to vary considerably, but probably average about 24 or 25 days. Males abandon their mates early in the incubation period, gathering in large flocks at traditional molting lakes and leaving other males to tend their females should renesting be necessary. In some years excessive flooding or cold weather results in massive nest desertion, and a resulting high degree of attempted renesting; additionally, predators such as raccoons, skunks, ravens, and crows are all locally serious sources of nest losses. Young canvasbacks have a relatively long fledging period of from 58 to 68 days, and females may abandon their young at a relatively early age to undergo their own molt, particularly among late-hatched broods. The flightless period for such birds is about three or four weeks, and a few females may remain unable to fly well into early fall, even into October in the areas of Delta, Manitoba (Hochbaum, 1944). Both juveniles and adult females suffer unusually high mortality rates for reasons that are still somewhat uncertain, but hunting is known to be an important factor in these losses.

Status. During two periods of the 1900s the canvasback has been known to be at perilously low population levels—during the dry years of the 1930s, and again during the 1960s and early 1970s, when a combination of unfavorable breeding years and excessive marsh destruction caused a serious and seemingly irreparable decline in their numbers. Bellrose (1976) reports that winter counts between 1955 and 1974 have indicated a reduction of more than 50 percent in numbers in all flyways but the Central Flyway, and that breeding-ground surveys over the same period suggest an average population of 560,000 birds for the major breeding areas. The badly unbalanced sex ratio of canvasbacks, their declining areas of prime breeding habitats, their sensitivity to oil or other pollution sources in their prime wintering areas, and their vulnerability to hunting all combine to make the future status of the canvasback a most uncertain one.

Relationships. In addition to its obvious close relationship to the redhead, the canvasback is even more similar to the Eurasian pochard in morphology and behavior. It is generally believed that the two North American forms must have evolved as a result of a double invasion of the continent from an ancestor similar to the Eurasian pochard, with one species (the redhead) adapting to more westerly and alkaline marsh conditions and the other becoming adapted to the deeper, less alkaline prairie marshes of the central plains and lowlands of North America. Although the two species seem to be effectively isolated from hybridization, they often forage in mixed flocks and probably compete to some degree for the same foods.

Suggested readings. Hochbaum, 1944; Erickson, 1948; Stoudt, 1971.

284 ◈ ◈ ◈

Eurasian Pochard

Aythya ferina (Linnaeus) 1758

Other vernacular names. European pochard; Tafelente (German); milouin d'Europe (French); zambullidor europeo (Spanish).

Subspecies and range. No subspecies recognized. Breeds in the British Isles, southern Scandinavia, and from eastern Russia through western Siberia to Lake Baikal, south to Holland, Germany, Romania, the Black Sea, and the Kirghiz steppes to the Yarkand River in western China. Winters from its breeding range south to northern Africa, the Nile Valley, Persian Gulf, India, Burma, southern China, and Japan. See map 98.

Measurements and weights. Folded wing: males, 207–24 mm; females, 201–12 mm. Culmen: males, 45–51 mm; females, 43–47 mm. Weights: males (in September) 930–1,100 g (av. 998 g); females, 900–995 g (av. 947 g) (Dementiev & Gladkov, 1967). Eggs: av. 62 x 44 mm, olive, 66 g.

Identification and field marks. Length 18–23″ (46–58 cm). Plate 47. *Males* in breeding plumage have a uniformly ruddy chestnut head and neck, and a black breast. The back, scapulars, flanks, and underparts are predominantly white, with black vermiculations of varying coarseness present except on the abdomen. The lower back, rump, and tail coverts are black, and the tail is nearly black. The upper wing coverts are a vermiculated gray, and the secondaries

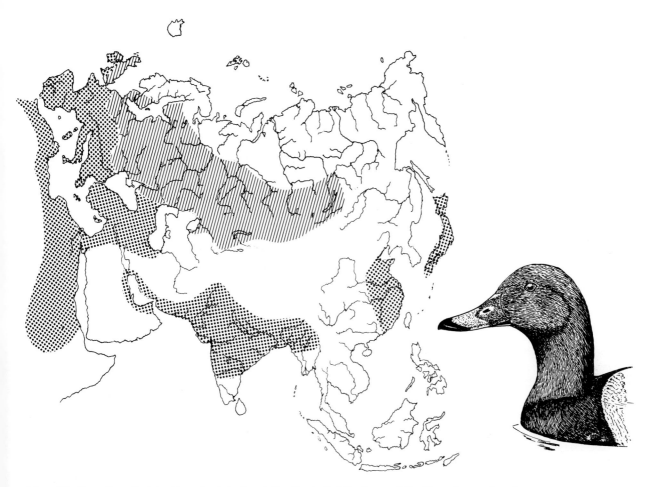

MAP 98. Breeding (hatched) and wintering (stippling) distributions of the Eurasian pochard.

are also gray, with no obvious speculum, while the primaries are more brownish, especially at their tips. The iris is red, the bill is mostly dark bluish, with a pale band between the nostrils and the blackish tip. The legs and feet are bluish gray, with darker webs. *Males in eclipse* resemble females, but retain a yellow to reddish iris color and show some vermiculations on the upperparts. *Females* have a brownish head, which is darker on the crown and hind neck and buff on the chin and throat, and also shows a buffy eye-ring and a vague pale stripe behind the eyes. The upperparts are brown or gray vermiculated with brown, while the rump, tail coverts, and tail are brownish black. The breast is brown, shading to pale gray on the abdomen and flanks. The upper wing coverts are grayish brown, without vermiculation; the secondaries are gray; and the primaries are dark brown. The iris is brown, the legs and feet are slate

gray with darker webs, and the bill is similar in color and pattern to that of the male. *Juveniles* resemble adult females, but have more mottled underparts, and young males have a ruddier head.

In the field, European pochards are sometimes found with and may be confused with red-crested pochards, but the reddish bill of the latter will separate them. In the unlikely event that they are seen in company with canvasbacks or redheads, the broad and conspicuous pale band across the bill is the best distinguishing mark (redheads may have a narrow and indefinite area behind the black bill tip, too); otherwise the birds look almost perfectly intermediate between these two species and might be regarded as possible hybrids. Females have the usual two general pochard calls, a growling *brerr* or *errr* associated with inciting, and an aggressive *pack* or *back*. Males produce a soft, breathing *wiwierrr* and a louder *kil-kil-kil* during display, but neither is particularly penetrating.

NATURAL HISTORY

Habitat and foods. The breeding habitat of this species apparently is identical to the habitats used by the redhead and canvasback in North America—namely, fairly fresh-water or alkaline marshes that are rich in submerged vegetation and are surrounded by emergent shoreline plants. Slowly flowing rivers and sometimes even eutrophic lakes are used for breeding also. Some open water areas are needed for landing and taking off, and most foraging is done at depths of only one to three meters, with the birds usually remaining submerged for periods of 15 to 30 seconds. The leafy parts and stems of aquatic plants such as muskgrass and pondweeds are the primary items of diet, although the seeds of these and other aquatic or shoreline species are also eaten, especially shortly after they have ripened. During the summer, insects are eaten in considerable quantities, not only by ducklings but also adults, with caddis fly larvae and the larvae of midges being particularly important. Birds that winter in coastal regions also consume considerable quantities of animal materials, such as cockles (*Cardium*) and other mollusks. However, Olney (1968) reported a very low incidence of animal foods among birds taken in inland waters of southeastern England and Northern Ireland, with nearly half of the volume of the food materials consisting of the spores and fruiting bodies of muskgrass.

Social behavior. Pochards become sexually mature their first winter of life, and presumably many females attempt to nest the following spring, but specific data are still lacking on this point. Social display occurs over an extended period, from December through early July in Bavaria according to Bezzel (1959, 1969), who noted that the incidence of paired females remained quite low until about April, when it sharply rose until June, when all females appeared to be paired. This was appreciably later than the other species present in the area. The calls and postures of this species are extremely similar to those of the canvasback. For example, inciting in the female consists of alternated lateral threatening movements and strong neck stretching, with associated *errr* notes. Males perform kinked-neck and head-throw displays much like those of the canvasback,

but the associated call is a soft, breathing *wiwierrr*. During the sneak posture, performed in nearly the same manner as in the canvasback, the same or a very similar call is also uttered. A louder *kil-kil-kil* sound is also produced by courting pochards, and perhaps corresponds to the coughing call of other pochard species. There is still very little information available on copulatory behavior in this species, but it possibly lacks display preening as a precopulatory display (Johnsgard, 1965a; Bauer & Glutz, 1969).

Reproductive biology. At least in the U.S.S.R., females often seek out nest sites that are located either in reed beds or on floating mats of reeds or other vegetation. In one area there, nearly half of the nests were among reed beds in shallow water of up to 30 centimeters in depth. During years of high water levels, when there are few emergent reed beds or floating mats, the birds may nest in sedge tussocks, in flooded fields, or under bushes on hummocks. The clutch typically numbers about 8 eggs, but often varies from 6 to 9, and multiple clutches may number as much as 25. Males depart at about the time incubation is initiated. Incubation requires from 24 to 28 days, averaging about 25 days. The fledging period is probably from 50 to 55 days, and females apparently remain with their offspring for most or all of this period (Dementiev & Gladkov, 1967; Bauer & Glutz, 1969).

Status. In contrast to the declining status of the redhead and canvasback in North America, the Eurasian pochard seems to be prospering, at least in northwestern Europe, where about 225,000 birds winter, in addition to some 750,000 that winter in the Mediterranean and Black sea area. The largest breeding concentrations occur in western U.S.S.R., representing some 200,000 pairs, but considerable numbers also breed in Finland, France, the Netherlands, Sweden, and East Germany, and a few hundred pairs occur in Denmark, Britain, and Spain (Ogilvie, 1975).

Relationships. The relationships of the canvasback, redhead, and Eurasian pochard have been considered by numerous writers; and although neither North American species is conspecific with the Old World form, it seems the Eurasian pochard is somewhat closer to the canvasback than to the redhead. This is suggested by its courtship behavior as well as its general ecological adaptations.

Suggested readings. Bezzel, 1969; Dementiev & Gladkov, 1967.

Redhead

Aythya americana (Eyton) 1838

Other vernacular names. Red-headed pochard; Rotkopfente (German); milouin américain (French); cabaza roja (Spanish).

Subspecies and range. No subspecies recognized. Breeds in North America from central Canada south to southern California, New Mexico, Nebraska, and Minnesota, with local breeding farther east and in Jalisco, Mexico (*Auk* 95: 152). Winters from Washington east to the middle Atlantic states and south to the Gulf Coast of Mexico and Guatemala. See map 99.

Measurements and weights. Folded wing: males, 230–42 mm; females, 210–30 mm. Culmen: males, 45–50 mm; females, 44–47 mm. Weights: adult males in fall average ca. 1,080 g, and females ca. 1,030 g, with maximum fall weights of 1,361 and 1,314 g, respectively. Eggs: av. 62 x 44 mm, white, 65 g.

Identification and field marks. Length 18–22" (40–46 cm). Plate 48. *Adult males* in breeding plumage have a bright reddish head and upper neck, with the lower neck, foreback, and breast black. The rest of the foreback and scapulars are vermiculated with black and white, producing a dark gray overall effect. The hind back is dusky, grading to brownish black on the upper and under tail coverts, and the tail is sooty brown. The sides and flanks are vermiculated, as is the back, while the rest of the underparts are white. The underwing surface is mostly white, and the upper wing coverts and tertials are fairly uniformly gray. The primaries and their coverts are brownish gray, and the secondaries are silvery gray with paler tips, the inner secondaries being narrowly margined with black. The bill is mostly pale bluish, with a whitish band behind the black tip, and sometimes with a black stripe behind the nostrils. The iris is lemon yellow, and the legs and feet are dark greenish gray, with darker webs. *Females* have a mostly reddish brown head and neck, darkest on the crown and palest around the base of the bill, with a faint pale eye-ring and postocular stripe. The chin is white, shading into a grayish brown throat and foreneck; the breast, sides, and flanks are brownish, the feathers having buffy tips. The rest of the underparts are whitish mottled with brown, while the upperparts are dark grayish brown, with ashy white speckling or tipping on the feathers. The tail is sooty brown. The upper wing coverts are mostly brownish gray, while the primaries and their coverts are light brownish gray and the secondaries are dark silvery gray tipped with whitish coloration and the inner secondaries are narrowly margined with black. The bill is grayish blue, with a darker tip and a faint pale band behind it, the iris is brown, and the legs and feet are greenish gray. *Males in eclipse* are similar to females, but have a more reddish brown head and a dull yellow orange iris. *Juveniles* resemble adult females but are more heavily mottled. Differences in the iris coloration of the sexes appear as early as eight to ten weeks after hatching, and males begin to show brownish red feathers in the cheeks at about this time.

NATURAL HISTORY

Habitat and foods. The preferred breeding habitat of redheads consists of nonforested environments with water areas that are sufficiently deep to provide permanent and fairly dense emergent vegetation for nesting. These include potholes and marshes that are usually somewhat alkaline and at least an acre in size, with an interspersion of open water covering from about 10 to 25 percent of the surface, and with emergent vegetation reaching about 20 to 40 inches in height. On the breeding grounds both aquatic vegetation such as the vegetative parts of pondweeds and various small invertebrates, including insect larvae, mollusks, and crustaceans, are consumed. A higher proportion of plant foods occurs on migration and wintering areas. On brackish or salt-water areas wigeon grass and shoal grass (*Diplantera*) are sometimes especially important, but the rootstalks, stems, leaves, and seeds of a variety of submerged aquatic plants seem to be taken, depending upon their availability (Johnsgard, 1975; Bellrose, 1976).

Social behavior. Like canvasbacks, redheads often form rather large flocks during fall migration where favored foraging areas occur, and such flocks often number in the thousands of birds. By midwinter the birds are in full nuptial plumage, and courtship behavior is apparently initiated on the wintering grounds. Probably all females form pair bonds during their first winter, even though it is believed that a substantial number of presumed yearling females may not attempt to nest. Pair-forming behavior is prevalent during the northward migration in spring,

and probably peaks about late April, when the first birds arrive on their breeding grounds. As in canvasbacks, most pair-forming behavior occurs on the water, although aerial chasing is prevalent late in the spring. The distinctive and rather catlike call of the male at this time is uttered with an extended and kinked-neck posture, or during a head-throw, in which the nape is brought all the way to the base of the bird's tail. Aggressive neck stretching is frequent, and a softer sound resembling a weak cough is also produced. Females perform the typical pochard inciting, with strong neck stretching and a rather soft growling note, and this is usually followed by the male's attempting to swim ahead of the female and turning the back of the head toward her, often with the crown feathers strongly depressed. Copulation is initiated by the male, through alternated bill-dipping and dorsal preening displays that may also be performed by the female. She then assumes a prone posture, and after treading, a single courtship call is produced by the male before he swims away from her in a rigidly held bill-down posture (Johnsgard, 1965a; Weller, 1967c).

Reproductive biology. When the paired birds have become established on their breeding grounds, they establish home ranges that overlap with those of other breeding pairs, although little or no intolerance between them results. Frequently two areas are used by breeding birds, a waiting-site pothole and a nesting-site pothole, which may be from about 50 to nearly 700 yards apart. Evidently a substantial proportion of females do not attempt to nest; Weller (1959) estimated that as many as half the females on the breeding grounds may fall into this category. At least some of these birds may be responsible for the high incidence of eggs that are dropped in the nests of breeding redheads, canvasbacks, and other duck species. These "parasitically laid" eggs have a low hatching success, and probably increase desertion and reduce hatching success in host nests. Females that do construct nests normally do so in thick beds of emergent vegetation, usually hardstem bulrush, but at times do select sites on dry land, particularly late in the breeding season. The average clutch size is difficult to establish, owing to parasitic nesting, but probably averages about 7 or 8 eggs. At least in some areas, renesting is fairly frequent as a result of nest failures through destruction or desertion caused by intruding females. Most males probably desert their mates shortly after incubation is underway, but in rare instances will remain until the time of hatching

MAP 99. Breeding (hatched) distribution of the redhead, including areas of major concentrations (crosshatching). Wintering distribution indicated by stippling.

or even beyond. Nesting success is evidently quite low among redheads, and the hatching success of parasitically laid eggs is particularly low (Weller, 1959). Olson (1964) believed that about half of the redheads hatched in Manitoba study areas were actually reared by canvasbacks, the redhead's most common host species. There is some controversy about how effective redhead females are at rearing their own young; brood mergers tend to make brood counts unreliable estimates of duckling mortality, but many older broods often appear to be untended by adults. The fledging period ranges from 56 to 73 days, and by the time the ducklings are eight weeks of age they will have been deserted by the female. The female then undergoes a relatively long flightless period of five to six weeks, which results in a quite late resumption of flight capabilities and in part might contribute to the relatively high vulnerability of females to hunting mortality in the fall.

Status. According to Bellrose (1976), the breeding populations of redheads during the twenty-year period 1955–74 averaged about 650,000, with substantial year-to-year variations. Winter surveys during the same period averaged 590,000 birds, with 80 percent of these birds located along the Gulf Coast from Florida to the Yucatán Peninsula. The low average nesting success of female redheads, their high vulnerability to hunting, and the concentration of the breeding populations in areas subject both to marshland drainage and periodic botulism outbreaks makes the long-term outlook for this species extremely pessimistic.

Relationships. As indicated in the section on the canvasback, these two species are clearly closely related, and presumably both are derived from an ancestral type not greatly different from the modern-day Eurasian pochard through a process of a double invasion of the North American continent.

Suggested readings. Low, 1945; Weller, 1959; Lokemoen, 1966.

Ring-necked Duck

Aythya collaris (Donovan) 1809

Other vernacular names. Ring-billed duck, ringneck; Halsringente (German); morillon à collier (French); pato de collar (Spanish).

Subspecies and range. No subspecies recognized. Breeds in North America from the Mackenzie District through the forested parts of southern Canada, south locally to California, Colorado, Nebraska, Iowa, Pennsylvania, and New York, and from New England to Nova Scotia. Winters along the Pacific coast from British Columbia to Baja California, in most of Mexico and adjoining Central America, along the Atlantic coast from Massachusetts southward, and in the West Indies. See map 100.

Measurements and weights. Folded wing: males, 195–206 mm; females, 185–95 mm. Culmen: males, 45–50 mm; females, 43–46 mm. Weights: adult males shot in the fall average ca. 790 g, and females ca. 690 g, with respective maximum weights of 1,087 and 1,178 g. Eggs: av. 58 x 41 mm, creamy, 51 g.

Identification and field marks. Length 15–18″ (40–46 cm). *Males* in breeding plumage have a black head, with a greenish gloss and a slight crest, terminated by an inconspicuous chestnut collar at the base of the neck. The upperparts are brownish black with a greenish gloss; the breast is black; the sides and flanks are finely vermiculated with black and white except immediately behind the black breast, where a white bar extends in front of the wing to the back. The tail coverts are black, and the tail is slate brown. The tertials and upper wing coverts are mostly dark grayish brown, with the larger coverts and tertials glossed with green, while the primaries and their coverts are dusky brown. The secondaries are dark gray, the outer ones usually tipped with dusky coloration and white, while the underwing surface is white. The iris is yellow; the bill is bluish gray, with a black tip, a white ring at the base of the mandible, and a pale band behind the black tip. The legs and feet are grayish blue, with darker webs. *Females* have a medium brown head with a blackish crown and whitish cheeks, throat, and chin, as well as a definite whitish eye-ring and postocular stripe. The back of the neck, foreback, sides, and flanks are brown, the feathers with lighter margins, while the rest of the back is blackish brown. The rump and upper tail coverts are blackish, the tail is slate brown, and the under tail coverts are white to dusky brown. The rest of the underparts and the underwing surface are white, and the wings are colored as in the male, except that the greenish gloss is lacking. The iris is brown, the bill is generally similar to the male's but duller (and not whitish at base as often depicted), and the legs and feet are grayish, with darker webs. *Males in eclipse* resemble females but are darker and retain a glossier upper wing surface and a yellowish eye. *Ju-*

veniles resemble adult females but are darker above and more mottled below.

In the field, male ring-necked ducks are most likely to be confused with scaup, while females are difficult to distinguish from female redheads. The white extension in front of the wing and the nearly black back coloration provide the best criteria for recognizing male ring-necked ducks, while females can usually be distinguished from female redheads by their more obvious eye-ring and postocular stripe, as well as their more grayish and blackish two-toned head pattern, as opposed to the more uniformly brownish head color of female redheads. In flight, ring-necked ducks lack the pale gray to white secondary markings typical of all other pochards, and their generally dark upper wing coloration, as contrasted to their white underwing surface, provides a very useful guide to identification. Females are relatively quiet but utter soft *rrrr* notes during inciting, while males have a soft

MAP 100. Breeding (hatched) distribution of the ring-necked duck, including areas of major concentrations (cross-hatching). Wintering distribution indicated by stippling.

breathing note and a louder whistling call that are also uttered on the water during display.

NATURAL HISTORY

Habitat and foods. The preferred breeding habitats of ring-necked ducks are sedge-meadow marshes, swamps, and bogs with waters ranging from fresh to somewhat acidic in pH, and especially those with surrounding cover of sweet gale (*Myrica*) or leatherleaf (*Chamaedaphne*). Floating-leaf aquatic plants that often are associated with nesting ring-necked ducks include water lilies (*Nymphaea* and *Nuphar*) and water shield (*Brasenia*). The seeds of water lilies and water shield and the seeds and vegetative parts of pondweeds are all important foods, as are the seeds of such emergences as bur reed (*Sparganium*), spike rush (*Eleocharis*) and bulrush, and the tubers of bulrush. In winter the birds occupy a variety of acidic, fresh-water and brackish habitats, including marshes, shallow lakes, estuarine bays, and coastal lagoons, but are generally found on less brackish waters than other pochards. Fall and winter foods include many of the same kinds of plants as consumed during summer and a small proportion of animal materials, such as mollusks and insects (Mendall, 1958; Johnsgard, 1975).

Social behavior. Ring-necked ducks are usually found in fairly small flocks during fall migration, and often show considerable sex segregation. A few birds appear to be paired even by October; presumably these are birds that had been mated the previous spring. Extensive social contacts begin on the wintering grounds, where pair-forming behavior becomes frequent, although it probably does not reach a peak until spring migration during March and April. Sexes renew their pair bonds annually, and evidently all females become paired their first year, since it is thought that at least most females breed as yearlings. Courtship behavior in ring-necked ducks is less conspicuous than in canvasbacks or redheads, since the associated vocalizations are weaker and the posturing is less evident. Males have a rapid head-throw display and kinked-neck call as in these species, associated with a soft whistling note, and often perform neck stretching with the crown feathers distinctively raised to form a triangular profile. A very inconspicuous sneak posture, with the head moved forward while the crown feathers are lowered, is also performed, and males also lower their head feathers when swimming ahead of inciting females and turning the back of the head toward them. Female inciting is the major display of that sex, and apparently plays an important role in stimulating courtship display as well as in forging pair bonds with a specific male. Copulation is preceded by the usual pochard bill-dipping and dorsal-preening behavior, and is followed by the equally typical single male call and bill-down posture (Johnsgard, 1965a).

Reproductive biology. As pairs return to their nesting areas, they establish home ranges that often overlap with those of other pairs and usually are near the area where the female was hatched. Breeding densities in favorable habitats tend to be higher than those of redheads or canvasbacks, perhaps reflecting less specific nesting requirements and a high degree of tolerance of other resident pairs in the same area. Nest sites are usually chosen on small floating mats of vegetation, amid clumps of emergent rooted vegetation, or on actual islands, with sedges, sweet gale, and leatherleaf being preferred nesting covers. Nests are usually placed but a short distance from open water, and additionally, space for landings and takeoffs is generally within 100 feet of the nest (Townsend, 1966). Clutches are laid at the rate of one egg per day and average 8 to 9 eggs for initial nesting attempts; renest clutches average 7 to 8 eggs. Females are apparently persistent renesters, and at least one case of a second renesting effort has been reported. Incubation is performed by the female alone and requires 26 to 27 days, with the male usually abandoning his mate during the last two weeks of incubation. Female ring-necked ducks are apparently more effective mothers than redheads or canvasbacks, and relatively few abandon their broods before they attain flight at the age of 7 to 8 weeks, even though they themselves may enter their flightless stage before that time. The duration of the flightless period is probably three or four weeks, with females molting about a month later than the males, which typically gather in molting areas some distance from the nesting grounds (Mendall, 1958; Erskine, 1972a).

Status. According to Mendall (1958), ring-necked ducks became much more abundant and extended their range in the northeastern states between the 1930s and 1950s, for reasons still not evident. Breeding-ground inventories between 1955 and 1973 suggest an average population of about 460,000 birds, with major annual fluctuations (Bellrose, 1976). Likewise, winter inventories have produced large annual variations in counts but no discernible

trends during this period. Ring-necked ducks are obviously less susceptible to loss of breeding habitat through marsh drainage than are canvasbacks and redheads, and breed in areas of minimal agricultural significance and human disturbance, so there is at present no reason for concern over the long-term outlook for this uniquely North American species.

Relationships. No doubt as a result of their superficial similarity to the scaups, ring-necked ducks have very often been thought of as close relatives of that group. However, comparison of the plumages of downy young and of females points out their close affinities with the typical pochards, and their foraging tendencies as well as their display patterns also confirm this relationship (Johnsgard, 1965a).

Suggested readings. Mendall, 1958; Townsend, 1966; Coulter & Miller, 1968.

Australasian White-eye

Aythya australis (Eyton) 1822

Other vernacular names. Hardhead, white-eyed duck; Australische Moorente (German); milouin d'Australie (French); pato ojos blancos de Australia (Spanish).

Subspecies and ranges. (See map 101.)
  A. a. australis: Australian white-eye. Resident in Australia, occurring throughout, as well as in Tasmania, with populations of uncertain status (possibly periodic occurrences) in East Java, the Celebes, New Guinea, New Caledonia, and formerly also New Zealand.
  A. a. extima: Banks Island white-eye. Resident on Banks Island.

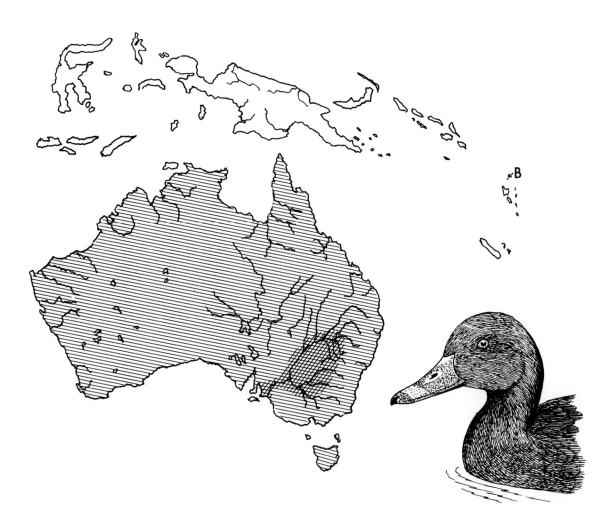

MAP 101. Breeding or residential range of the Australian white-eye, including major breeding areas (cross-hatching) and peripheral range (hatching).

Measurements and weights. Folded wing: males, 183–243 mm; females, 186–234 mm. Culmen: males, 35–50 mm; females, 39–49 mm. Weights (of *australis*): males, 525–1,100 g (av. 902 g); females, 530–1,060 g (av. 838 g). Eggs (of *australis*): av. 54 x 42 mm, pale cream, 60 g.

Identification and field marks. Length 18″ (46 cm). *Adult males* are generally dark brown throughout, with the head, neck, breast, and mantle nearly uniformly rich brown, while the scapulars and flanks are slightly edged or barred with buffy or whitish coloration. The lower breast and abdomen are white to brownish, and the under tail coverts are white. The tail and upper wing surface are brown, except for the secondaries and the inner primaries, which are white

with brown tips. The bill is dark gray, with a black nail and a pale grayish bar near the tip, the legs and feet are gray, and the iris nearly white. *Females* are similar to males, but are generally lighter and have a brown iris. The bill pattern is also more evident than that of the male. *Juveniles* resemble adult females, but are paler and have more mottled underparts.

In the field, the pochardlike body shape separates this from other Australian ducks, since no other member of this group occurs in Australia. In addition, the white eye of the male is distinctive, and both sexes have relatively conspicuous white under tail coverts. In flight, the white speculum, extending well out into the primaries, is highly conspicuous. Like that of other pochards, the voice of the female is relatively harsh and not very loud, while the male evidently produces only a soft whistling note and a whirring call, both uttered during display.

Natural History

Habitat and foods. According to Frith (1967), Australian white-eyes exhibit a preference for deep water areas having abundant emergent vegetation such as is present in cattail swamps and lignum-lined creeks. They sometimes also occur on coastal swamps and fresh-water lakes such as those at the mouth of the Murray River, and although they are sometimes found on the deep mountain lakes of the southern tablelands, they rarely if ever breed in these areas. Like all pochards, these birds forage primarily by diving, but they have been observed upending as well, in addition to dabbling in shallow water and stripping seeds from overhead plants. Extensive samples of food remains from birds collected in New South Wales indicate a strongly vegetarian diet, with the seeds and flowers of grasses and sedges alone accounting for more than half of the volume. In a similar number of samples from northern Queensland, grass seeds were nearly lacking but those of water lilies and smartweeds (*Polygonum*) were correspondingly more important. In both areas animal foods were relatively insignificant, and consisted mainly of aquatic insects and some surprisingly large mussels (Frith, 1967). In a sample of over 300 gizzards from a permanent swamp in New South Wales, pondweeds were found to be the most important single food source, and various mussels and snails were the primary animal foods represented (Frith et al., 1969).

Social behavior. In spite of this species' tendency to inhabit permanent swamp areas, it does at times spread out considerably, particularly in summer, when these swamps become drier and force the birds to move elsewhere. With extended periods of drought very extensive movements to deeper waters may occur, resulting in concentrations of 50,000 to 80,000 birds in extreme cases. On the other hand, during years of interior flooding the birds can also take advantage of this situation and begin breeding rapidly, shortly after the gray teal and black ducks initiate their nesting activities (Frith, 1967). Thus either pair bonding must be relatively permanent, or the birds must have the ability to establish pairs quite rapidly when breeding conditions become favorable. This latter is more probably the case. Pair-forming behavior has not been studied in the wild, and obser-

vations of captive birds indicate that the same array of displays is present as in other pochards. The inciting by females consists of direct threatening movements alternated with partial retreats, and is accompanied by a rather harsh and rattling call. Strangely, females also fairly regularly perform the same head-throw display as do males, and infrequently utter the kinked-neck call as well. In the males this latter display is very common and especially conspicuous because it is usually uttered in rapid succession several times, with a strong retraction of the neck each time. The head-throw is strongly developed, and is distinctly asymmetrical, with the bill being tilted in the direction of the courted female as it is brought forward from the back. In both of these displays a soft whirring sound is produced. Males also perform a threatlike sneak display, and at times nod-swim. Display preening is frequent and is especially conspicuous because of the white speculum pattern that is thus exhibited. During precopulatory display the male alternates bill-dipping and dorsal-preening displays with slight head-pumping movements, to which the female apparently makes no overt response. After treading, the usual kinked-neck call is uttered, followed by the male's swimming away in a rigid bill-down posture (Johnsgard, 1965a).

Reproductive biology. Breeding in this species varies somewhat with location in Australia. In the winter rainfall area of southwestern Australia, nesting occurs in the spring, between October and November. In the permanent swamps of inland New South Wales, most clutches probably occur between September and December. Progressively farther north the influence of the summer rains becomes more evident, and in Northern Australia nesting may occur as late as April and May. Finally, in the arid interior where flooding is unpredictable, breeding can occur at any favorable time. Nests are usually constructed in water that is several feet deep, and often are in cattail cover, in reeds, or even in lignum or in the butts of flooded trees. Observed clutches have ranged from 6 to 18 eggs, and generally vary between 9 and 12 (Frith, 1967). Incubation is performed by the female only, and has been established to require 25 days. The fledging period is still not known, but there is no current evidence that the male participates in any way in brood care in this species, so that multiple nesting is not likely to occur.

Status. This species is an important game duck in Australia, and in addition to hunting losses, it has suffered greatly from swamp drainage associated with agricultural development. Frith (1967) comments that it has declined from being the most abundant coastal species of waterfowl to one that is now rare on the coast and elsewhere is only locally common. Like many of the pochard species around the world, its population could rapidly be endangered by drainage and other habitat alterations of permanent swamps and marshes.

Relationships. This species is obviously a typical member of the white-eye group of pochards, which are all associated with Africa, Europe, and Asia, and have no direct ecological counterparts in the Americas. Probably the species' nearest relative is the Siberian white-eye, with which it shares a number of behavioral similarities (Johnsgard, 1965a).

Suggested readings. Frith, 1967.

Siberian White-eye

Aythya baeri (Radde) 1863

Other vernacular names. Baer's pochard, Baer's white-eye, Asiatic white-eyed pochard; Schwarzkopfmoorente (German); milouin de Baer (French); pato ojos blancos de Baer (Spanish).

Subspecies and range. No subspecies recognized. Breeds from Transbaikalia to the lower Ussuri River and the Amur, possibly also on Kamchatka. Winters in southeastern China, occasionally in upper Assam and Burma, and rarely in Japan. See map 102.

Measurements and weights. Folded wing: males, 210–33 mm; females, 186–203 mm. Culmen: males, 48–50 mm; females, 47–48 mm. Weights: males, ca. 880 g; females, ca. 680 g (Palmer, 1976). Eggs: av. 51 x 38 mm, cream, 43 g.

Identification and field marks. Length 18″ (46 cm). *Adult males* in breeding plumage have a glossy greenish black head except for a white chin-spot, a dark reddish brown breast that grades into brown flanks, and upperparts that are blackish brown to chestnut brown. The lower flanks, abdomen, and under tail coverts are white, while the tail and upper tail coverts and lower back are black. The upper wing surface is brown, except for a white speculum formed mainly by the secondaries, which are tipped with black. The inner webs of the primaries are pearl gray, producing an extension of the speculum. The iris is white, the legs and feet gray, with darker webs, and the bill is dark bluish, with a lighter tip and a black nail. *Females* are generally like the male, but have little or no iridescence on the head and have a brown iris, a duller brown breast color, and usually a light brownish spot between the eye and the bill. *Males in eclipse* resemble females, but retain a white or nearly white iris. *Juveniles* resemble adult females, but have russet brown on the abdomen.

In the field, this species closely resembles the other white-eyes, but is the only pochard with iridescent color on the head. They do not overlap with any other white-eyed pochard except on their wintering range; there they occur with the smaller and more chestnut ferruginous white-eye. Females utter a coarse *gaaaak* call, and the males also have a very similar, harsh *krraaaa* note, uttered during display, that is quite different from the male calls of the other white-eyes.

NATURAL HISTORY

Habitat and foods. In the breeding season, this species is reported to occupy small lowland lakes, preferably those containing aquatic vegetation and reeds. Treeless rather than forested habitats are preferred as well, and it has been said that on migration the birds may be found on rapidly flowing rivers, although this is certainly not typical pochard habitat.

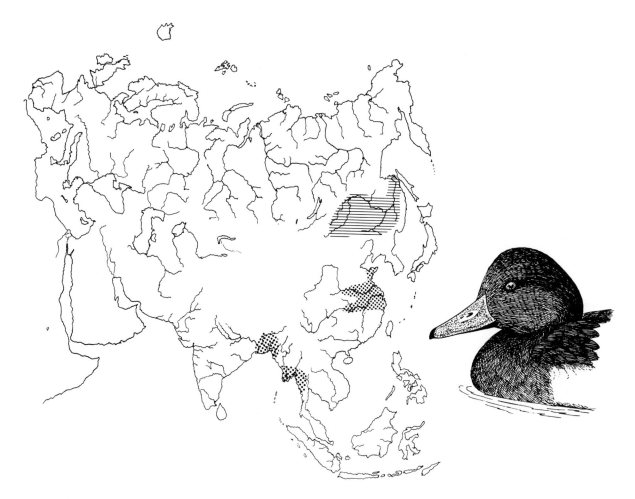

MAP 102. Breeding (hatched) and wintering (stippling) distributions of the Siberian white-eye.

Virtually nothing is known of the foods taken, but in India the birds are said to be fishy and almost inedible, presumably as a result of animal materials in their diet (Dementiev & Gladkov, 1967). Two birds shot in October had the remains of water plants in their stomachs.

Social behavior. Very little has been recorded on the pair-forming or pair-bonding behavior of this species. It is said that the birds arrive in their nesting areas of the Argun River region already paired, while in the Ussuri region and on the lower Iman they are still in small groups after spring arrival, and pair later. The only descriptions of display in this species are my own observations (1965a) on captive birds, which suggest different affinities from those implied in the general belief that this species is simply an eastern variant of the ferruginous white-eye. Females

have a harsh inciting call that is very much like those of the Australian and ferruginous white-eyes, and like these two species also performs a head-throw display almost identical to that of the males. Females also have a display preening behind the wing that is directed to preferred males. The male displays are surprisingly similar to those of the Australian white-eye. For example, the kinked-neck call is the most frequent of the male displays, and is performed repeatedly, with strongly exaggerated neck movements and a harsh *krraaaa* call. Also in common with the Australian species, the male often stretches his head and neck forward over the water in a sneak posture, followed immediately by the kinked-neck call. The most elaborate display, the head-throw, is accompanied by the same harsh call given during the kinked-neck display, and as the head is swung forward from its extreme position on the rump it is tilted in the direction of the courted female. Display preening of the white speculum is a common display by males, and is often performed mutually with females. Males also typically respond to female inciting by swimming ahead of them and turning the back of the head. Copulatory behavior is of the typical pochard type; but as in the Australian white-eye, it is preceded by slight head-pumping movements by males, in addition to mutual bill-dipping and dorsal-preening displays by both birds. After treading, the usual call and bill-down displays are performed by the male (Johnsgard, 1965a).

Reproductive biology. Very few observations are available on the nesting of this species. The birds are said to nest along the shorelines of lakes or on stream banks, and one nest with 10 eggs has been described from the lower Iman Valley, the nest being situated in dense sedge cover about seven meters from water. In captivity, the clutch size ranges from 6 to 9 eggs, and the latter figure is probably close to the normal clutch size. Johnstone (1965) reported that the incubation period was 27 days in the first case of captive breeding of this species, and that the downy young are about midway in appearance between those of the ferruginous and Australian white-eyed pochards.

Status. Apparently this species is common in the southern Maritime Territory of the U.S.S.R. and in the eastern half of Manchuria, but is relatively rare throughout the rest of its range. In winter it is most common in the central part of China's coastal areas south to the mouth of the Yangtze River (Dementiev & Gladkov, 1967). It also winters uncommonly and erratically in Manipur, Assam, western Bengal, and Bangladesh, but its similarity to the ferruginous white-eye makes its status in this area difficult to determine (Ali & Ripley, 1968). The species is probably not so rare as the little information available on it would suggest.

Relationships. The display repertoire and the pattern of the downy young both suggest somewhat different affinities from those implied by the generally held position that this species is a very close relative, possibly only a subspecies, of the ferruginous white-eye. Instead, its nearest affinities are probably with the Australian white-eye, or at least it seems to provide a real phyletic link between the ferruginous and Australian white-eyes (Johnsgard, 1965a).

Suggested readings. Dementiev & Gladkov, 1967.

Ferruginous White-eye

Aythya nyroca (Güildenstädt) 1770

Other vernacular names. Common white-eye, white-eyed pochard; Moorente (German); milouin nyroca (French); pato ojos blancos comun (Spanish).

Subspecies and range. No subspecies recognized. Breeds in southern Europe (mainly East Germany, Hungary, Czechoslovakia, and Poland) east to western Siberia (Ob Valley), south to northern Africa (Morocco and Algeria), Iran, Turkestan, Kashmir, the Pamirs, and southern Tibet. Winters in the Mediterranean region, the Nile Valley, Persian Gulf, and Burma. See map 103.

Measurements and weights. Folded wing: males, 178–93 mm; females, 172–85 mm. Culmen: males, 40–43 mm; females, 36–40 mm. Weights: males in winter, 500–650 g (av. 583 g); females in winter, 410–600 g (av. 520 g) (Bauer & Glutz, 1969). Eggs: av. 50 x 37 mm, deep cream, 43 g.

Identification and field marks. Length 16″ (41 cm). *Adult males* in breeding plumage have a rich reddish chestnut head and neck, with a small white chin-patch, which is separated by a blackish collar from a breast of the same chestnut color. The breast is separated rather sharply from a white abdomen, duller brown flanks, and a uniformly greenish black mantle. The under tail coverts are white, and bounded anteriorly by blackish coloration; the upper tail coverts and tail are also black, with a greenish gloss. The upper wing surface is mostly dark brown, except for a white speculum formed by the bases of the secondaries and most of the primaries, which are tipped with black. The iris is white, the bill is grayish black, paler toward the tip and with a black nail, and the legs and feet are lead-colored. *Males in eclipse* are

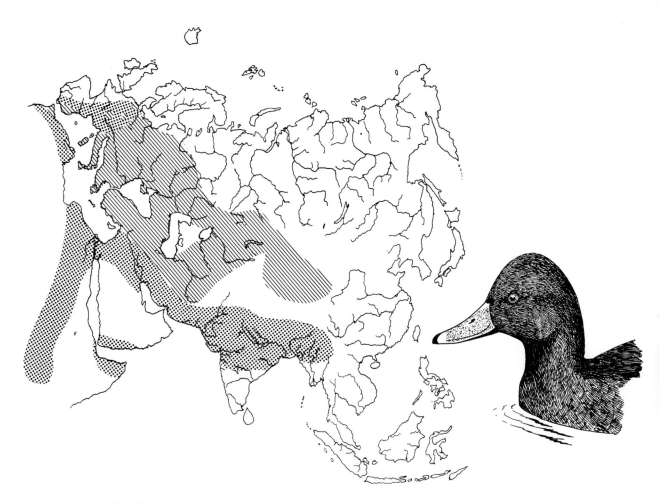

Map 103. Breeding (hatched) and wintering (stippling) distributions of the ferruginous white-eye.

a duller reddish brown color on the head and neck, and the lower neck feathers are edged with sandy or white coloration. *Females* are generally similar to males, but lack the chestnut tones of the head, neck, and breast, which are instead dull brown, like the flanks and mantle. The iris is brown, but the bill and leg coloration is like the male's. *Juveniles* resemble adult females, but have silvery brown underparts and blackish barring on the breast, and lack the white under tail coverts of adults.

In the field, these white-eyes might be confused with female tufted ducks or perhaps female scaup, but the uniformly colored face and the conspicuous white under tail coverts of white-eyes should separate the species readily. In flight they closely resemble both of these species, as the wing patterning is nearly identical, and the dark brown head of the male may appear blackish in poor light. The female is known to utter a harsh *gak* call during display and

also an inciting *errr* note, while the male produces soft *wheeeoooo* and *wee'-whew* calls while displaying.

NATURAL HISTORY

Habitat and foods. The breeding habitats of this pochard species are largely shallow pools of either fresh or brackish water, or rivers that have dense shoreline vegetation and an abundance of submerged aquatic plants. It particularly seems to prefer areas with extensive reed beds, and needs a smaller area of open water than does the Eurasian pochard, with which it overlaps in range appreciably (Voous, 1960). It also seems to prefer shallower waters than does that species, frequently foraging in water from one to four meters in depth, but often remaining under water for nearly a minute. Like Eurasian pochards, the birds are predominantly vegetarians, and feed on the leafy parts and roots of a variety of submerged plants, including pondweeds, duckweeds, and musk grass. Additionally, they consume seeds of various grasses (*Setaria, Echinochloa, Hordeum*) and herbs such as smartweeds (*Polygonum*) and pondweeds. Aquatic insects and insect larvae constitute a small part of their diet (Dementiev & Gladkov, 1967; Bauer & Glutz, 1969).

Social behavior. It is somewhat uncertain whether this species normally breeds in its first year, although Ogilvie (1975) reports that this is the case. Courtship displays may be seen among captive birds over a rather prolonged period in the winter, sometimes as early as January. Females take an active part in the display, performing an inciting that alternates between overt threats or attacks and swimming back toward the preferred males, neck stretching and repeatedly uttering the *gak-gak-gak* call. Another surprising component of female behavior is their rather frequent performance of a head-throw display and a kinked-neck call that are essentially identical to those same displays of males, except for their associated vocalizations. Displaying males often swim about in a distinctive jerky manner called nod-swimming, usually depressing their tail and partially erecting their crown feathers at the same time. They also often utter the kinked-neck call repeatedly, a soft *wheeoooo* note that is the same or very similar to a call given in a head-forward or sneaking posture. From this posture the bird quickly lifts his head and performs a neck-stretching display, so that the two phases seem to constitute a single display. Males also perform a head-throw display, uttering the same vocalization as made during the kinked-neck call, and another high-pitched and double-noted call is uttered during coughing, as the bird inconspicuously flicks its folded wings. Preening behind the wing and turning the back of the head toward inciting females have also been observed in this species. Precopulatory behavior apparently consists of the male's making bill-dipping and dorsal-preening movements, and the postcopulatory behavior is of the usual pochard type (Johnsgard, 1965a).

Reproductive biology. Apparently female ferruginous white-eyes have a strong tendency to place their nests very close to water; in the U.S.S.R. they typically nest in reed beds, on floating mats of vegetation, on islets or hummocks, and along the bank in emergent shoreline reeds. The usual clutch size is from 7 to 11 eggs, and in rare instances as

many as 14 may be present. Clutches of from 16 to 20 eggs have also been found, evidently the result of two females' efforts, and mixed clutches of this species and the Eurasian pochard have also been reported. Presumably the male deserts at about the time incubation gets underway, or at least before the end of the incubation period, which requires from 25 to 27 days. Another 55 to 60 days are needed for the young to attain flight, but the timing of the molts of adult birds with respect to the development of the young is still essentially unestablished (Dementiev & Gladkov, 1967; Bauer & Glutz, 1969).

Status. Population estimates for this southerly and easterly distributed species are quite limited, but it has been estimated that about 140,000 pairs may breed in the western parts of the U.S.S.R. and a very few additional birds in East and West Germany, Czechoslovakia, Hungary, France, Italy, and Spain. A few hundred birds winter in northwestern Europe, but about 75,000 are to be found in the Mediterranean and Black sea area (Ogilvie, 1975).

Relationships. Behavioral evidence suggests that this species might be somewhat apart from the other more typical white-eyes and have some affinities with the scauplike ducks. Information on the Madagascan white-eye is still extremely scant, but it presumably is the nearest relative of this species.

Suggested readings. Dementiev & Gladkov, 1967; Bauer & Glutz, 1969; Owen, 1977.

Madagascan White-eye

Aythya innotata (Salvadori) 1894

Other vernacular names. Madagascan pochard; Madagaskar-Moorente (German); milouin de Madagascar (French); pato ojos blancos de Madagascar (Spanish).

Subspecies and range. No subspecies recognized. Limited to Madagascar (Malagsi), primarily or entirely on the northern and eastern parts of the island. See map 105.

Measurements and weights. Folded wing: males, 190–201 mm; females, 188–95 mm. Culmen: males, 46–49 mm; females, 44–46 mm. Weights: no record. Eggs: 55 x 40 mm, buffy gray.

with extensive marsh areas on two sides, and with large beds of papyrus and reeds in the middle. Delacour (1954–64) says that in their behavior the birds appeared to be typical pochards, but no specific information on foraging behavior or foods is available.

Social behavior. Delacour (1954–64) was reminded of redheads by this species' display, which he observed in June of 1929. No further accounts of this species' behavior are available, but it would be of interest to have a comparison of its displays with those of the ferruginous white-eye.

Reproductive biology. Delacour (1954–64) judged, from the age of immature specimens he obtained, that the breeding season is probably from October to January. Nests have not been found in the wild, but a 26- to 28-day incubation period has been established for captive birds. Unfortunately, none of the captive stock that was imported into Europe survived World War II, and no more birds have been imported since.

Status. No recent information is available on the status of this species.

Relationships. As indicated in the account of the ferruginous white-eye, this species is presumably an insular derivative of that more widely ranging form, but no detailed studies of anatomy or behavior are available to provide evidence on this possibility.

Suggested readings. Delacour, 1954–64.

Tufted Duck

Aythya fuligula (Linnaeus) 1758

Other vernacular names. None in general English use; Reiherente (German); canard morillon (French); pato de copete (Spanish).

Subspecies and range. No subspecies recognized. Breeds in Iceland, the British Isles, and most of Europe and Asia north to 70° latitude and south to central Europe, the Balkan Peninsula, Kirghiz steppes, Lake Baikal, the Amur River, Sakhalin, and the Commander Islands. Winters from its breeding range south to northern Africa, the Nile Valley, the Persian Gulf, India, southern China, and the Philippines. See map 104.

Measurements and weights. Folded wing: males, 198–208 mm; females, 189–202 mm. Culmen:

Identification and field marks. Length 18″ (46 cm). *Adults* closely resemble adults of the ferruginous white-eye but are generally darker in color. The male's head and chest are a darker chestnut than in that species, and there is no white chin-patch, nor is there a distinct collar around the neck. Further, the scapulars are not freckled with brown, nor is the white abdomen so sharply defined. The secondaries have distinct black edges on the outer margins of the white speculum patch. The iris is white or nearly white, while the bill, legs, and feet are blackish. *Females* closely resemble female ferruginous white-eyes, and like that species have a brown iris and lack chestnut coloration on the head and breast. *Juveniles* are much like those of common white-eyes, but are darker on the scapulars and mantle.

In the field, this is the only species of pochard likely to be encountered in Madagascar, and thus its body shape alone should separate it from other ducks there.

NATURAL HISTORY

Habitat and foods. This species is confined to the plateau region of eastern Madagascar, at elevations of about 3,000 to 4,000 feet, from Lake Alaotra south to Antsirabe. Lake Alaotra, where the birds are most common, is about 25 miles long and 7 miles wide, males, 38–42 mm; females, 38–41 mm. Weights: males (in February), 1,000–1,400 g (av. 1,116 g); females, 1,000–1,150 g (av. 1,050 g) (Dementiev & Gladkov, 1967). Eggs: av. 59 x 41 mm, greenish gray, 56 g.

Identification and field marks. Length 17–18″ (43–46 cm). *Males* in breeding plumage have a black head and neck, with a long, narrow crest that nearly touches the back. The upperparts are blackish, the scapulars having a greenish cast and faint vermiculations. The breast, tail coverts, and tail are all black, while the abdomen and flanks are white. The upper wing coverts are dark brown, and the secondaries are white with black tips. The primaries are dark brown, with the inner ones having gray or white on the inner webs. The iris is yellow, the bill is pale blue with a black tip, and the legs and feet are lead blue, with darker webs. *Males in eclipse* are much like females, but are more grayish throughout, with some vermiculations showing on the flanks. *Females* are similar to scaup females, but are darker dorsally, have a small occipital crest, and show little or no white at the base of the bill. Some females also exhibit a white area on the under tail coverts, but most are dull brown in this region. The soft-part colors are similar to those of the male, but the bill is more grayish and the iris coloration is less bright. *Juveniles* are similar to adult females, but young males are somewhat vermiculated dorsally and have darker heads.

In the field, tufted ducks are most likely to be confused with greater scaups, but the crest of the male will normally allow for separation, and even the female exhibits a slight crest. Females also have less white in front of the eyes than do scaup females, but this trait is not useful in summer, when female scaup acquire a brownish face pattern. In flight, tufted ducks are extremely difficult to separate from scaup, and the head characteristics just mentioned provide

the best clues. The calls of the male tufted duck consist of a mellow *whee'oo* and a rather windy *wha'wa-whew*, which are very similar to the corresponding calls of the greater scaup. A low growling call and a *gack* or *quack* note are produced by females.

NATURAL HISTORY

Habitat and foods. The breeding-habitat needs of tufted ducks and greater scaup appear to be quite similar; both species are generally associated with larger and deeper bodies of water that have an abundance of invertebrate foods. Tufted ducks also breed on artificial reservoirs, and even on ponds in cities, especially where islands for nesting are available. They also breed along slowly flowing rivers with abundant foods, and throughout the year they concentrate on mollusks, crustaceans, and insect larvae that they probe for among the mud and silt of the bottom debris. They probably usually dive only to depths of about two to four meters, and probably feed on whatever is present in greatest quantities. Olney (1963a) studied the foods taken from tufted ducks in various areas of England and Northern Ireland and found considerable dietary differences, although mollusks or crustaceans predominated in all the samples, with insect remains and plant seeds occurring in minor quantities. In the U.S.S.R. mollusks are also the predominant food in most areas and most times of the year, with crustaceans occurring most frequently among birds using inland reservoirs. Both inland and coastal waters are used by wintering birds, but the birds probably rarely are found foraging where the water is more than six meters in depth (Dementiev & Gladkov, 1967; Bauer & Glutz, 1969).

Social behavior. Probably many, but not all, tufted ducks breed at the end of the first year of their life, and females in particular may not nest until their second year. The pair-bond system is one of seasonal monogamy, with pairing beginning in January and reaching a peak in April, when over 90 percent of the females are likely to have obtained mates. This pattern of pairing is later than that typical of the *Anas* species wintering in Bavaria, but earlier than that of the Eurasian pochard (Bezzel, 1959). As might be expected, the social displays of tufted ducks are very much like those of scaups. Females incite with a scauplike neck-stretching movement while uttering soft *karrrr* notes, and additionally perform display

MAP 104. Breeding (hatched) and wintering (stippling) distributions of the tufted duck.

preening behind the wing as well as a head-throw display comparable to that of the male. Male displays include a three-noted coughing call in a neck-stretching posture, performed with a conspicuous wing- and tail-flicking movement. The head-throw is performed quite rapidly, and is accompanied by a mellow whistling *whee'oo* call, the same sound associated with the inconspicuous kinked-neck posture and call. Males quite frequently preen behind the wing toward females, and in addition they quickly respond to inciting by turning the back of the head toward inciting birds. In common with the New Zealand scaup, but not the greater or lesser scaups, males also perform a nod-swimming display. Copulation is typically initiated by the male, through bill-dipping, dorsal preening, and preening behind the wing. Females may respond in the same way, but often assume the receptive pose quite suddenly. The postcopulatory behavior is of the usual pochard type, with the male calling once with the kinked-neck call, then swimming away in a bill-down posture (Johnsgard, 1965a).

Breeding biology. Tufted duck females exhibit a moderately strong tendency to return to the same area in which they nested the year before, and in Loch Leven, Scotland, they often nest within 50 meters of the previous nest location (Newton &

Campbell, 1976). They are relatively late nesters, and in several areas including Loch Leven have been found to favor nesting among gull or tern colonies, which seems to improve their hatching success. Additionally they are prone to nest on islands when they are available, and at least in Iceland show a strong preference for selecting low shrubs, broad-leafed herbs (*Angelica* and *Archangelica*) and to some extent sedges for nesting cover (Bengtson, 1970), while in Loch Leven most nests are in tall perennial grasses (*Deschampsia* and *Phalaris*). Females lay relatively large clutches, generally averaging 10 eggs in Iceland and Finland, but having been found in Scotland to decline from about 14 eggs at the onset of the laying season to about 6 at the end of it, with no indication of attempted renesting. The incubation period is normally 24 days, probably varying from 23 to 25 under natural conditions. The higher nesting success found in nests among gull colonies in Scotland has been attributed to the birds keeping jackdaws away from the nesting area. However, when unusual crowding of nests in such areas occurs, the incidence of nest losses from desertion associated with multiple use of the same nests is increased (Newton & Campbell, 1976). Tufted duck broods are raised by females on habitats rather intermediate between the relatively open waters favored by most pochard species and the vegetation-lined potholes used by most dabbling ducks, and ducklings of this species are more prone to hide in vegetation when disturbed than are scaups and other pochard ducklings (Bengtson, 1971b). The fledging period is probably 45 to 50 days, and females apparently remain with their broods for most of this period before becoming flightless.

Status. In general, the tufted duck seems to be increasing through much of its European breeding range, and Ogilvie (1975) estimated the wintering population of northwestern Europe to be 525,000 birds, concentrated especially in Denmark, while southern Europe supports about 300,000 wintering birds. In terms of breeding, Finland alone is believed to support about 40,000 pairs, and the birds also occur in the thousands in Iceland, the Baltic states, and Britain. The size of the populations breeding and wintering farther east cannot be estimated at present.

Relationships. As I have pointed out earlier (1965a), this species might well be called the "tufted scaup," as it is clearly a member of that group, and in particular seems to have strong behavioral similarities to the New Zealand scaup. It overlaps appreciably in its ecologic adaptations with the greater scaup and presumably competes with it locally or seasonally, but like the lesser scaup of North America is less prone to use marine habitats in the nonbreeding season.

Suggested readings. Hilden, 1964; Bengtson, 1970; Owen, 1977.

New Zealand Scaup

Aythya novae-seelandiae (Gmelin) 1789

Other vernacular names. Black teal, black scaup; Neuseeland-Tauchente (German); morillon noir (French); pato de Nueva Zelandia (Spanish).

Subspecies and range. No subspecies recognized. Resident in New Zealand (both islands), and on the Auckland and Chatham islands. See map 105.

Measurements and weights. Folded wing: males, 175–87 mm; females, 170–81 mm. Culmen: males, 38–41 mm; females, 35–38 mm. Weights: males, 630–760 g (av. 695 g); females, 545–690 g (av. 610 g). Eggs: 64 x 41 mm, cream, 63 g.

Identification and field marks. Length: 16–18" (40–46 cm). *Adult males* are almost uniformly dark, with a black to iridescent green head, breast, and back, the

back feathers obscurely vermiculated with brownish coloration. The abdomen is whitish, grading to brown on the flanks, and the tail and tail coverts are black. The upper wing surface is generally dark brown to greenish, except for the secondaries, which form a white speculum with a black trailing edge. The bill is bluish, with a black nail, the iris is yellow, and the legs and feet are gray and blackish. *Females* are generally dark brown in the areas where the male is black, except for a seasonally developed small white patch behind the base of the bill. The soft-part colors are like those of the male, except for a brown iris and somewhat darker bill. *Juveniles* are similar to females, but lack the white facial patch and have nearly white abdomens.

In the field, the diving-duck profile and extremely dark body coloration should identify this species. In flight, the white wing speculum and a grayish white underwing coloration contrast with the otherwise dark body coloration. Courting males utter a clear, three- or four-noted whistle, with the last note more prolonged, while females have a fairly high-pitched growling call.

NATURAL HISTORY

Habitat and foods. The preferred habitat of this diving duck is on inland lakes and lagoons near the sea, particularly those with clean waters and that are not extremely deep. They also occur on mountain lakes up to about 3,000 feet elevation, and on hydroelectric reservoirs of the North Island, often colonizing only shortly after they are flooded (Falla et al., 1967). They often forage in water five or six feet deep, and when diving often remain submerged for 20 seconds or more. At such times they keep their wings closed and use their legs and feet as rudders and for locomotion (Oliver, 1955). Like the other scaups, they probably feed predominantly on invertebrate life, but no studies of their food intake have yet been made.

Social behavior. These birds are fairly gregarious, and even during the breeding season some flocks are apparent, suggesting that breeding may be delayed until the second year of life. Social displays have been observed only among captive birds, but seem to be very much like those of the other scaups. Females produce inciting displays with strong neck stretching and associated *errrr* calls, to which the males respond with a variety of postures and calls. The most frequent male display is coughing, in which the multi-

noted whistle described earlier is uttered. They also utter a second *whe-whe* call while kinking their extended neck, and rather infrequently perform a head-throw display. A sneaking display, with the head and neck extended toward a female or another male, and accompanied by a soft note, is yet another of the male displays. Males often turn the back of the head toward inciting females, but although females have been observed display-preening, this has not yet been reported among males. Copulation is preceded by alternated bill dipping and dorsal preening by the male, to which the female may or may not respond in the same fashion. After treading, the male calls once and then swims away from the female in a characteristic bill-down posture (Johnsgard, 1965a).

Reproductive biology. The nesting season of this species is said to extend from October through March, and the nests are often situated at the very edge of a lake or other water, in dense cover. Sometimes the nest is placed under an overhanging bank, and at other times it is nearly buried among flax roots. The nest is usually extensively lined with down, and the clutch size varies from 5 to 8 eggs. Normally the eggs are laid on alternate days, and incubation requires 27 to 30 days (Reid & Roderick, 1973). Nests are frequently placed close together, in a colonial fashion, and shortly after hatching the ducklings begin to dive for their food, often to considerable depths (Oliver, 1955; Falla et al., 1967). Within 75 days the young are fully feathered and presumably fledged (Reid & Roderick, 1973).

Status. Since 1934 the New Zealand scaup has been protected in New Zealand, and since that time it has perhaps begun to increase and started to reoccupy its drastically reduced original range. It is probable that the establishment of new reservoirs has also helped its status (Williams, 1964). Like some other pochards, it is extremely sensitive to hunting and, if it is to survive, will probably continue to require protection.

Relationships. As suggested earlier (Johnsgard, 1965a), this species is closely related to the tufted duck and to the other two species of scaups. Interestingly, it is the only Southern Hemisphere species of scaup, and it seems strange that neither southern Africa nor southern South America has a counterpart present.

Suggested readings. Oliver, 1955.

Greater Scaup

Aythya marila (Linnaeus) 1761

Other vernacular names. Bluebill, broadbill; Bergente (German); milouinan (French); costero grande (Spanish).

Subspecies and ranges. (See map 105.)

A. m. marila: European greater scaup. Breeds in Iceland, Scandinavia, and northern Eurasia across much of Siberia, mainly north of latitude 60°, with the eastern limits still uncertain. Winters on the coast of western Europe, the eastern Mediterranean, the Black Sea, the Persian Gulf, and in northwestern India.

A. m. mariloides: Pacific greater scaup. Breeds on the mainland of eastern Asia perhaps as far west as the Lena River, on the islands of the Bering Sea, and in North America from the Bering coast eastward to Hudson Bay and Ungava Bay, and on Newfoundland. Winters along the coast of China, Korea, and Japan, and on the coast of North America from Alaska to California as well as from the Gulf of St. Lawrence to South Carolina.

Measurements and weights. Folded wing: males, 215–33 mm; females, 210–20 mm. Culmen: males, 43–47 mm; females, 41–46 mm. Weights: mixed-age males of *mariloides* shot in the fall average about 1,000 g; adult males of *marila* in winter average 1,250 g; while respective female weights are about 900 and 1,200 g. Eggs: av. 62 x 40 mm, brown to olive, 67 g.

Identification and field marks. Length 16–20″ (40–51 cm). Plate 49. *Adult males* in breeding plumage have a head and neck that are black with a greenish gloss, the breast and extreme foreback are also black, while the rest of the back and scapulars are finely vermiculated with black and white. The rump and tail coverts are brownish black, the tail is blackish brown, while the abdomen, sides, and flanks are white, with slight vermiculations often present on the flanks. The upper wing coverts are blackish brown, with some white speckling; the secondaries are white, with a gray band near the tip and narrowly tipped with white; the tertials are dusky, with a greenish gloss; while the primaries are dark grayish brown, with the inner six or seven having a whitish area on the inner webs. The bill is pale blue with a relatively wide (8 mm or more) black nail, the iris is orange yellow, and the legs and feet are grayish blue to greenish. *Females* have a brownish head with a darker crown and a white facial patch that extends around the base of the bill and across the forehead; a pale whitish area is present in the region of the ear during summer. The neck, back, scapulars, tail, and upper tail coverts are brown, the rump is blackish brown. The breast, sides, and flanks are grayish brown, the feathers tipped with whitish, while the underparts are mostly white except for a brownish area near the vent. The wing coverts and tertials are dark brown, the tertials often faintly glossed with greenish coloration, the primaries are grayish brown, and the secondaries are white, with a brown posterior border. At least some of the inner primaries are distinctly whitish on the inner webs. The bill is grayish blue with a darker nail, the iris is dark yellow to brownish, and the legs and feet are lead-colored, with darker webs. *Males in eclipse* resemble females but have a darker head with little or no white on the forehead and cheeks, and a more blackish breast. *Juveniles* resemble adult females but have less white in the face and are generally duller.

In the field, greater scaups are most likely to be confused with lesser scaups. Besides inhabiting more marine environments, greater scaups exhibit a lower head profile produced by a straighter culmen profile and an uncrested head, which in males is green-glossed and in females has more white on the face than occurs in lesser scaups. In flight, both sexes exhibit a greater amount of white on the primaries than do lesser scaups. Female greater scaups have a slightly louder growling *arrr* call, and males during

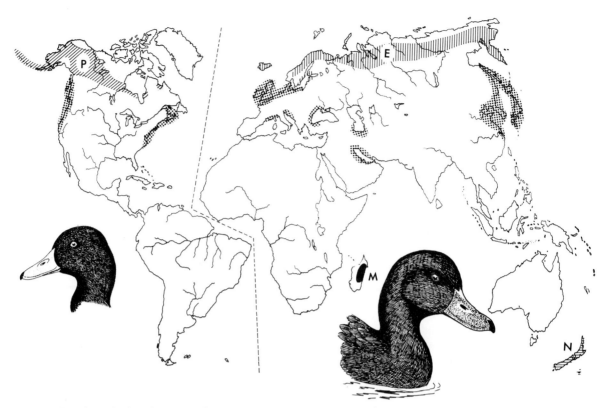

MAP 105. Residential distributions of the New Zealand scaup ("N") and Madagascan white-eye ("M") and breeding distributions of Pacific ("P") and Eurasian ("E") greater scaups. The wintering distributions of the two latter races are indicated by stippling.

display utter a very soft, cooing *wa'hooo* note and a triple-noted *week* whistle.

NATURAL HISTORY

Habitat and foods. The greater scaup's preferred breeding habitat consists of tundra or low forest adjacent to tundra, where shallow water areas occur that have grassy shorelines and a high density of invertebrate life such as amphipods and aquatic insect larvae. Summer foods have not been specifically investigated, but probably are much like those of the lesser scaup. During the fall and winter the birds are found on appreciably larger and deeper fresh-water lakes and on more brackish and salt-water habitats than are lesser scaup, and during that time they concentrate largely on bivalve and univalve mollusks such as *Mytilus, Mulinia,* and *Littorina,* and to a smaller degree on crustaceans. Their favored winter habitats in coastal areas are brackish and salt-water estuarine bays and coastal bays with foraging areas available where they can feed by diving to about 25

feet, but preferably where the water is only about five feet deep (Cronan, 1957; Stewart, 1962).

Social behavior. To a degree greater than most other pochards, scaups seem to exhibit a deferred maturity, with the result that most of the females do not nest until they are two-year-olds, although pair bonds are seemingly formed by yearlings and at least some hand-raised birds breed their first year in captivity. Scaup are highly gregarious on their migratory and wintering areas, and social display probably begins in winter, judging from observations of captive individuals. The fact that the birds often winter well away from shore and that their calls and postures are not highly conspicuous makes information on this phase of the life cycle difficult to obtain. However, their display patterns are very much like those of the lesser scaup, differing mainly in vocalizations and quantitively in posturing. There is a soft, whistled coughing call, an even weaker *wa-hoo* note associated with a rapid head-throw or a kinked-neck posture, and several silent displays such as turning the back of the head toward inciting females and display preening of the wing, exposing the white speculum pattern. Copulation is preceded by the male's performing bill-dipping, dorsal-preening, and wing-preening movements, to which the female may respond with the same displays. Treading is followed by the male's uttering a single kinked-neck call and then swimming away from the female in a stereotyped bill-down posture (Johnsgard, 1965a).

Reproductive biology. Greater scaup are fairly late nesters, and usually arrive on their breeding ground the latter half of May. In favored areas the breeding pairs tend to establish overlapping home ranges that are not defended in any manner approaching typical territoriality. Bengtson (1970) reported on such nesting aggregations in Iceland, and Weller et al. (1969) found a similar condition on Great Slave Lake. In both areas, island nesting seems to be prevalent. In the latter region, the birds often placed their nests in the grassy cover of the past year's growth, often in cracks in rocks or under woody vegetation. In Iceland, perennial herbaceous vegetation or shrubs under half a meter in height provided cover for most of the nests; there is also a strong tendency for scaup to place their nests among colonies of nesting terns or gulls (Hilden, 1964). Normal clutch sizes seem to be rather variable, ranging from about 8 to 10, but the clumped nesting tendencies of scaup often result in females laying eggs in the nests of other scaup as well as those of different species nesting in the vicinity. The incubation period is probably 24 to 25 days, with some estimates of up to 28 days. Males usually abandon their mates fairly early in the nesting period and gather offshore in large flocks, or may migrate some distance for molting. Females also tend to move their young broods to fairly open water, exposing them to chills and higher predation rates, but there is little tendency for brood mergers as in eiders. The fledging period has not been established, but is likely to be similar to the 45- or 50-day period established for lesser scaups in Alaska.

Status. In North America, the greater scaup is easily confused with the lesser scaup during aerial surveys, and thus its actual numbers are hard to estimate. Along the coastal portions of Alaska, the breeding population of greater scaup was estimated to average about half a million birds between 1957 and 1973, and Bellrose (1976) further estimated that the total Alaska population may be about 550,000 birds, with an additional 200,000 nesting in Canada. Ogilvie (1975) estimated that about 150,000 winter in northwestern Europe and another 50,000 in the Mediter-

ranean and Black sea area. No estimates for the segments wintering in central and eastern Asia are available.

Relationships. This species has the largest and most coastal distribution of all the scaups. It is clearly a very close relative of the lesser scaup, and is slightly less closely affiliated with the New Zealand scaup. It is rather surprising that no other Southern Hemisphere scauplike species exist or that the greater scaup has not colonized the temperate parts of South America.

Suggested readings. Weller et al., 1969; Hilden, 1964; Bengtson, 1970.

Lesser Scaup

Aythya affinis (Eyton) 1838

Other vernacular names. Bluebill, broadbill, little bluebill; Veilchenente (German); petit milouinan (French); costero chico (Spanish).

Subspecies and range. No subspecies recognized. Breeds in North America from central Alaska eastward to western Hudson Bay and southward locally to Idaho, Colorado, Nebraska, and the Dakotas. Winters from British Columbia south along the Pacific coast to Mexico, Central America, and Colombia, on the Atlantic coast from the mid-Atlantic states south to Colombia, and in the West Indies. See map 106.

Measurements and weights. Folded wing: males, 190–201 mm; females, 185–98 mm. Culmen: males, 38–42 mm; females, 36–40 mm. Weights: adult males shot in the fall average ca. 850 g, and females ca. 800 g, with respective maxima of 1,087 g and 951 g. Eggs: av. 56 x 40 mm, stone to olive, 44 g.

Identification and field marks. Length 15–19" (38–48 cm). *Adult males* in breeding plumage closely resemble males of the greater scaup, but have a purplish-glossed and slightly crested head, more extensive vermiculations on the sides and flanks, coarser vermiculations on the back and scapulars, and no white present on the inner vanes of the primaries, although the innermost ones may be quite pale. The soft-part colors are likewise the same, but the bill has a narrower (under 7 mm wide) black nail, which is more concave in profile and slightly smaller throughout. *Females* are very much like female greater scaup, but tend to have a smaller white facial patch, and their primaries lack white in their inner webs. *Males in eclipse* resemble females but retain darker heads and have little or no white in the cheek area. *Juveniles* resemble adult females but are generally duller and have less white in the face.

In the field, lesser scaups are difficult to separate from greater scaups, and the criteria mentioned in that species' account should be used. The display vocalizations of the two species differ slightly, with the male lesser scaup uttering a relatively faint *whee-ooo* note and a single-noted *whew* whistle. The tendency of the lesser scaup to inhabit fresh-water

areas and marshes rather than deeper lakes and tide-water areas also is a useful aid to identification.

NATURAL HISTORY

Habitat and foods. Lesser scaup typically breed in the vicinity of interior lakes and ponds with associated low islands and moist sedge meadows and surrounding environments of prairies or partially wooded

MAP 106. Breeding (hatched) distribution of the lesser scaup, including areas of major concentrations (cross-hatched). Wintering distribution indicated by stippling.

parklands. Lakes of moderate depth, especially those with bulrushes near shore and brushy coves, are also favored, particularly if they have abundant populations of amphipods and aquatic insect larvae, which are major foods during the summer. When on migration and during the winter, fresh to brackish waters are favored habitats, but unlike greater scaup, the birds are not often found in salt-water areas. At this time they also feed on a preponderance of animal foods, including fishes and, particularly, mollusks. In their greater dependence on invertebrate foods than on specific aquatic plants, scaup thus differ considerably from most other pochards, and they thus concentrate in areas of abundant animal life at depths that are usually less than 20 feet below the surface (Johnsgard, 1975; Rogers & Korschgen, 1966).

Social behavior. A high degree of sociality is typical of fall and winter flocks of lesser scaup, which often "raft" in the thousands wherever food and protection allow. During the winter pair-forming behavior is

initiated, and although apparently only a small proportion of yearling females actually nest, it is probable that all of them form pair bonds during their first year of life. Social display begins relatively late in lesser scaup, probably normally in January or February, or about the time the northward migration is initiated. It continues at a high level of activity through the migration period, and by late April the migrating females passing through eastern Washington are mostly paired. By the time they have arrived on breeding grounds in the Northwest Territories in mid-May, all of the females are paired (Trauger, 1971). Most courtship occurs on the water surface, but it is also marked by frequent diving by the entire courting party, as well as by aerial chases. The calls of the courting male includes a soft *whee-ooo* note, uttered with a head-throw or a kinked-neck posture, and a louder whistle associated with a convulsive coughlike movement of the body. A rudimentary sneaking display, with the head lowered and pointed toward the female, is sometimes performed, and males very often swim ahead of inciting females with their crown feathers strongly depressed. Display preening behind the wing, exposing the white speculum pattern, is a fairly common display during pair formation in scaup, unlike most other pochards. This display as well as the usual bill-dipping and dorsal-preening postures are performed before copulation. Postcopulatory displays include the male's typical stereotyped swim away from the female with the bill pointed downward, probably accompanied by the kinked-neck call (Johnsgard, 1965a).

Reproductive biology. As noted earlier, probably only a minority of yearling females attempt to nest, and Trauger (1971) estimated that less than 15 percent of them succeed in hatching broods, while second-year and older females were successful in progressively greater degrees, with more than 40 percent of the latter age group producing broods. Females have a moderately strong tendency to return to their natal areas for nesting, whereas males do not. Upon returning to their breeding grounds, each pair typically occupies a fairly large and poorly defined home range, which is usually shared with other pairs in apparent harmony. Typically the home range is centered on a permanent pothole or marsh from two to five acres in size, at least ten feet deep and surrounded by trees and ungrazed grassland. Nest sites are often in the grasslands areas well away from shore, but usually are within 50 yards of shore and in herbaceous cover between a foot and two feet in height. Islands, floating sedge mats, and tern colonies are also preferred areas of nesting in some locations. Nesting in association with terns or gulls is not unique to scaup and indeed is fairly common among ground-nesting waterfowl, which apparently gain added protection from some predators by this device (Vermeer, 1970). Eggs are probably laid on a daily basis until the clutch is complete. Early clutches typically number about 10 eggs, while later ones or renesting efforts average progressively fewer. As many as four nesting attempts by a single female have been documented. Males usually abandon their mates at about the time that incubation gets underway, but may stay until about the midpoint of the incubation period, which averages 23 to 25 days. Renesting is often prevalent in spite of the relatively late initiation of nesting in scaup, but in general scaup appear to be relatively inefficient nesters, with most studies indicating successful hatching by no more than about 30 percent of the breeding pairs in typical years. The fledging period of the young is probably between 45 and 50 days, but females often desert their broods well in advance of fledging, frequently when they are but a few weeks old. Motherless broods often merge, further confusing estimates of productivity, and no good information is yet available on mortality of juveniles during their first fall of life (Johnsgard, 1975; Bellrose, 1976).

Status. In marked contrast to the situation with the canvasback and redhead, continental populations of lesser scaup seem to have remained fairly stable over the past few decades. Surveys by federal biologists do not attempt to separate the two scaup species, which between 1955 and 1975 have averaged nearly 7 million birds during the breeding season (Bellrose, 1976). These two species are thus the most abundant of the North American pochards, and in spite of seemingly low nesting effectiveness by females the populations have shown no measurable reduction in recent years. As with the ring-necked duck, adaptations for dryland nesting and the use of habitats in western and northern Canada that are still fairly free of human disturbance or development probably help to maintain the scaup populations in a favorable situation.

Relationships. The greater and lesser scaups are obviously very closely related, although hybridization under wild conditions has never been proven to occur. It seems likely that the lesser scaup, like the tufted duck in Eurasia, represents a North American offshoot from earlier stock that gave rise to the greater scaup, and the two scaups currently occupy largely complementary breeding and wintering habitats and ranges, although their food intakes are very similar.

Suggested readings. Gehrman, 1951; Trauger, 1971.

Tribe Mergini (Sea Ducks)

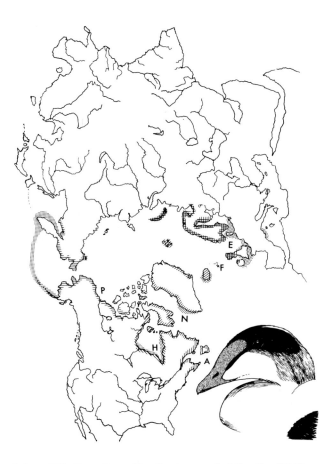

MAP 107. Breeding distributions of Atlantic ("A"), European ("E"), Faeroe ("F"), Hudson Bay ("H"), northern ("N"), and Pacific ("P") common eiders. Wintering distribution indicated by stippling.

Drawing on preceding page: Long-tailed Ducks

Eider (Common Eider)

Somateria mollissima (Linnaeus) 1758

Other vernacular names. Eider duck; Eiderente (German); eider ordinaire (French); eidero comun (Spanish).

Subspecies and ranges. (See map 107.)

S. m mollissima: European eider. Breeds on the coasts of Scotland and northern Eurasia from Scandinavia to Novaya Zemlya and Vaigach Island. Largely sedentary, wintering as far south as France and occasionally to Spain and Italy.

S. m. faeroeensis: Faeroe eider. Resident on the Faeroe Islands.

S. m. dresseri: American eider. Breeds along the coasts of Labrador, Newfoundland, Quebec, Nova Scotia, New Brunswick, and Maine. Winters mostly from the Gulf of St. Lawrence to Massachusetts and eastern Long Island.

S. m. sedentaria: Hudson Bay eider. Breeds on the coast and islands of Hudson Bay from Cape Fullerton to the east coast, south of Southampton, Coats, and Mansel islands, and south to James Bay. Winters mostly in the vicinity of the Belcher Islands.

S. m. borealis: Northern eider. Breeds from Somerset and Ellsemere islands south to Southampton Island, Hudson Strait, and northern Labrador, and probably also on Greenland and Iceland (the Iceland population is sometimes considered *mollissima,* and sometimes is separated, along with birds from Greenland, Spitsbergen, and Norway, as a distinct form, *islandica*). Winters from southern Greenland and Labrador south to Nova Scotia and rarely to New England.

S. m. v-nigra: Pacific eider. Breeds in eastern Siberia, on the islands of the Bering Sea, and in northwestern North America from Cook Inlet northwestward around the Bering Sea coast and along the coast of the Arctic Ocean to Victoria Island and Coronation Gulf. Winters in the Bering Sea, primarily around the Aleutian Islands.

Measurements and weights. Folded wing: males, 269–328 mm; females, 266–95 mm. Culmen: males, 49–61 mm; females, 44–57 mm. Weights: adult males of *sedentaria,* 2,450–2,725 g (av. 2,500 g); females, 1,575–1,825 g (Freeman, 1970). Males of *dresseri* shot in the fall average ca. 2,000 g, and

females ca. 1,500 g. Males of *v-nigra* shot in the fall average ca. 2,600 g, and females ca. 2,500 g. Eggs (of *mollissima*) average 77 x 50 mm, olive green, 110 g.

Identification and field marks. Length 22–28" (56–71 cm). Plate 50. *Adult males* in breeding plumage have a black forehead and black area around the eye and crown, which is divided in the middle by a white streak to the hind neck. The rest of the head is white except for a sea green patch extending from the ear to the occiput, interrupted by a narrow white line near the back of the head, and (in *v-nigra* only) a black V-mark passing from the chin back along the sides of the throat. The neck, breast, and mantle are white or creamy white, the rump, tail coverts, and tail are black, and the rest of the body is black except for a white patch on each side of the rump. The sickle-shaped tertials and most upper wing coverts are white, except for the greater secondary coverts, which are black. The secondaries are black, and the primaries and their coverts are grayish black. The iris is dark brown, the feet are olive green to yellowish, with blackish webs, and the bill varies from greenish or olive to bright orange (in *v-nigra*), with a variably wide upward extension toward the eyes and a grayish nail. *Females* have a grayish brown head and neck that is finely streaked with black. The breast is dusky brown with blackish barring, and the sides, flanks, back, and upper tail coverts are brown (grayish in *sedentaria*) with darker brown barring, as are the upper wing coverts, except for the greater secondary coverts, which are unbarred and tipped with white. The primaries and their coverts are brown, the secondaries are brownish tipped with buff, and the underparts are unmarked brown. The iris is brown, the bill gray, and the legs and feet greenish yellow, with darker webs. *Males in eclipse* are mostly dark-colored except for the white wings, the body being generally blackish brown except for a few white feathers on the breast, a light brown crown, and usually some white feathers on the back. *Subadult males* resemble adults, but have gray tertials and wing coverts and duller black underparts. *Juveniles* do not resemble adult females, but are dull brown, lacking strong barring or white markings on the feathers. Males develop a whitish breast during their first winter and generally resemble an adult in eclipse plumage.

In the field, the large size and marine habitat of eiders separates them from most waterfowl except scoters, and no scoter species exhibits white on the back and breast. Common eiders are the only eiders in which the males have a black crown stripe that effectively hides the eye, and they also are more extensively white on the breast and back than are the other large eiders. Females have strong vertical barring on their sides and flanks, and thus differ appreciably from female king eiders. They also lack the eye-ring markings of the spectacled eider. In flight, eiders are slow and ponderous in spite of a rapid wingstroke, and males may be easily identified by the black crown markings and their more extensively white upper wing and back coloration. During display, males utter several courtship notes that are all dovelike cooing sounds, and females produce several loud and rather hoarse-sounding calls. Calling during flight is apparently infrequent.

Natural History

Habitat and foods. During the breeding seasons, eiders seek out low-lying rocky coastlines with numerous islands, with more limited usage of sandy islands and coastal fresh-water lakes or rivers. Boulder-covered islands are preferred to gravel- or rock-covered ones, and grassy islands are chosen over shrubby or wooded ones. In some areas the birds nest well away from the coast, near tundra ponds, but usually they remain close to marine foods. Even during the summer, foods consist mostly of such invertebrates as amphipods, isopods, and bivalve mollusks, especially mussels (*Mytilis*) and periwinkles (*Littorina*), and smaller amounts of univalve mollusks and echinoderms. During fall and winter the birds are found well away from shore, but apparently continue to concentrate on mollusks and crustaceans as dietary staples, usually foraging over mussel beds between two and ten meters below the surface. They evidently rarely dive to depths of more than 16 meters when foraging, and most dives are probably of no more than a few meters. At times the birds dig in muddy shorelines with their feet, evidently for small clams or worms (Palmer, 1976).

Social behavior. So far as is known, eiders become sexually mature in their second winter of life, when the males attain their nuptial plumage for the first time. However, some females may not nest until their third year of life. Pair formation presumably begins when the birds are still at sea, but most observations stem from the breeding grounds or from captive individuals. Spurr and Milne (1976) reported that in the European eider most of the pairs that were formed before midwinter were of older birds that probably remated with earlier mates, while later pairs were formed by younger females that did not nest or laid late in the season. Many of the birds are obviously already paired when seen on spring migration approaching their breeding grounds, but displays often continue well after arrival there. Inciting by the female is the common and apparently important display of that sex, and the call is accompanied by strong bill-pointing and chin-lifting movements. There are several male displays, and the major postures are associated with cooing sounds and thus are called cooing movements. There are three major cooing movements, as well as some compound cooing movements caused by linkage of the single display elements. In addition, some silent displays, such as neck stretching, lateral head turning, dorsal preening, bathing, and wing flapping, are commonly performed. Many of these same male displays occur in the precopulatory situation, after the female has assumed a prone position in the water, with a moderate degree of predictability. After treading, the male performs a single display (third cooing movement) and then swims away while performing lateral head turning (Johnsgard, 1965a; McKinney, 1961).

Reproductive behavior. When the birds arrive at their nesting grounds they spend a good deal of time seeking out and contesting ownership of suitable nest sites, with females occasionally engaging in severe fights. Many birds use old sites of their own or of other females. In particular they tend to favor locations with rocky overhangs and which are well drained and become snow-free early in the nesting season. In most areas the birds nest in distinct colonies, presumably as an antipredator adaptation, and on occasion will nest in tern colonies, apparently for the same reason. Nest densities in such favored locations are often extremely high, with "territories" averaging no more than 100 to 300 square feet per nest not uncommon in such locations. The birds use communal loafing areas, and males typically visit the nest site only during the egg-laying period. Shortly afterwards they desert their mates and begin to gather before molting. Females lay their eggs at the rate of one per day, and clutch sizes tend to be strongly correlated with the timing of nesting, with early clutches often averaging about 5 eggs and late clutches considerably smaller. There is also a relationship between clutch size and latitude, with birds nesting in midlatitudes for the species having the largest average clutches (Johnsgard, 1973). Estimates of the normal incubation period range from 25 to 30 days, and perhaps 26 days represents the most typical span. Females quickly take their broods to water, and in many areas considerable brood merging occurs, with dozens of ducklings and several adult females associated in single groups. An eight-week fledging period has been estimated, and apparently females begin their own flightless period before the young birds reach their flight stage, so that adults and young probably gain their powers of flight at about the same time. In some areas a long migration, involving all males and a substantial number of the females, is undertaken prior to molting (Cooch, 1965; Hilden, 1964).

Status. Without attempting to consider the subspecies separately, Bellrose (1976) suggested that a total North American population of this species might be from 1.5 to 2.0 million birds, the bulk of which are presumably of the American race. The Pacific race is probably next most abundant, and may consist of about a third of a million birds. The size of the Greenland population is unknown, but Iceland supports about half a million pairs, Great Britain about 10,000 pairs, and the Baltic area, including Sweden and Finland, about 300,000 pairs (Ogilvie, 1975). Beyond that, there is the U.S.S.R. population. Of the three subspecies represented there, Dementiev and Gladkov (1967) provided estimates for only *mollissima*, which they estimated at a minimum of 90,000 nests. Palmer (1976) provided some summaries of historical eider down harvests for various countries, and suggested that at one time down from over half a million nests was obtained annually in the U.S.S.R., but the colonies there now are greatly reduced in size.

Relationships. This is the most widespread and geographically variable of the eider species, and in at least some respects seems to be the least specialized in behavior and morphology of the *Somateria* group. Its nearest relative is certainly the king eider, and some wild hybrids with it have been reported.

Suggested readings. McKinney, 1961; Cooch, 1965.

King Eider

Somateria spectabilis (Linnaeus) 1758

Other vernacular names. None in general English use. Prachteiderente (German); eider royal (French); eidero rey (Spanish).

Subspecies and range. No subspecies recognized. Breeds in a circumpolar distribution in Greenland, northern Russia, Siberia, northern Alaska, and the arctic coasts of Canada including most of the Arctic islands and perhaps also the northern coast of Labrador. Winters on the north Pacific, especially around the Aleutian Islands, but occasionally as far south as California, and on the Atlantic coast from Greenland to Newfoundland, with stray individuals occurring south to Georgia and inland to the Great Lakes. See map 108.

Measurements and weights. Folded wing: males, 275–90 mm; females, 260–82 mm. Culmen: males, 28–34 mm; females, 30–35 mm. Weights: males in spring, 1,530–2,010 g (av. 1,830 g); females, 1,500–1,870 g (av. 1,750 g) (Brandt, 1943). Eggs: av. 64 x 43 mm, bright olive, 73 g.

Identification and field marks. Length 19–25" (43–63 cm). Plate 51. *Males* in breeding plumage have the crown and back of the head a pale bluish gray, forming a smooth crest, which is separated from greenish cheeks by a whitish line. There is a black spot below the eye, a black V-mark on the sides of the throat, and black feathers bordering the enlarged base of the bill. The neck, throat, and foreback are white, the breast is creamy white, and the rest of body and tail are brownish black except for a large white patch on each side of the rump. The middle and lesser wing coverts are white, the marginal coverts are dusky, and the greater coverts, secondaries, and sickle-

shaped tertials are black. The primaries and their coverts are brownish black. The bill is bright orange to red, with a whitish nail and a seasonally enlarged orange to reddish forehead knob. The iris is dark brown, and the legs and feet are yellowish to dull orange, with darker webs. *Females* have a cinnamon buff head and neck, finely streaked with black, averaging darker on the crown and lighter on the throat, and a rudimentary crest evident on the hind neck. The neck is brownish black, the feathers edged with tawny and buff coloration; and the rump, tail coverts, breast, sides, and flanks are cinnamon buff, with darker U-shaped markings, especially on the flanks. The tail is dark brown, the upper wing coverts and tertials are brown, while the primaries and secondaries are blackish brown, with the secondaries and their coverts tipped with white. The iris is brown, the bill is gray, and the legs and feet are greenish gray to dull yellow, with darker webs. *Males in eclipse* are dark brownish black and generally lack the crescentic markings of the female, and

additionally retain the white upper wing coverts and usually some white feathers on the breast or foreback. *Subadult males* in their second year resemble adults, but the median wing coverts are margined or shaded with dusky coloration. *Juveniles* of both sexes are initially quite brownish, but during late winter the males gradually acquire a darker back, scapulars, and flanks, and develop a varying amount of white on the breast and rump.

In the field, male king eiders may be separated from common eiders by their more extensive black coloration, especially on the back, and by their orange red bill-knob. Females lack the vertical barred pattern of the common eider, but instead have more *Anas*-like crescentic markings and lack the definite eye-ring of the spectacled eider. In flight, the more extensively black upperparts (back and tertials) are evident, and the red bill and enlarged knob at its base are conspicuous for great distances. Females utter distinctive hollow-sounding notes similar to those made by a hammer striking a hollow wall, while during courtship on water the males utter tremulous cooing sounds.

NATURAL HISTORY

Habitat and foods. Preferred breeding habitats of the king eider include fresh-water ponds, lakes, and streams on arctic tundra, usually near coastlines, with occasional nesting just above the high-tide line of seacoasts. For the rest of the year the birds are at sea, often resting on drift ice or feeding some distance off the coastline in fairly deep waters. During the summer months the birds probably feed largely on the larvae of aquatic insects such as caddis flies and particularly on midges, supplemented by small quantities of vegetable materials such as grasses, sedges, and some broad-leaved herbs. During the rest of the year they concentrate on invertebrates, particularly mollusks, sea urchins, sand dollars, and crustaceans.

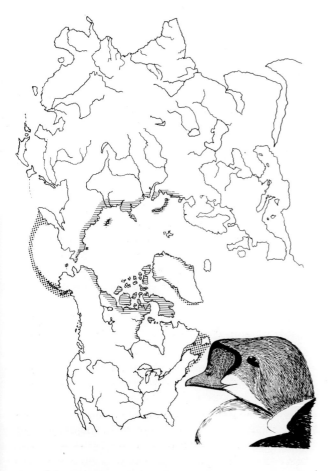

MAP 108. Breeding (hatched) and wintering (stippling) distributions of the king eider.

Practically the only plants consumed while the birds are at sea are small quantities of algae, eelgrass (*Zostera*), and wigeon grass (*Ruppia*) (Palmer, 1976; Johnsgard, 1975).

Social behavior. Little is known of the sociality of eiders during the nonbreeding period, but they are known to concentrate in enormous flocks in some areas, such as off St. Lawrence Island, where one flock in February was estimated to include 15,000 birds. Presumably pair formation must begin in these flocks; but as the birds begin to move back to the coastlines for breeding, the earliest arrivals are preponderantly males, with flocks of nearly equal sex ratios arriving only later. Pair bonds are thought to be formed in the second winter or second spring of life, and social display is frequently seen among spring migrants as well as on the breeding grounds as soon as they have arrived. May and June are thus presumably the major period of courtship activity, but this would mean that pair bonds in this species may last no more than about two months, or until the time the female has begun incubation, usually by early July. Displays of king eiders, based on observations of captive birds and supplemented by some observations in the wild, are rather similar to those of the common eider. Beside a number of activities that are only slightly modified from their corresponding comfort movements, such as head rolling, bathing, and body shaking, a more stereotyped form of wing flapping occurs in the king eider than in the other eiders, and it seems to expose the black abdomen and throat markings effectively. Two displays are associated with cooing calls, reaching (the homologue of the common eider's third cooing movement), and a repetitive display called pushing (the homologue of the common eider's second cooing movement). Head turning is performed more conspicuously and ponderously than in the common eider, and often follows one of the other displays. Displays often occur in fairly predictable sequences, with the time intervals separating them often as fixed as the duration of the displays themselves. The major female display is inciting, to which males often respond by swimming rapidly ahead of her while performing head turning. Females solicit copulation by extending themselves prone on the water, while the male performs all of the same displays used during pair formation, but primarily those derived from comfort movements. After treading, he performs a single display, then swims rapidly away while performing head-turning movements (Johnsgard, 1965a).

Reproductive biology. Females often do not arrive on their high-arctic nesting grounds until about mid-June, and tend to spread out over the tundra habitats except in certain locations, such as where river islands provide protection from arctic fox predation. Nests are usually well away from water, often being situated on dry and rocky slopes as far as a quarter mile from the nearest water. Egg laying often begins in late June, with one egg per day being laid until a clutch averaging 5 eggs has been completed. In spite of frequent high rates of nest predation by foxes, there is little or no indication of renesting efforts, which are probably precluded by the short breeding season. Incubation is by the female alone; and as soon as the clutch is completed, the males begin to leave the nesting area, migrating in vast flocks to molting areas at sea, often more than a thousand miles from the nesting areas. Incubation requires 22 to 24 days, and shortly after hatching, broods begin to merge, with crèches of up to 100 or more ducklings often being thus formed, with several females in attendance. Unsuccessful females and females that are displaced from their broods once the nurseries are formed soon flock and also undergo a long molt migration (Salomonsen, 1968). About 100,000 molting eiders concentrate off the coast of western Greenland, probably including most of the population of eiders from eastern Canada, while those from the remainder of Canada and Alaska migrate west over Point Barrow to molt in waters of the Bering Sea. Probably at least a million eiders, including common eiders, cross Point Barrow in late summer during this migration (Thompson & Person, 1963).

Status. The North American population of king eiders is probably between 1 and 1.5 million birds (Bellrose, 1976), and in addition there is an obviously substantial but unestimable Asian population. Dementiev and Gladkov (1967) indicate that, together with the long-tailed duck, the king eider is the commonest duck species of coastal tundras, especially in the eastern U.S.S.R.

Relationships. The king eider and common eider are clearly close relatives, as indicated by the many similarities in their behavior and their plumage patterns. Additionally, wild hybrids between them have been reported on several occasions, especially in Iceland, where male king eiders sometimes form bonds with female common eiders (Palmer, 1976).

Suggested readings. Parmelee et al., 1967; Salomonsen, 1968; Johnsgard, 1975.

Spectacled Eider

Somateria fischeri (Brandt) 1847

Other vernacular names. None in general English use. Plüschkopfente (German); eider de Fischer (French); eidero de Anteojos (Spanish).

Subspecies and range. No subspecies recognized. Breeds in eastern Siberia from the Chukot Peninsula to the Yana River delta and probably sporadically to the Lena River. In North America it occurs sporadically from the Baird Inlet north and east to Demarcation Point, but is locally distributed and nesting is common only on the lower Kuskokwim delta. Probably winters in the Bering Sea, but is rarely observed during that season. See map 109.

Measurements and weights. Folded wing: males, 255–67 mm; females, 240–50 mm. Culmen (to feathering): males, 21–26 mm; females, 20–25 mm. Weights: adults of both sexes average ca. 1,630 g, with a maximum of 1,850 g reported. Eggs: av. 64 x 45 mm, olive, 73 g.

Identification and field marks. Length 20–23" (51–58 cm). Plate 52. *Adult males* in breeding plumage have a somewhat shaggy crest of greenish feathers on the sides and back of the head, and a velvetlike area of similarly colored feathers in front of the eye and extending forward into the bill. These two areas are separated by a large white area around the eye, framed narrowly with black so as to suggest spectacles. The throat, neck, and back are creamy white, the hind back and rump are dark brown, and the tail coverts and tail are grayish brown. Except for large white patches on either side of the rump, the rest of the underparts are dark grayish brown to grayish black, with a somewhat silvery bloom on the breast and sides. The sickle-shaped tertials and lesser and middle wing coverts are creamy white, the marginal coverts are dusky, and the secondaries and their coverts are blackish. The primaries and their coverts are dark grayish brown. The iris is whitish, ringed with light blue, the bill is orange with a paler nail, and the legs and feet are dull yellowish to olive brown. *Females* are generally like female common eiders, but differ in having dark brown feathering extending out on the bill from its base diagonally upward to a point just above the nostrils and a light brown ring around the eyes of the same size and shape as the "spectacles" in the male, and they

average somewhat more rust-colored throughout. The iris is brown and pale bluish, the bill is gray, and the legs and feet are yellowish brown. *Males in eclipse* are predominantly gray to grayish black, the head is grayish white where it is normally green, and the white "spectacles" are a darker gray. The body is mostly grayish, except for the white upper wing coverts, and the scapulars are also gray. *Subadult males* in their second winter resemble adults but have a somewhat less well developed crest, light gray scapulars, and darker gray tertials. The white patch on the sides of the rump is poorly developed, and there is some dark gray mottling on the sides of the neck, based on observations of captive birds. Some first-year males may also reach this stage, while others remain very femalelike. *Juveniles* are quite different from adults in that the juvenile male has only slightly developed "spectacles" and somewhat resembles the adult female, but is darker above, with the underparts faintly but uniformly barred with dusky coloration, while the wings are brownish black. In juvenile females the wings are more like those of adult females, but the underparts are spotted rather than barred.

In the field, the distinctive "spectacles" of both sexes provide the best field mark; in males the silvery black underpart coloration that extends part way up the breast is also distinctive. The dark brown area ahead of the female's "spectacles" contrasts strongly with the "spectacles" and the rest of her head plumage; this is also evident on flying birds. In flight these eiders are swifter and more agile than their larger relatives, and the blackish color of the underparts of males extends well forward of the leading edge of their wings. Both sexes are relatively quiet, and the cooing calls of the males carry but a short distance. The female's vocalizations are very much like those of the other large eiders.

NATURAL HISTORY

Habitat and foods. The preferred breeding habitat of this eider in Alaska consists of rather luxuriant lowland tundra having small ponds and reasonable proximity to salt water. The birds nesting in the U.S.S.R. occupy similar habitats of moist tundra, especially low areas that are flooded in June and are

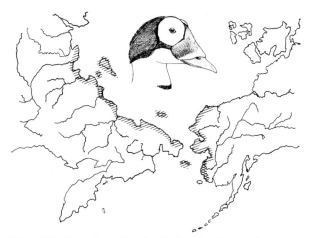

MAP 109. Breeding (hatched) distribution of the spectacled eider, including areas of major concentrations (cross-hatched).

vegetated with sedge and mare's-tail (*Hippurus*), sometimes extending as far as 50 kilometers from the coast. During the summer the birds concentrate on coastal shallows that are inhabited by the larvae of crane flies and caddis flies, which provide the bulk of the summer foods for adults and probably also for juveniles. Mare's-tail is one of the few plant foods of importance to young birds, along with pondweeds and crowberries (*Empetrum*). When the birds return to the sea, they again resort to an almost exclusively animal diet, with mollusks apparently one of the major foods. Echinoderms such as sand dollars and sea urchins are very minor components of the spectacled eider's diet, which is markedly different from the king eider's in this respect (Palmer, 1976; Kistchinski & Flint, 1974).

Social behavior. Almost nothing is known of the social behavior of spectacled eiders away from their breeding grounds, as their whereabouts during winter remains essentially unknown. In captivity they seem to be much like the other eiders in their general gregariousness, and as wild birds return in spring to their nesting areas they are found in small flocks composed mainly of already paired birds, plus a small percentage of unmated males. Few courtship activities have been seen on the breeding grounds; my (1964a) observations are virtually the only ones of wild birds, and they have subsequently been supported by some observations on captives. Display observed on the breeding grounds appeared to result when unpaired males located untended females, and was thus sporadic and often broken up when the female's mate approached. Male displays consist of

several that are common to the other species of *Somateria* and include as well some elements found in the Steller eider, thus providing a link between these two groups of eiders. The usual ritualized comfort movements such as wing flapping, preening, bathing, body shaking, and head rolling occur, plus a few more specialized displays. One is a rapid backward movement of the head on the shoulders, much like rearing in *Polysticta*, but usually preceded by a movement of the head forward and downward, as in the king eider's reaching display. The other is a pushing display much like that of the king eider, and a backward bill-toss that is followed, and at times also preceded, by a forward neck-jerk similar to one of the display combinations of the common eider. Not many copulatory sequences have been seen, but evidently the precopulatory behavior consists primarily of the ritualized comfort movements already mentioned, particularly preening movements. Treading is often preceded by a single body-shaking movement, and in the cases observed has always been followed by a single head-forward-rearing display, which is in turn followed by a few head-turning movements (Johnsgard, 1964a).

Reproductive biology. These eiders typically arrive on their Alaskan nesting grounds in May, and often must wait several weeks for the lowland tundra to become sufficiently snow-free for nesting. The nests are most commonly located around the edges of tundra ponds in Alaska (Dau, 1974; Johnsgard, 1964b), but observations in the U.S.S.R. indicate that small islets of such ponds are also favored. In both instances the past year's growth of tall grasses and sedges is the usual nest cover. Nests are sometimes also located in colonies of small gulls and terns, whose behavior tends to ward off larger gulls and probably also foxes from the nest vicinity. Clutches are laid at the rate of one egg per day, and although the earliest clutches may exceed 5 eggs, normal and late clutches are usually noticeably smaller and may average about 4 eggs. Males leave their females early in the incubation period, and most observations indicate that females rarely if ever leave their nests willingly once incubation is underway. The incubation period is normally 24 days. After hatching, the young are not immediately led to salt water, but instead broods tend to remain separate and are reared to fledging on fresh to slightly brackish ponds within a mile or two of the nest. The fledging period is about 50 days, or at most no more than 53 days (Dau, 1974; Johnsgard, 1964b).

Status. It has been estimated that the Alaskan breeding population of this species may number about 100,000 birds (Johnsgard, 1975), and it is also believed that this population is the bulwark of the world's population. In Siberia the species occurs over a fairly wide area, but is almost nowhere extremely abundant. Kistchinski and Flint (1974) estimated about 17,000 pairs in the Indigirka delta, where it is perhaps more common than anywhere else, except possibly for the Kolyma delta. It is thus likely that the world population of breeders is no more than 200,000 birds, with an equal or greater number of immatures. This estimate is fairly close to one of a half-million birds made earlier by Uspenski (1972).

Relationships. Although there seems to be no justification for maintaining a separate genus (*Lampronetta*) for the spectacled eider, it does in many ways appear to provide an evolutionary bridge between the typical large eiders and the genus *Polysticta*.

Suggested readings. Johnsgard, 1964a, 1964b; Dau, 1972, 1974, 1976.

Steller Eider

Polysticta stelleri (Pallas) 1769

Other vernacular names. None in general English use. Scheckente (German); eider de Stellar (French); eidero de Steller (Spanish).

Subspecies and range. No subspecies recognized. Breeds in arctic Siberia from Novaya Zemlya (rarely northern Scandinavia) eastward to the Bering Sea (but only locally common), on St. Lawrence Island, and in North America locally on the Kuskokwim delta and in northern Alaska from Wainwright east at least to Pitt Point, and possibly to Humphrey Point. Winters off northern Finland and Norway, the Kamchatka Peninsula, the Commander, Kurile, and Aleutian islands, and the Kenai Peninsula. See map 110.

Measurements and weights. Folded wing: males, 209–17 mm; females, 208–15 mm. Culmen: males, 36–40 mm; females, 35–40 mm. Weights: both sexes average ca. 860 g, with maximums of 951 and 907 for males and females, respectively (Nelson & Martin, 1953). Additional weight data summarized by Palmer (1976) indicate maximums of 1,000 and 999 g for the two sexes. Eggs: av. 59 x 41 mm, pale olive buff, 58 g.

Identification and field marks. Length 17–19″ (43–48 cm). Plate 53. *Adult males* in breeding plumage have a shiny white head and neck, except for black on the chin and throat, around the eyes, and on the occiput, and a plushlike greenish group of feathers forming an occipital tuft, as well as a less distinct area of green in front of the eyes. The hind neck and sides of the neck are black, and the black extends around the foreneck as a narrow collar and down the middle of the back to the tail. The brown tail is long and pointed, with black under tail coverts; a black rounded spot is also present in front of and below the base of the wings. The longer scapulars are decurved, pointed, and iridescent bluish black striped with white. Most of the upperparts are cinnamon buff, deepening to orange on the chest and underparts. The upper wing coverts are white, except for the primary coverts and the primaries, which are brown. The secondaries and outer webs of the tertials are iridescent blue, the secondaries tipped with white and the tertials white on their inner webs. The iris is dark brown, the bill bluish gray, lighter toward the tip, and the legs and feet are dark bluish gray. *Females* have a dusky brown and brownish buff head and neck, with an indistinct whitish eye-ring and pale cheeks. The body is predominantly dark brown, with paler edges on the feathers, especially the flanks and scapulars, and darker on the breast, shading to blackish on the abdomen. The tail is dark brown, the upper wing coverts are dusky brown with grayish tips, except for the greater secondary coverts, which are dusky with broad white tips. The primaries and their coverts are brown, the secondaries are iridescent blue tipped with white, and the sickle-shaped tertials have grayish blue outer webs and dusky brown inner webs. The iris is brown, the bill dusky blue, and the legs and feet are bluish gray. *Males in eclipse* closely resemble females, but retain their white upper wing coverts. *Subadult males* in their second year resemble adults, but may have a few scattered brown feathers. *Juveniles* resemble the adult female, but are lighter-colored, more reddish, and mottled below. First-year males gradually develop a black neck-ring and dusky throat.

In the field, Steller eiders, particularly the rather *Anas*-like females, are most likely to be mistaken for

species other than the larger eiders. Males in nuptial plumage have a distinctive white head and cinnamon body coloration, and in flight the white forewings and distinctive speculum pattern are quite evident. Like the other eiders, this species is not highly vocal, but females do utter a loud *qua-haaa'* call during inciting, while the males are almost totally silent during display.

NATURAL HISTORY

Habitat and foods. No detailed studies of breeding habitat needs have yet been made, but it is clear that the Steller eider nests in lowland tundra near the coast, often in tidewater flats. They also often nest in lacustrine basins on mossy tundra, but in general are probably to be found somewhat closer to the coastline than are spectacled eiders, which have a very similar breeding distribution. Late spring and summer foods of the birds are primarily amphipod crustaceans and bivalve mollusks, which presumably are from salt-water or at least brackish-water environments. Fresh-water animal materials include the

larvae of midges and caddis flies. Vegetative materials are taken in small amounts, and include such items as pondweeds and crowberries (*Empetrum*). The winter foods are evidently mainly such rather soft-bodied invertebrates as amphipod and isopod crustaceans, univalve and bivalve mollusks, and barnacles (Johnsgard, 1975; Palmer, 1976).

Social behavior. This small eider is perhaps the most sociable of all the eiders, and in some areas such as Nelson Lagoon and Izembek and Bechevin bays at the tip of the Alaskan peninsula the flock sizes are overwhelmingly large. This, however, occurs only during the fall molting period and in early spring, before waters to the north become ice-free and allow for northward migration. Observations at Izembek Bay by McKinney (1965b) indicate that much pairforming behavior occurs during the spring assembly period of April and is marked by extensive display on the water, aerial flights, and massed diving behavior. Typically several males display around a single female, with frequent chases and fighting among them, and also with short flights toward the females. Inciting is the major female display, and is marked by loud calling and strong chin lifting. Males perform a variety of displays, all of which are evidently done silently and which include several ritualized comfort movements such as body shaking, head rolling, and preening. The most spectacular display, rearing, consists of a rapid raising of the head and body, momen-

MAP 110. Breeding (hatched) and wintering (stippling) distributions of the Steller eider.

tarily exposing the chestnut-colored underparts, and is often preceded and followed by rapid lateral head movements as the males first swim toward and then away from a female. All of the displays are performed much more rapidly than their counterparts among the larger eiders, and the overall level of activity is also much greater. Copulation takes a form similar to that of the other eiders, in that the female becomes prone after little if any preliminary posturing. Thereafter the male performs a long series of displays that primarily are a sequence of bathing, bill-dipping, and preening movements in rapid succession. Mounting is apparently always preceded by a single body shake, after which the male quickly steams over the water and begins treading. After copulation, he performs a single rearing display, then steams away in an erect posture, usually while head-turning (Johnsgard, 1964a; McKinney, 1965b).

Reproductive biology. Relatively little is known of the breeding biology of this species, which in Alaska nests in only a few localities. There it arrives later than the other three eiders in the Hooper Bay area, about three weeks before nesting gets underway in late June. The nests are well scattered over lowland tundra, and often are placed a few meters from a tundra pond, either on a slight hummock or in a depression between hummocks, and usually are well concealed in grass cover. Studies on the lower Kashunuk River indicate a very low nesting density there in recent years—only about one nest per 100 acres in the early 1960s. However, the species was apparently at one time more common there, so this may not be an accurate reflection of spacing behavior. The average

clutch size is of 7 or 8 eggs, which are presumably laid at daily intervals. Evidently the male remains in the vicinity of the nest and female for a longer period than most other eiders, perhaps until about the time of hatching. The incubation period is still unestablished. Evidently gulls and other predators are serious sources of mortality among eggs and ducklings; and shortly after hatching, the females move their broods to brackish inlets or even salt water, where they form "herds" and forage in the litter of the tidal flats. The fledging period is not definitely established, but studies by Brandt (1943) indicated that hatching began in early July and some birds were on the wing by the end of August, so an approximate 50-day fledging period seems likely. About the time that hatching is underway the males begin to gather in large flocks off the coastline near the nesting grounds, but the species has an unusually late wing-molt period, allowing a substantial migration to molting areas that may be as far as 3,000 kilometers from the nesting grounds. There are marked yearly differences in the timing and degree of this migration to the tip of the Alaska Peninsula; sometimes it occurs as early as August, while in other years the birds have not arrived until early November, after having completed their molt in some other area (Jones, 1965).

Status. The winter population of Steller eider in the vicinity of the Alaska Peninsula has been estimated at about 200,000 birds (Jones, 1965), and in addition there is a second major wintering area from the Kamchatka Peninsula southward to the Kurile Islands. Uspenski (1972) estimated a total world population of about half a million birds. Probably nearly all of these nest on the Siberian mainland; there are now virtually none nesting along the lower Yukon and Kuskokwim deltas; and although a nesting population is now known to occur in the Prudhoe Bay area (Lubbock, 1976), it is not likely to be extremely large.

Relationships. Behavioral studies (Johnsgard, 1964b) and anatomical analyses (Woolfenden, 1961) both indicate not only that *Polysticta* should be recognized as distinct, but also that it has affinities with such sea duck genera as *Clangula* and *Polysticta*. However, Brush (1976) found that electrophoretic patterns of the feather proteins of *Somateria* and *Polysticta* were alike, and differed from those of the other sea ducks, thus favoring tribal separation of the eiders.

Suggested readings. Brandt, 1943; Jones, 1965; McKinney, 1965b.

Labrador Duck

Camptorhynchus labradorius (Gmelin) 1789

Other vernacular names. Pied duck; Labradorente (German); canard du Labrador (French); pato del Labrador (Spanish).

Subspecies and range. No subspecies recognized. Extinct since about 1875. Originally occurred along the Atlantic coast of North America, possibly breeding in Labrador or farther north, but the nesting grounds were never definitely established. Wintered south to Chesapeake Bay, but mainly along Long Island.

Measurements and weights. Folded wing: males, 210 mm; females, 206–9 mm. Culmen: males, 43–45 mm; females, 40–42 mm. Weights: males, 864 g; females, 482 g. Eggs: some probable Labrador duck eggs are pale olive to yellowish brown and average 62.7 x 42.5 mm, weight unknown.

Identification and field marks. Length about 20–23" (51–56 cm). *Adult males* have a head, neck, and scapulars that are white except for a black stripe extending from the crown to the nape, a black collar around the base of the neck, and a yellowish area of stiffened feathers on the cheeks. The back, rump, upper tail coverts, tail, primaries, and entire underparts are black, and the scapulars and tertials are edged with black. The upper wing surface is white, except for the primaries, which are black. The iris color was probably yellow or reddish hazel, and the legs and feet were probably grayish blue, but there is disagreement on this. The unusually soft-edged bill was black for most of its length, but the basal portion behind the nostrils was probably pale grayish blue and was separated from the black portion by a yellow, orange, or flesh-colored band. The eclipse plumage of males, if it existed, is unknown. *Females* are generally uniformly brownish gray, grading toward bluish slate dorsally, to sandy brown on the rump and tail coverts, and to light grayish brown on the underparts. The tail is very dark brown, the wing coverts are the same bluish-slate as the mantle, except for the greater secondary coverts, which, with the secondaries, form a white speculum. The primaries and their coverts are blackish brown, as in the male. The soft-part colors are probably similar to those of the male. *Juveniles* apparently resembled adult females for most of their first year, with young males probably beginning to get white feathers on the

head, throat, and upper breast by the end of their first winter of life.

NATURAL HISTORY

Habitat and foods. Evidently this little sea duck had a highly specialized diet, to judge from its unusual bill structure, that may have involved both dabbling at the surface in a shovelerlike manner and also diving for its food. Since the birds were sometimes caught by fishermen on trotlines that had been baited with blue mussels (*Mytilus*), it may be imagined that mollusks were a part of their diet, and the birds apparently often fed close to shore along sandy bays or in estuaries where mussels might be abundant. It is quite possible that the Labrador duck occupied much the same habitat and consumed the same type of foods as the Steller eider, which also has a soft-edged bill and is of very similar bodily proportions.

Social behavior. Nothing specific was ever recorded on this subject.

Reproductive biology. Nothing is known of this. It has been suggested (Phillips, 1922–26) that the birds may have nested on a few islands in the Gulf of St. Lawrence or in southern or eastern Labrador, in which case it would have been highly susceptible to nest robbing by fisherman or "eggers."

Status. Extinct since the 1870s, with the last known specimen taken in the fall of 1875, probably along Long Island. A less likely final record is for 1878, when a bird was reputedly shot near Elmira, New York, but the inland location makes this record suspect, and the specimen no longer exists. No convincing reasons for the bird's disappearance have ever been advanced; it was not an important sport species, nor was it sought after by market hunters. It seems most likely that a breeding-grounds disturbance, such as perhaps the arrival of effective mammalian predators in a previously isolated nesting area, may have left the species defenseless.

Suggested readings. Phillips, 1922–26; Humphrey & Butsch, 1958.

Harlequin Duck

Histrionicus histrionicus (Linnaeus) 1758

Other vernacular names. None in general English use. Kragenente (German); canard harlequin (French); pato arliquín (Spanish).

Subspecies and range. No subspecies currently recognized. Breeds in northern and eastern Asia, on the islands of the Bering Sea, in Greenland and Iceland, and in continental North America from Alaska and the Yukon south through the western mountains to central California and Wyoming (originally to Colorado), and from Baffin Island and Labrador to the Gaspé Peninsula and perhaps Newfoundland. Winters in or near its breeding grounds of the arctic islands, south to California and Long Island. See map 111.

Measurements and weights. Folded wing: males, 200–10 mm; females, 190–97 mm. Culmen: males, 25–28 mm; females, 24–26 mm. Weights: males average ca. 680 g and females ca. 540 g, with the respective reported maximums being 750 and 562 g (Johnsgard, 1976). Eggs: 54 x 38 mm, creamy, 53 g.

Identification and field marks. Length 15–21" (38–51 cm). Plate 55. *Males* in breeding plumage have a dark slate blue head and neck with a purplish sheen, a black crown and throat, and a white cheek patch extending above the eye, where it becomes dull chestnut and passes back to the nape. There is a second rounded spot behind the eye, and a third white vertical stripe along the back of the head, with all of these narrowly margined with black. The back, breast, and underparts are generally slate blue, shading to dusky on the abdomen and to bluish black or black on the rump and tail coverts. The black tail is long and pointed, and white is present only as a small

rounded spot on each side of the rump. There is a black-bordered white stripe around the base of the neck, and a similar vertical stripe from the base of the wing down the side of the breast; the scapulars are also tipped with black and white. The sides and flanks are chestnut brown, and the primaries and their coverts are dark brown. The secondaries from an iridescent blue to purplish speculum and the tertials are white, margined with black on the outer webs, while the other upper coverts are dark slate with a purplish gloss, except for rows of white spots on the outer middle coverts and inner greater coverts. The iris is brown, the bill bluish gray with a yellowish nail, and the legs and feet are grayish blue, with dusky webs. *Females* have an olive brown head and neck, with a darker crown and a lighter chin and throat. There is a small rounded whitish area above the eye and a larger one behind the eye, and the cheek area below the eye is also whitish, with darker spotting increasing toward the chin and throat. The upperparts are dark olive brown to brownish black, the tail is dark purplish brown, the under tail coverts are olive brown, and the sides and flanks are medium brown, becoming paler on the abdomen and mottled with grayish white. The upper wing surface is dark grayish brown to brown, the secondaries being glossed with a purplish sheen. The iris is dark brown, the bill is dusky, and the legs and feet are dull grayish blue. *Males in eclipse* resemble females, but have darker brown underparts and the brighter upper wing coloration typical of males. *Immature males* have the white head markings bordered with grayish brown, but the cinnamon stripe is absent or poorly developed, and the lateral neck and breast markings are smaller than in adults. *Juveniles* resemble adult females but have paler upperparts and more spotted underparts.

In the field, the habitat of rocky shores or rushing mountain streams tends to eliminate most other species of ducks, and the unique spotted plumage of the male is also distinctive. Females have a close resemblance to female or juvenile scoters, but are smaller and have a much smaller bill. In flight the birds appear swift, and closely follow the course of mountain streams. They appear quite dark-bodied when in flight, and only the white spotting of the male may help to identify them as harlequins. The male's bluish coloration is surprisingly inconspicuous in the water, and the white spotting may even be a concealing adaptation in a white-water environment. Neither sex is very vocal, and the calls that have been described are mostly high-pitched piping notes or stac-

cato sounds quite unlike those of most other sea ducks.

Habitat and foods. The presence of cold, rapidly flowing waters that are rich in aquatic insect life is central to the habitat needs of harlequin ducks, and additionally the birds seem to favor forested over nonwooded environments when they are available. In the summertime the birds forage in rapid streams for the larvae and pupae of midges (*Chironomida*) and black flies (*Simulium*) and for caddis fly larvae (*Trichoptera*). They feed much in the manner of torrent ducks, and are equally well adapted to swimming in torrential currents as that species. During winter, many harlequin ducks move to coastal waters, where they congregate along rocky headlands and dive in the surf for such foods as crustaceans, mollusks, and echinoderms. They are well adapted to prying such mollusks as chitons from submerged rocks, and also obtain a variety of univalve mollusks in much the same manner. During foraging dives, the birds tend to remain under water longer than most other diving ducks, but in general they probably do not dive to such great depths as many of the others which feed in still or less rapidly flowing waters (Bengtson, 1966b).

Social behavior. Perhaps because the species is adapted to foraging in a specialized manner on a very restricted food supply, harlequins usually are not found in large flocks, nor are they highly gregarious. Birds raised in captivity acquire adult plumage and begin sexual display activity in their second winter of life, and presumably two-year-old females attempt to nest under natural conditions. Little is known of the timing or mechanism of pair formation in this species, but Bengtson (1966b) observed a very low incidence of mated birds among flocks in December, when some display activity was seen. A few displays

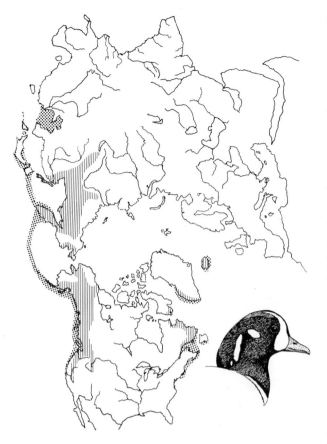

MAP 111. Breeding (hatched) and wintering (stippling) distributions of the harlequin duck.

of head-nodding behavior by the male and often by both sexes, with lateral bill shaking and bill dipping frequent between head nods. The male also performs a series of short rushes toward the female, which end with him nibbling the side of her face. Mounting is done deliberately and is followed by male aggressive behavior (Kuchel, 1977).

Reproductive biology. In Iceland, the birds move from coastal wintering areas to territories located along stretches of interior rivers during May and become well dispersed and spaced in such areas. In that area, Bengtson (1966b) found that the density was about two pairs per mile of river, being highest near lake outlets, and territorial boundaries were either indistinct or even lacking, although males exhibited possessive behavior toward their mates. Probably females tend to use the same nesting site year after year; such sites are usually extremely well hidden locations under dense bushes, in rock crevices, or on islands. Surface-nesting is apparently more typical than hole-nesting, but surface nests are always well concealed from above by dense vegetation. Eggs are evidently laid at the rather slow rate of one every two days, and clutch sizes are relatively small, probably averaging about 6 and ranging from 4 to 8. Males desert their mates as soon as incubation begins, and within a few days they leave for coastal habitats. The incubation period is not yet fully established, but probably is 28 to 29 days, rather than the earlier estimates of up to 34 days. Likewise, the fledging period is rather uncertain, but may be about 40 days, although estimates of as much as 60 to 70 days have also been made. Apparently some females molt in their breeding areas, while others evidently leave their broods and fly elsewhere to undergo their flightless period. Brood mergers have been reported common in Iceland and elsewhere (Bengtson, 1966b, 1972b; Palmer, 1976).

Status. The size of the Icelandic population of harlequin ducks is not known with certainty but probably is not more than 5,000 birds (Gudmundsson, 1971), and likewise, no estimate is available for Greenland's population. The vast majority of the North American population is associated with the Aleutian Islands, and Bellrose (1976) reports that during the fall and spring from 600,000 to a million harlequin ducks may be present on the Aleutian Islands National Wildlife Refuge. Additionally, there is an unknown but presumably rather small population in Siberia and the adjoining Asian islands.

have also been seen on the nesting grounds involving presumably paired birds. The most commonly seen display is a head nodding rather like the elliptical head movements of goldeneyes, and likewise performed by both sexes. It is seemingly silent, and also evidently of hostile derivation, as in goldeneyes. Bill dipping, lateral bill shaking, wing flapping, and various preening movements often occur during apparent courtship and some or all of these may represent actual displays. Females perform what seems to be a typical inciting movement with an associated harsh call, but descriptions are still meager for both male and female displays. Some observers have described a definite head-throw display by males, while others who have observed the birds at considerable length have not reported anything more elaborate than vigorous head-nodding displays. Various trilled, squeaking, or whistling notes have been attributed by various observers to displaying males as well, but again there is little agreement on this point. Copulatory behavior apparently consists

Relationships. Although a distinctive genus and one that has clearly evolved in conjunction with strong selective pressures associated with specialized foraging behavior, *Histrionicus* appears in many ways to be a typical sea duck. Woolfenden (1961) believes that this genus and *Clangula* come close to the core of this group of birds, with *Histrionicus* having some clear affinities with the scoters as well. What little is known of the species' displays also suggests typical sea duck affinities. Unfortunately, Brush (1976) was unable to test its feather proteins against those of the other sea duck genera.

Suggested readings. Bengtson, 1966b, 1972b; Kuchel, 1977.

Long-tailed Duck

Clangula hyemalis (Linnaeus) 1758

Other vernacular names. Oldsquaw; Eisente (German); canard nequelon (French); pato de cola largo (Spanish).

Subspecies and range. No subspecies recognized. Breeds in a circumpolar belt that includes arctic North America, Greenland, Iceland, northern Europe and Asia, and the islands of the Bering Sea. Winters in salt-water and deep fresh-water habitats, including the coastlines and larger lakes of Europe, Asia, and North America, the Caspian Sea, the Great Lakes, and coastal Greenland and Japan. See map 112.

Measurements and weights. Folded wing: males, 219–36 mm; females, 202–10 mm. Culmen: males, 26–29 mm; females, 23–28 mm. Weights: males shot in the fall average ca. 800 g and females ca. 650 g, with maximum weights of 1,042 and 815 g, respectively. Eggs: av. 54 x 38 mm, olive buff, 43 g.

Identification and field marks. Length 15–23″ (38–58 cm). Plate 54. *Males in winter plumage* have the forehead and sides of the head a pale brownish gray, with a white eye-ring, the sides of the neck blackish brown shading to buffy brown below, and the rest of the head, neck, and extreme foreback white. The scapulars are gray to white, including several ornate-

ly long and pointed ones. The breast, rest of the upperparts, and the upper wing surface are brown to brownish black, the primaries lighter brown and the secondaries dark chestnut brown. The underparts are white, shading to pearly gray on the upper sides and flanks; the outer tail feathers are white to dusky, edged with white; and the central pair is black and greatly elongated. The iris is yellow to reddish brown, the bill is dusky at the base and pink toward the tip, with a black nail, and the legs and feet are bluish gray, with darker webs. *Males in summer* have a brownish black head, neck, and breast, except for a large white eye-ring that extends backward as a postocular stripe, and an ashy gray area between this white patch and the bill, which gradually diminishes in size as the white area enlarges. The back and scapulars are blackish brown, the feathers broadly edged with reddish brown to buffy coloration, and the longest scapulars shorter than in the winter plumage. The rest of the plumage is as in the winter plumage, but is generally faded and worn. *Males in late summer and autumn* lose the brown and ashy gray head feathers from the summer plumage and replace them with white, so that the head is entirely white except for a dusky patch in the region of the ears. The scapulars are again molted and replaced with still shorter buff-edged feathers. *Adult females in winter* have the forehead, crown, and a patch at the side of the neck brownish black, while the rest of the head is white except for a dusky chin and throat. The upper breast and foreback are blackish brown, the feathers, especially the scapulars, having broad reddish brown edges and markings. The hind back, wings, and tail are blackish brown, except for the outer tail feathers, which are ashy brown; the secondaries, which are dull chestnut brown; and the middle coverts, which are mostly buff. The iris is brown, the bill is blackish to lead-color, and the legs and feet are grayish. *Adult females in summer* are generally more brownish, with the brown crown and ear-patch increasing in size and grayish brown feathers appearing over most of the face, except immediately around the eyes, and on the sides of the neck. The black becomes more brownish, with the buff-tipped scapulars replaced by less contrastingly patterned ones, and the sides and flanks become considerably browner or grayer. There is a late-summer molt of the scapulars, which are replaced by ones that are more uniformly olive brown. *Immature males* lack the fully elongated central tail feathers and ornamental scapulars during their first winter of life. *Juveniles* resemble females in summer, but lack the russet edges on the scapulars and are light brownish gray on the head and neck.

In the field, the open-water habitat, fairly small size, and elongated tail of the males help identify these ducks, which are seasonally so variable in plumage as to make any single field mark almost useless. In the winter both sexes exhibit more white than almost any of the other sea ducks, and the loud calls of the males are very conspicuous at this time.

MAP 112. Breeding (hatched) and wintering (stippling) distributions of the long-tailed duck.

They include several "yodeling" calls, often sounding like *ugh, ugh, ah-oo-gah*, or *a-oo, a-oo, a-oo'-gah*. The calls of the female are soft and infrequent by comparison. In flight, the birds exhibit white abdomen coloration that contrasts with their generally brownish black wing surfaces, and both sexes also have relatively dark breasts in winter.

NATURAL HISTORY

Habitat and foods. The breeding habitat of this most northerly ranging duck species consists of arctic tundra in the vicinity of lakes, ponds, or coastlines. Wooded country is avoided, but shrubby areas are preferred to sedges or grassy cover for nesting. The species nests virtually throughout the high arctic of the Northern Hemisphere, often at considerable distances from the coastline, but even during the summer the birds seem to prefer marine foods to freshwater ones. During the summer, larval insects and crustaceans are important sources of food for adults, while juveniles concentrate to a large extent on crustaceans such as cladocerans. Mollusks are also an important food for long-tailed ducks, but not to the extent that crustaceans are. In the winter the birds are found both on the open sea and on deep fresh-water lakes such as the Great Lakes; in the Chesapeake Bay area they also regularly extend into salt and brackish estuarine bays and sometimes into fresh-water estuaries. They then forage in waters that are deeper than those used by nearly all other diving ducks, having been caught by fishermen at depths of up to about 200 feet below the surface of Lake Michigan, and regularly foraging at depths of 50 feet or more (Ellarson, 1956).

Social behavior. Long-tailed ducks are highly social and gregarious birds, and spend a great deal of time during winter, spring, and early summer in pair-forming or territorial activities. They become sexually mature and usually form initial pair bonds in their second year of life; and although pair bonds are renewed annually, they tend to be quite strong, with females often remating with their mate of the previous year. Social displays are marked by a great deal of vocal activity, towering flights followed by a zigzag descent to the water, and short display flights by the drake toward the female. Females utter gutteral calls during a chin-lifting display that seems to represent inciting behavior. Males produce two distinctive courtship calls, the *ah-har-lik* call, uttered with or without a bill-toss, and a quite different call associated with the rear-end display, in which the head and neck are lowered over the water, the tail is raised vertically, and both feet are kicked slightly. Males also turn the back of the head toward females, and in addition perform a neck-stretching display. Various other postures that may represent displays have also been described (Alison, 1970; Myres, 1959). Copulation is evidently preceded by such male displays as bill tossing and lateral head shaking, and sometimes bill dipping, neck stretching, and porpoising. Many of these same displays have been observed in the postcopulatory situation as well (Alison, 1970).

Reproductive biology. Long-tailed ducks arrive on their high arctic breeding grounds already paired, and soon begin to establish territories that are strongly defended from incursions by other pairs, resulting in a high degree of dispersion. There is also a high degree of homing by females to ponds used in the past year. Although males establish and defend territories of varying sizes, females apparently rarely nest within the territorial limits, and instead often nest near other females, in a colonial manner. They particularly seek out islands in tundra ponds or lakes, and are thus usually quite close to the water's edge, but at times may place their nests several hundred feet from water. In Iceland, low shrubs provide the most frequent nest cover, followed by high shrubs, sedges, perennial herbs, and meadow cover (Bengtson, 1970). In some areas the birds also nest among tern colonies, presumably as an antipredator adaptation. Nesting is initiated rapidly by the female, with eggs being laid on an approximately daily basis and the clutch size averaging about 6 or 7 eggs. The incubation period, during which the male deserts his mate, lasts 24 to 26 days. Perhaps because of the short available breeding season, females seem quite prone to abandon their broods while they are still quite young and begin the postnuptial molt, and the young ducklings typically gather to form large, often parentless groups. They grow very rapidly, and reportedly require only 35 days to attain flight, perhaps the shortest fledging period of any of the arctic ducks (Alison, 1975; Bengtson, 1972a).

Status. The long-tailed duck is perhaps the most abundant of the arctic-nesting ducks, but its numbers are hard to estimate owing to its wide range and dispersal tendencies. Bellrose (1976) estimated an early summer (breeding) population of 3 to 4 million birds in North America. Estimates for Greenland,

Iceland, and most of Scandinavia are not available, but the western parts of the U.S.S.R. support about 740,000 breeding pairs, and about 100,000 birds normally winter in the Baltic area (Ogilvie, 1975). In the U.S.S.R. this species is the most abundant duck of tundra areas, especially in northern Siberia (Dementiev & Gladkov, 1967). Presumably its worldwide population might be in the neighborhood of 10 million birds.

Relationships. Although a well-defined genus, *Clangula* clearly shows structural and behavioral affinities with such other sea ducks as the scoters, the harlequin duck, and even the goldeneyes. Woolfenden (1961) regarded it as a part of the core of the sea duck group, with its closest links to *Histrionicus* and the scoters, while Brush (1976) found it has feather protein electrophoretic patterns identical to those of *Melanitta, Bucephala*, and *Mergus*.

Suggested readings. Alison, 1970, 1975; Ellarson, 1956.

Black Scoter

Melanitta nigra (Linnaeus) 1758

Other vernacular names. American scoter; Trauerente (German); macreuse noire (French); anade negro marino común (Spanish).

Subspecies and ranges. (See map 113.)
  M. n. nigra: European black scoter. Breeds in Iceland, Scotland, Spitsbergen, and northern Europe and Asia east at least to the Khatanga River. Winters mainly off the coast of western Europe, and on the Mediterranean, Black, and Caspian seas.
  M. n. americana: Pacific black scoter. Breeds in northern Asia from the Lena-Yana watershed to the Anadyr Basin and the Kamchatka Peninsula, on the Kurile Islands, and in North America from Bristol Bay north to about Kotzebue Sound and Mt. McKinley. Breeding records in Canada few and scattered. Winters on the Asian and North American coastlines of the Pacific Ocean, on the Great Lakes, and on the Atlantic coast of North America.

Measurements and weights. Folded wing: males, 228–42 mm; females, 220–29 mm. Culmen: males, 45–49 mm; females, 42–46 mm. Weights: adult males in fall and winter average ca. 1,100 g and females ca. 950 g, with respective maximums of 1,268 g and 1,087 g. Eggs: av. 66 x 45 mm, creamy to buff, 72 g.

Identification and field marks. Length 17–21" (43–51 cm). *Adult males* have a head, neck, and body that is entirely black, glossy above and less glossy below; a black and rather pointed tail; under wing coverts that are brownish black and silvery gray; and a glossy black upper wing surface. The bill is blackish, with a yellow-orange basal enlargement (larger in *americana* than in *nigra*) that terminates at the nostrils. The iris is brown, and the legs and feet brownish black, with darker webs. *Females* have the top of the head down to the level of the eyes and back of the neck dark brown, while the rest of the head and neck is whitish spotted with dusky coloration. The tail and upper wing surface are dusky brown. The iris is brown, the bill blackish, and the legs and feet are dark greenish brown. *Immature males* gradually become black except on the breast and wings during their first winter, and the bill begins to assume the adult shape and color. Maximal bill enlargement may

not occur until the third year. *Juveniles* resemble adult females, but are paler, especially on the underparts and lower half of the head.

In the field, the totally black color of male black scoters is the best field mark, with the conspicuous yellow bill marking (smaller in the European form) the only contrasting color present. Females are best identified on the basis of their two-toned head coloration that is reminiscent of some pochards. In flight, the birds also appear very dark, and often fly low but swiftly over the surf, with their wings producing a strong whistling noise. The male utters a clear, mellow whistle during courtship, and the female's voice is quite grating and similar to the sound of a door swinging on rusty hinges.

NATURAL HISTORY

Habitat and foods. The breeding habitat of the black scoter consists of fresh-water ponds, lakes, and rivers in tundra or wooded country, especially where shrubs are present for nesting cover. Yet, in spite of this seemingly generalized requirement, fitting much of northernmost North America, the species is relatively rare nearly everywhere, and is apparently

MAP 113. Breeding distributions of the European ("E") and Pacific ("P") black scoters. Wintering distributions are indicated by stippling.

common only in eastern Asia, especially around Kamchatka. Little more can be said of breeding requirements, except that lakes or ponds with islets seem to be preferred, and presumably those providing a source of summer foods such as crustaceans, mollusks, insects such as midges and caddis flies, and edible aquatic plants, including pondweeds. During fall and winter the birds occupy primarily salt-water habitats; in the Chesapeake Bay region they concentrate in the littoral zone of the ocean, usually just outside the zone of breakers, and forage in mussel-rich areas less than 25 feet in depth. In addition, they also feed on other mollusks such as periwinkles, crustaceans such as barnacles and shrimp, and to a very limited extent on echinoderms and fish. Although they are the smallest of the scoters, they are not particularly deep divers, and their dives probably rarely exceed 40 feet (Cottam, 1939).

Social behavior. Black scoters, like other sea ducks, do not reach adult plumage or sexual maturity until their second winter, and at that time presumably form initial pair bonds. However, during the winter the birds are usually well away from shore and hard to study, so that the intensity of pair-forming behavior at that time is uncertain. Most observations are for the spring migration period April through June, at or near the breeding grounds. Courtship occurs in small flocks which usually consist of a single female and five or more males, with the numbers of males increasing as the spring progresses. Displays of the males are diverse and often consist of direct aggressive actions or comfort movements such as head shaking, wing flapping, preening, body shaking, and the like. The most elaborate display sequence involves neck stretching, repeated calling, a vertical tail-snap, and a low rush toward or past the female. No definite inciting behavior has been described, but it almost certainly exists as in the other scoters. Evidently behavior associated with copulation is very simple, and involves preening movements by both sexes. The male mounts after performing a single body shake, and afterwards may swim away from the female while calling or may perform other postcopulatory behavior (Johnsgard, 1965a; Myres, 1959; Bengtson, 1966a).

Reproductive biology. Black scoters are relatively late nesters, arriving on their Alaskan breeding grounds in late May and not nesting until late June, but nesting somewhat earlier in Iceland. In Alaska the birds often nest on grass-covered islands, usually in large grass clumps of the past year's growth, while in Iceland they seem to favor dense stands of willow scrub or birch, where shrubby growth provides concealment for the nest. Island nesting is infrequent. In Iceland, initial nesting efforts average about 9 eggs, while renesting attempts average 6 eggs, but little can be said of average clutches in the North American population. Evidently the drake deserts his mate at about the time that incubation begins, and moves back to the coast or undertakes a molt migration of some distance. Incubation lasts about 27 or 28 days, with a few estimates of periods as long as 33 days. There are rather few observations on brood life, but apparently brood mergers are not typical of this species. The fledging period is still not established, but has been estimated as six to seven weeks. Molt migrations of adults are well developed in this species, and in the North Sea up to 150,000 birds, including adult males, juveniles of both sexes, and some adult females, gather in late summer (Salomonsen, 1968).

Status. The North American population of the black scoter is very difficult to judge with certainty, but it is thought that the Alaskan breeding population may consist of about 235,000 birds, and a presumably much smaller additional number breed in Canada. Wintering-ground counts suggest that the Aleutian Islands support about 250,000 birds, and another 155,000 winter south of Alaska (Bellrose, 1976). Late-summer and winter counts in western Europe suggest a total of about 150,000 birds (Ogilvie, 1976), and additionally there is a substantial wintering area off the east coast of Asia. Dementiev and Gladkov (1967) state that this species is second only to the long-tailed duck in abundance during spring migration along the Gulf of Penzhina (west coast of Kamchatka), but specific estimates of numbers were not provided.

Relationships. I concluded (1965a) that this is the most isolated of the three scoter species, but that it should not be generically separated from the other two. It may also provide the closest link with *Clangula* of the three scoters.

Suggested readings. Bengtson, 1966a, 1970.

Surf Scoter

Melanitta perspicillata (Linnaeus) 1758

Other vernacular names. Skunk-headed coot; Brillente (German); macreuse à lunnettes (French); anade marino de las rompientes (Spanish).

Subspecies and range. No subspecies recognized. Breeds in North America from western Alaska east through the Yukon and the Northwest Territories to southern Hudson Bay, and in the interior of Quebec and Labrador. Winters on the Pacific coast south to the Gulf of California, on the Atlantic coast south to Florida, and to some extent in the interior, especially on the Great Lakes. See map 114.

Measurements and weights. Folded wing: males, 240–56 mm; females, 223–35 mm. Culmen: males, 34–38 mm; females, 33–37 mm. Weights: males average ca. 1,000 g and females ca. 900 g, with maximum weights of about 1,130 g reported for both sexes. Eggs: av. 67 x 53 mm, creamy to pinkish buff, ca. 80 g.

Identification and field marks. Length 18–22" (46–55 cm). *Adult males* have a black head and neck except for a white forehead patch and a long, triangular patch on the nape, the point extending down the back of the neck. The rest of the body is black, except for the abdomen, which is mottled with lighter brown. The tail is short and black, and the upper wing surface is black. The iris is white, and the legs and feet are bright red on the outer side, orange-red on the inner side, with blackish webs. The bill has a pale yellow tip, and is otherwise white on the sides and red at the base above a swollen black oval area, shading to yellowish toward the tip. The lower mandible is yellow toward the base and flesh-colored elsewhere. *Females* have a dusky brown head and neck, with a darker crown and sometimes with a whitish patch on the back of the head comparable to that of the male, as well as two indistinct whitish patches on the face, one between the eye and the bill and the other in the ear region. The body is blackish brown except for a dusky abdomen; the tail and upper wing surface are blackish brown. The iris is dark brown, the bill is blackish with a black oval patch at the base that is surrounded by pale gray and generally less swollen than the male's. The legs and feet are dull yellowish to brownish, with darker webs. *Immature males* gradually acquire black feathers and white napes during their first winter, but not the white forehead. *Juveniles* resemble adult females but have a paler breast and lack the whitish nape patch. The whitish markings on the head are more conspicuous, and the head is dark brown from the eyes upward, resembling the black scoter pattern.

In the field, male scoters may be readily recognized by the white forehead and nape marks and the garishly colored bill. Females closely resemble female white-winged scoters, and may be difficult to recognize unless the white secondaries of that species can be seen. Like black scoters, this species shows no white in the wings when in flight, but its wings produce a humming rather than whistling noise, and the white head markings of males are usually readily apparent.

NATURAL HISTORY

Habitat and foods. No detailed analysis of breeding habitats of the surf scoter is available, but like the other scoters, it evidently prefers to nest around fresh-water lakes, ponds, or rivers with shrubby or low woodland cover in the vicinity. Wintering habitats include the littoral zone of the ocean and adjoining coastal bays, sometimes extending up into brackish estuaries. Very infrequently stray birds appear on fresh-water lakes or larger rivers in the interior during winter; they are usually females or immature birds. Not much is known of food preferences either, except that mollusks, and particularly mussels, are apparently the major dietary item, with crustaceans, insects, and plant materials being minor components except perhaps among juveniles (Cottam, 1939).

Social behavior. Apparently pair bonds are reestablished each winter and spring, following the first year of life. There is a prolonged period of social display among wintering birds, which has been best documented by Myres (1959). Flock sizes in early winter are fairly large, but courting parties frequently consist of a single female and a small number of active males, with much fighting or threatening behavior typical. Females have few obvious displays; chin lifting associated with a crowlike call is the most obvious one and seems to represent a form of inciting behavior. Many of the male displays consist of variably ritualized aggressive postures, such as crouch and threat postures, often interspersed with underwater chases. There is also a stretched-neck posture called the sentinel, and from this posture breast scooping is performed, which seems to be a combination of head-shaking and breast-preening movements, accompanied by a gurgling call. A chest-

lifting display that has a strong similarity to the rearing of the Steller eider is likewise performed by males, as well as a short display flight that is terminated by holding the wings in an upraised position as the bird skids to a stop. Copulation takes the same form as in the other sea ducks, with the female assuming a prone position while the male performs an extended series of bill-dipping, preening, and drinking movements. After treading, the male typically performs a single chest-lifting display (Myres, 1959).

Reproductive biology. Very few nests of surf scoters have been described, but it appears that they are often located some distance from water, and are always very well concealed from view. Nests are reportedly often concealed under low branches of a conifer, under bushes, or in grass. Clutch sizes of from 5 to 7 eggs are apparently typical, and up to 9 eggs have been reported. The incubation period is still unknown, but is likely to be no more than the 27 or 28 days characteristic of the other scoters. Several broods have been seen with two females in attendance, suggesting that brood mergers may be common in this species. Evidently males undertake molt migrations to the coastline before hatching and together with nonbreeding females molt in shallow bays and inlets. The fledging period has not been established, but it is possible that females leave their broods in advance of fledging and also undertake a migration prior to molting (Bellrose, 1976; Johnsgard, 1975).

Status. Bellrose (1976) has estimated a wintering population of 765,000 surf scoters and a breeding population of 257,000 birds, but these figures are highly uncertain, inasmuch as little effort is made by federal biologists to distinguish the scoter species during their annual surveys. In any case, the surf scoter is clearly the least common of the three scoter species on a worldwide basis. Like some of the other sea ducks, its major threat to survival is perhaps the possibility of massive losses due to oil spills on areas of concentration, since its breeding areas are diffuse and still fairly undisturbed.

Relationships. I have suggested (1965a) that the surf scoter and white-winged scoter are probably fairly closely related but in addition show some similarities to other sea ducks such as the Steller eider and the goldeneyes, emphasizing the central position of the scoters among the sea ducks.

Suggested readings. Myres, 1959; Bellrose, 1976; Johnsgard, 1975.

White-winged Scoter

Melanitta fusca (Linnaeus) 1758

Other vernacular names. Velvet scoter, white-winged sea coot; Samtente (German); macreuse à ailes blanches (French); anade marino de alas blancas (Spanish).

Subspecies and ranges. (See map 115.)

  M. f. fusca: European white-winged scoter. Breeds across northern Europe and Asia from Scandinavia to at least the Yenisei and perhaps to the mouth of the Khatanga River. Winters in the Atlantic Ocean from Norway to Spain, and on the Caspian Sea.

  M. f. stejnegeri: Asiatic white-winged scoter. Breeds in eastern Asia from the Altai to Kamchatka Peninsula, and perhaps on the Commander Islands. Winters on the coast of eastern Asia south to Japan and China.

  M. f. deglandi: Pacific white-winged scoter. Breeds in North America from northwestern Alaska east to Hudson Bay and south to southern Manitoba. Winters on both coastlines, south to Baja California and to South Carolina.

Measurements and weights. Folded wing: males, 269–93 mm; females, 251–66 mm. Culmen: males, 37–50 mm; females, 38–43 mm. Weights: males (of *fusca* and *deglandi*) in fall and winter average from 1,500 to 1,700 g and females from 1,200 to 1,600 g,

MAP 114. Breeding (hatched) and wintering (stippling) distributions of the surf scoter.

with considerable age and seasonal variation. Eggs: av. 72 x 48 mm, creamy to buff, 92 g.

Identification and field marks. Length 19–24" (48–61 cm). *Adult males* have a black head and neck except for a small crescent-shaped white spot around and behind the eye; otherwise the entire plumage is black, except for white secondaries and their greater coverts and a brownish tinge on the sides and flanks. The iris is white; the legs and feet are orange vermillion on the inner sides and purplish pink on the outer sides, with dusky webs. The bill is enlarged at the base with a black knob (largest in *stejnegeri*, smallest in *fusca*), a yellow to whitish tip and nail, and reddish to yellow or orange striping on the sides of the bill, varying with the subspecies. *Females* have a brownish black head and neck, with paler spots between the eye and bill and in the region of the ear. Otherwise the body is entirely blackish brown, except for whitish edging on some of the body feathers and white secondaries. The iris is brown, the bill is dull black, mixed with whitish and sometimes also pink on the upper mandible, and the legs and feet are light brownish red, with darker webs. *Immature males* lose their femalelike facial markings the first winter and begin to assume the first black feathers of their adult plumage. The iris becomes white by the second autumn of life. *Juveniles* resemble adult females but are paler on the breast, and sometimes are even whitish there. The pale head markings are also much larger and more whitish than in adult females.

In the field, white-winged scoters are the bulkiest of all the scoter species, and at any distance males ap-pear nearly black, since the white eye markings are easily overlooked and the white wing speculum is often hidden in swimming birds. Females lack the white nape markings sometimes shown in female surf scoters, and additionally show less overall contrast in their head markings. The best field mark is the white speculum of both sexes, which is readily visible in flying birds, plus the ponderous flight of this species, which often flies at low altitude in long strings or loose flocks. Females sometimes utter a thin whistling note, and males have a more bell-like call that is ap-parently associated with courtship display.

NATURAL HISTORY

Habitat and foods. The breeding habitat of this rather widespread species of scoter consists of coast-lines and lakes in the northern coniferous forest, especially where there are boulder-covered islets with shrubs and low trees present, but also extensive her-baceous vegetation. The birds often nest well away from coastal areas, even in the interior of continents, and thus access to salt water is not necessary. Sum-mer foods are not well studied, but at least juvenile birds seem to concentrate on crustaceans such as am-phipods, while the larvae of stone flies (*Perlidae*) and caddis flies (*Trichoptera*) have been found in some adult males. Fall and winter foods are mainly ob-tained in coastal regions, where the birds concentrate in waters that are usually under 20 feet deep, in the littoral zone just beyond the breakers. They also ex-tend into brackish-water estuaries, and stragglers sometimes winter on larger rivers and reservoirs.

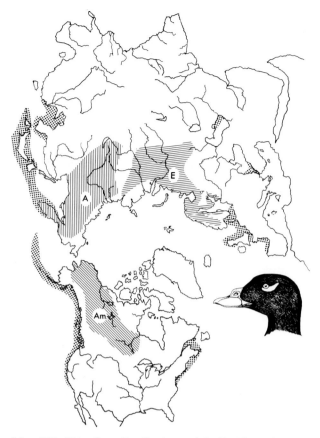

MAP 115. Breeding distributions of the Pacific or American ("Am"), Asian ("A"), and European ("E") white-winged scoters. Wintering distributions indicated by stippling.

Mollusks, primarily bivalves such as mussels, clams, and scallops, predominate in the winter foods, and to a much smaller extent such crustaceans as crabs are consumed. The birds often remain submerged for considerable durations, frequently for a minute or more, when foraging in deep water, and even ducklings only a few days old can remain submerged for up to 30 seconds when being chased (Johnsgard, 1975; Bellrose, 1976).

Social behavior. During most of the year this species is on coastal waters, where flocks of considerable size often gather. The birds are not known to nest before their second year of life, and some may not breed until even later, but probably all second-year females form pair bonds. This probably occurs in late winter and spring, since nearly all birds are paired when they arrive on their nesting grounds. Social displays include several aggressive postures such as the crouch, and a more extreme threat or attack posture with the head and neck stretched forward in the

water. There is also a neck-erect-forward posture with the neck outline greatly thickened, which may also serve as a threat display. Males are known to produce a whistling call, but the associated posturing has not been described, and females also utter a thin whistle, apparently in conjunction with inciting behavior. Several other rather simple displays have been described, such as wing flapping, drinking, body shaking or upward-stretch, and preening behind the wing, but at least some of these are probably more directly associated with copulatory behavior. Evidently a combination of display preening and water-twitching, preceded by display drinking, provide the major precopulatory displays of males, but no specific postcopulatory posturing has yet been described (Myres, 1959).

Reproductive biology. These scoters arrive on the nesting grounds relatively late, and furthermore a period of a month or more may elapse before the female begins egg laying. The ecological significance of this adaptation is obscure, but probably restricts the potential for arctic breeding as well as for renesting. In the Gulf of Bothnia, Hilden (1964) found that females usually selected junipers or bushes for nest cover, with broad-leaved herbs, herb-shrub mixtures, or boulders accounting for most of the rest of the sites. Nests were generally placed in woodland areas well away from the shore and almost invariably were well concealed from above, with exposed nests typically found only where the birds nested in association with gulls or terns. The interval between successive eggs has been estimated at about 40 hours, and clutch sizes seem to average 9 to 10 eggs, with later clutches slightly smaller than earlier ones. The incubation period is usually between 26 and 29 days, averaging 27.5 days, and probably males abandon their mates at about the time incubation gets underway. Likewise, females tend to abandon their young at a relatively early stage, probably in order to complete their own molt before bad weather begins; and as a result a good deal of brood merging usually occurs in the first week or two of life. The fledging period has not been established with certainty, but indirect evidence suggests that it may be between 63 and 77 days. In contrast to the males, which often undertake molt migrations of considerable length, breeding females probably normally molt on their breeding grounds (Rawls, 1949; Hilden, 1964; Koskimies & Routamo, 1953).

Status. Using the only available evidence, which is rather questionable, Bellrose (1976) indicated that the

North American population of breeding white-winged scoters might be as high as 675,000 birds, but that this is likely to be an inflated estimate. However, about 250,000 apparently winter in the Aleutian Islands alone. Also, the wintering range on the Pacific coast extends from Alaska to Baja California, and nearly 150,000 birds winter from southern Alaska southward. Since as many as 100,000 white-winged scoters have been reported in a single day off the Massachusetts coast, and as many as 400,000 birds have been estimated in winter flocks of Cape Cod and Nantucket, there may indeed be close to a million white-winged scoters in North America. In Europe, about 30,000 birds winter in the Baltic area, and some winter south to Spain and around the Black and Caspian seas as well (Ogilvie, 1975). The size of the East Asian population of white-winged scoters is totally unknown.

Relationships. Plumage similarities of the downy young and females of the surf and white-winged scoters indicate that they are very close relatives, and in addition, some surprising behavioral similarities occur between *Melanitta* and *Bucephala* (Johnsgard, 1965a; Myres, 1959).

Suggested readings. Hilden, 1964; Rawls, 1949; Koskimies & Routamo, 1953.

Bufflehead

Bucephala albeola (Linnaeus) 1758

Other vernacular names. Butterball; Büffelkopfente (German); garrot albéole (French); pato cabeza clara (Spanish).

Subspecies and range. No subspecies recognized. Breeds in North America from southern Alaska and northern Mackenzie District through the forested portions of Canada east to James Bay and south into the western United States to northern California and northwestern Wyoming. Winters along the Pacific coast from the Aleutian Islands to central Mexico, along the Gulf and Atlantic coasts from southern Canada to Texas, and in the interior where open water occurs. See map 116.

Measurements and weights. Folded wing: males, 163–80 mm; females, 150–63 mm. Culmen: males, 25–29 mm; females, 23–26 mm. Weights: males average ca. 450 g and females ca. 330 g, with maximum weights of about 590 g in both sexes occurring during fall. Eggs: av. 52 x 37 mm, creamy to pale olive buff, 37 g.

Identification and field marks. Length 13–16″ (33–40 cm). Plate 56. *Adult males* in breeding plumage have a large white patch extending across the back of the head from cheek to cheek, forming a bushy crest, with the rest of the head iridescent with shades of purple blue, bronze, and green. The neck, scapulars, breast, underparts, and sides are white; the abdomen is slightly grayish; and the flank feathers are white, narrowly edged with black, as are the outer tail coverts. The tail is dark gray, the upper wing surface is blackish, except for a white speculum that extends from the inner and middle secondaries diagonally forward across the greater and middle coverts to the outer lesser and marginal coverts, which are grayish to dusky. The iris is dark brown, the bill is dusky, with a darker tip, and the legs and feet are flesh pink. *Females* have a brownish head and neck except for an oval white patch that extends from below the eye back toward the nape. The back is dull blackish; the sides, flanks, breast, and rear underparts are dark sooty gray; the abdomen is whitish; and the tail is grayish brown. The upper wing surface is blackish brown, except for the inner and middle secondaries and inner greater coverts, which are white, tipped with blackish. The iris is dark brown, the bill is lead-color, and the legs and feet are bluish gray to pinkish. *Males in eclipse* resemble females, but have more white on the sides of the head and on the upper wing coverts. *Immature males* begin to acquire a large white nape patch in their first winter, but the rest of the head remains largely brownish. Males at this age have darker heads than females, and their upper tail coverts are paler grayish brown. *Juveniles* resemble adult females, but are more brownish and generally duller, with smaller white patches on the sides of the head.

In the field, the predominance of white in the body plumage of males makes them highly conspicuous and distinctive; only the smew has as much or more white. The large white crest marking of the male is also distinctive; unlike that of the hooded merganser, it is not tipped with black. The female is an extremely small duck that might be mistaken for a small grebe; its small teardrop-shaped white head marking pro-

vides the best field mark. In flight, the birds flash their white speculum pattern and are smaller than other North American diving ducks except ruddy ducks. Both sexes are unusually silent, even during courtship display.

NATURAL HISTORY

Habitat and foods. Favored breeding habitats of this North American hole-nesting duck consist of lakes and ponds in or near open temperate woodland, particularly those lakes that are moderately fertile, with large areas of open water and maximum depths of at least three meters. Trees that are located either in water or very close to it, and which have cavities such as those formed by flickers (*Colaptes*) are obviously an important breeding habitat component. Lakes that have an abundance of such summer foods as nymphs, water boatmen, aquatic beetles and their larvae, and other aquatic invertebrates are also favored. Snails are also eaten in quantity when the birds are on fresh-water areas, but in winter they are as likely to be found in brackish-water habitats such as estuarine bays, where they feed on crustaceans and mollusks to a much larger degree. Even fish are consumed to some extent by wintering birds, and in some areas aquatic vegetation is also consumed in considerable quantities during that time of year. Buffleheads are evidently opportunistic foragers, and probably can utilize considerably smaller items as major foods, as can goldeneyes, which might account for the importance of snails and small bivalve mollusks present in most food samples (Erskine, 1972b; Wienmeyer, 1967).

Social behavior. Buffleheads become sexually mature during their second winter of life, and as early as late January may begin the social displays that result in pair bonding. The flock sizes of adult wintering birds tend to be rather small, a fact which is perhaps related to the surprisingly high level of aggressiveness typical of courting birds. Most of the male displays are clearly aggressive in nature, and are sometimes alternated with actual male-to-male attacks, often by underwater approach. Probably the most common male display is oblique-pumping, which is directed both to other males and toward females. Crest erection is also common, but no calling is performed by males during display activity. Short flights, followed by a complex and rapid sequence of wing flapping, head bobbing, and lifting of the folded wings over the back, are frequent. Females regularly swim behind displaying males in a following display, and also raise their crests in a head display, both of which are important stimuli to male display activity. Copulation is preceded by a very short prone posture on the part of the female, while the male typically performs preening and bill-dipping movements combined with lateral head movements. Postcopulatory behavior is evidently quite variable, but often consists of vigorous bathing or diving by the male (Myres, 1959).

Reproductive biology. On their return to the breeding grounds, females often seek out areas where they were raised, and often return to nest sites that they used in a previous year. Often many pairs will share a lake; Erskine (1972b) found a few cases of single trees having two simultaneously occupied nests. If females do not use the nest site of a previous year, they usually nest nearby, and will accept tree holes from as low as a meter above the water to more than 15 meters, and with openings as small as six centimeters in diameter. Eggs are laid at rather variable intervals, but average about 38 hours apart, and clutch sizes for initial nesting attempts are 8 or 9 eggs. Dump-nesting tendencies may inflate observed clutch

MAP 116. Breeding (hatched) distribution of the buffle-head, including areas of major concentrations (cross-hatched). Wintering distribution indicated by stippling.

sizes, and renesting efforts tend to produce smaller-than-average clutches. Apparently most males leave their territories as soon as the female begins incubation, and at least in some areas probably undergo a migration to favored molting areas. The incubation period averages about 30 days under natural conditions; and shortly after hatching, the young jump out of their nesting holes and are led to water. Although brood territories are established by the females, transfers of young from one brood to another are not uncommon, and at times single broods have been seen with two females in attendance. It is estimated that fledging requires from 50 to 55 days, and probably the mother abandons the ducklings before this time so that she can undergo her own flightless period, but relatively little is known of this phase of the reproductive cycle (Erskine, 1972b).

Status. Erskine (1972b) estimated a breeding-ground population of 500,000 buffleheads in the early 1960s, and Bellrose (1976) produced more recent evidence suggesting spring populations approaching 750,000 birds. Evidently the species has increased somewhat in the midwestern and eastern states in recent years,

but has declined in the far western states, for reasons that are not now apparent.

Relationships. Most taxonomists now support the view that this species should be included with the goldeneyes in the single genus *Bucephala*. Erskine (1972b) believes that *Mergus* represents the nearest generic relative of *Bucephala*, but it is significant that a number of the behavioral traits of *Bucephala* also occur in *Melanitta*, and the scoters must also be considered closely related to the goldeneyes (Johnsgard, 1960c).

Suggested readings. Myers, 1959; Erskine, 1972b.

Barrow Goldeneye

Bucephala islandica (Gmelin) 1789

Other vernacular names. None in general English use. Spatelente (German); garrot d'islande (French); pato ojos dorados de Barrow (Spanish).

Subspecies and range. No subspecies recognized. Breeds in Iceland, southwestern Greenland, northern Labrador, and from southern Alaska and Mackenzie District southward through the western states and provinces to California and Wyoming. Winters primarily along the Pacific coast south to central California, also along the Atlantic coast to the mid-Atlantic states, and around Greenland and Iceland. See map 117.

Measurements and weights. Folded wing: males, 232–48 mm; females, 205–24 mm. Culmen: males, 31–36 mm; females, 28–31 mm. Weights: males in fall average ca. 1,100 g and females ca. 800 g, with respective maximums of 1,314 and 907 g. Eggs: av. 62 x 45 mm, bluish green, 70 g.

Identification and field marks. Length 16–20" (40–51 cm). Plate 57. *Adult males* in breeding plumage are similar to those of the common goldeneye, differing in the following major ways: the white patch on the head is crescent-shaped, and the iridescence is glossy purple. The head has a flatter crown profile and longer nape feathers, and the nail of the bill is distinctly raised above the culmen profile. The body is more extensively black, especially on the flanks, which are more heavily margined with black, and on

the scapulars, which are margined with black in such a way as to produce a pattern of oval white spots in linear series. The upper wing surface is also more extensively black, especially on the lesser and middle coverts. The soft-part colors are the same. *Females* likewise resemble female common goldeneyes, but have a slightly darker brown head with a more flattened crown profile, a broader and more pronounced ashy brown breast band, and more black on the middle and lesser wing coverts. The soft-part colors are the same, except that the bill is more extensively yellowish, and often is entirely bright yellow in breeding females. *Males in eclipse* resemble females,

but have darker heads and often retain a trace of the white cheek pattern. The middle wing coverts are also more extensively white. *Immature males* are like those of the common goldeneye, with the white face markings appearing late in the first winter or spring, and the head remaining brownish. *Juveniles* resemble adult females, but initially have a brownish iris.

In the field, this species is most likely to be confused with the common goldeneye, but the differences in head shape and in the white cheek markings on the male provide for fairly ready identification. Lone females are much more difficult to identify. If the bill is entirely yellow, the bird can confidently be regarded as a Barrow goldeneye; and when the two species are side by side, the differences in head shape and overall darkness of head and breast are evident. In flight, males show less white in the forewing than do common goldeneyes, and the white speculum of the secondaries is separated from the forewing white patch by a narrow black line that is lacking in the common goldeneye. Males lack the loud whistling notes of courting common goldeneyes, and instead utter a variety of grunting and clicking sounds. Calls of the female are soft, and even during courtship females are relatively silent birds.

NATURAL HISTORY

Habitat and foods. The breeding habitat of this species is rather more flexible than that of the common goldeneye or bufflehead, inasmuch as it is not dependent upon tree cavities for nest sites, and instead can utilize rock crevices or even surface nest sites in treeless habitats. Thus, its distribution may be more closely related to the availability of food, mostly in the form of amphipods and other aquatic invertebrates associated with fresh-water or alkaline lakes. In Iceland, where large trees are lacking, the birds seem to prefer to breed near running water rather than lakes, and in most areas deep lakes with little

MAP 117. Breeding (hatched) and wintering (stippling) distributions of the Barrow goldeneye.

aquatic vegetation present are apparently avoided. Summer foods include not only amphipods, but also many aquatic insects, especially the nymphs of dragonflies and damsel flies. After the breeding season the birds typically move to brackish water in estuaries or to coastal lakes, with salt-water environments seemingly avoided. At that time of the year mollusks, especially mussels, become important foods, and are supplemented by crustaceans, fish, and marine algae. Many birds also winter in the interior, on fresh-water lakes and rivers that remain unfrozen, and at such times may even resort to feeding on grain provided by humans when it is available (Cottam, 1939; Johnsgard, 1975).

Social behavior. Like the other sea ducks, Barrow goldeneyes become sexually mature during their second winter of life, and at that time begin an extended period of social display that persists for several months through the spring migration. Yearling birds are often present in the flocks of wintering adults but do not participate in social display, and during such display the courting groups typically consist of several adult males and one or two females. There is a great deal of overt and ritualized aggression among the males and a considerable degree of aggression by females toward males. The major female display is a strong side-to-side movement of the head, performed silently but certainly equivalent to inciting behavior, as well as rotary pumping movements of the head, also silently performed. The major displays of the males consist of a head-throw associated with a kick and splash of water, an aggressive crouch posture, a neck-withdrawing movement often performed in response to female inciting, and rotary pumping movements generally given in conjunction with the same display by females. Precopulatory behavior consists of the female's assuming a prone posture on the water, usually after mutual drinking by both birds, and remaining in that posture while the male performs a long series of displays that are largely derived from comfort movements. Treading is always preceded by a stereotyped sequence of rapid movements and a quick rush to the female. After copulation the male swims rapidly away from the female with his neck extended and the crest fluffed, while uttering grunting sounds and performing lateral head-turning movements (Johnsgard, 1965a; Myres, 1959).

Reproductive biology. One of the few studies of the nesting biology of this species is that of Bengtson (1971b), who found that among ten species that he studied, this one was unique in its establishment of a true territory before egg laying, as well as its formation of defended brood territories after hatching. In Iceland, where tree-cavity nesting is not possible, the birds nested primarily in other kinds of cavities or even under dense shrubs, with the density of the nesting birds ranging from 30 to 600 pairs per square kilometer. Most of the nests were located within ten meters of water, and studies of tree nesting in British Columbia indicated that the birds usually selected holes with entrances between 7 and 10 centimeters in diameter, and apparently preferred cavities with vertical rather than horizontal openings (Erskine, 1960). Clutch sizes of initial nesting efforts are usually between 9 and 11 eggs, while renesting attempts have appreciably smaller clutch sizes. Males apparently desert their mates shortly after incubation gets underway, and the incubation period under natural conditions is fairly long, averaging about 32 days. Females establish strongly defended brood territories for their young, and not only attack the young of other species but also strange young of their own species, which in densely inhabited areas may be a serious cause of duckling mortality. The fledging period is approximately eight weeks, but the adult females usually abandon their young before fledging and either fly up to 25 miles to a molting area, or undergo their flightless period in the same vicinity (Bellrose, 1976; Munro, 1939).

Status. The two major populations of this species occur in Iceland and in western North America. The Icelandic population is believed to number only about 1,000 pairs, mostly concentrated in the Lake Mývatn area (Hudson, 1975). Bellrose (1976) has made a "crude" estimation of the prebreeding population in western North America as between 125,000 and 150,000 birds. Beyond these populations, the Labrador and Greenland birds add a small additional component to the total, which is perhaps no more than 200,000 breeding birds.

Relationships. It is interesting that in spite of the remarkable similarities between the females of common and of Barrow goldeneyes, natural hybrids between the species have been reported only a few times. This might be largely related to the substantial differences in male display behavior and vocalizations, although the behavior associated with copulation is nearly the same in both forms (Johnsgard, 1965a). Another curious feature of the Barrow goldeneye is its failure to colonize any part of Asia, in spite of its rather more flexible nesting requirements than those characteristic of the far more

cosmopolitan common goldeneye. Perhaps competition with that species has restricted its opportunities for range expansion.

Suggested readings. Munro, 1939; Myres, 1959.

Goldeneye (Common Goldeneye)

Bucephala clangula (Linnaeus) 1758

Other vernacular names. Whistler; Schellente (German); garrot ordinaire (French); pato ojos común (Spanish).

Subspecies and ranges. (See map 118.)

B. c. clangula: European goldeneye. Breeds in Iceland, northern Europe, and Asia from Norway to Kamchatka and Sakhalin. Winters from the British Isles, southern Scandinavia, and the southern limits of its breeding range south to the Mediterranean, the Persian Gulf, northern India, Burma, southern China, and Japan.

B. c. americana: American goldeneye. Breeds from Alaska to southern Labrador and Newfoundland, and south through the forested portions of Canada plus northern and northeastern parts of the United States. Winters from the Alaskan coast south to California, in the interior wherever open water is present, and on the Atlantic coast from Newfoundland to Florida.

Measurements and weights. Folded wing: males, 215–35 mm; females, 188–220 mm. Culmen: males, 35–43.5 mm; females, 28–35 mm. Weights: males (of both races) average ca. 1,000 g in fall and females ca. 800 g, with maximums of 1,406 and 1,133 g, respectively. Eggs: av. 60 x 42 mm, greenish, 57 g.

Identification and field marks. Length 16–20" (40–51 cm). *Adult males* in breeding plumage have a white circular to oval patch between the eye and the base of the bill, and the rest of the head blackish with an iridescent greenish gloss. The neck, breast, anterior underparts, sides, flanks, and some scapulars are white, the flanks and longer scapulars having narrow black margins. The shorter scapulars are entirely black. The back, rump, and upper tail coverts are

buffleheads, but are appreciably larger, and might also be confused with goosanders, which have a similar white wing speculum but are much longer in body outline. Males utter loud whistling notes during aquatic courtship, and the wings of birds in flight produce a strong whistling noise (thus the vernacular name "whistler"), but otherwise both sexes are relatively silent.

NATURAL HISTORY

black; the tail is grayish brown; and the under tail coverts and rear underparts are grayish to dusky. The upper wing surface is blackish to blackish brown, except for a white patch that includes the exposed webs of about six secondaries, eight greater coverts, and the adjoining middle and lesser coverts. The iris is bright yellow, the bill is black, and the legs and feet are bright yellow to orange. *Females* have a head that is entirely dark chocolate brown, separated from a more grayish body by a white neck band. The upper breast, anterior back, sides, and flanks are gray to brownish gray, the feathers edged with lighter tips, and the back becoming darker toward the rump and upper tail coverts. The tail is grayish brown, the under tail coverts are sooty gray, and the abdomen becomes white anteriorly. The upper wing surface is brownish black to black and darkest on the outer secondaries, tertials, and adjoining secondaries, while the middle five secondaries are white, with their greater coverts also white but tipped with blackish coloration. The middle coverts are mostly grayish white or gray tipped with white. The iris is pale yellow to greenish yellow, the bill is dusky, sometimes tipped with yellow, and the legs and feet are dull orange with darker webs. *Males in eclipse* resemble females but usually have a darker head color, and a trace of the white cheek spot is often evident. The wing coverts also remain more extensively white at this time. *Immature males* during the first fall and winter assume a plumage that is darker on the back than are females, and a white spot on the cheeks may be evident by spring. *Juveniles* resemble adult females, but have a brown iris and are generally darker on their upper wing coverts.

In the field, common goldeneyes are most often confused with Barrow goldeneyes; distinctions are provided in the account of that species. They might also be confused with various mergansers, but no merganser species has such distinctively yellow eyes or a short and stout bill. In flight the birds resemble

Habitat and foods. This goldeneye breeds over a very wide area of the Northern Hemisphere, primarily where lakes or deep marshes having abundant invertebrate life are located near forests or stands of hardwood or coniferous trees of moderate size. Rock crevices are apparently not suitable substitutes for nest sites in this species, but a variety of tree species are utilized for nesting. Foods of juveniles during the breeding season probably consist largely of inverte-

MAP 118. Breeding distributions of the American ("A") and Eurasian ("E") common goldeneyes. Wintering distributions indicated by stippling.

brates such as larval stages of dragonflies, damsel flies, and May flies, as well as adult aquatic insects. Probably much the same applies to adults, but little information on summer foods is yet available. During the rest of the year the birds are highly opportunistic and forage on a wide array of animal life, including crustaceans, mollusks, fish, frogs, tadpoles, worms, and virtually anything else that is available in interior and coastal waters. They are widely distributed ecologically at such times, but seem to prefer brackish estuarine bays as well as saltwater bays, with nonsaline waters utilized mainly during migration periods. They also forage at depths ranging from a few to 20 feet or more, but prefer the shallower areas (Johnsgard, 1975; Bellrose, 1976).

Social behavior. Goldeneyes are known normally to mature in their second winter of life, and occasionally even yearling females attempt to breed but probably only rarely do so successfully. Pair bonds are renewed annually during a prolonged period of courtship in winter and spring, which normally peaks about March, by which time about 80 percent of the females appear to be paired. Goldeneye courtship is perhaps the most spectacular and complex of that of any North American duck, with males showing a bewildering array of postures and calls that seem to have little predictability or obvious differences in function. Clearly, many are derived from hostile patterns, and direct threats or attacks among the males are not uncommon. Likewise, the inciting behavior of females, although highly ritualized, is clearly of hostile origin. Three different types of head-throw occur, one of which lacks an associated kicking of the feet, while the other two are slow and fast versions of a simultaneous head-throw and backward kicking movement that throws water behind the displaying bird. Several other silent displays are also performed, all of which have been analyzed carefully as to their time and sequential characteristics (Dane & Van der Kloot, 1964). Copulation is likewise preceded by an extended and diverse series of male displays, performed while the female lies prone in the water. Treading is preceded by a rigidly stereotyped sequence that is terminated by a rapid steaming to the female, and afterward the male retreats from her in an equally stereotyped manner (Johnsgard, 1965a; Lind, 1959).

Reproductive biology. Goldeneyes are relatively early migrants, and pairs begin to return to their breeding grounds soon after they have thawed. Like most hole-nesting ducks, females exhibit a strong homing tendency and very often nest in the same cavity as they used the previous year. If natural cavities are used, they are generally in trees that are large enough to provide internal cavity diameters of about 20 centimeters, although the height of the opening and its dimensions seemingly are not so important. Cavities with lateral openings are used more often than those with vertical openings, and trees that are in open stands or near the edges of marshes seem to be preferred to those in dense stands, perhaps for ease of entry by flying birds. Females deposit their eggs at the rate of one per 1.3 to 2.0 days and have clutch sizes that average about 10 eggs. In areas where cavities are limited, considerable competition for available nest sites occurs and several females may attempt to use the same site, often resulting in desertion and reduced hatching success. The incubation period averages 30 days, but ranges from 27 to 32 days under natural conditions. Females usually leave the nest the day after the eggs have hatched and quickly lead their broods to water. Females are seemingly rather careless mothers, often losing several of the ducklings in the process of moving them from one area to another. Such abandoned ducklings combine to form rather large groups, especially as they grow older. The fledging period has been variously estimated to require 56 to 66 days, and probably in most cases the female will have abandoned her brood before fledging in order to begin her own flightless period, but she usually remains in the same general area to molt. On the other hand, males typically abandon their mates about the time incubation gets underway, and at least in some areas often move to river mouths or coastal inlets to undergo their molts (Carter, 1958; Johnson, 1967; Johnsgard, 1975).

Status. Bellrose (1976) suggested that a breeding-grounds estimate of 1.25 million birds in Alaska and Canada represents a "crude" index to this species' abundance in North America. Ogilvie (1975) indicated that the wintering population in western Europe is about 150,000, with the majority located in the Baltic. However, in the western U.S.S.R. alone there are an estimated 120,000 pairs, plus an additional 50,000 in Finland and unknown but considerable numbers breeding in Scandinavia. Likewise, the population of eastern Asia must be considerable.

Relationships. As noted in the account of the Barrow goldeneye, these species are obviously very close relatives, as are the bufflehead and, perhaps surprisingly, the smew and hooded merganser. Bill shape cannot confidently be used to estimate relationships among the members of this group, which show strong similarities in behavior, the patterns of their downy young, skeletal characteristics, and electrophoretic patterns of the feather profiles (Johnsgard, 1965a; Woolfenden, 1961; Brush, 1976).

Suggested readings. Carter, 1958; Dane & Van der Kloot, 1964.

Hooded Merganser

Mergus cucullatus (Linnaeus) 1758

Other vernacular names. Fish duck, sawbill; Kappensäger (German); harle couronné (French); mergo capuchino (Spanish).

Subspecies and range. No subspecies recognized. Breeds in North America from southeastern Alaska and adjacent Canada east through the southern and middle wooded portions of the border provinces to New Brunswick and Nova Scotia, and south to Oregon, Idaho, and Montana. East of the Great Plains it breeds from Minnesota through the Great Lakes states to the Atlantic coast, with less frequent breeding southward through the Mississippi Valley to the Gulf Coast. Winters along the Pacific coast south to Mexico, along the Gulf Coast, and on the Atlantic coast north to New England. See map 119.

Measurements and weights. Folded wing: males, 195–201 mm; females, 184–98 mm. Culmen: males, 38–41 mm; females, 35–39 mm. Weights: males in fall average ca. 680 g and females ca. 540 g, with respective maximums of ca. 900 and 680 g. Eggs: av. 52 x 47 mm, white, 60 g.

Identification and field marks. Length 16–19″ (40–48 cm). Plate 58. *Adult males* in breeding plumage have a blackish brown forehead and the rest of the head and neck glossy black, except for a large, erectile white patch extending from behind the eyes in a triangular fashion to the tip of a rounded crest, which is margined with black. The back is blackish brown, shading to grayish brown on the rump and upper tail coverts, and the tail is dark sooty brown. The scapulars are black, with black also extending down the sides of the breast in two vertical bands; the rest of the underparts are white, except for reddish brown sides and flanks, which are vermiculated with black. The lesser and middle wing coverts are ashy gray; the greater coverts are blackish, some of them being tipped with white; while the primaries, their coverts, and outer secondaries are grayish brown to blackish. The inner secondaries are black, with white outer margins, and the black tertials are pointed and elongated, with white central stripes. The iris is yellow, the bill is blackish, and the legs and feet are dull brown, with darker webs. *Females* have a head and neck that are grayish brown except for the crest, which is lighter and brighter, and a whitish chin and throat. The upperparts are generally dark brown, with the tips of the scapulars and upper tail coverts lighter, and the tail is dark sooty brown. The breast is ashy brown, shading to whitish on the abdomen and darker brownish on the sides and flanks. The upper wing surface is similar to the male's but the lesser and middle wing coverts are not lighter than the other coverts, the secondaries and greater coverts have less white present, and the tertials are less pointed and are brownish with white stripes. The iris is yellow-brown, the bill is blackish, with dull orange on the lower mandible and basal half of the upper mandible, and the legs and feet are dull yellow to brown, with darker webs. *Males in eclipse* resemble females, but the head and neck are more mottled, the flanks are plain grayish brown, and more white is present on the upper wing surface. *Immature males* are femalelike most of their first year, with a white head patch developing in winter or spring. *Juveniles* resemble adult females but have no crest, and the entire plumage is more brownish.

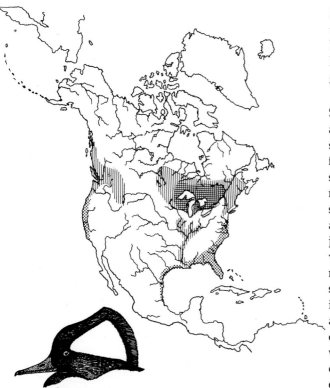

MAP 119. Breeding (hatched) distribution of the hooded merganser, including areas of major concentrations (cross-hatched). Wintering distribution indicated by stippling.

In the field, the merganserlike bill and strongly crested head will serve to separate both sexes from other species, although the male's crest can be strongly depressed and fairly inconspicuous. The white tertial stripes on the lower back of both sexes are usually quite evident, and in flight a limited amount of white is visible on the secondaries. Neither sex is highly vocal, but during courtship males utter a strange, froglike *crooooooo* note that carries some distance.

NATURAL HISTORY

Habitat and foods. During the breeding period, hooded mergansers are usually found in wooded habitats that contain clear-water streams or equally clear lakes. Such waters provide foraging for small fish and invertebrates, and tree cavities for nesting sites. Streams that have sandy or cobble bottoms are preferred to mud-bottom streams, and fairly shallow, fast-moving waters are preferred to slower or deeper rivers. Besides fish, frogs, tadpoles, crayfish, other crustaceans, snails, and some other mollusks are consumed, but little plant material is ingested.

Probably much the same foods are taken during winter, when the birds concentrate on fresh or brackish waters of rivers, estuaries, and interior lakes, generally avoiding salt-water habitats (Johnsgard, 1975; Stewart, 1962).

Social behavior. Perhaps because of the problems associated with competition for limited food resources, flock sizes of hooded mergansers tend to be fairly small during the fall and winter, usually consisting of no more than 15 birds. Courtship begins in midwinter among birds more than one year old, and such displaying groups often consist of from 3 to about 10 birds, with up to 3 females present. The major display activity of the female consists of a rather vigorous inciting behavior, marked by head-bobbing movements and repeated uttering of a hoarse *gak* sound. Females also perform an elliptical head-pumping movement almost exactly like that of the males, and usually perform it simultaneously with that display. Most of the male displays are associated with crest raising, which may occur independently of other displays or in conjunction with them. Silent elliptical head pumping, similar to that of goldeneyes and likewise apparently hostile in function, is common. The only display that has a moderately loud call is a head-throw that is preceded by several increasingly strong head-shaking movements. Swimming ahead of an inciting female with the depressed crest directed toward her is another common male display. Copulation is preceded by the female's assuming a prone position in the water, often after both birds have performed drinking movements. The male then begins a series of body-shaking and drinking movements, interspersed with jerky back-and-forth head movements, and terminates his precopulatory behavior with a stereotyped display sequence that includes a ritualized approach to her. After treading he swims away from the female with a fully erected crest, and finally dives or bathes (Johnsgard, 1961f, 1965a).

Reproductive biology. Pair formation is probably completed by the time the birds arrive on their breeding grounds, and there is a strong tendency for females to return to their natal homes or to where they nested the previous year. Females frequently reoccupy the nest site of the previous year if it is still available, but otherwise look for suitable sites in the near vicinity. Cavities adjacent to water are preferred to those some distance from it, and there is probably some competition with goldeneyes and other hole-nesting species for suitable sites. Females lay eggs in such sites at a rate of about one egg every two days, and initial nesting efforts have average clutch sizes of about 10 eggs, with experienced breeders averaging slightly larger clutches than hens nesting for the first time. Males usually desert their females at the time that incubation gets underway, and a relatively long incubation period of 32 to 33 days is typical. As in many hole-nesting species, dump nesting is a frequent cause of desertion of clutches, but otherwise fairly high nesting success seems to be characteristic. Females move their newly hatched ducklings out of the nest within a day of hatching, and usually go to shallow waters quite close to timber. Rivers with high levels of food resources are preferred for brood rearing, but sometimes beaver ponds or other standing-water habitats are used. The fledging period of the ducklings is about 70 days, but it is still uncertain how long females remain with their broods before leaving them to begin their own flightless period, and there is little available information on the activities of adults during the postbreeding period (Morse et al., 1969).

Status. Although the method of calculation was indirect, Bellrose (1976) estimated that the prebreeding population of hooded mergansers may average about 76,000 birds. The dependence of these birds on clear and unpolluted streams, and on nesting sites in fairly large trees, places them at a substantial disadvantage relative to many other North American ducks in terms of survival probabilities. Artificial nesting boxes may be used by the birds, but inasmuch as they are not significant sporting birds they have not received the attention that the wood duck has in this regard, and most hunters tend to regard all mergansers as "trash ducks," or, even worse, as destroyers of fish that should be shot on sight.

Relationships. As indicated by a study of its sexual patterns (Johnsgard, 1961c), as well as its postcranial skeletal structure (Woolfenden, 1961), the hooded merganser has strong evolutionary affinities with the genus *Bucephala*, and Woolfenden has argued for a retention of the genus *Lophodytes* on this basis. However, Brush (1976) found no obvious differences in the electrophoretic patterns of the feather proteins among all of the mergansers, the goldeneyes, scoters, and the long-tailed duck, which further suggests that they are a fairly close-knit group, and that very few genera should be recognized.

Suggested readings. Morse et al., 1969; Johnsgard, 1961f; Bouvier, 1974.

Smew

Mergus albellus Linnaeus 1758

Other vernacular names. None in general English use. Zwergsäger (German); harle piette (French); bech de serra petit (Spanish).

Subspecies and range. No subspecies recognized. Breeds in forested areas from northern Scandinavia across northern Russia to eastern Siberia, south to the lower Volga, Turkestan, and the Amur. Winters on the coasts and lakes of Europe south to the Mediterranean, the Caspian, Iran, northern India, southern China, and Japan. See map 120.

Measurements and weights. Folded wing: males, 192–205 mm; females, 178–86 mm. Culmen: males, 28–30 mm; females, 25–28 mm. Weights: males (fall), 540–935 g; females, 515–650 g (Dementiev & Gladkov, 1967). Eggs: av. 52 x 38 mm, creamy buff, 39 g.

Identification and field marks. Length 14–16″ (35–40 cm). *Males* in breeding plumage have a white head and neck except for a greenish black patch extending from the bill to around the eyes, and another blackish V-shaped mark around the nape, forming a short crest. The breast and foreback are white, with two black bands extending down the sides of the breast in front of the wings. The tail is gray, the outer scapulars white, some with black margins, and the underparts are white, except for the sides and flanks, which are vermiculated with gray. The middle and lesser wing coverts are white, while the marginal and greater coverts are mostly black, and the secondaries are dark brown to black, tipped with white. The primaries are dark brown, and the tertials are silvery gray. The iris is dark gray, the bill bluish to blue-gray, with a darker nail, and the feet and legs are bluish gray. *Males in eclipse* are very similar to

Map 120. Breeding (hatched) and wintering (stippling) distributions of the smew.

females, but have whiter upper wing coverts and darker upperparts. *Females* have cinnamon brown on the upper head extending from the base of the bill to the nape, which is more grayish, and the rest of the head and neck is white. The breast, sides, flanks, and upperparts are gray, the scapulars being mostly light gray; while the rump, tail coverts, and tail are brownish to silvery gray. The underparts are mostly pure white, and the upper wing surface is like that of the male, but has less white on the coverts and the tertials are shorter and browner. The iris is dark brown, the bill dark gray, and the legs and feet bluish gray. *Juveniles* resemble adult females, but the central wing coverts have brownish edges.

In the field, the predominantly white pattern of the male is highly conspicuous and distinctive; no other European duck exhibits so much white on the head and body. The female's sharply two-toned head pattern, with bright cinnamon above and white below, is also apparent from great distances. The slim, typical merganser body form is evident in flight; the head, neck, and body are held in a straight line. Calling by both sexes is largely limited to display; females utter a harsh, rattling *krrrr, krrrr* sound during inciting, and males produce a very soft rattling call that sounds much like the noise made when winding a timepiece.

Natural History

Habitat and foods. This is a forest-nesting species which is similar in its ecological needs to the hooded merganser and bufflehead. It apparently avoids rapidly flowing streams, and instead favors the bottomlands of larger rivers, flooded riverside woods, and lakes that are associated with river bottomlands. During the nonbreeding period the birds are usually found on larger lakes and impoundments, and at the mouths of rivers, but in general they prefer fresh-water to marine habitats. Besides the usual merganser fare of small fishes, they also eat mollusks and crustaceans and substantial quantities of insects such as beetles, dragonflies, and caddis fly larvae. The fish that are taken are generally roach (*Rutilus*), minnows, small carp (*Cyprinus*), and other nongame species, but the birds have also been known to consume trout and salmon. They typically feed in rather shallow water, and remain under water for rather short periods of 15 to 20 seconds. Pairs often dive synchronously, and at times flocks forage in a cooperative manner (Nilsson, 1974). Their bill shape is merganserlike in its serrated edges, but is shorter and stouter than those of the other mergansers, and presumably is better adapted for crushing than are the others (Dementiev & Gladkov, 1967; Ogilvie, 1975).

Social behavior. There is evidently some separation of the sexes during the winter, with females and immature birds wintering farther south than adult males, and tending to use inland rather than coastal habitats to a greater degree than adult males. In southern Sweden the birds spend about half of the daylight hours foraging, interspersed with resting and preening activities. Social display was not observed by Nilsson (1974) until late January, and by March about a third of the birds seemed to have become paired. The courting groups usually consisted of one or two females and up to eight males, with a good deal of shifting of numbers typical. Females have a conspicuous and energetic inciting movement and associated call, to which males often respond by swimming rapidly ahead and orienting their V-shaped markings toward her. Males also erect the front part of their crown into a rather shaggy crest and while swimming in this posture perform pouting, a bridlinglike movement of the head backward along the back while uttering a soft mechanical rattling sound. Males also perform a silent neck stretching, and sometimes also do a sudden head-fling, tossing the head upward and backward and uttering the same call as that produced during pouting, suggesting that this is a more elaborate version of that display. Copulation has been observed in wild birds as early as January, and probably is usually initiated by mutual drinking movements. The female then assumes a distinctive prone posture, with her bill on the water but her tail well elevated above it. The male then performs an extended series of drinking and preening displays in no obvious sequence.

There is evidently no ritualized approach to the females as in the hooded merganser or goldeneyes; but after treading is completed, the male performs a single head-fling display and then swims rapidly away from the female while turning the back of the head toward her (Johnsgard, 1965a; Nilsson, 1974).

Reproductive biology. Apparently female smew breed in their second year of life, judging from what is known of related species. There is not a great deal known of the nesting biology of smews, but they evidently prefer to nest in hollow broad-leaved trees, including oaks, willows, and aspens, often in cavities so low that they can readily be looked into by humans, and at times the birds also accept nesting boxes. The cavity is lined with abundant whitish down, and from 6 to 9 eggs are laid, with up to 14 reported, presumably from multiple nesting efforts. Additionally, the birds often have mixed clutches with goldeneyes, and wild hybrids between these two species have been reported. The incubation period is 28 days, during which time the male deserts his mate and moves to a molting area. Females are reportedly very "tight" sitters, and near the end of the incubation period can sometimes be picked up. There is no specific information on sources of mortality among eggs and ducklings, and the fledging period under wild conditions is not known, but some young raised in captivity were reportedly fully fledged by ten weeks of age (Dementiev & Gladkov, 1967; Bauer & Glutz, 1969).

Status. Although no estimates of breeding populations are available, Ogilvie (1975) reports that about 10,000 smews winter in northwestern Europe, and another 30,000 occur in the Mediterranean and Black sea vicinity. The Netherlands supports the largest numbers of wintering smews in northwestern Europe. Most smews presumably nest in the U.S.S.R., where they are most common in western Siberia and in the open areas among the lowland coniferous forests of European U.S.S.R., but numerical estimates for the U.S.S.R. and China are not available. The birds are uncommon in Japan during winter (Austin, 1948).

Relationships. Similarities in female plumages and those of downy young, and the occurrence of wild hybrids, all suggest a close relationship between smews and goldeneyes. Additionally, smews and hooded mergansers seem to be close relatives, but this in part is a reflection of the fact that they are ecological counterparts. The inciting behavior of female smews is more *Mergus*-like than *Bucephala*-like, further indicating the intermediate status of this species between these two seemingly distinctive genera (Johnsgard, 1965a).

Suggested readings. Nilsson, 1974; Owen, 1977.

Brazilian Merganser

Mergus octosetaceus Vieillot 1817

Other vernacular names. None in general English use. Dunkelsäger (German); harle du Brésil (French); mergánsar pico serrucho (Spanish).

Subspecies and range. No subspecies recognized. Resident in southern Brazil in the states of Goiás, São Paulo, Santa Catarina, and Paraná, and also in the Paraná River drainage of eastern Paraguay and northeastern Argentina. See map 121.

Measurements and weights. Folded wing: males, 183–88 mm; females, 180–84 mm. Culmen: males, 49–51 mm; females, 38–40 mm. Weights: no record. Eggs: light cream colored, 66 x 45 mm (Kolbe, 1972).

Identification and field marks. Length: ca. 20″ (50 cm). *Adults* have a black head and upper neck, with some green iridescence, and generally dark greenish brown upperparts. The lower neck, breast, and flanks are gray, finely vermiculated with hoary white, while the abdomen is barred irregularly with brown and white. The upper wing surface is generally blackish, except for the secondaries and their coverts, which form a white speculum bordered anteriorly with black and white bars. The legs and feet are rosy red, the iris is brown, and the bill is black. *Females* differ from males in their smaller size, shorter bill, and more poorly developed crest. *Juveniles* closely resemble females.

In the field, the slim, merganserlike shape separates this species from all other South American waterfowl. The birds fly swiftly, following the river courses closely, and hold their necks stiffly outstretched in line with the body, in the usual merganser fashion. The major calls so far described

are a simple *queek* uttered in flight, and various sounds made while defending the young.

NATURAL HISTORY

Habitat and foods. The contribution by Partridge (1956) constitutes nearly the sum total of our knowledge of this rare species, which for some time was believed to be extinct. The species is now largely or entirely limited to the tributaries of the Paraná River, which are wild and torrential streams that flow through tropical forests. Evidently the upper reaches of these streams are inaccessible to the predatory dorado fish (*Salmimus*), regarded as a major enemy of the downy young. The streams do support other fish, and the remains of 11 stomachs indicate that the merganser's primary food consists of fish (cichlids, characinids, etc.) from 6 to 19 cm long, supplemented by the larvae of dobson flies (*Corydalis*) and a few snails. The birds feed during the daytime, often perching on rocks among rapids and diving for food. Like torrent ducks, they often feed in areas of rapids where a large stone emerges, and feed near such rocks, remaining submerged for periods of about 15 or 20 seconds.

Social behavior. Evidently these mergansers are almost always found in pairs; Partridge noted only a single exception to this in his observations, where two pairs occupied the same stretch of rapids. During August he observed some apparent mating displays among these two pairs of birds, which included the male chasing a female in circles, paddling with the wings in the water but not leaving the water surface. At times both pairs would engage in this behavior, after which they would leave the water and preen. Judging from the behavior of other waterfowl, it seems unlikely that this was typical display; more of the cavity, which was three meters deep, there was a great deal of fine, rotten wood present but no down or other nesting materials. While the female incubated the male remained near the nest site, resting or foraging. Once a day the female left the nest and joined her mate in foraging, spending an hour to an hour and a half in this activity. After resting for a time in the river, the female would fly directly back to the nest cavity, while the male would fly past the nest entrance and then return to the river. Hatching occurred about a week after the nest was originally discovered, on August 30, and four ducklings were subsequently observed on the river with both parents. No information on the fledging period is yet available.

Status. Partridge (1956) believed that this merganser is not rare where suitable habitat occurs, but that its total range is limited to the tributary streams of the Alto Paraná river system. Because the birds are distributed as pairs along those tributary rivers, the total population must be greatly restricted. Since these rivers are still largely inaccessible, he believed that the species will continue to survive there indefinitely.

Relationships. I have suggested (1965a) that this species is not very closely related to any single one of the Northern Hemisphere mergansers, and must have had an earlier origin. Like the Auckland Island merganser, it seems to be a relatively isolated form.

Suggested readings. Partridge, 1956.

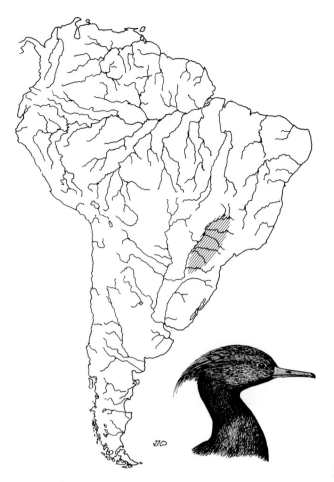

MAP 121. Residential distribution of the Brazilian merganser.

probably it represented the "dashing and diving" behavior of many ducks when bathing energetically. Partridge saw one copulation, which was preceded by the female's becoming prone and motionless in the water. Both birds were completely submerged during treading, and afterward the female uttered a long cry, after which they bathed and perched on a rock.

Reproductive biology. Breeding in this species begins in June, with incubation in July and the young emerging in the first part of August, according to Partridge. This corresponds to the rainy period and presumably to an abundance of available food. Only a single nest has been described, which was found by Partridge in a hollow cavity of a leguminous tree (*Peltophorum*) adjacent to the river. The cavity was located about 25 meters above the level of the river, and the entrance measured 35 by 15 centimeters. At the bottom

Red-breasted Merganser

Mergus serrator Linnaeus 1758

Other vernacular names. Fish duck, sawbill; Mittelsäger (German); harle huppé (French); mergansár de pecho rojo (Spanish).

Subspecies and ranges. (See map 122.)

M. s. serrator: Common red-breasted merganser. Breeds in Iceland, the British Isles, northern Europe and Asia from Scandinavia to Kamchatka, the Aleutian Islands, and from Alaska east across nearly all of arctic Canada except for the northern part of Keewatin District and the arctic islands, south to northern British Columbia and Alberta, central Saskatchewan and Manitoba, southern Ontario, the Great Lakes states, New York, New England, and the eastern Canadian provinces to Newfoundland. Winters mostly on salt water from the southern parts of its breeding range southward along the coastlines of Europe, Asia, and North America.

M. s. schiøleri: Greenland red-breasted merganser. Resident in Greenland. (This is a poorly characterized subspecies that might not deserve recognition.)

Measurements and weights. Folded wing: males, 224–60 mm; females, 217–30 mm. Culmen: males, 53–62 mm; females, 48–55 mm. Weights: males in fall and winter average ca. 1,200 g and females ca. 925 g, with respective maximums of 1,314 and 1,268 g. Eggs: av. 63 x 45 mm, deep buff, 72 g.

Identification and field marks. Length 19–26″ (48–66 cm). *Adult males* in breeding plumage have the entire head and neck black, with a green gloss, the feathers on the back of the head elongated to form a long, shaggy, and rather double-pointed crest. The foreback, sides of breast, and inner scapulars are black, shading to vermiculated green on the rump and upper tail coverts. The tail is ashy gray, the underparts are whitish, and the sides and flanks are white, with black vermiculations. The breast is cinnamon brown streaked with black, separated from the blackish

neck by a wide white collar that is incomplete behind. The outer scapulars are white, and a group of black-margined white feathers extend down from the base of the wing to separate the brownish breast from the sides. The marginal and tertial coverts are gray, while the others are white or tipped with white, the greater coverts having exposed black bases. The primaries and their coverts are slaty brown. The secondaries have the exposed webs black outwardly and white or white margined with black inwardly. The longer tertials are white margined with black, and the inner ones are blackish. The iris is red, the legs and feet are dull red to orange, and the bill is carmine red, with a more dusky culmen and a black nail. *Females* have a grayish brown crown and crest, are darker around the eyes, have whitish lores, and whitish coloration on the chin and throat that gradually merges with cinnamon brown on the cheeks, sides of the head, and the neck. The upperparts are grayish brown, the feathers having lighter edges and darker shaft-lines; the inner scapulars are blackish brown; the breast is whitish to brownish gray, and the sides and flanks are grayish brown, as is the tail. The underparts are white, the primaries and their coverts are dark brown, the outer secondaries are black, and the remainder are white with black exposed basally. The tertials are dark brownish gray, the greater coverts are black or black tipped with white, and the other coverts are brownish gray. The iris is reddish brown, the bill is dull red, and the legs and feet are dull red to orange. *Males in eclipse* resemble females, but have less white on the chin and have white on the middle and lesser coverts, a darker back, and more reddish eyes. *Immature males* re-

semble females but in their first spring begin to develop black feathers on the head, back, and sides of the breast. *Juveniles* resemble females but have a smaller crest, grayer upperparts, less white on the foreneck, and less blackish coloration around the eyes.

In the field, this species is most likely to be confused with the goosander, although the shaggy crest of males and their brownish breast coloration should easily separate them. Females have a distinctly less contrasting facial pattern, with the whitish areas of the chin and throat not so distinctly contrasting with the darker cheeks and neck feathers; in addition, their brown head and neck coloration gradually merges with the more grayish breast. Females of both species utter harsh calls during courtship display, but male red-breasted mergansers produce a distinctive catlike *yeow-yeow* call during elaborate posturing. In flight, the long, thin body profile, with the head and neck held in line with the body, marks the birds as mergansers, and the brownish breast of males, bordered in front and behind with white, provides the best field mark for distinction from goosanders and hooded mergansers.

NATURAL HISTORY

Habitat and foods. During the breeding season, red-breasted mergansers are usually found around inland lakes and streams that are not far removed from the coast. Although they extend into tundra areas, they are more often found around deeper lakes than tundra ponds. However, by their ground-nesting behavior they are not dependent on trees like many mergansers. Nonetheless, habitats having natural cavities such as those provided by boulder fields are favored over more exposed areas. During the summer the adult birds apparently remain fish eaters, but the young feed on insects such as aquatic beetles and the larvae of May flies as well as small fish and shrimp. During the fall and winter the birds tend to move to coastal areas, where they concentrate in such habitats as the open ocean, salt-water and brackish estuarine bays, and to a limited extent on fresh and slightly brackish waters as well. In any case, they prefer clear and shallow waters not affected by heavy wave action, where they can see to forage. Fish make up the majority of the diet at such times, with limited amounts of shrimps, crayfish, and other crustaceans also being consumed. The

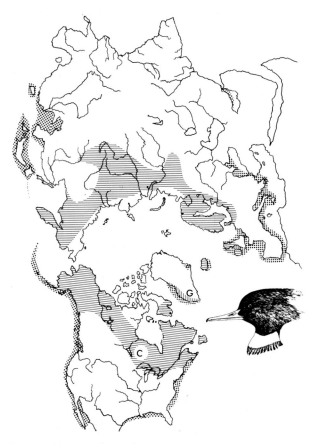

MAP 122. Breeding distributions of the Greenland ("G") and common ("C") red-breasted mergansers. Wintering distributions indicated by stippling.

birds sometimes forage in groups, apparently co-operating in chasing fish until they are trapped and caught (Johnsgard, 1975; Bellrose, 1976).

Social behavior. Although first-year males have at times been observed in courtship display, full male plumage and presumably full sexual maturity are not attained before the second winter. Likewise, pair bonds are reestablished each winter during a rather prolonged period of display, but apparently they are fairly loose and some divergences from typical monogamy have been reported. By as early as February active display can be seen among coastal wintering birds, and courtship in this species is both conspicuous and rather bizarre in terms of the male posturing. Female inciting with a strong and harsh double note occurs occasionally but does not seem to be the major stimulus for display, which at times seems to be independent of specific female activities. The most obvious and complex male display consists of a series of movements including diagonal neck stretching (the salute), followed by a quick lowering of the neck and part of the head into the water (the curtsy) as a catlike note is simultaneously uttered. Males also sprint over the water, sending up a spray on both sides. Copulatory behavior consists of the female's assuming a prone posture on the water, often after both birds have performed drinking movements. Thereafter the male begins an extended series of drinking, preening, wing-flapping, and shaking movements, without clear sequential connections. After copulation, the male performs a single knicks display (the combination salute and curtsy mentioned above), and then both birds begin to bathe (Johnsgard, 1965a).

Reproductive biology. Red-breasted mergansers are usually paired by the time of their arrival on the nesting grounds, although some display occurs after arrival as well. Little is known of home ranges or possible territoriality in this species, but in general the nests tend to be well scattered, with some tendency for concentration on islands. In Iceland, Bengtson (1970) found that island nesting concentrations were roughly twice as dense as mainland ones, and that the majority of nests were placed in holes or cavities, with shrubs providing cover for most of the rest. Studies on the Gulf of Bothnia by Hilden (1964) indicated a strong tendency for nesting under boulders or amid thick bushes. Where tree cavities or artificial nesting boxes are utilized, the birds favor those with entrances about 10 centimeters in diam-

eter and with internal diameters of about 30 to 40 centimeters. Clutches of this species average about 9 to 10 eggs for initial nesting attempts, but dump nesting resulting from competition for limited nesting sites often produces larger observed clutch sizes. Incubation periods usually average about 32 days but range from 29 to 35 days under natural conditions. Males desert their mates early in incubation; and although they may tend to move to brackish or saline waters, there is no evidence for a major migration in this species to a common molting area. As in some other sea ducks, especially late nesters, brood aggregations are common in this species, and frequently one or more females may be seen in company with several dozen young of various ages. The fledging period probably is about 60 days, which would suggest that females are likely to become flightless before the young are able to fly, and would seemingly also tend to limit the success of northerly-breeding birds in late springs (Curth, 1954; Hilden, 1964; Johnsgard, 1975).

Status. Bellrose (1976), using a combination of data sources, judged that the summer population of the red-breasted merganser in North America might consist of about 237,000 birds. This compares with an estimate of about 40,000 wintering birds in northwestern Europe, including Great Britain, and another 50,000 in the Mediterranean and Black sea region (Ogilvie, 1975). This would exclude the populations wintering off Iceland and Greenland; and in addition, the large population that winters along the coast of China and Siberia cannot be estimated. Although not a game species, the red-breasted merganser has suffered population losses in North America as a result of presumed pesticide poisoning as well as local control efforts by fishing interests.

Relationships. The evolutionary relationships of this species to the goosander and the Chinese merganser are of interest; and although no behavioral information is available on the latter species, it seems on anatomical grounds to occupy an evolutionary position somewhere between those of the red-breasted merganser and goosander. By comparison with the goosander, the red-breasted merganser has a somewhat narrower and longer bill structure, and also has somewhat broader nesting adaptations.

Suggested readings. Curth, 1954; Johnsgard, 1975; Bellrose, 1976.

Chinese Merganser

Mergus squamatus Gould 1864

Other vernacular names. Scaly-sided merganser; Schuppensäger (German); harle écaillé (French); mergánsar barreado (Spanish).

Subspecies and range. No subspecies recognized. Breeding range uncertain, but includes Ussuriland and perhaps eastern and northern Manchuria. In Ussuriland, breeding known only in the central and southern Sikhote-Alin range, from the Khor River basin to the Iman River basin. Winters mainly in eastern and central China, in the Yangtze River valley west to Szechwan and south to Fukien, Kwangtung, and northern Tonkin (Vaurie, 1965). See map 123.

Measurements and weights. Folded wing: males, 250–65 mm; females, 240–45 mm. Culmen: males, 46–54 mm; females, 43–46 mm. Weights: no record. Eggs: no description.

Identification and field marks. Length: ca. 22″ (55 cm). *Adult males* in breeding plumage closely resemble the red-breasted merganser in the color and shape of the head, which is shaggy-crested and greenish. The lower neck, breast, and abdomen are salmon pink, while the flanks are distinctively scalloped in a scaly pattern. The upper back and inner scapulars are black, bounded by white laterally, and the lower back and rump are gray, edged with black in the same manner as the flanks. The tail is gray, and the upper surface of the wings is patterned as in *serrator*, but the anterior coverts are much greater. The bill is dull red, with a black nail and a

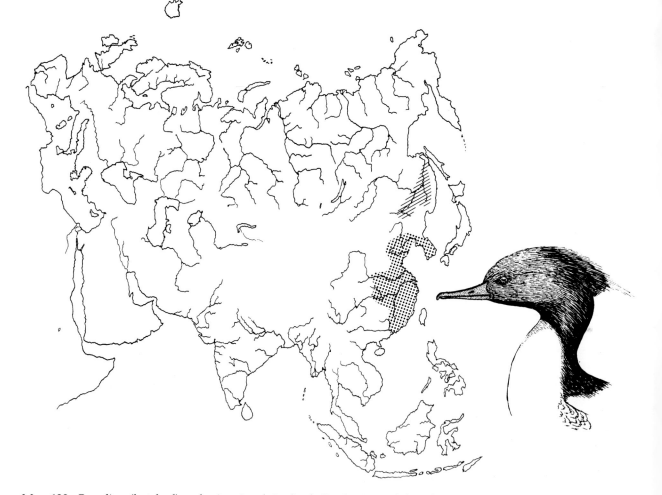

MAP 123. Breeding (hatched) and wintering (stippling) distributions of the Chinese merganser.

dusky stripe along the culmen, the legs and feet are orange, and the iris is brown. *Females* are very much like female goosanders but are more whitish on the breast and sides and have a trace of the scalelike markings on the flanks. The soft-part colors are as in the male. *Males in eclipse* closely resemble females but have more white on the wings. *Juveniles* apparently resemble females, but too few have been described to be certain of this.

In the field, the combination of a shaggy head crest, a white to pink breast color, and grayish flanks with crescent-shaped markings are distinctive characteristics of the male. Females cannot safely be separated in the field from the other two species of Asian mergansers.

NATURAL HISTORY

Habitat and foods. This species breeds in the rapidly flowing mountain streams of the Sikhote-Alin mountains and occupies the middle and upper taiga (coniferous forest) zones. It reportedly avoids extremely narrow creeks where diving may be difficult, and additionally is found only in areas with forested banks. It is largely limited to the area between 35 and 50 kilometers inland. In the winter, however, the birds move downstream and may occur between 9 and 19 kilometers from the sea. Habitats in the wintering areas of southern China must differ considerably from these but presumably are clear rivers or lakes with abundant fishes. Almost nothing is known of the foods other than that they include various fishes (*Salvelinus, Oncorhynchus*).

Social behavior. Almost nothing can be said with certainty of this species' social behavior. It evidently is relatively rare everywhere and never is encountered in large flocks. "Large numbers" have been reported only during fall at Lake Chingpo Hu in

Kirin province in southern Manchuria, where local residents reported them breeding (Dementiev & Gladkov, 1967). Nothing has been reported yet on the timing of pair formation, the displays, or the strength of the pair bond.

Reproductive biology. Although no eggs have been described, two nests were reportedly found in a tributary of the Sitsa River on the eastern slope of the Sikhote-Alin range. Both were in hollow trees, one about 1.5 meters above the water in a broken trunk overhanging the river, the other about 3 to 3.5 meters above the water in a linden tree. In common with the goosander, and in contrast to the red-breasted merganser, females have white-colored down, suggestive of normal tree-cavity nesting. The number of eggs in the clutch is unreported, but observed broods of from 8 to 12 young suggest a normal clutch of 10 or more eggs. Broods have been seen in July, August, and even September.

Status. This species' population status is extremely uncertain, but it is apparently quite rare, with an extremely restricted breeding area.

Relationships. The Chinese merganser has plumage similarities to both the red-breasted merganser and the goosander, and additionally has a tracheal anatomy very similar to that of the goosander. In the absence of further information, it must be assumed that the species occupies an evolutionary position somewhere between these two species (Johnsgard, 1965a).

Suggested readings. Dementiev & Gladkov, 1967; Delacour, 1954–64.

Goosander (Common Merganser)

Mergus merganser Linnaeus 1758

Other vernacular names. American merganser, fish duck, sawbill; Gänsesäger (German); harle bièvre (French); mergánsar (Spanish).

Subspecies and ranges. (See map 124.)

M. m. merganser: Eurasian goosander. Breeds in Iceland, Scotland, Scandinavia, and across Rus-

sia and Siberia to Kamchatka, on the Kurile and Commander islands, and south across Europe and Asia to Switzerland, Poland, Romania, and central Russia, with the southern limits in Asia ill-defined. Winters south to the Mediterranean, Black, and Caspian seas, northern India, Assam, and China.

M. m. comatus: Oriental goosander. Breeds in the Pamirs and northeastern Afghanistan, eastward through Ladakh and Tibet to Sikang, Tsinghai, western Kansu, and adjacent Szechwan. Winters in the Himalayan foothills and south to Burma and Yunnan.

M. m. americanus: American goosander. Breeds in North America from southern Alaska east across central Canada to James Bay and across the Labrador Peninsula to Newfoundland, south in the western mountains to California, Arizona, and New Mexico, and east to the Great Lakes states and New England. Winters on fresh water and salt water from its breeding grounds to southern California and Florida.

Measurements and weights. Folded wing: males, 275–95 mm; females, 244–75 mm. Culmen: males, 55–61 mm; females, 45–50 mm. Weights: males (of *americanus*) average ca. 1,600 g and females ca. 1,200 g during fall, with respective maximums of 1,859 and 1,769 g. Eggs: av. 66 x 46 mm, creamy, 82 g.

Identification and field marks. Length 21–27" (51–68 cm). *Adult males* in breeding plumage have the entire head and neck black, with a green gloss; the feathers of the back of the head are elongated to form a short, bushy crest. The foreback and inner scapulars are black, grading to ashy gray on the hind back, rump, and upper tail coverts. The tail is gray; the neck, outer scapulars, breast, sides, flanks, and underparts are white, often with a salmon pink tint. The flanks are lightly vermiculated with gray. The marginal and lesser wing coverts are black; the primaries and their coverts are dark slate brown; the outer secondaries and their greater coverts are blackish; while the remaining secondaries and tertials are white or white

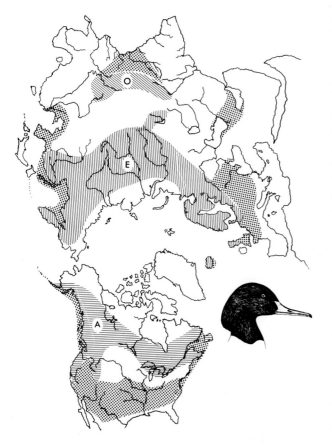

MAP 124. Breeding distributions of the American ("A"), European ("E"), and Oriental ("O") goosanders. Wintering distributions indicated by stippling.

narrowly edged with black, the inner tertials becoming gray or black. The other coverts are white, except for the inner greater coverts, which have exposed black bases (in *americanus*). The iris is dull reddish to brown, the bill is red, with a black culmen and nail, and the legs and feet are orange red with darker webs. *Females* have a grayish brown crown and crest, with a white throat, lower portion of the cheeks, and a narrow line from the eye to the base of the bill; these areas all contrast rather sharply with the cinnamon red of the rest of the head and neck. The upperparts are ashy, the feathers having darker shaft streaks and lighter edges. The inner scapulars are blackish brown; the foreneck and upper breast are whitish, merging abruptly with the cinnamon of the neck; the sides and flanks are grayish brown; the tail is ashy gray; and the underparts are white, sometimes faintly tinted with salmon. The primaries and their coverts are dark brown, the outer secondaries are black, and the others are white, with little

or no black exposed basally. The tertials are dark brownish gray, the greater coverts are black or black tipped with white, and the other coverts are brownish gray. The iris is dull reddish to brown, the bill is purplish red, with a darker nail, and the legs and feet are orange red, with darker webs. *Males in eclipse* resemble females, but have white upper wing coverts. *Immature males* begin to acquire black feathers on the head and neck during their first winter and spring, and develop a rather mottled appearance. *Juveniles* resemble adult females but have a shorter crest, and the white on the throat extends down to the chest.

In the field, its large size and merganserlike bill separate this species from all others except the somewhat smaller red-breasted merganser. Males in breeding plumage appear mostly white (or pinkish), except for a black head and back, and their red bill is also quite widely visible. Females are best separated from red-breasted mergansers by their more grayish body and brownish head, the two color areas sharply separated from each other, and by the more sharply defined white areas on the cheeks and throat. The call of the displaying male is also distinctive, consisting of a guitarlike *uig-a* note, while the female produces harsh *karrr* sounds. In flight, the very long-bodied profile and uniformly white underpart coloration of both sexes helps to distinguish this species from other species.

NATURAL HISTORY

Habitat and foods. This widespread and largest species of merganser breeds in temperate forested habitats of the Northern Hemisphere that usually consist of clear fresh-water lakes, rivers, or ponds associated with the upper portions of rivers in forested regions. It is an inland rather than coastal breeder, and is limited almost exclusively to areas with forests nearby, although sometimes cavities in boulders provide alternative nest sites where tree cavities are unavailable. Throughout the year the birds concentrate on fish as their major source of food, although ducklings consume large quantities of insects such as May flies until they become adept at catching fish. During the fall and winter the birds spread out over a variety of habitats, but usually concentrate on clear rivers and ice-free lakes. They also sometimes feed in estuaries at the mouths of rivers or on bay marshes that are slightly brackish. They typically forage in fairly shallow waters but have been known to dive as deep as 30 feet, and sometimes a flock will forage cooperatively, driving fish into areas where they can be readily captured. In some areas this species is persecuted because of its reputed damage to salmon or trout fisheries, but it tends to feed on what is readily available and can easily be caught, which in most areas is predominantly roughfish (Johnsgard, 1975; Bellrose, 1976).

Social behavior. So far as is known, goosanders mature the second year of life, since no records of year-old birds breeding have yet come to light. Pair bonds are formed during an extended period of display lasting from winter through the spring migration period, until the arrival of the birds at their nesting areas. Display is marked by a great deal of chasing on the water surface, underwater attacks, and some aerial chases. Females incite with a loud, harsh call and quick forward lunges in the water, and the male often attempts to swim ahead of such a female while directing the back of his head toward her and ruffling his crown feathers into a short but distinctive crest. The males produce a distinctive courtship call, a curious twanging note similar to that produced by a guitar, and in addition utter a clear bell-like call during a sudden vertical stretching of the head and neck in a salute posture. They also at times throw out a jet of water far behind them with a backward kicking movement. As in the red-breasted merganser, copulation is often preceded by both birds performing drinking movements, after which the female becomes prone and the male continues an extended series of drinking, preening, shaking, and other movements that seem to have no sequential predictability. No specific postcopulatory posturing has been observed, other than the male swimming away from the female while calling repeatedly and directing the back of his head toward her (Johnsgard, 1965a).

Reproductive biology. Shortly after the spring flocks arrive on their nesting grounds, they break up into pairs that take up home ranges along stretches of river that support populations of fish suitable for them and their ducklings. Pairs occupying such stretches of rivers seem to be well isolated from one another, but in lake or coastal situations with islands that provide suitable nesting sites a more concentrated nesting population may develop. When nest sites are available in trees, the birds seem to prefer

cavities with openings about 12 centimeters wide and internal diameters of about 25 centimeters. Artificial nesting boxes are used in some areas, and these are usually 85 to 100 centimeters high, with openings 50 to 60 centimeters from the base. In Iceland and on the Gulf of Bothnia few trees are large enough to provide such sites, and there the birds often nest under boulders, under bushes, or even in buildings. Concealment from above, with an associated darkness of the nesting cavity, seems to be a paramount consideration for nest sites. Eggs are deposited on a nearly daily basis until a clutch averaging 9 to 10 eggs has been completed, and at about this time the male leaves the area, sometimes moving to coastal situations for molting. Incubation requires between 32 and 35 days, a relatively long period, and in addition there is a quite long fledging period of some 60 to 70 days. Not surprisingly, the female often deserts her brood well before fledging to begin her own postnuptial molt, at which time the ducklings often begin to form rather large assemblages. About a month is required for the females to attain the power of flight again, which in the Maritime Provinces means that it may be early October before they are able to fly (Erskine, 1971; White, 1957).

Status. On the basis of rather uncertain assumptions, Bellrose (1976) calculated that the summer population of goosanders in North America may be as high as 641,000 birds, although winter counts average only about 165,000 of these conspicuous birds. Most of these are on large impoundments in the interior, where they can easily be counted, and thus it seems unlikely that the breeding population figure cited above is a reasonable one. In northwestern Europe about 70,000 birds represent the average wintering population, most of which occurs in the Baltic area. There are also about 10,000 that winter in the Mediterranean and Black sea region (Ogilvie, 1975). No estimates of the population of the oriental race are possible, nor is the size of the population of the Eurasian merganser that winters in eastern Asia known.

Relationships. As noted earlier, the goosander is probably a quite close relative of the Chinese merganser and a slightly less close relative of the red-breasted merganser, judging from available data.

Suggested readings. Erskine, 1971; White, 1957.

Auckland Island Merganser

Mergus australis Hombron and Jacquinot 1841

Other vernacular names. None in general English use. Aucklandsäger (German); harle austral (French); mergánsar de la Isla Auckland (Spanish).

Subspecies and range. No subspecies recognized. Extinct; was originally limited to the Auckland Islands and (as subfossil remains) New Zealand.

Measurements and weights. Folded wing: males, 186–220 mm; females, 176–80 mm. Culmen: males, 60–61 mm; females, 53–55 mm. Weights: no record. Eggs: no record.

Identification and field marks. Length 23″ (58 cm). *Adults* have a very dark brown head, crest, and neck, but with lighter chin and throat. The back, scapulars, upper tail coverts, and tail are very dark bluish black, while the breast is dull gray, with some lighter crescent-shaped markings. The underparts are mottled gray and white, while the flanks are uniformly dark bluish gray. The upper wing surface is slate gray to black, except for the middle secondaries, which are white on the outer vanes, and the greater secondary coverts, which are tipped with white. The iris is dark brown, the legs and feet orange, with dusky webs and joints, and the bill is yellowish orange, with a black culmen and nail. *Females* have shorter crests and less reddish coloration in the crown, and have one white wing bar instead of two.

In the field, the merganserlike shape alone would separate this species from any other bird in the Auckland Islands or New Zealand.

NATURAL HISTORY

Habitat and foods. The review by Kear and Scarlett (1970) indicates that this bird occupied interior rivers for the most part and occurred along the coast only on estuarine creeks and along sheltered harbors. Most of the specimens, however, appear to have come from coastal locations, and the few substantiated food remains, including a fish (*Galaxias*) and a polychaete (*Nereis?*), are apparently salt-water forms.

Social behavior. Nothing specific has been recorded about the behavior of this species. Birds in pairs were taken in October, November, January, and perhaps May, so that the birds were obviously monogamous.

Reproductive biology. The review by Kear and Scarlett (1970) suggests that the pair bond may have been a long-term one and that egg laying may have occurred in late November or early December, based on observations of young ducklings that were collected from a brood tended by both parents. This brood evidently consisted of four ducklings, but no more information is available on clutch size or brood size.

Status. The review by Kear and Scarlett indicates that the last record of this species was probably in 1902, and that only 26 skins exist in the world's museums. These include 4 ducklings and at least 12 males and 9 females. They suggest that the original population on the Aucklands was probably not great, and perhaps consisted of no more than a few hundred birds. When the species became extinct in New Zealand is uncertain, but it was probably after 1800 (Williams, 1964). Hunting by Polynesian moa hunters was presumably a factor in their extinction there, since skeletal remains of this merganser have been found associated with their refuse heaps. Introduction of various mammals on the Auckland Islands probably contributed to their extinction there.

Relationships. As noted in an earlier review (Johnsgard, 1965a), this species is perhaps derived from one of the Northern Hemisphere mergansers such as the Chinese merganser or the common merganser (goosander), although it does possess a number of unique plumage characteristics. There is no good evidence to favor the view that it is closely related to the Brazilian merganser.

Suggested readings. Kear & Scarlett, 1970.

Tribe Oxyurini (Stiff-Tailed Ducks)

Drawing on preceding page: Australian Blue-billed Duck (female diving)

Black-headed Duck

Heteronetta atricapilla (Merrem) 1841

Other vernacular names. None in general English use. Schwarzkopfente (German); canard à tête noire (French); pato rinconero or pato sapo (Spanish).

Subspecies and range. No subspecies recognized. Breeds in central Chile from Santiago to Valdivia, in Argentina, including the provinces of Buenos Aires, Santa Fe, and Santiago del Estero, and in central Paraguay. It also occurs, without evidence of breeding, in Uruguay, Brazil, and Bolivia, but apparently winters through much of its breeding range. See map 125.

Measurements and weights. Folded wing: males, 157–78 mm; females, 154–82 mm. Culmen: males, 40–47 mm; females, 41–48 mm. Weights: males, 434–680 g (av. 513 g); females, 470–720 g (av. 565 g). Eggs: 59 x 44 mm, white, 60 g.

Identification and field marks. Length 14–15″ (35–38 cm). *Adult males* have a black head and neck, sometimes with a white throat patch, and the mantle and scapulars are black, with reddish vermiculations. The breast, flanks, and under tail coverts are reddish to tawny, vermiculated with black, while the abdomen is silvery white, mottled with brown. The upper surface of the wings is dark brown, except for the coverts, which are speckled with reddish, and the secondaries, which are tipped with white. The iris is brown, the legs and feet are lead gray, with greenish on the edges of the tarsi, and the bill is grayish blue to black except during the breeding season, when it becomes rosy red between the nostrils and at the base of the bill. *Females* are generally brownish on the head, with a pale buff streak through the eyes and a buff chin and throat. The upperparts are blackish brown, speckled with reddish, while the flanks, abdomen, and wing coloration are nearly like those of the male. The soft-part colors are also like the male's, except that the female never develops red at the base of the bill, and instead this area becomes yellowish orange or yellowish pink.

In the field, this species at times resembles a dabbling duck, such as a cinnamon teal, more closely than a typical stifftail. This is especially true of females; males can usually be readily identified by their black heads and the seasonally brilliant reddish bill coloration. The birds are able to take off quickly, and fly with strong and rapid wingbeats and with their heads held fairly low. Females have not been heard to utter any vocalizations, and the calls of the males are generally low and grunting, although a soft whistle is part of the display call.

Natural History

Habitat and foods. Weller (1967b) reviewed the habitats of this species, and stated that the birds occupy semipermanent to permanent fresh-water marshes that are dominated by extensive stands of emergent bulrushes. This plant not only provides escape cover, but its seeds were found by Weller (1968b) to be a major food item as well. He noted that they also consume snails, some other plant seeds, and duckweeds. The foods are obtained by dabbling at the water's surface, by straining mud in shallow water, by upending in somewhat deeper water, and occasionally by diving in water several feet deep. The bill shape is slightly spatulate and well adapted for straining small materials, but evidently few small invertebrates are thus obtained. Inasmuch as black-headed ducks were at times found by Weller in partly wooded country and at times on artificial impoundments, he concluded that the nature of the emergent vegetation was the most important component of the habitat affecting these birds.

Social behavior. Weller (1968b) observed paired birds between September and December, but during the postbreeding period of January to March only a small proportion of the birds were paired, suggesting a temporary and rather weak pair-bonding situation. During much of the postbreeding period the birds were in groups at times numbering 10 to 15 birds, but courtship was observed only during July and August, or only slightly before the onset of breeding. This is in strong contrast to many South American dabbling ducks, which re-form bonds shortly after the breed-

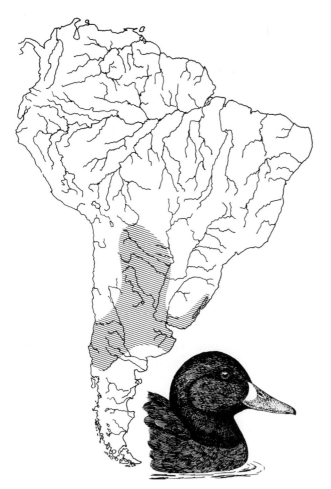

MAP 125. Breeding or residential distribution of the black-headed duck.

ing season is completed. Displays by lone males were directed to paired or unpaired females that they encountered, but no inciting behavior or other obvious courtship displays on the part of females were seen by Weller. They instead tended to attack the displaying males, in a manner also characteristic of female *Oxyura*. The primary display of unpaired males is the toad call (called gulping in my 1965a review), which consists of the male's inflating the neck and cheeks, suddenly raising and opening the bill, and uttering a triple-noted call as the folded wings are lifted twice and the tail is lifted and shaken. This is apparently the only highly ritualized display of males, other than a rather poorly developed turning of the back of the head. Copulatory behavior has not yet been described in detail, but apparently occurs without obvious preceding display (*Wildfowl News* 77: 9).

Reproductive biology. The black-headed duck is unique among the family *Anatidae* in that it is believed to be wholly parasitic in its reproduction. Weller's (1968b) studies support this view; he summarized evidence as to the wide variety of host species' nests in which eggs have been found (totaling 12, plus 2 more reported in 1975 by Höhn), and found that 3 species of coots are the most frequent host species. The long egg-laying period, lasting from September until December, encompasses the major nesting periods of several potential host species, and over half of the 133 red-fronted coot (*Fulica rufifrons*) nests that he found contained one or more duck eggs. The white egg of the black-headed duck is poorly adapted to coot parasitism, but does approach in size and color the eggs of the rosybill, which is also frequently parasitized. The incubation period of the black-headed duck is apparently about 21 days, and thus is shorter than that of coots or rosybills, an obvious advantage for a social parasite. Hatching success was, however, relatively low, with about 80 percent of the red-fronted coot nests that were parasitized successfully hatching, but only about 20 percent of the parasitically laid eggs successfully hatching in one study area. Weller found that many of the duck eggs that failed to hatch became buried under the clutch of coot eggs, and others were deposited too late in the host's incubation period to hatch at the time that the host eggs hatched. Weller also found that the young black-headed ducklings are highly precocial and able to forage for themselves shortly after hatching, a condition that also favors survival in a socially parasitic species. Weller believed that the success of the species' parasitism is based on its wide spectrum of host choice and exploitation, rather than on the evolution of close mimicry and dependency on a single species for its reproductive success. Observations at the Wildfowl Trust in 1977 indicated that two eggs are normally laid in each host's nest, the eggs being deposited at daily intervals at about the time the host's clutch is complete (Michael Lubbock, pers. comm.).

Status. Although apparently still fairly common within its limited breeding range, at least on its preferred marshy habitats, the black-headed duck has the same close dependence on permanent marshes that *Oxyura* species do, and would be very vulnerable to the loss of these areas by drainage or pollution. The special biological interest in this uniquely parasitic species of waterfowl should dictate particular care that its population not be allowed to become endangered.

Relationships. Weller (1968b) reviewed the long and varied taxonomic history of this form and summarized the evidence as to its two likeliest close affinities, the dabbling duck group and the stifftails. In agreement with my analysis of the species' behavior (1965a), Woolfenden's anatomical analysis (1961), and Delacour's (1954–64) judgment, he concluded that the species should be placed in the stifftail tribe, but emphasized its somewhat intermediate nature in many respects. My own conclusion was that this species is the least specialized of all the stifftails, and was almost certainly directly derived from a dabbling duck-like ancestor during the early separation of these two now distinctive tribes. Brush (1976) has found that the feather proteins of the black-headed duck indicate that it should be grouped with the dabbling ducks rather than the stifftails, whose electrophoretic profiles are quite distinctive but probably most like those of the sea ducks.

Suggested readings. Weller, 1968b; Johnson, 1965.

Masked Duck

Oxyura dominica (Linnaeus) 1766

Other vernacular names. None in general English use. Maskenruderente (German); canard masqué (French); pato domínico (Spanish).

Subspecies and range. No subspecies recognized. Breeds from coastal Texas (rarely) southward to Mexico (locally), Central America, and the West Indies, and in the lowlands of South America from Colombia to northern Argentina. Probably resident in most areas, but distribution is poorly known and movements are unreported. See map 126.

Measurements and weights. Folded wing: males, 135–42 mm; females, 133–40 mm. Culmen: males, 33–35 mm; females, 32–34 mm. Weights: males, 369–449 g (av. of 9 406 g); females, 298–393 (av. of 6 339 g) (Haverschmidt, 1972). Eggs: av. 54 x 41 mm, white, 52 g.

Identification and field marks. Length 12–14″ (30–35 cm). *Adult males* in breeding plumage have the forehead, crown, sides of the head, and cheeks black, the chin usually white, and the back of the head, neck, and upper breast rusty cinnamon. The back, scapulars, sides, and flanks are rusty cinnamon, marked with black in the feather centers, producing a spotted effect. The lower breast, abdomen, and under tail coverts are dark brown, marked with black and white, and the tail is blackish. The upper wing surface is mostly dark brown, except for the outer eighth or ninth secondary, which is white, tipped with dark brown, and its adjoining greater and middle coverts, which are white. The iris is brown; the bill is cobalt blue, with a black anterior culmen ridge and nail; and the legs and feet are dark brown or black on the inner side and brown spotted with black on the outer side, inner toe, and webs. *Females* have a dark brown crown and similar-colored stripes through the eye and from the lore through the ear region, with whitish coloration between these stripes and on the chin and throat. The neck is light brown, streaked with darker brown; the breast, sides, and flanks are dusky brown, the feathers barred and tipped with whitish coloration; the upperparts are brownish black, the feathers tipped with buff; and the abdomen is grayish buff to silvery white. The tail and upper coverts are brownish black; the under tail coverts are brownish buff; and the upper wing surface is dark brown, except for the middle secondaries, which are white basally, and their adjoining greater coverts and a few middle coverts, which are also white. The iris is brown, with a pale blue ring around it, the bill is brown with a black tip, and the legs and feet are as in the male. *Males in eclipse* (winter) plumage are very much like the female and may perhaps be best separated by the greater amount of white on their wing

MAP 126. Breeding or residential distribution of the masked duck.

coverts and their less contrasting facial striping. *Juveniles* closely resemble females, but perhaps have more mottled underparts.

In the field, masked ducks are most likely to be seen on ponds overgrown with water lilies, and they often remain motionless and half concealed under such floating leaves. The strongly striped head of females provides the best distinction from ruddy ducks, and males lack the white cheek markings of that species. In flight, white appears on the secondaries of both sexes, which is not true of any of the other stiff-tailed ducks. Various sounds have been attributed to masked ducks, but their descriptions have been inadequate for usefulness in field identification.

NATURAL HISTORY

Habitat and foods. Throughout most of its range, the masked duck is associated with tropical to subtropical marshes or swamps that are densely vegetated with both emergent vegetation and extensive

areas of floating-leaf plants such as water lilies and water hyacinths. Mangrove swamps are apparently used occasionally, but probably not by breeding birds. Leaf-covered aquatic habitats used by jacanas are also often used by masked ducks, which probably rely on the floating leaves for escape cover by hiding under them. Plant seeds are apparently the major food of masked ducks, and their mandibles are clearly more adapted for consuming plant materials than are those of typical *Oxyura* species, which are obviously modified for sieving debris in mud-bottom habitats. Foraging is done by diving, often in rather shallow waters, but the birds also at times come up on dry land, a practice which may account for the presence of wild millet (*Echinochloa*) and other terrestrial plant seeds in gizzards of these birds (Johnsgard, 1975).

Social behavior. Virtually nothing is known of the social behavior or display characteristics of masked ducks, which are generally extremely inconspicuous and difficult to observe in their tropical marsh habitats. Flocks seem to be quite small, rarely consisting of more than ten birds, and only a few fragmentary records of social display have been obtained. It is known that the male is able to inflate its neck and cock its tail, and it performs a kind of breast-beating display that seems similar in most respects to the bubbling display of the ruddy duck. Yet, inasmuch as other *Oxyura* species have rather different kinds of repeated or convulsive head movements in conjunction with neck inflation, it is still impossible to judge the similarities of the masked duck's displays to those of the other *Oxyura* forms. The copulatory behavior is likewise still totally unknown.

Reproductive biology. Very few nests have been found in North America, but unpublished studies by Dale Crider in Argentina provide some information on the nesting behavior of this species. In that area the birds breed in the fall, when water levels in rice fields are rising. Evidently the favored location for nesting there as well as in Cuba is in such rice fields, but in Panama nests have been found in rushes. Nests in rice clumps are usually immediately beside fairly deep water, into which the female can readily escape, and the nests are typically well hidden by roofed-over vegetation, with a lateral entry. Apparently in common with the ruddy duck there is a strong tendency toward dump nesting, so that clutch sizes are generally rather larger than is typical of a single female. Thus, estimates of 4 to 6 eggs are probably closer to normal than the nests having from 8 to 18 eggs which have been described for Cuba. The eggs are evidently laid at the rate of one per day, and lack the chalky surface that other *Oxyura* eggs typically have. The incubation period is still rather uncertain, but Dale Crider estimates it to be about 28 days. There is likewise no information on the fledging period. Crider reports that males were never seen in association with any broods, and observers in Texas have seen only femalelike adults with broods. Males do assume a distinctly femalelike plumage after breeding, and in Argentina some males were found to be flightless at a time when others were still in their full breeding plumage, suggesting a low degree of breeding synchrony (Johnsgard, 1975).

Status. The status of this species in the United States was extensively reviewed by Johnsgard and Hagemeyer (1969), and it is clear that it must be regarded as an accidental breeding anywhere north of Mexico. In Mexico it is probably a rare permanent resident along much of the Gulf Coast, and at least at one time was fairly common in Cuba. It seems to be generally uncommon throughout most of its South American range, with northern Argentina apparently representing its only real area of relatively common occurrence.

Relationships. Although obviously one of the stiff-tailed ducks, the masked duck differs in some respects from the typical *Oxyura* species, and in its morphology seems to be rather less specialized, especially in its bill structure and diving adaptations. Woolfenden (1961) listed a total of 20 such osteological features that separated this species from the others, and thus believed that the genus *Nomonyx* should be retained for it. However, Brush (1976) found its electrophoretic feather protein patterns to be identical to those of the other *Oxyura* forms. In the absence of additional information, I have included it in *Oxyura*.

Suggested readings. Johnsgard & Hagemeyer, 1969; Johnsgard, 1975.

Ruddy Duck

Oxyura jamaicensis (Gmelin) 1789

Other vernacular names. Butterball, stifftail; Schwarzkopfruderente (German); erismature à joues blanches (French); pato rojo americano (Spanish).

Subspecies and ranges. (See map 127.)

  O. j. jamaicensis: North American ruddy duck. Breeds from central British Columbia and southwestern Mackenzie District southeast across the Canadian prairies to the Red River Valley and south through the Great Plains to Colorado, with local or sporadic breeding farther south to Arizona and Texas and limited breeding in central Mexico and the West Indies. Winters primarily along the Pacific, Gulf, and Atlantic coasts, from British Columbia and Delaware southward to Guatemala.

  O. j. andina: Colombian ruddy duck. Resident in the lakes and marshes of the Andes of central and eastern Colombia.

O. j. ferruginea: Peruvian ruddy duck. Resident in the Andean lakes of South America from southern Colombia to Chile. Also breeds on the lowland lakes and marshes of central Chile and Argentina south to Tierra del Fuego, with these latter populations wintering farther north.

Measurements and weights. Folded wing: males, 142–63 mm; females, 135–54 mm. Culmen: males, 39–45 mm; females, 37–45 mm. Weights: males of *jamaicensis* average ca. 550 g in fall, and females ca. 500 g, with respective maximums of 815 g and 794 g. Males of *ferruginea* range from 817 to 848 g (Niethammer, 1953). Eggs: *jamaicensis* av. 62 x 46 mm, chalky white, 73 g; *ferruginea*, av. 73 x 52 mm.

Identification and field marks. Length 14–19" (35–48 cm). *Adult males* in breeding plumage have the crown and nape glossy black, with the rest of the head also black (*ferruginea*), mottled black and white (*andina*), or completely white (*jamaicensis*). The neck, upperparts other than the rump, and the sides and flanks are glossy reddish chestnut. The rump is dark brown to blackish, shading into dark chestnut on the upper tail coverts; the tail is brownish black; the wings are dark brown, the under tail coverts and abdomen are silvery white to mottled. The iris is dark brown, the bill cobalt blue, with pinkish edges, and the legs and feet are bluish gray, with darker webs. *Females* have the crown and sides of the head to a point below the eyes dull rufous, finely barred or freckled with black, with a darkish and indistinct stripe of black and chestnut running from the base of the upper mandible back through the ear region. The rest of the face, throat, and neck are buffy white to ashy. The hind neck, upperparts, sides, and flanks are dull brown, vermiculated and freckled with chestnut. The tail and wings are grayish brown with some chestnut, and the underparts are silvery white. The iris is brown, the bill is dusky, and the legs and feet are bluish gray. *Males in eclipse* (winter) plumage closely resemble females but (in *jamaicensis*) have pure white cheeks and chin. *Juveniles* resemble females but have the breast more mottled with dusky coloration.

In the field, the lengthened tail, which is often held partially raised, provides the best field mark, along with the generally dumpy body profile. Males of the North American population always have white cheeks, but females could readily be confused with female masked ducks where the two sometimes occur together, as on the Gulf Coast. There, the single, rather than double, dark streak through the face provides the easiest means of separating the species. Neither sex is very vocal, although males in spring produce a distinctive drumming sound by beating their lower mandible on the breast. Ruddy ducks rarely fly, but when in flight they have a very rapid wingbeat and typically fly very low over the water, with little or no veering or flaring.

NATURAL HISTORY

Habitat and foods. Ruddy ducks, at least in the North American population, prefer habitats that encompass fresh-water and alkaline permanent marshes, with extensive areas of emergent vegetation, stable water levels, and enough open water for landing and taking off. Such marshes are usually mud-bottomed and not very deep, providing ample foraging opportunities for probing in the bottom debris for the larvae of midges and other flies, as well as abundant vegetable foods in the form of pondweeds, sedges, and bulrushes. As the birds leave their breeding grounds they tend to move to coastal areas, where they concentrate in brackish or slightly brackish coastal lagoons and shallow estuaries.

MAP 127. Breeding or residential distributions of the North America ("N"), Andean ("A"), and Peruvian ("P") ruddy ducks.

There they feed on a variety of submerged plants and also on small mollusks and crustaceans, particularly bivalves, amphipods, and ostracods. Feeding in turbid waters is probably done without many visual clues, and the tip of the bill is rich in sensory endings, while its slightly recurved nail may be useful for tearing leaves or stems from underwater plants (Johnsgard, 1975; Siegfried, 1973).

Social behavior. It is generally believed that ruddy ducks become sexually mature their first winter and that females breed as yearlings, although in captivity females often do not breed until their second year. The birds remain in fairly large flocks through the winter, without obvious pairing, and it is not until well into their spring migration northward and the assumption of breeding plumage by the males that pair-forming activities become conspicuous. These persist after arrival on the nesting areas; Siegfried (1976b) indicates that a loose pair-bonding social system exists in this species, and that most courtship actually occurs after arrival on the nesting grounds. Unlike the maccoa duck, males do not establish specific territories, and male displays are thus concerned with pair-bond establishment rather than territorial defense. Females evidently do little to stimulate display, and instead respond to male approaches by bill-threatening gestures. Tail cocking (tail flashing) and neck enlargement by means of inflation of a tracheal air sac are the most obvious male-to-female displays, while male-to-male displays are of an aggressive nature, and may terminate in chases or flights. The most conspicuous male display is bubbling, in which the bill is repeatedly struck downward against the inflated chest, producing a drumming sound and a ring of bubbles around the breast as air is forced out from under it. A weak call terminates this display, but all of the other displays are either silent or depend on nonvocal sounds such as splashing of water. A short and low display flight over the water, with an associated sound made by the feet repeatedly hitting the water surface, is one example of these displays. Copulation is preceded by bill-dipping and head-shaking movements on the part of the male, who then suddenly mounts the female before she assumes a receptive posture. Treading is typically followed by the male's performance of the bubbling display several times in rapid succession (Johnsgard, 1965a).

Reproductive biology. As noted, birds often arrive on their nesting grounds still unpaired, and during the first few weeks a good deal of display activity is

thus typical. Females seek out areas that combine open water where foraging and taking off and landing are possible with adjacent beds of emergent vegetation such as bulrushes. Such plants, when in water about a foot deep, relatively dense, and pliable enough to be bent down to form a nest support, are favored. In addition, the presence of muskrat runs through the beds of emergent vegetation allow easy and inconspicuous access for the female. The eggs are laid at a daily rate; and in nests not affected by dump-nesting, clutches average about 8 eggs. However, some nests contain considerably more eggs than this as a result of dump nesting, and ruddy ducks often also drop eggs in the nests of other marsh-nesting waterfowl. Siegfried (1976b) attributed the "parasitism" of this species to a lack of attunement between environmental clues such as quality and quantity of available nesting cover and the female's physiological and behavioral response system. The hatching success of parasitically laid eggs is relatively low (Joyner, 1975). Males have but loose pair-bonding tendencies, and may court other females after their mates have begun their incubation period. The usual incubation period is from 23 to 26 days under natural conditions, and about 21 days in an incubator, suggesting that females are not very persistent "sitters." They likewise are relatively poor mothers, and often within a few days after hatching the brood begins to become broken and scattered. This is in part a result of the ducklings' high precocity and tendency to stray from their mothers, often joining other ducklings that also have become separated. The fledging period is probably between 52 and 66 days, and rarely if ever does the female remain in contact with her brood for this entire period.

Status. Bellrose (1976) suggests that a breeding population of about 600,000 birds exists in North America, judging from survey data of federal biologists. Additionally, the West Indian population and those of South America add to the total world population, but none of these apparently small populations can effectively be estimated. At least in North America the ruddy duck population has clearly suffered greatly in recent decades, with the extensive marsh destruction that has occurred in the middle of its favored breeding grounds, and the periodic losses of large numbers of birds on wintering areas as a result of oil-spill disasters.

Relationships. The relationships of the South American "ruddy duck" forms (*andina*, *ferruginea*, and *vittata*) to one another and to *jamaicensis* have been a source of some controversy and even today remain a subject of speculation (Siegfried, 1976a). I have suggested (1965a) that *andina* and *ferruginea* are fairly recent derivatives of a more northerly ancestral form, while Siegfied argues that *ferruginea* may be a long-established form that was derived from *vittata*. Social display patterns in these two forms are so different that it is difficult to accept this argument, particularly since the obviously more isolated *australis* has retained such *vittata*-like display patterns. Perhaps a study of *andina* will help to resolve this question.

Suggested readings. Joyner, 1975; Siegfried, 1976a, 1976b.

White-headed Duck

Oxyura leucocephala Scopoli 1769

Other vernacular names. White-headed stifftail; Weisskopfruderente (German); erismature à tête blanche (French); pato de cabeza blanca (Spanish).

Subspecies and range. No subspecies recognized. Breeds in the Mediterranean region, primarily the western half, at the mouth of the Danube, in the steppes around the Caspian Sea, and on the lower Volga north to Stalingrad and east to the foothills of the Altai range. Also breeds on the upper Yenisei. The southern breeding limits are at the Iran-Afghanistan boundary. Winters from the north coast of Africa through the Nile Valley, Turkey, the Persian Gulf, and most of northern India. See map 128.

Measurements and weights. Folded wing: males, 155–72 mm; females, 150–67 mm. Culmen: males, 46–48 mm; females, 43–45 mm. Weights: males, 558–865 g (av. 737 g); females, 539–631 g (av. 593 g) (Savage, 1965). Eggs: av. 66 x 50 mm, white, 97 g.

Identification and field marks. Length 18″ (46 cm). Plate 59. *Adult males* in breeding plumage have a black crown and an otherwise white head, bounded below by a black neck that grades into a rusty chestnut breast and a more grayish chestnut back color.

The breast also merges with grayish chestnut flanks and underparts that are silvery white. The tail is black and the upper tail coverts chestnut. The upper wing surface is gray, with whitish vermiculations on the coverts and secondaries. The iris is dark brown, the legs and feet brown to lead-colored, and the bill is bright cobalt blue and very swollen at the base, with whitish pink margins. *Females* lack white on the head, and are generally reddish throughout, with a dark chestnut crown and cheek-stripe, and are buffy to whitish elsewhere on the cheeks and throat. The iris is dark brown, the bill is lead-colored and only slightly less swollen than in males, and the legs and feet are black. *Males in eclipse* plumage approach the female in appearance but retain a white head and black crown. *Juveniles* resemble adult females, but young males have a dirty-white face and perhaps are slightly ruddier on the back. Chestnut feathers appear on the male when it is nearly a year old, followed by a white face, but full plumage is not attained until the second spring of life.

In the field, the swollen bill, which is blue in breeding males, and the lengthened tail feathers are good field marks, along with the almost grebelike diving behavior. Like other stifftails, the birds rarely fly and make almost no vocal sounds.

NATURAL HISTORY

Habitat and foods. On their breeding grounds, white-headed ducks occupy permanent brackish or fresh-water marshes that have dense beds of emergent vegetation, open pools, and submerged aquatic plant life. Probably brackish rather than fresh-water areas are preferred for breeding, and in addition the birds typically winter in moderately saline lakes (from about 2,000 to 8,000 parts per million dissolved salts). These support little emergent vegetation but are rich in algae and submerged aquatics such as wigeon grass and pondweeds. The seeds of at least the former are consumed (Savage, 1965), probably so too are the leaves and seeds of pondweeds and other aquatics. Invertebrate foods include the larvae of midges (Chironomidae), which are probably the most important single invertebrate food, but in addition the remains of mollusks and crustaceans have been reported (Dementiev & Gladkov, 1967). Birds in captivity spend much time diving for food and evidently do not depend on grain that is the staple fare of most captive waterfowl. They average about 20 seconds per dive on such

foraging dives, and have resting intervals averaging about six to ten seconds (Matthews & Evans, 1974).

Social behavior. Until recently almost nothing was known of the displays and social behavior of this species, but the account by Matthews and Evans (1974) has remedied much of that gap in our knowledge of the species. In three males and three females observed over two years by these persons (and myself), display occurred from late March until the summer nesting period, and reappeared for a short period in September (with the birds in eclipse plumage), but no pair bonds ever appeared to be established. On the small pond where the birds were placed no obvious territorial limits were established, and display seemed to be associated simply with chance contacts while swimming. Female displays were limited to open-bill threats or direct attacks, in the usual manner of stifftails, but the male displays were varied and numerous. No obvious neck inflation occurs among males, but an extended neck with tail cocking is a commonly performed display. Males also often perform a hunched-rush display much like that of other *Oxyura* males, and additionally have a sideways-

hunch posture in which they orient themselves broadside before a female. This display is usually associated with a "tickering-purr" sound, tail vibrating, and rapid foot paddling in place. Sometimes this posture is followed by an energetic kickflap, which strongly resembles the sousing of the Australian blue-bill and probably even more closely conforms to the extreme form of the maccoa duck's vibrating trumpet posture. This display is always followed by sideways-piping, in which a double-noted piping call is uttered as the tail is twisted and vibrated and the folded wings are slightly raised and lowered. Copulatory behavior has not yet been described.

Reproductive biology. The nesting of this species is evidently unusually late; captive birds at the Wildfowl Trust in England laid between late July and mid-August, and egg or small duckling records of wild birds range from late May through July. The nests are typical of stifftails, being constructed of heavy stands of reeds, but always immediately adjacent to open water for escape purposes. Frequently the nests of other waterbirds (coots, tufted ducks, white-eyes) are exploited by the females. The remarkably large eggs, about one-seventh the weight of the female, are laid at daily intervals, and so a clutch of 7 eggs, weighing as much as the female herself, may be deposited in a week. Clutch sizes range from about 4 to as many as 13, but such large clutches are almost certainly the result of two females laying in the same nest. Almost no down is placed in the nest, contrary to early reports. During the incubation period of 25 days the female incubates most of the time, also contrary to earlier writers, with occasional breaks of from a few to about 40 minutes for preening and bathing. Males play no role in parental care, but Matthews and Evans did observe a male approach and display to a very newly hatched duckling. Ducklings begin to forage by diving on their first day after hatching and gradually increase their mean diving times through the first seven weeks of life, when the durations are about equal to those of adults. Females closely tend their ducklings while they are very young, and the mother and brood may return to their old nest periodically for brooding. By the third week of life, however, the independence of the ducklings was noticeable in one case, and by the fifth

MAP 128. Breeding (hatched) and wintering (stippling) distributions of the white-headed duck.

week of life the mother and duckling were totally independent. Fledging of one duckling occurred on the 58th day (Matthews & Evans, 1974).

Status. This species is considered endangered in Europe and has now probably been eliminated as a breeding species in Spain. It is possibly gone from Sardinia, and is virtually extirpated from Yugoslavia. The last definite nesting in Romania was in 1957, in Corsica it was in the 1960s, and in Hungary, Italy, and Sicily it was in the 1950s. Perhaps fewer than 30 pairs are now breeding in all of Europe outside Russia, and there too the numbers are probably quite small. Currently some 6,000 to 9,000 birds winter in Turkey (Hudson, 1975). It has been estimated by Ogilvie (1975) that the total world population may be about 15,000 birds, with Turkey and Russia holding most of these, and the remainder in Iran, Pakistan, and North Africa.

Relationships. Although there are some plumage similarities between the North American ruddy duck and the white-headed duck, these are rather superficial ones, and recent observations on behavior do not support the view I advanced earlier (1965a) that these two species are quite closely related. The swollen bill is suggestive of the maccoa duck's similar one (but both might be independently associated with large nasal gland development for drinking brackish water), and additionally there seem to be some display similarities in these two species.

Suggested readings. Dementiev & Gladkov, 1967; Matthews & Evans, 1974; Cramp & Simmons, 1977.

Maccoa Duck

Oxyura maccoa Eyton 1838

Other vernacular names. None in general English use. Afrikanische Ruderente (German); erismature maccoa (French); pato maccoa (Spanish).

Subspecies and range. No subspecies recognized. Breeds in eastern Africa from the highlands of Ethiopia and Kenya to eastern Zaire and Uganda, and in southern Africa from Rhodesia and probably Botswana south to the Cape. See map 129.

Measurements and weights. Folded wing: males, 160–72 mm; females, 166–72 mm. Culmen: males, 38–41 mm; females, 36–39 mm. Weights: both sexes 450–700 g. Eggs: ca. 68 x 50 mm, chalky and bluish white, 96 g.

Identification and field marks. Length 19–20″ (48–51 cm). *Adult males* in breeding plumage have the entire head and upper neck black, the chin sometimes grayish. The lower neck, upper breast, mantle, and upper tail coverts are bright chestnut, while the flanks are light chestnut and the lower breast, abdomen, and vent are grayish brown to silvery. The upper wing surface is grayish brown, with some ocher-colored freckling. The blackish tail feathers are pointed and narrow. The iris is brown, the legs and feet slate gray, and the bill is cobalt blue. *Females* and nonbreeding males resemble the female North American ruddy duck, but have less distinctive facial striping. They are grayish brown on the upperparts, with tan and buff freckling, and their sides and flanks are ashy brown, barred and slightly freckled with buffy white. The soft-part colors are like those of the breeding male, except for the bill, which is slate gray. *Juveniles* resemble the adult female, but have a more uniform coloration, lacking the spotting or barring on the upperparts and showing more sepia than grayish on the underparts.

In the field, the long tail, rotund body, and frequent diving will separate this from all other African species except perhaps the white-backed duck, which is more fulvous-colored and lacks a long tail. Although females are nearly mute, the male utters a frog- or trumpet-like call during display. Like the other stifftails, the bird takes flight with difficulty and rarely attains a height far above the water.

NATURAL HISTORY

Habitat and foods. Although the maccoa duck is sometimes seen in other habitats, its preferred en-

vironment for breeding consists of waters having extensive, tall emergent vegetation such as *Phragmites* and associated open stretches of water of medium depth, with a relative absence of floating or partially submerged vegetation (Macnae, 1959). During the breeding season the birds also may be found on mine impoundments and sewage-farm basins, particularly the latter, where decomposing organic matter provides much invertebrate food, especially *Tubifex* worms, the eggs of *Daphnia*, and similar materials that the ducks are able to strain with their bills and extract from the debris. Such foods would be expected to be major parts of this species' diet, although virtually no food analyses are available yet. One report cited by Clancey (1967) indicated that the seeds and roots of aquatic plants are consumed, as well as algae, larvae, and small fresh-water mollusks.

Social behavior. Macnae (1959) has suggested that the males of this species are highly territorial, and Clark (1964) supports this view, indicating that areas as large as 1,000 square yards may be defended. Incursions into such areas by other males may lead to intense fighting, but females may or may not be closely associated with such territory holders. Clark states that territories are established by July, but that peak territorial activity does not occur until October or November. More than one female may occupy the territory of a single male, and Siegfried (1976a) states that up to eight may nest in a male's territory. It is thus clear that monogamous pair bonds are lacking and the role of the male is primarily one of defending a suitable nesting area rather than forming a pair bond with and defending a specific female. The male's displays were originally described in part by various persons (Johnsgard, 1968c; Clark, 1964), but only recently have been fully documented and compared with related species (Siegfried & Van der Merwe, 1975). The most conspicuous of the male displays is the independent vibrating trumpet call, uttered with the head extending forward and the tail variably raised. The rather froglike call that is uttered serves as an important response to the presence of other males, and is probably the most important territorial proclamation signal. When males are displaying in the presence of females, a modified and more complex version, called the vibrating trumpet call, is performed, which consists of a preliminary rapid approach to the female in a stretched swim or more rapid ski posture, and the call is preceded by a strong upward and slightly backward neck jerk prior to the utterance of the call. Bill dipping and lateral

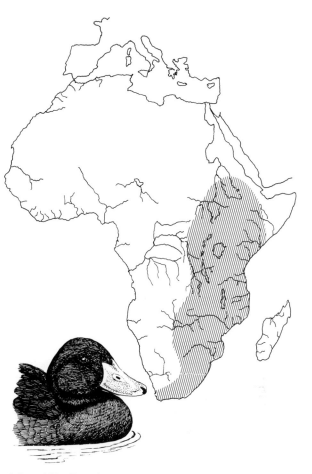

MAP 129. Breeding or residential distribution of the maccoa duck.

flicking of water with the bill (water flicking) may be performed before or after the vibrating trumpet display, and thus are not specific precopulatory displays as they are in some other *Oxyura* species. Dab-preening, wing flapping, cheek rolling, head dipping, wing shuffling, and a general shake of the body during swimming are also displays, and males and females sometimes dive in synchrony. The last complex male display of the maccoa duck is one that has strong similarities to sousing in the Australian species, and appears to be homologous with it, although it is evidently less common and of less sexual significance in the maccoa duck. Copulatory behavior is still incompletely described, but precopulatory display by the female apparently consists of bill dipping and water flicking, following the male's vibrating trumpet call and preceding her assumption of a prone position. Bathing follows copulation, but the postcopulatory behavior of both sexes does not appear to be definitely ritualized (Siegfried & Van der Merwe, 1975).

Reproductive biology. The nesting season of the maccoa duck is considerably prolonged, with South African records extending from June until April, but with a peak between September and December. Females construct their nests in beds of emergent vegetation, usually among cattails but sometimes among other emergents and infrequently on dry land among low scrub. The typical nests are constructed by pulling down leaves of cattails and bending them down to form a nest basin. Later, reed stems may be pulled over the nest to form a partial dome as well. The average clutch size has been determined to be 5 eggs, with a range of 4 to 8. Evidently one or two days are skipped in the egg-laying sequence, so that eight days are normally needed to complete a clutch of 6 eggs. Only the female incubates, and the incubation period is probably 25 to 27 days. The young are watched attentively by the female for varying periods, but often for no more than a few weeks, after which they largely fend for themselves. Although males sometimes approach females with broods, they are typically chased away by the females (Clark, 1964). The fledging period is apparently not yet established.

Status. Although fairly widespread in Africa, the maccoa duck is generally uncommon throughout most of its range, with the greatest numbers being found on certain reedy lakes in Kenya and lakes west of the Rift Valley farther south. In southern Africa it is most common on fresh waters in the southwestern Cape, and in the gold-mining areas of the Transvaal and Orange Free State, where dam construction favors its survival. It is not a significant game species, and is in no apparent present danger.

Relationships. I once suggested (1961a) that the Argentine, Australian, and African species of *Oxyura* constitute an evolutionary group that is relatively distinct from the other *Oxyura* forms of the Northern Hemisphere (*leucocephala* and *jamaicensis*). More recent observations suggest that *maccoa* shares a few similarities with these latter two species and at least behaviorally may help to link the genus *Oxyura* into a single unit. Additional anatomical and behavioral evidence would be desirable to test this idea.

Suggested readings. Clancey, 1967; Johnsgard, 1968c; Siegfried & Van der Merwe, 1975.

Argentine Blue-billed Duck

Oxyura vittata (Philippi) 1860

Other vernacular names. Argentine ruddy duck; Argentinische Ruderente (German); erismature tacheté (French); pato rana or pato rojo de la Argentina (Spanish).

Subspecies and range. No subspecies recognized. Breeds in Chile from Atacama to Llanquihue, in Argentina from La Rioja and San Juan south to Tierra del Fuego, in Uruguay, and in extreme southeastern Brazil. Winters north to Paraguay and south-central Brazil. See map 130.

Measurements and weights. Folded wing: males, 137–55 mm; females, 132–40 mm. Culmen: males, 38–45 mm; females, 36–38 mm. Weights: both sexes 550–675 g (Kolbe, 1972); 1 male weighed 610 g, 1 female 560 g (Weller, 1968a). Eggs: 65 x 47 mm, white and chalky, 86 g.

Identification and field marks. Length 16″ (46 cm). *Adult males* in breeding plumage have black heads and necks, usually with no white on the chin, and deep chestnut-red breast, flank, and upperpart coloration, the abdomen grading to silvery white, barred with gray. The under tail coverts may be whitish or rusty, and the tail is black, as is the upper wing surface, although rusty feathers may be interspersed. The iris is dark brown, the bill is cobalt blue, and the legs and feet are blackish to olive gray. *Females* (and nonbreeding males) closely resemble females of the ruddy duck, but the head is more distinctly striped with buffy white below the eyes, and the chin and throat are nearly pure white. Nonbreeding males are likely to have a few rusty-colored feathers present on the body, which are lacking in females. *Juveniles* also closely resemble adult females, but the blunt-tipped tail feathers should identify them.

In the field, Argentine blue-bills are most likely to be confused with Peruvian ruddy ducks, which are appreciably larger and lighter in their chestnut tones, and females and nonbreeding males of which have less distinctly striped facial markings. Peruvian ruddy duck males also often have white throats or mottled white markings on the lower cheeks, and considerably wider bills. Neither sex is sufficiently

vocal to provide useful field clues, and when the birds make their rare flights they have the same labored and buzzing flight of other *Oxyura* species.

NATURAL HISTORY

Habitat and foods. In Chile, this species is said to occupy lowland lakes that are also used by the Peruvian ruddy duck, while in eastern Argentina they were reported by Weller (1967a) to utilize large lakes and semiopen marshes with large pools. Where deep open roadside marshes were connected with larger marshes, they too were occupied. On windy days the birds were sometimes also seen on pools covered with duckweeds and azolla (*Azolla*). The foods of this species have not yet been studied but presumably are much like those of the North American ruddy duck.

Social behavior. Evidently there is some off-season flocking by this species; Weller (1967a) noted a group of nearly 400 birds on one lake during late May and early June. Evidently in eastern Argentina the bright breeding plumage is molted over a period from about late January until April or May, while the nesting season extends from mid-October until early January (Weller, 1967a, 1968a). Pair-bonding patterns are not well studied in any of the stifftails, but are almost certainly temporary and relatively weak. Weller observed courtship, but does not indicate its timing or frequency of occurrence. The few published descriptions of the displays are at considerable variance with one another, and it is clear that the displays differ greatly from those of the North American species and

instead more closely approximate those of the Australian blue-billed duck (Johnsgard, 1967a). Probably the most common display given by males is a form of head jerking ("head-pumping" in Weller's terminology) that is directed toward other males as well as to females. The male also performs a choking display (that corresponds exactly to the sousing of the Australian species) in which the head and greatly expanded neck are held forward over the water and a convulsive series of jerking movements are made. This display was filmed by Bradbury and Bradbury (1968), and one sequence that they filmed contained a series of 18 such jerking movements, lasting about 12 seconds. A low display flight over the water, called the ringing rush in the North American species, was also filmed by them. Females participate very little in display, other than to threaten males

MAP 130. Breeding (hatched) and wintering (stippling) distributions of the Argentine blue-billed duck.

that approach too closely. Behavior associated with copulation is still undescribed, although I have observed some possible precopulatory display consisting of bill dipping and head rolling on the part of a male (Johnsgard, 1965a).

Reproductive biology. Johnson (1965) reported that although both species of "ruddy ducks" nest in the same area of central Chile, this species builds a much smaller nest than the larger Peruvian duck and lays a smaller clutch of only 3 to 5 eggs. Nests are hidden in emergent vegetation and are nearly flat, scarcely reaching above the water level. Earlier reports indicate that the average clutch may be larger, of 6 to 12 eggs, but at least some of the larger clutches that have been reported are probably the result of dump nesting by two or more females. Incubation is by the female alone, and the length of the incubation period has still not been established. No doubt only the female cares for the young, and even she is likely to desert them to fend for themselves at a relatively early age (Weller, 1967a).

Status. The population size of such a secretive duck as this is impossible to judge, but doubtless it is controlled by the availability of large and permanent marshes supporting the vegetation and invertebrate life on which stifftails depend. Drainage of such marshes, rather than hunting or other pressures, are certain to be the controlling factors affecting the survival of these specialized species.

Relationships. In the past, this species has often been confused with *ferruginea*, and even has been considered conspecific with it at various times. Johnson (1965) established that the two species are indeed distinct, and behavioral evidence supports the position that they are not even very closely related, at least within the genus *Oxyura* (Johnsgard, 1967a). Most probably South America was invaded by an ancestral stifftail that also gave rise to the modern species of Australian *Oxyura*, and secondly by a northern form that was also ancestral to the North American ruddy duck, and the two forms came into contact in southern South America. This is the only area where two typical *Oxyura* species overlap in breeding, and the possible ecological interactions between them have yet to be studied.

Suggested readings. Johnson, 1965; Johnsgard, 1967a.

Australian Blue-billed Duck

Oxyura australis Gould 1836

Other vernacular names. Australian stifftail, stifftail; Australische Ruderente (German); erismature d'Australie (French); pato pico azul (Spanish).

Subspecies and range. No subspecies recognized. Resident in southern Australia, primarily in the Murray and Darling river systems and adjacent coastal areas of Victoria and South Australia, as well as in coastal portions of Western Australia. See map 131.

Measurements and weights. Folded wing: males, 150–73 mm; females, 142–63 mm. Culmen: males, 37–48 mm; females, 32–37 mm. Weights: males, 610–965 g (av. 812 g); females, 476–1,300 g (av. 852 g). Eggs: av. 66 x 48 mm, light green, 90 g.

Identification and field marks. Length 14–16″ (30–35 cm). *Adult males* in breeding plumage have a black head, while the foreneck, breast, flanks, and back are dark chestnut and the underside is silvery gray to brown, with darker flecking. The upper tail coverts are brownish black, with chestnut flecking, while the under tail coverts are mottled with gray and brown. The tail is stiffened and elongated, and is black above. The upper wing surface is generally brown, with chestnut flecking on the coverts. The iris is brown, the bill cobalt blue with a black nail, and the legs and feet are gray. *Females* and males in nonbreeding plumage are barred and mottled with tones of blackish brown, black, and light brown, grading toward silvery gray on the undersides, and have obscure facial striping. The bill is slate gray, but the other soft-part colors are like those of the breeding

MAP 131. Breeding or residential range of the Australian blue-billed duck.

male. *Juveniles* resemble adult females but are paler and have grayish green bills, and gray-brown feet.

In the field, the chunky body shape, elongated and frequently cocked tail, and expert diving behavior mark this species, which is appreciably smaller than the musk duck, the only other stifftail in Australia. The blue bill of the male in breeding plumage is also distinctive, as is its chestnut red coloration at that time. Neither sex is very vocal; most of the sounds made by displaying males are splashing noises or mechanical sounds.

NATURAL HISTORY

Habitat and foods. Two rather different habitats are frequently used by blue-billed ducks: fairly large and relatively shallow lakes during the winter period, when large flocks often congregate on such waters, and the permanent *Typha*-dominated fresh-water marshes of the interior. Two other habitats are sometimes also used for breeding, lignum swamps and tea tree thickets in coastal regions. Lakes such as Kangaroo Lake in northern Victoria are greatly favored and evidently are regular wintering areas for birds that breed farther north in the Murray River drainage; here the birds mingle with musk ducks, coots, and other waterfowl, usually fairly near shore or *Typha* beds, and dive for their food. Foods on the wintering areas have not been studied, but probably are much the same as those taken on the breeding grounds. A sample of 546 stomach remains from a breeding-ground swamp in New South Wales indicated that nearly equal amounts of plant and animal materials are consumed (Frith et al., 1969). The plants represented were primarily hornwort (*Ceratophyllum*), smartweeds (*Polygonum*), water milfoil (*Myriophyllum*), and azolla (*Azolla*). Adult materials were primarily insects, although small arthropods, mollusks, and crustaceans also were present in small numbers. As with other *Oxyura* species, the larvae of midges (*Chironomidae*) are the single most important source of animal foods, and are obtained by sieving bottom debris. In contrast to the musk duck, this species consumes few crustaceans or mussels, and the size of the plant and animal foods is appreciably smaller.

Social behavior. On wintering areas such as Lake Kangaroo, the majority of the blue-bills are in non-breeding condition and obviously unpaired. In late July, I counted over 1,300 birds along a mile of Lake Kangaroo's shoreline, and the vast majority of the males were partly or entirely femalelike in coloration, with no pair bonds evident. By late July, a few males in breeding plumage were beginning to display, although the earliest record for ducklings in that area is November. Thus, sexual behavior probably extends over a period of several months in that region, and probably is associated largely with the breeding grounds themselves. Males are quite aggressive toward other males in breeding condition, and much of their display seems to be concerned with territorial advertisement and defense. A great number of male displays have been seen in this species (Johnsgard, 1966b), including tail cocking, neck inflation, and several ritualized comfort movements such as dab-preening, head rolling, and wing flapping. The most elaborate of the male displays is sousing, a convulsive series of movements made with the head and neck held low over the water, the tail strongly cocked, and air apparently being expelled forcibly from the neck. The display seems to have strong motor similarities to the choking display of the Argentine blue-bill, and lesser affinities with the extreme form of the vibrating trumpet display of the maccoa duck (Johnsgard, 1968c). Males also rush over the water in a motor boat display, and sometime perform short display flights (or ring rushes) toward females. In the only described case of observed copulation, a male chased a female at high speed until he caught her, and copulation occurred with the birds entirely submerged (Wheeler, 1953). Probably no real pair bonds are established in this species, but instead effective reproduction by males depends on their ability to obtain and advertise a territory in which females can nest successfully.

Reproductive biology. In the Lake Kangaroo area and other parts of northern Victoria, the nesting season apparently extends from October to February or March, while in Western Australia the nesting time coincides with the winter rainfall, and extends from September to November. Nesting sites are usually built in dense marsh or swamp vegetation, often in beds of cattails, but sometimes also in lignum, reeds, or tea trees. Rarely the old nest of a coot or some other marsh bird is utilized. In most instances the clutch consists of 5 or 6 eggs, but the reported range is from 3 to 12, with the latter almost certainly being the result of several females' efforts. Incubation is entirely by the female, and requires 26 to 28 days. There is only one record of a male being seen with a female and brood, and no reason to believe that male participation in brood care is likely. By the eighth week of life the young are fully feathered and presumably fledged, and within a year the full adult condition has been attained (Frith, 1967).

Status. The blue-billed duck is protected from hunting throughout its range, and in any case hunting would not be a likely conservation problem because of the infrequent flights of this species and its relatively inaccessible habitats. The most serious factor affecting its status is the availability of the permanent marshes and swamps that it needs for breeding. Without preservation of an adequate number of these, its future is in doubt (Frith, 1967).

Relationships. Few taxonomists have tried to resolve the intrageneric affinities of the typical stifftails, and even the species limits have been the subject of confusion. The best behavioral evidence favors the view that the nearest relative of the Australian blue-bill is the Argentine blue-bill and that it is probably somewhat less closely related to the maccoa duck (Johnsgard, 1968c).

Suggested readings. Johnsgard, 1966b; Frith, 1967; Braithwaite & Frith, 1969.

Musk Duck

Biziura lobata (Shaw) 1796

Other vernacular names. None in general English use. Lappenente (German); canard mosqué (French); pato almízclero de Australia (Spanish).

Subspecies and range. No subspecies recognized. Resident in southern and western Australia, plus Tasmania, with the northern limits at Fraser

Island, Queensland, and North-west Cape, Western Australia, but most common in the permanent swamps of New South Wales, Victoria, and in Tasmania. See map 132.

Measurements and weights. Folded wing: males, 205–40 mm; females, 165–202 mm. Culmen: males, 36–47 mm; females, 31–41 mm. Weights: males, 1,811–3,120 g (av. 2,398 g); females, 993–1,844 g (av. 1,551 g). Eggs: av. 79 x 54 mm, pale greenish white, 128 g.

Identification and field marks. Length 24–29″ (61–73 cm). *Adults* of both sexes are generally blackish brown, with lighter brown to grayish penciling over most of the head and body except the upper part of the head, which is nearly black. The lower breast and underparts are whitish brown, flecked or barred with blackish coloration. The tail and upper wing surface are black. The bill is black and has a large pendulous lobe beneath, the feet and legs are dark gray, and the iris is brown. *Females* are identical in plumage to males, but are much smaller and have only an incipient lobe. *Juveniles* resemble adult females, but exhibit yellow on the anterior half of the lower mandible.

In the field, the large body size, grayish black overall coloration, and superb diving abilities of the musk duck make them almost impossible to confuse with other species. Females approach blue-billed ducks in size, but are predominantly grayish black rather than brownish in color. Like other stifftails, the birds rarely fly and are almost mute, although a whistling sound is uttered by males in full display.

NATURAL HISTORY

Habitat and foods. The range and preferred habitats of the musk duck closely conforms to those of the blue-billed duck, although the musk duck is perhaps somewhat the more widespread. It is especially numerous in Tasmania and the large swamps of the Murray and Darling river basins, while during the nonbreeding season the birds spread out to deeper lakes, estuaries, and even marine environments some distance from the shore. Evidently more mobile than the blue-billed duck, the musk duck is able to locate and breed in the temporarily flooded lakes of Australia's dry interior, including Lake Eyre. However, their favored breeding habitats are the permanent swamps, where they forage in moderately deep water. They often remain submerged for a half minute or more, and doubtless extract much of their food from the muddy bottoms of these marshes. Frith (1967) and Frith et al. (1969) report that nearly all the food is of animal origin and consists primarily of the larvae and adults of aquatic insects. Additionally, mollusks and crustaceans, including large crayfish, are regularly eaten. Unlike *Oxyura* species, the musk duck concentrates on larger food materials, such as large bugs and the larvae of dragonflies and May flies. Occasionally even small ducklings may be consumed. Fish and frogs are apparently also rarely taken by these large birds. Almost no food is taken from the water surface; this difference and the difference in size of foods taken seem to reduce competition between musk ducks and blue-billed ducks where the two occur together.

Social behavior. During much of the fall and winter musk ducks are found in southern New South Wales in flocks of varying sizes which probably consist of both adult and immature birds. The period of time to sexual maturity in males has not been established, but might be more than a year, judging from the number of males with small lobes that may be seen in company with adult males and females. Braithwaite and Frith (1969) suggest that females probably breed at one year but that yearling males may be unable to compete effectively for territories with older birds. As spring approaches, the adult males spend increasing amounts of time establishing their territories, which consist of stretches of shoreline and adjacent reedbeds. During all hours of the day and night males can be heard performing one or more of their three primary displays, the paddling kick, the plonk kick, and the whistle kick. These displays, which seem to represent progressive levels of ritualization, complexity, and time-interval constancy (Johnsgard, 1966b), produce sounds that carry for a half mile or more under favorable conditions. The sounds of the paddling kick are nothing more than the noise of water splashed upward and backward by the feet. The noise made by the webs of the feet as they leave and strike the water are responsible for the plonk-kick sounds in the birds of eastern Australia (Johnsgard, 1966b), while in Western Australia a distinct vocalization accompanies this call (Robinson & Robinson, 1970). Last, the whistle kick is primarily a vocally produced noise. In extreme cases of this display the tail is cocked so that it presses against the back, the neck and cheeks are greatly inflated, and the pendent lobe on the bill is expanded to form a turgid wedge. In this posture the male splashes water laterally with both feet as it utters a clear whistle at approximately three-second intervals. Females and other males are attracted to such displaying males; males that approach too closely are chased away, while if a female approaches closely, the male suddenly attempts to copulate with her in a rapelike manner. Seemingly no association between the pairs occurs; fertilized females presumably seek out a nesting site within the male's large territory and quite probably more than one female will occasionally nest there. There appears to be in the musk duck a greater development of promiscuous mating systems than in perhaps any other species of this family, although some *Oxyura* species approach it, and likewise the genera *Cairina* and *Sarkidiornis* of the perching ducks seem to be similar.

Reproductive biology. Nesting usually is done in dense beds of cattails, and in Victoria usually occurs between September and November, with some exceptional earlier and later records. In Western Australia nesting occurs at much the same time, between August and November, in spite of the climatic differences that prevail. During years of unusual flooding in the interior, nesting may occur at other times, and at Lake Kangaroo territorial males have been

MAP 132. Breeding (hatched) range of the musk duck, including areas of major concentrations (cross-hatched).

heard displaying in all months of the year. Besides nesting in cattail beds, musk ducks sometimes also place their nests on low branches of tea trees where they touch the water, on low stumps, or even rarely on the ground on islands. The average clutch is of fewer than 3 eggs, and the observed range is from 1 to 10, but these large clutches are clearly the efforts of two or more females (Frith, 1967). The period of incubation has not yet been established. After hatching, the young are tended for a time by the female, and have even been reported riding on the back of the female. Additionally, females directly feed their ducklings by diving for food and passing it to them; such behavior may account for the unusually low clutch size in this species. This strong dependency of the young on their mother is quite different from the usual situation in stifftails; the ducklings of the other species are unusually precocial and soon become independent.

Status. Although of no importance as game birds, and probably inedible because of their musky odor, some musk ducks are drowned by fish nets in which they become entangled. Like the blue-billed duck, their dependence on the permanent swamps that are in danger of being drained for agricultural purposes may ultimately pose conservation problems for the species (Frith, 1967).

Relationships. In a review of this species' behavior, I concluded (1966b) that *Biziura* is a close relative of *Oxyura,* and that nearly all of the unique features of this genus can be attributed to the species' social behavior, with attendant selective pressures that have promoted the evolution of large body size, extreme sexual dimorphism, and elaborate territorial signals. In many ways the musk duck is the predictable end point of a trend that is already evident in the genus *Oxyura.* Brush (1976) reported that the feather proteins of *Oxyura* and *Biziura* formed identical electrophoretic patterns, supporting this position.

Suggested readings. Johnsgard, 1966b; Frith, 1967; Lowe, 1966.

Sources Cited

Alder, L. P. 1963. The calls and displays of African and Indian pygmy geese. In *Wildfowl Trust, 14th Annual Report*, pp. 174–75.

Ali, S. 1960. The pink-headed duck *Rhodonessa caryophyllacea* (Latham). *Wildfowl Trust, 11th Annual Report*, pp. 55–60.

Ali, S., & Ripley, D. 1968. *Handbook of the birds of India and Pakistan, together with those of Nepal, Sikkim, Bhutan and Ceylon*. Vol. 1. Bombay: Oxford University Press.

Alison, R. 1970. The behaviour of the oldsquaw (*Clangula hyemalis*) in winter. M.S. thesis, University of Toronto.

———. 1975. *Breeding biology and behavior of the oldsquaw (Clangula hyemalis L.)*. A.O.U. Monographs, no. 18.

Austin, O. L., Jr. 1948. *Waterfowl of Japan*. G.H.Q. Supreme Cmdr. Allied Powers, Nat. Res. Sec., Tokyo.

Baldwin, P. M. 1945. The Hawaiian goose, its distribution and reduction in numbers. *Condor* 47:27–37.

———. 1947. Foods of the Hawaiian goose. *Condor* 49:108–20.

Balham, R. W. 1952. Gray and mallard ducks in the Manawatu District, New Zealand. *Emu* 52:163–91.

Balham, R. W., & Miers, K. H. 1959. *Mortality and survival of gray and mallard ducks banded in New Zealand*. N.Z. Dept. Int. Affairs Wildlife Pub. No. 5.

Banko, W. 1960. *The trumpeter swan: Its history, habits, and population in the United States*. U.S. Dept. of Interior, Fish and Wildlife Service, North American Fauna No. 63, 214 p.

Barry, T. W. 1966. Geese of the Anderson River delta, Northwest Territories, Canada. Ph.D. dissertation, University of Alberta.

Bauer, K. M., & Glutz von Blotzheim, Urs. N. 1968, 1969. *Handbuch der Vögel Mitteleuropas*. Vols. 2, 3. Frankfurt: Akademische Verlagsgesellschaft.

Beames, I. 1969. Welcome immigrants: Mandarin duck. *Animals* 12:315–17.

Beckwith, S. L., & Hosford, H. J. 1957. A report on seasonal food habits and life history notes of the Florida duck in the vicinity of Lake Okeechobee, Glades County, Florida. *American Midland Naturalist* 57:461–73.

Bell, H. 1969. Birds of Ok Tedi, New Guinea. *Emu* 69:193–211.

Bellrose, F. C. 1976. *Ducks, geese and swans of North America*. 2d ed. Harrisburg, Pa.: Stackpole.

Bellrose, F. C., & Hawkins, A. S. 1947. Duck weights in Illinois. *Auk* 64:422–30.

Bengtson, S. A. 1966a. [Observation on the sexual behavior of the common scoter, *Melanitta nigra*, on the breeding grounds, with special reference to courting parties.] *Vår Fågelvärld* 25:202–26. (In Swedish, with English summary.)

———. 1966b. Field studies on the harlequin duck in Iceland. In *Wildfowl Trust, 17th Annual Report*, pp. 79–94.

———. 1970. Location of nest-sites of ducks in Lake Mývatn area, north-east Iceland. *Oikos* 21:218–29.

———. 1971a. Variations in clutch-size in ducks in relation to their food supply. *Ibis* 113:523–26.

———. 1971b. Habitat selection of duck broods in Lake Mývatn area, north-east Iceland. *Ornis Scandinavica* 2:17–26.

———. 1972a. Reproduction and fluctuations in the size of duck populations at Lake Mývatn, Iceland. *Oikos* 23:35–58.

———. 1972b. Breeding ecology of the harlequin duck *Histrionicus histrionicus* (L.) in Iceland. *Ornis Scandinavica* 3:1–19.

Bennett, L. J. 1938. *The blue-winged teal, its ecology and management*. Ames, Iowa: Collegiate Press.

Benson, C. W.; Brooke, R. K.; Dowsett, R. J.; & Irwin, M.P.S. 1971. *The birds of Zambia*. London: Collins.

Berger, A. J. 1972. *Hawaiian birdlife*. Honolulu: University Press of Hawaii.

Bezzel, E. 1959. Beiträge zur Biologie der Geschlechter bei Entenvögeln. *Anzeiger der Ornithologischen Gesellschaft in Bayern* 5:269–355.

———. 1969. *Die Tafelente*. Neue Brehm-Bücherei 405. Wittenberg, Lutherstadt: A. Ziemsen Verlag.

Bolen, E. G. 1964. Weights and linear measurements of black-bellied tree ducks. *Texas Journal of Science* 16:257–60.

———. 1967. The ecology of the black-bellied tree duck in southern Texas. Ph.D. dissertation, Utah State University.

———. 1971. Pair-bond tenure of the black-bellied tree duck. *Journal of Wildlife Management* 35:385–88.

———. 1973. Copulatory behavior in Dendrocygna.

Southwestern Naturalist 18:348–50.

Bolen, E. G., & Forsyth, B. J. 1967. Foods of the black-bellied tree duck in south Texas. *Wilson Bulletin* 79:43–49.

Bolen, E. G.; McDaniel, B.; & Cottam, C. 1964. Natural history of the black-bellied tree duck (*Dendrocygna autumnalis*) in southern Texas. *Southwestern Naturalist* 9:78–88.

Bolen, E. G., & Rylander, M. K. 1973. Copulatory behavior in *Dendrocygna*. *Southwestern Naturalist* 18:348–50.

Bouvier, J. M. 1974. Breeding biology of the hooded merganser in southwestern Quebec, including interactions with common goldeneye and wood duck. *Canadian Field-Naturalist* 88:323–30.

Boyd, H. 1953. On encounters between wild white-fronted geese in winter flocks. *Behaviour* 5:85–129.

Bradbury, J., & Bradbury, I. 1968. Duck displays. *Animal Kingdom* 71(2):13–15.

Braithwaite, L. W. 1976. Notes on the breeding of the freckled duck in the Lachland River valley. *Emu* 76:127–32.

Braithwaite, L. W., & Frith, H. J. 1969. Waterfowl in an inland swamp in New South Wales. III. Breeding. *C.S.I.R.O. Wildlife Research* 14:65–109.

Braithwaite, L. W., & Miller, B. 1975. The mallard, *Anas platyrhynchos*, and mallard-black duck, *Anas superciliosa rogersi*, hybridization. *Australian Wildlife Research* 2:47–61.

Brand, D. J. 1961. A comparative study of the Cape teal (*Anas capensis*) and the Cape shoveller (*Spatula capensis*), with special reference to breeding biology, development and food requirements. Ph.D. dissertation, University of South Africa.

———. 1966. Nesting studies of the Cape shoveler *Spatula capensis* and the Cape teal *Anas capensis* in the western Cape Province, 1957–1959. *Ostrich* supplement no. 6:217–21.

Brandt, H. 1943. *Alaska bird trails.* Cleveland: Bird Research Foundation.

Bruggers, R. L. 1974. Nesting biology, social patterns and displays of the mandarin duck, *Aix galericulata*. Ph.D., dissertation, Bowling Green University.

Brush, A. H. 1976. Waterfowl feather proteins: Analysis of use in taxonomic studies. *Journal of Zoology* 179:467–98.

Bryant, D. M., & Leng, J. 1975. Feeding distribution and behaviour of shelduck in relation to food supply. *Wildfowl* 26:20–30.

Carter, B. C. 1958. *The American goldeneye in central New Brunswick.* Canadian Wildlife Service, Wildlife Management Bulletin, ser. 2, no. 9.

Cawkell, E. M., & Hamilton, J. E. 1961. The birds of the Falkland Islands. *Ibis* 103a:1–27.

Chapin, J. P. 1932. Birds of the Belgian Congo. *American Museum of Natural History Bulletin* 65:1–756.

Clancey, P. A. 1967. *Gamebirds of southern Africa.* New York: American Elsevier Publishing Co.

Clark, A. 1964. The maccoa duck (*Oxyura maccoa* (Eyton)). *Ostrich* 35:264–76.

———. 1966. The social behaviour patterns of the southern pochard *Netta erythrophthalma brunnea. Ostrich* 37:45–46.

———. 1969a. The breeding of the Hottentot teal. *Ostrich* 40:33–36.

———. 1969b. The behaviour of the white-backed duck. *Wildfowl* 20:71–74.

———. 1971. The behaviour of the Hottentot teal. *Ostrich* 42:131–36.

———. 1976. Observations on the breeding of whistling ducks in southern Africa. *Ostrich* 47:59–64.

Cooch, F. G. 1965. *The breeding biology and management of the northern eider (Somateria mollissima borealis) in the Cape Dorset area, Northwest Territories.* Canadian Wildlife Service, Wildlife Management Bulletin, ser. 2, no. 10.

Cooke, F., & McNallay, C. M. 1975. Mate selection and colour preferences in lesser snow geese. *Behaviour* 53:151–70.

Cottam, C. 1939. *Food habits of North American diving ducks.* USDA Technical Bulletin 643.

Coulter, M. W., & W. R. Miller. 1968. Nesting biology of black ducks and mallards in northern New England. *Vermont Fish and Game Dept. Bulletin,* no. 68–2.

Cramp, D., & Simmons, K. E. L. (eds.). 1977. *The birds of the western Palearctic.* Vol. 1. Oxford: Oxford University Press.

Cronan, J. M., Jr. 1957. Food and feeding habits of the scaups in Connecticut waters. *Auk* 74:459–68.

Curth, P. 1954. *Der Mittelsäger.* Neue Brehm-Bücherei 126. Wittenberg, Lutherstadt: A. Ziemsen Verlag.

Dane, B., & van der Kloot, W. G. 1964. An analysis of the displays of the goldeneye duck (*Bucephala clangula* L.). *Behaviour* 22:282–328.

Dane, C. W. 1966. Some aspects of breeding biology of the blue-winged teal. *Auk* 83:389–402.

Dau, D. P. 1972. Observations on the natural history of the spectacled eider (*Lampronetta fischeri*). Dept. Wildlife and Fish, University of Alaska. (Mimeo.)

———. 1974. Nesting biology of the spectacled eider *Somateria fischeri* (Brandt) on the Yukon-Kuskokwim delta, Alaska. M.S. thesis, University of Alaska.

———. 1976. Clutch sizes of the spectacled eider on the Yukon-Kuskokwim delta, Alaska. *Wildfowl* 27:111–13.

Delacour, J. 1954–64. *The waterfowl of the world.* 4 vols. London: Country Life.

Delacour, J., & Mayr, E. 1945. The family Anatidae. *Wilson Bulletin* 57:4–54.

Dementiev, G. P., & Gladkov, N. A. (eds.). 1967. *Birds of the Soviet Union.* Vol. 2. Trans. by Israel Program for Scientific Translations, U.S. Dept. Interior & Natl. Sci. Foundation, Washington, D.C.

Duebbert, H. F. 1966. Island nesting of the gadwall in North Dakota. *Wilson Bulletin* 78:12–25.

Dzubin, A.; Miller, H. W.; & Schilman, G. V. 1964. White-fronts. In *Waterfowl Tomorrow,* pp. 135–43. U.S. Dept. Interior, Washington, D.C.

Edkins, D., & Hansen, I. A. 1972. Wax esters secreted by the uropygial gland of some Australian waterfowl, including the magpie goose. *Comparative Biochemistry and Physiology* 41B:105–12.

Einarsen, A. S. 1965. *Black brant: Sea goose of the Pacific coast.* Seattle: University of Washington Press.

Eisenhauer, D. I., & Frazer, D. A. 1972. Nesting ecology of the emperor goose (*Philacte canagica* Sewastianov) in the Kokechik Bay region, Alaska. Dept. Forestry & Conservation, Purdue University.

Eisenhauer, D. I., & Kirkpatrick, C. M. 1977. Ecology of the emperor goose in Alaska. *Wildlife Monographs* 57.

Eisenhauer, D. I.; Strang, C. A.; & Kirkpatrick, C. M. 1971. Nesting ecology of the emperor goose (*Philacte canagica* Sewastianov) in the Kokechik Bay region, Alaska. Dept. Forestry & Conservation, Purdue University.

Elder, W. R., & Woodside, D. H. 1958. Biology and management of the Hawaiian goose. In *Transactions of the 23rd North American Wildlife Conference,* pp. 198–214.

Eldridge, J. L. 1977. A preliminary ethogram of the torrent duck. M.S. thesis, University of Minnesota.

Elgas, R. 1970. Breeding populations of tule white-fronted geese in northwestern Canada. *Wilson Bulletin* 82:420–26.

Ellarson, R. S. 1956. A study of the old-squaw duck on Lake Michigan. Ph.D. dissertation, University of Wisconsin.

Eltringham, S. K., & Boyd, H. 1960. The shelduck population in the Bridgwater Bay area. In *Wildfowl Trust, 11th Annual Report,* pp. 107–17.

Erickson, R. C. 1948. Life history and ecology of the canvas-back, *Nyroca valisineria* (Wilson), in southeastern Oregon. Ph.D. dissertation, Iowa State College.

Erskine, A. J. 1960. A discussion of the distributional ecology of the bufflehead (*Bucephala albeola;* Anatidae: Aves) based on breeding biology studies in British Columbia, M.S. thesis, University of British Columbia.

_____. 1971. Growth and annual cycle of weights, plumages and reproductive organs of goosanders in eastern Canada. *Ibis* 113:42–58.

_____. 1972a. Postbreeding assemblies of ring-necked ducks in eastern Nova Scotia. *Auk* 89:399–400.

_____. 1972b. *Buffleheads.* Canadian Wildlife Service Monograph Series, no. 4.

Falla, R. A.; Sibson, R. B.; and Turbott, E. G. 1967. *A field guide to the birds of New Zealand.* Boston: Houghton Mifflin.

Ferguson, W. H. 1966. Will my birds nest this year? *Modern Game Breeding* 2:18–20, 34–35.

Ferns, P. N., & Green, G. H. 1975. Observations of pink-footed and barnacle geese in the Kong Oscar Fjord region of north-east Greenland, 1974. *Wildfowl* 26:131–38.

Fischer, H. 1965. Das Triumphgeschrei der Graugans (*Anser anser*). *Zeitschrift für Tierpsychologie* 22:247–304.

Fisher, J.; Simon, N.; & Vincent, J. 1969. *The red book: Wildlife in danger.* London: Collins.

Freeman, M. M. R. 1970. Observations on the seasonal behavior of the Hudson Bay eider. *Canadian Field-Naturalist* 84:145–53.

Frith, H. J. 1964. The downy young of the freckled duck, *Stictonetta naevosa. Emu* 64:42–47.

_____. 1965. Ecology of the freckled duck, *Stictonetta naevosa* (Gould). *C.S.I.R.O. Wildlife Research* 10:125–39.

_____. 1967. *Waterfowl in Australia.* Honolulu: East-West Center Press.

Frith, H. J.; Braithwaite, L. W.; & McKean, J. L. 1969. Waterfowl in an inland swamp in New South Wales. II. Food. *C.S.I.R.O. Wildlife Research* 14:17–64.

Frith, H. J., & Davies, S. J. J. F. 1961. Ecology of the magpie goose, *Anseranas semipalmata* Latham (Anatidae). *C.S.I.R.O. Wildlife Research* 6:91–141.

Gates, J. M. 1957. Autumn food habits of the gadwall in northern Utah. *Proc. Utah Acad. Science, Arts & Letters* 34:69–71.

_____. 1962. Breeding biology of the gadwall in northern Utah. *Wilson Bulletin* 74:43–67.

Geis, A. D.; Smith, R. I.; & Rogers, J. P. 1971. *Black duck distribution, harvest characteristics and survival.* U.S. Fish & Wildlife Service, Special Scientific Report—Wildlife No. 139.

Gehrman, K. H. 1951. An ecological study of the lesser scaup duck (*Aythya affinis* Eyton) at West Medical Lake, Spokane County, Washington. M.S. thesis, Washington State College.

Gibson, E. 1920. Further ornithological notes from the neighbourhood of Cape San Antonio, Buenos Ayres. *Ibis* 11:1–97.

Girard, G. L. 1939. Notes on the life history of the shoveller. In *Transactions of the 14th North American Wildlife Conference,* pp. 364–71.

Gisenko, A. I., & Mischin, I. P. 1952. [New data on the geographical distribution and biology of *Cygnopsis cygnoides* in Sakhalin.] *Zoologicheskii Zhurnal* 31:312–14. (In Russian.)

Gladstone, P., & Martell, C. 1968. Some field notes on the breeding of the greater kelp goose. *Wildfowl* 19:25–31.

Glover, F. A. 1956. Nesting and production of the blue-winged teal (*Anas discors* Linnaeus) in northwest Iowa. *Journal of Wildlife Management* 20:28–46.

Grice, D., & Rogers, J. P. 1965. *The wood duck in Massachusetts.* Mass. Div. Fisheries & Game, Final Report, Project No. W-19-R.

Gudmundsson, F. 1971. Stradumendur (*Histrionicus histrionicus*) á Islandi. Fyrri hluti. *Natturufraedingurinn* 41:1–28. (In Icelandic.)

Guiler, E. R. 1966. The breeding of the black swan (*Cygnus atrata* Latham) in Tasmania with special reference to some management problems. *Papers and Proceedings of the Royal Society of Tasmania* 100:31–52.

———. 1967. The Cape Barren goose, its environment, numbers and breeding. *Emu* 66:211–35.

———. 1974. The conservation of the Cape Barren goose. *Biological Conservation* 6:252–57.

Hansen, H. A.; Shepherd, P. E. K.; King, J. G.; & Troyer, W. A. 1971. *The trumpeter swan in Alaska.* Wildlife Monographs, no. 26.

Hanson, H. C. 1965. *The giant Canada goose.* Carbondale: Southern Illinois University Press.

Haverschmidt, F. 1968. *Birds of Surinam.* Edinburgh: Oliver and Boyd.

———. 1972. Bird records from Surinam. *Bulletin of the British Ornithologists' Club* 92:49–53.

Hawkes, B. 1970. The marbled teal. *Wildfowl* 21:87–88.

Headley, P. C. 1967. *Ecology of the emperor goose.* Report of Alaska Cooperative Wildlife Unit, University of Alaska.

Hester, F. F., & Quay, T. L. 1961. A three-year study of the fall migration and roosting-flight habits of the wood duck in east-central North Carolina. In *Proceedings of the 15th Annual Conference, S.E. Association Game & Fish Commissioners,* pp. 55–60.

Hilden, O. 1964. Ecology of duck populations in the island group of Valassaaret, Gulf of Bothnia. *Annales Zoologici Fennici* 1:1–279.

Hochbaum, H. A. 1944. *The canvasback on a prairie marsh.* Harrisburg, Pa.: Stackpole; and Washington, D.C.: Wildlife Management Institute.

Höhn, E. O. 1975. Notes on black-headed ducks, painted snipe, and spotted tinamous. *Auk* 92:566–75.

Hori, J. 1964. The breeding biology of the shelduck *Tadorna tadorna.* *Ibis* 106:333–60.

———. 1969. Social and population studies in the shelduck. *Wildfowl* 20:5–22.

Hudec, K., & Rooth, J. 1970. *Die Graugans.* Neue Brehm-Bücherei 429. Wittenberg, Lutherstadt: A. Ziemsen Verlag.

Hudson, R. (ed.). 1975. *Threatened birds of Europe.* London: Macmillan.

Humphrey, P. S., & Butsch, R. S. 1958. The anatomy of the Labrador duck, *Camptorhynchus labradorius* (Gmelin). *Smithsonian Miscellaneous Collections* 135(7):1–23.

Humphrey, P. S., & Ripley, S. D. 1962. The affinities of the pink-headed duck. *Postilla* 61:1–21.

Hyde, D. O. (ed.). 1974. *Raising wild ducks in captivity.* New York: E. P. Dutton.

Jackson, E. E.; Ogilvie, M. A.; & Owen, M. 1974. The Wildfowl Trust expedition to Spitsbergen, 1973. *Wildfowl* 25:102–16.

Jacob, J., & Glaser, A. 1975. Chemotaxonomy of Anseriformes. *Biochemical Systematics and Ecology* 2:215–20.

Johnsgard, P. A. 1960a. The systematic position of the ringed teal. *Bulletin British Ornithologists' Club* 80:165–67.

———. 1960b. Pair-formation mechanisms in *Anas* (Anatidae) and related genera. *Ibis* 102:616–18.

———. 1960c. Classification and evolutionary relationships of the sea ducks. *Condor* 62:426–33.

———. 1961a. The taxonomy of the Anatidae—A behavioural analysis. *Ibis* 103a:71–85.

———. 1961b. The tracheal anatomy of the Anatidae and its taxonomic significance. In *Wildfowl Trust, 12th Annual Report,* pp. 58–69.

———. 1961c. The breeding biology of the magpie goose. In *Wildfowl Trust, 12th Annual Report,* pp. 92–103.

———. 1961d. Evolutionary relationships among the North American mallards. *Auk* 78:1–43.

———. 1961e. The systematic position of the marbled teal. *Bulletin of the British Ornithologists' Club* 81:37–43.

———. 1961f. The sexual behavior and systematic position of the hooded merganser. *Wilson Bulletin* 73:226–36.

———. 1964a. Comparative behavior and relationships of the eiders. *Condor* 66:113–29.

———. 1964b. Observations on the breeding biology of the spectacled eider. In *Wildfowl Trust, 15th Annual Report,* pp. 104–7.

———. 1965a. *Handbook of Waterfowl Behavior.* Ithaca: Cornell University Press.

———. 1965b. Observations on some aberrant Australian Anatidae. In *Wildfowl Trust, 16th Annual Report,* pp. 73–83.

———. 1966a. The biology and relationships of the torrent duck. In *Wildfowl Trust, 17th Annual Report,* pp. 66–74.

———. 1966b. Behavior of the Australian musk duck and blue-billed duck. *Auk* 83:98–110.

———. 1967a. Observations on the behavior and relationships of the white-backed duck and the stiff-tailed ducks. In *Wildfowl Trust, 18th Annual Report,* pp. 98–107.

———. 1967b. Sympatry changes and hybridization incidence in mallards and black ducks. *American Midland Naturalist* 77:51–63.

———. 1968a. *Waterfowl: Their biology and natural history.* Lincoln: University of Nebraska Press.

———. 1968b. Some putative mandarin duck hybrids. *Bulletin British Ornithologists' Club* 88:140–48.

———. 1968c. Some observations on maccoa duck behavior. *Ostrich* 39:219–22.

———. 1973. Proximate and ultimate determinants of clutch-size in the Anatidae. *Wildfowl* 24:144–49.

———. 1974a. The taxonomy and relationships of the northern swans. *Wildfowl* 25:155–61.

_____. 1974b. *Song of the north wind: A story of the snow goose.* New York: Doubleday.

_____. 1975. *Waterfowl of North America.* Bloomington: Indiana University Press.

Johnsgard, P. A., & DiSylvestro, R. 1976. Seventy-five years of changes in mallard–black duck ratios in eastern North America. *American birds* 30:904–9.

Johnsgard, P. A., & Hagemeyer, D. 1969. The masked duck in the United States. *Auk* 84:691–95.

Johnson, A. W. 1963. Notes on the distribution, reproduction and display of the Andean torrent duck, *Merganetta armata. Ibis* 105:114–16.

_____. 1965. *The birds of Chile, and adjacent regions of Argentina, Bolivia and Peru.* Vol. 1. Buenos Aires: Platt Establecemientos Gráficos.

Johnson, L. L. 1967. The common goldeneye duck and the role of nesting boxes in its management in north-central Minnesota. *Journal of the Minnesota Academy Science* 34:110–13.

Johnstone, S. T. 1960. First breeding of the spotted whistling duck (*Dendrocygna guttata*). In *Wildfowl Trust, 11th Annual Report,* pp. 11–12.

_____. 1965. 1964 breeding results at the Wildfowl Trust. *Avicultural Magazine* 71:20–23.

_____. 1970. Waterfowl eggs. *Avicultural Magazine* 76:52–55.

Jones, R. D., Jr. 1965. Returns from Steller's eiders banded in Izembek Bay, Alaska. In *Wildfowl Trust, 16th Annual Report,* pp. 83–85.

Jones, R. D., Jr., & Jones, D. M. 1966. The process of family disintegration in black brant. In *Wildfowl Trust, 17th Annual Report,* pp. 75–78.

Joyner, D. E. 1975. Nest parasitism and brood-related behavior of the ruddy duck (*Oxyura jamaicensis rubida*). Ph.D. dissertation, University of Nebraska–Lincoln.

Kear, J. 1967. Notes on the eggs and downy young of *Thalassornis leuconotus. Ostrich* 38:227–29.

_____. 1972. The blue duck of New Zealand. *Living Bird* 11:175–92.

_____. 1973. The magpie goose *Anseranas semipalmata* in captivity. *International Zoo Yearbook* 13:28–32.

_____. 1975. Salvadori's duck of New Guinea. *Wildfowl* 26:104–11.

_____. 1976. Good news on endangered wildfowl. *Wildfowl Trust Bulletin* 74:13.

Kear, J., & Burton, P. J. K. 1971. The food and feeding apparatus of the blue duck *Hymenolaimus. Ibis* 113:483–93.

Kear, J., & Murton, R. K. 1973. The systematic status of the Cape Barren goose as judged by its photoresponses. *Wildfowl* 24:141–43.

Kear, J., & Scarlett, R. J. 1970. The Auckland Islands merganser. *Wildfowl* 21:78–86.

Kear, J., & Steel, T. H. 1971. Aspects of social behaviour in the blue duck. *Notornis* 18:187–98.

Keith, L. B. 1961. *A study of waterfowl ecology on small impoundments in southeastern Alberta.* Wildlife Monographs, no. 6.

Kerbes, R. H.; Ogilvie, M. A.; & Boyd, H. 1971. Pink-footed geese of Iceland and Greenland: A population review based on an aerial survey of pjórsárver in June, 1970. *Wildfowl* 22:5–17.

King, J. G. 1970. The swans and geese of Alaska's Arctic slope. *Wildfowl* 21:11–17.

Kistchinski, A. A. 1971. Biological notes on the emperor goose in northeast Siberia. *Wildfowl* 22:29–34.

Kistchinski, A. A., & Flint, V. E. 1974. On the biology of the spectacled eider. *Wildfowl* 24:5–15.

Koepcke, H. W., & Koepcke, M. 1965. *Las silvestres de importancia económica del Perú.* Lima.

Kolbe, H. 1972. *Die Entenvogel der Welt.* Redebuel: Neumann Verlag.

Koskimies, J., & Routamo, E. 1953. Zur Fortpflanzungsbiologie der Samtente *Melanitta f. fusca* (L.). 1. Allgemeine Nistokologie. *Papers on Game Research* 10:1–105.

Krapu, G. 1974. Foods of breeding pintails in North Dakota. *Journal of Wildlife Management* 38:408–17.

Krechmar, A. V., & Leonovich, V. V. 1967. [Distribution and biology of the red-breasted goose in the breeding season.] *Problemy Severa* 11:220–34. (In Russian, translated by National Research Council, Ottawa.)

Kuchel, C. R. 1977. Some aspects of the behavior and ecology of harlequin ducks breeding in Glacier National Park, Montana. M.S. thesis, University of Montana.

Kumari, E. 1971. Passage of the barnacle goose through the Baltic area. *Wildfowl* 22:35–43.

Lack, D. 1968. *Ecological adaptations for breeding in birds.* London: Methuen.

Lavery, H. J. 1972. The gray teal at saline drought-refuges in north Queensland. *Wildfowl* 22:56–63.

Leopold, A. S. 1959. *Wildlife of Mexico: The game birds and mammals.* Berkeley: University of California Press.

Lind, H. 1959. Studies on courtship and copulatory behaviour in the goldeneye (*Bucephala clangula* (L). *Dansk Ornithologisk Forenings Tidsskrift* 53:177–219.

Lokomoen, J. T. 1966. Breeding ecology of the redhead duck in western Montana. *Journal of Wildlife Management* 30:668–81.

Lorenz, K. Z. 1951–53. Comparative studies on the behaviour of Anatinae. *Avicultural Magazine* 57:157–82; 58:8–17, 61–72, 86–94, 172–84; 59:24–34, 80–91.

Lorenz, K. Z., & von de Wall, W. 1960. Die Ausdrucksbewegungen der Sichelente, *Anas falcata* L. *Journal für Ornithologie* 101:50–60.

Low, J. B. 1945. Ecology and management of the red head, *Nyroca americana,* in Iowa. *Ecological Monographs* 15:35–69.

Lowe, V. T. 1966. Notes on the musk duck. *Emu* 65:279–90.

Lubbock, M. 1976. The Wildfowl Trust expedition to

Alaska. *Wildfowl Trust Bulletin* 74:10–11.

McCartney, R. B. 1963. The fulvous tree duck in Louisiana. M.S. thesis, Louisiana State University.

McKelvey, S. D. 1977. The Meller's duck on Mauritius: Its status in the wild and captive propagation. *Game Bird Breeders', Aviculturists', Zoologists' and Conservationists' Gazette*, May–June 1977, pp. 11–13.

McKenzie, H. R. 1971. The brown teal in the Auckland Province. *Notornis* 18:280–86.

Mackenzie, M. J. S., & Kear, J. 1976. The white-winged wood duck. *Wildfowl* 27:5–17.

McKinney, F. 1961. An analysis of the displays of the European eider *Somateria mollissima mollissima* (Linnaeus) and the Pacific eider *Somateria mollissima v. nigra* (Bonaparte). *Behaviour*, supplement no. 7.

———. 1965a. The displays of the American green-winged teal. *Wilson Bulletin* 77:112–21.

———. 1965b. The spring behavior of wild Steller's eiders. *Condor* 67:273–90.

———. 1967. Breeding behaviour of captive shovelers. In *Wildfowl Trust, 18th Annual Report*, pp. 108–21.

———. 1970. Displays of four species of blue-winged ducks. *Living Bird* 9:29–64.

Macnae, W. 1959. Notes on biology of maccoa duck. *Bokmakierie* 11:49–52.

Marriott, R. W. 1970. The food and water requirements of Cape Barren geese (*Cereopsis novaehollandiae* Latham). Ph.D. dissertation, Monash University, Melbourne.

Matthews, G. V. T., & Evans, M. E. 1974. On the behaviour of the white-headed duck with especial reference to breeding. *Wildfowl* 25:56–66.

Meanley, B., & Meanley, A. G. 1959. Observations on the fulvous tree duck in Louisiana. *Wilson Bulletin* 71:33–45.

Mehta, K. L. 1960. A pinkhead duck (*Rhodonessa caryophyllacea* (Latham)) at last? *Journal of the Bombay Natural History Society* 57:417.

Mendall, H. L. 1958. *The ring-necked duck in the northeast.* University of Maine Bulletin 60 (16).

Mickelson, P. G. 1973. Breeding biology of cackling geese (*Branta canadensis minima* Ridgway) and associated species on the Yukon-Kuskokwim delta, Alaska. Ph.D. dissertation, University of Michigan.

Middlemiss, E. 1958. The southern pochard *Netta erythropthalma brunnea. Ostrich*, supplement no. 2.

Miller, A. H. 1937. Structural modifications in the Hawaiian goose (*Nesochen sandvicensis*): A study in adaptive evolution. *University of California Publications in Zoology* 42:1–80.

Minton, C. D. T. 1968. Pairing and breeding of mute swans. *Wildfowl* 19:41–60.

Moffett, G. M., Jr. 1970. A study of nesting torrent ducks in the Andes. *Living Bird* 9:5–28.

Moisan, G.; Smith, R. I.; & Martinson, R. K. 1967. *The green-winged teal: Its distribution, migration and population dynamics.* U.S. Fish & Wildlife Service, Special

Scientific Report—Wildlife no. 100.

Morse, T. E.; Jakabosky, J. L.; & McCrow, V. P. 1969. Some aspects of the breeding biology of the hooded merganser. *Journal of Wildlife Management* 33:596–604.

Moynihan, M. 1958. Notes on the behavior of the flying steamer duck. *Auk* 75:183–202.

Munro, J. A. 1939. Studies of waterfowl in British Columbia: Barrow's goldeneye, American goldeneye. *Transactions of the Royal Canadian Institute* 22:259–318.

———. 1949a. Studies of waterfowl in British Columbia: Green-winged teal. *Canadian Journal of Research,* sec. D, 27:149–78.

———. 1949b. Studies of waterfowl in British Columbia: Baldpate. *Canadian Journal of Research,* sec. D, 27:289–307.

Murphy, R. C. 1916. Anatidae of South Georgia. *Auk* 33:270–77.

———. 1936. *Oceanic birds of South America.* Vol. 2. New York: American Museum of Natural History.

Murton, R. K., & Kear, J. 1975. The role of daylength in regulating the breeding seasons of wildfowl. In *Light as an ecological factor,* 2:337–57. Oxford: Blackwell.

Myres, M. T. 1959. The behavior of the sea-ducks and its value in the systematics of the tribes Mergini and Somateriini, of the family Anatidae. Ph.D. dissertation, University of British Columbia.

Nelson, A. D., & Martin, A. C. 1953. Gamebird weights. *Journal of Wildlife Management* 17:36–42.

Newton, I., & Campbell, C. R. G. 1976. Breeding of ducks at Lock Leven, Kinross. *Wildfowl* 26:83–102.

Newton, I., & Kerbes, R. H. 1974. Breeding of greylag geese (*Anser anser*) on the Outer Hebrides, Scotland. *Journal of Animal Ecology* 43:771–83.

Niethammer, G. 1952. Zur Anatomie und systematischen Stellung der Sturzbach-Ente (*Merganetta armata*). *Journal für Ornithologie* 93:357–60.

———. 1953. Zur Vogelwelt Boliviens. *Bonner Zoologische Beiträge* 4:195–303.

Nilsson, L. 1974. The behaviour of wintering smew in southern Sweden. *Wildfowl* 25:84–88.

Ogilvie, M. A. 1967. Population changes and mortality of the mute swan in Britain. In *Wildfowl Trust, 18th Annual Report,* pp. 64–73.

———. 1975. *Ducks of Britain and Europe.* Berkhamsted: T. & A. D. Poyser.

———. 1976a. *The winter birds.* London: Michael Joseph.

———. 1976b. Dark-bellied brent geese in Britain and Europe, 1955–76. *British Birds* 69:422–39.

Ogilvie, M., & Boyd, H. 1976. The numbers of pink-footed and greylag geese wintering in Britain: Observations, 1969–1975, and predictions, 1976–1980. *Wildfowl* 27:63–75.

Oliver, W. R. B. 1955. *New Zealand birds.* 2nd ed. Wellington: A. H. and A. W. Reed.

Olney, R. J. S. 1963a. The food and feeding habits of the tufted duck, *Aythya fuligula. Ibis* 105:55–62.

_____. 1963b. The food and feeding habits of the teal, *Anas crecca* L. *Proceedings of the Zoological Society of London* 140:169–210.

_____. 1965. The food and feeding habits of the shelduck, *Tadorna tadorna. Ibis* 107:527–32.

_____. 1968. The food and feeding habits of the pochard, *Aythya ferina. Biological Conservation* 1:71–76.

Olrog, C. C. 1968. *Las aves Sudamericanas.* Vol. 1. Tucumán: Instituto Miguel Lillo.

Olson, D. P. 1964. A study of canvasback and redhead breeding populations, nesting habitats and productivity. Ph.D. dissertation, University of Minnesota.

Oring, L. W. 1964. Behavior and ecology of certain ducks during the postbreeding period. *Journal of Wildlife Management* 28:223–33.

_____. 1969. Summer biology of the gadwall at Delta, Manitoba. *Wilson Bulletin* 8:44–54.

Owen, M. 1977. *Wildfowl of Europe.* London: Macmillan.

Palmer, R. S. (ed.). 1976. *Handbook of North American birds.* Vols. 2, 3. New Haven: Yale University Press.

Parmelee, D. F.; Stephens, H. A.; & Schmidt, R. H. 1967. *The birds of Victoria Island and adjacent small islands.* National Museums of Canada Bulletin no. 222.

Partridge, W. H. 1956. Notes on the Brazilian merganser in Argentina. *Auk* 73:473–88.

Pearse, R. J. 1975. *Cape Barren geese in Tasmania; Biology and management to 1975.* National Parks and Wildlife Service, Tasmania, Wildlife Division Technical Report 71/1.

Pengelly, W. J., & Kear, J. 1970. The hand-rearing of young blue duck. *Wildfowl* 21:115–21.

Penkala, J. M.; Applegate, J. E., & Wolfast, L. J. 1975. Management of Atlantic brant: Implications of existing data. In *Transactions of the 40th North American Wildlife and Natural Resources Conference,* pp. 325–33.

Pettingill, O. S., Jr. 1965. Kelp geese and flightless steamer ducks in the Falkland Islands. *Living Bird* 4:65–78.

Petzold, H.-G. 1964. Beiträge zur vergleichenden Ethologie der Schwäne (Anseres, Anserini). *Beiträge zur Vogelkunde* 10:1–123.

Philippona, J. 1972. *Die Blessgans.* Neue Brehm-Bücherei 457. Wittenberg, Lutherstadt: A. Ziemsen Verlag.

Phillips, J. C. 1922–26. *A natural history of the ducks.* 4 vols. Boston: Houghton Mifflin.

Platz, F. 1974. Untersuchungen über die Ontogenese von Ausdrucksbewegungen und Lautäusserungen bei der Kolbenente (*Netta rufina* Pallas) mit einem Beitrag zur Anatomie des Stimmapparates. *Zeitschrift für Tierpsychologie* 36:293–428.

Ploeger, P. L. 1968. Geographical differentiation in arctic Anatidae as a result of isolation during the last glacial. *Ardea* 56:1–159.

Poston, H. J. 1969. *Home range and breeding biology of the shoveler.* Canadian Wildlife Service Report Series, no. 25.

Prestwich, A. A. 1974. The pink-headed duck (*Rho-donessa caryophyllacea*) in the wild and in captivity. *Avicultural Magazine* 80:47–52.

Prevett, J. P. 1973. Family behavior and age-dependent breeding biology of the blue goose, *Anser caerulescens.* Ph.D. dissertation, University of Western Ontario.

Raikow, R. J. 1971. The osteology and taxonomic position of the white-backed duck, *Thalassornis leuconotus. Wilson Bulletin* 83:270–77.

Rand, A. L., & Gilliard, E. T. 1967. *Handbook of New Guinea birds.* London: Weidenfeld and Nicolson.

Rand, A. L., & Rabor, D. S. 1960. Birds of the Philippine Islands: Siquijor, Mount Malindag, Bohol, and Samar. Chicago Museum of Natural History, *Fieldiana* (Zoology) 35:221–441.

Raveling, D. G. 1969. Preflight and flight behavior of Canada geese. *Auk* 86:671–81.

Rawls, C. K., Jr. 1949. An investigation of the life history of the white-winged scoter (*Melanitta fusca deglandi*). M.S. thesis, University of Minnesota.

Reid, B., & Roderick, C. 1973. New Zealand scaup *Aythya novaeseelandiae* and brown teal *Anas aucklandica chlorotis* in captivity. *International Zoo Yearbook* 13:12–15.

Riggert, T. L. 1977. *The biology of the mountain duck on Rottnest Island, Western Australia.* Wildlife Monographs, no. 52.

Robinson, F. N., & Robinson, A. H. 1970. Regional variation in the visual and acoustic signals of the male musk duck, *Biziura lobata. C.S.I.R.O. Wildlife Research* 15:73–78.

Rogers, J. P., & Korschgen, L. J. 1966. Foods of lesser scaups on breeding, migration and wintering areas. *Journal of Wildlife Management* 30:258–64.

Rowan, M. K. 1963. The yellowbill duck *Anas undulata* Dubois in southern Africa. *Ostrich,* supplement no. 5.

Ryan R. A. 1972. Body weight and weight changes of wintering diving ducks. *Journal of Wildlife Management* 36:759–65.

Ryder, J. P. 1967. *The breeding biology of Ross' goose in the Perry River region, Northwest Territories.* Canadian Wildlife Service Report Series, no. 3.

_____. 1970. Timing and spacing of nests and breeding biology of Ross' goose. Ph.D. dissertation, University of Saskatchewan.

_____. 1972. Biology of nesting Ross's geese. *Ardea* 60:185–215.

Rylander, M. K., & Bolen, E. G. 1970. Ecological and anatomical adaptations of North American tree ducks. *Auk* 87:72–90.

Rylander, M. K., & Bolen, D. C. 1974. Feeding adaptations in whistling ducks (*Dendrocygna*). *Auk* 91:86–94.

Sage, B. L. 1958. Hybrid ducks in New Zealand, *Bulletin of the British Ornithologists' Club* 78:108–13.

Salomonsen, F. 1968. The moult migration. *Wildfowl* 19:5–24.

Salvan, J. 1970. Remarques sur l'evolution de l'avifauna

Malgache depuis 1945. *Alauda* 38:191–203.

Savage, C. 1952. *The mandarin duck.* London: A. & C. Black.

———. 1965. White-headed ducks in West Pakistan. In *Wildfowl Trust, 16th Annual Report,* pp. 121–23.

Schmidt, C. R. 1969. Preliminary notes on breeding the Falkland Island flightless steamer duck, *Tachyeres brachypterus,* at Zurich Zoo. *International Zoo Yearbook* 9:125–27.

Schönwetter, M. 1960. *Handbuch der Oologie.* Vol. 2. Berlin.

Schulten, G. G. M. 1974. The food of some duck species occurring at Lake Chilwa, Malawi. *Ostrich* 45:224–26.

Scott, D. 1971. The Auckland Island flightless teal. *Wildfowl* 22:44–45.

Scott, D., & Lubbock, J. 1974. Preliminary observations on waterfowl of western Madagascar. *Wildfowl* 25:117–20.

Scott, P. 1954. South America—1953. In *Wildfowl Trust, 6th Annual Report,* pp. 55–69.

———. 1960. BBC/IUCN Darwin Centenary Expedition. In *Wildfowl Trust, 11th Annual Report,* pp. 61–76.

———. 1970. Redbreasts in Rumania. *Wildfowl* 21:37–41.

Scott, P., & Boyd, H. 1957. *Wildfowl of the British Isles.* London: Country Life.

Scott, P.; Boyd, H.; & Sladen, W. J. L. 1955. The Wildfowl Trust second expedition to central Iceland, 1953. In *Wildfowl Trust, 7th Annual Report,* pp. 63–98.

Scott, P., & Fisher, J. 1953. *A thousand geese.* London: Collins.

Scott, P., & Wildfowl Trust. 1972. *The swans.* London: Michael Joseph.

Sherwood, G. A. 1965. Canada geese of the Seney National Wildlife Refuge. Seney National Wildlife Refuge.

Siegfried, W. R. 1962a. Nesting behavior of the redbill teal *Anas erythrorhyncha* Gmelin. *Investigational report,* Dept. of Nature and Conservation (Republic of South Africa) 2:19–24.

———. 1962b. Observations on the post-embryonic development of the Egyptian goose *Alopochen aegyptiacus* (L.) and the redbill teal *Anas erythrorhyncha* (Gmelin). *Investigational report,* Dept. of Nature and Conservation (Republic of South Africa) 2:9–17.

———. 1965. The Cape shoveler *Anas smithii* (Hartert) in southern Africa. *Ostrich* 37:155–98.

———. 1968. The black duck in the south-western Cape. *Ostrich* 39:61–75.

———. 1973. Summer feed and feeding of the ruddy duck in Manitoba. *Canadian Journal of Zoology* 51:1293–97.

———. 1976a. Social organization in ruddy and maccoa ducks. *Auk* 93:560–70.

———. 1976b. Breeding biology and parasitism in the ruddy duck. *Wilson Bulletin* 88:566–74.

Siegfried, W. R., & van der Merwe, F. 1975. A description and inventory of the displays of the maccoa duck. *Zeitschrift für Tierpsychologie* 37:1–23.

Singh, L. P. 1966. The pinkheaded duck (*Rhodonessa caryophyllacea* (Latham)) again. *Journal of the Bombay Natural History Society* 63:440–41.

Skead, D. M. 1976. Social behaviour of the yellow-billed duck and red-billed teal in relation to breeding. M.S. thesis, University of Natal.

———. 1977. Pair-forming and breeding behaviour of the Cape shoveller at Barberspan. *Ostrich,* supplement no. 12.

Smith, R. I. 1968. The social aspects of reproductive behavior in the pintail. *Auk* 85:381–96.

Sowls, L. K. 1955. *Prairie ducks: A study of their behavior, ecology and management.* Harrisburg, Pa.: Stackpole; Washington, D.C.: Wildlife Management Institute.

Spencer, H. E., Jr. 1953. The cinnamon teal (*Anas cyanoptera* Vieillot): Its life history, ecology, and management. M.S. thesis, Utah State Agricultural College.

Springer, P. 1975. *Report on observations of Aleutian Canadian geese in northern coastal California, spring 1975.* U.S. Fish & Wildlife Service, Migratory Bird Research Station, Humboldt State University, Arcata, Calif.

Springer, P. F., et al. 1978. Reestablishing Aleutian Canada geese. In *Endangered birds: Management techniques for threatened species.* Madison: University of Wisconsin Press, pp. 331–38.

Spurr, E., & Milne, H. 1976. Adaptive significance of autumn pair formation in the common eider *Somateria mollissima* (L.). *Ornis Scandinavica* 7:85–89.

Stewart, R. 1958. *Distribution of the black duck.* U.S. Fish & Wildlife Service Circular 51.

———. 1962. *Waterfowl populations in the upper Chesapeake region.* U.S. Fish & Wildlife Service, Special Scientific Report—Wildlife no. 65.

Stoudt, J. H. 1971. *Ecological factors affecting waterfowl production in the Saskatchewan Parklands.* U.S. Fish & Wildlife Service, Resource Publication no. 99.

Swedburg, G. E. 1967. *The koloa: A preliminary report on the life history and status of the Hawaiian duck* (Anas wyvilliana). Wildlife Branch, Division of Fish and Game, Department of Lands and Natural Resources, Hawaii.

Thompson, D. Q., & Person, R. A. 1963. The eider pass at Point Barrow, Alaska. *Journal of Wildlife Management* 27:348–56.

Townsend, G. H. 1966. A study of waterfowl nesting on the Saskatchewan river delta. *Canadian Field-Naturalist* 80:74–88.

Trauger, D. L. 1971. Population ecology of lesser scaup (*Aythya affinis*) in subarctic taiga. Ph.D. dissertation, Iowa State University.

Tso-hsin, C. (ed.). 1963. [*China's economic fauna: Birds.*] Science Publishing Society, Peiping. Trans. 1964 by U.S. Department of Commerce, Washington, D.C.

Uspenski, S. M. 1965. *Die Wildgänse Nordeurasiens*. Neue Brehm-Bücherei 352. Wittenberg, Lutherstadt: A. Ziemsen Verlag.

———. 1972. Die Eiderenten (Gattung *Somateria*). Neue Brehm-Bücherei 452. Wittenberg, Lutherstadt: A. Ziemsen Verlag.

Vaurie, C. 1965. The birds of the Palearctic fauna: Non-passeriformes. London, H. F. & G. Witherby.

Vermeer, K. 1970. Some aspects of the nesting of ducks on islands in Lake Newell, Alberta. *Journal of Wildlife Management* 34:126–29.

Veselovsky, Z. 1973. The breeding biology of Cape Barren geese *Cereopsis novaehollandiae*. *International Zoo Yearbook* 13:48–55.

Voous, K. H. 1960. *Atlas of European birds*. London: Elsevier, Nelson.

Warham, J. 1958. The nesting of the pink-eared duck. In *Wildfowl Trust, 9th Annual Report*, pp. 118–27.

Warner, R. E. 1963. Recent history and ecology of the Laysan duck. *Condor* 63:3–23.

Watson, G. E. 1975. *Birds of the Antarctic and Sub-Antarctic*. Washington, D.C.: American Geophysical Union.

Weller, M. W. 1959. Parasitic egg laying in the redhead (*Aythya americana*) and other North American Anatidae. *Ecological Monographs* 29:333–65.

———. 1967a. Notes on some marsh birds of Cape San Antonio, Argentina. *Ibis* 109:391–416.

———. 1967b. Distribution and habitat selection of the black-headed duck. *Hornero* 10:299–306.

———. 1967c. Courtship of the redhead (*Aythya americana*). *Auk* 84:544–59.

———. 1968a. Notes on some Argentine anatids. *Wilson Bulletin* 80:189–212.

———. 1968b. The breeding biology of the parasitic black-headed duck. *Living Bird* 7:169–207.

———. 1972. Ecological studies of Falkland Island's waterfowl. *Waterfowl* 23:25–44.

———. 1974. Habitat selection and feeding patterns of brown teal (*Anas castanea chlorotis*) on Great Barrier Island. *Notornis* 21:25–35.

———. 1975a. Habitat selection by waterfowl of Argentine Isla Grande. *Wilson Bulletin* 87:83–90.

———. 1975b. Ecological studies of the Auckland Islands flightless teal. *Auk* 92:280–97.

———. 1975c. Ecology and behavior of the South Georgia pintail *Anas g. georgica*. *Ibis* 117:217–31.

———. 1976. Ecology and behaviour of steamer ducks. *Wildfowl* 27:45–53.

Weller, M. W.; Trauger, D. L.; & Krapu, G. L. 1969. Breeding birds of the West Mirage Islands, Great Slave Lake, N.W.T. *Canadian Field-Naturalist* 83:344–60.

Wetmore, A. 1926. Observations on the birds of Argentina, Paraguay, Uruguay and Chile. *U.S. National Museum Bulletin* 133.

Wheeler, J. R. 1953. Notes on the blue-billed ducks at Lake Wendouree, Ballarat. *Emu* 53:280–82.

White, H. C. 1957. Food and natural history of mergansers on salmon waters in the Maritime Provinces of Canada. *Fisheries Research Board of Canada Bulletin* 116:1–63.

Wienmeyer, S. N. 1967. Bufflehead food habits, parasites, and biology in northern California. M.S. thesis, Humboldt State College, Arcata, Calif.

Williams, C. S. 1967. *Honker: A discussion of the habits and needs of the largest of our Canada geese.* Princeton: D. Van Nostrand.

Williams, G. R. 1964. Extinction and the Anatidae of New Zealand. In *Wildfowl Trust, 15th Annual Report*, pp. 140–46.

Williams, S. O. III. 1978. The Mexican duck in Mexico: Natural history, distribution, and population status. Ph.D. dissertation, Colorado State University.

Wilmore, S. B. 1974. *Swans of the world*. New York: Taplinger.

Winterbottom, J. M. 1974. The Cape teal. *Ostrich* 45:110–32.

Woods, R. W. 1975. *The birds of the Falkland Islands.* Oswestry: Anthony Nelson.

Woolfenden, G. E. 1961. Postcranial osteology of the waterfowl. *Bulletin of the Florida State Museum, Biological Sciences* 6:1–29.

Wright, B. S. 1954. *High tide and an east wind: The story of the black duck.* Harrisburg, Pa.: Stackpole; Washington, D.C.: Wildlife Management Institute.

Würdinger, I. 1970. Erzeugung, Ontogenie and Funcktion der Lautäusserungen bei vier Gänsearten (*Anser indicus, A. caerulescens, A. albifrons* and *Branta canadensis*). *Zeitschrift für Tierpsychologie* 27:257–301.

———. 1973. Breeding of bar-headed goose *Anser indicus* in captivity. *International Zoo Yearbook* 13:43–47.

Young, C. M. 1970. Territoriality in the common shelduck *Tadorna tadorna*. *Ibis* 112:330–35.

Young, J. G. 1972. Breeding biology of feral graylag geese in south-west Scotland. *Wildfowl* 23:83–87.

Zaloumis, E. A. 1976. Incubation period of the African pygmy goose. *Ostrich* 47:231.

Zimmerman, D. R. 1974. Return of the nene. *Animal Kingdom* 77(3):22–28.

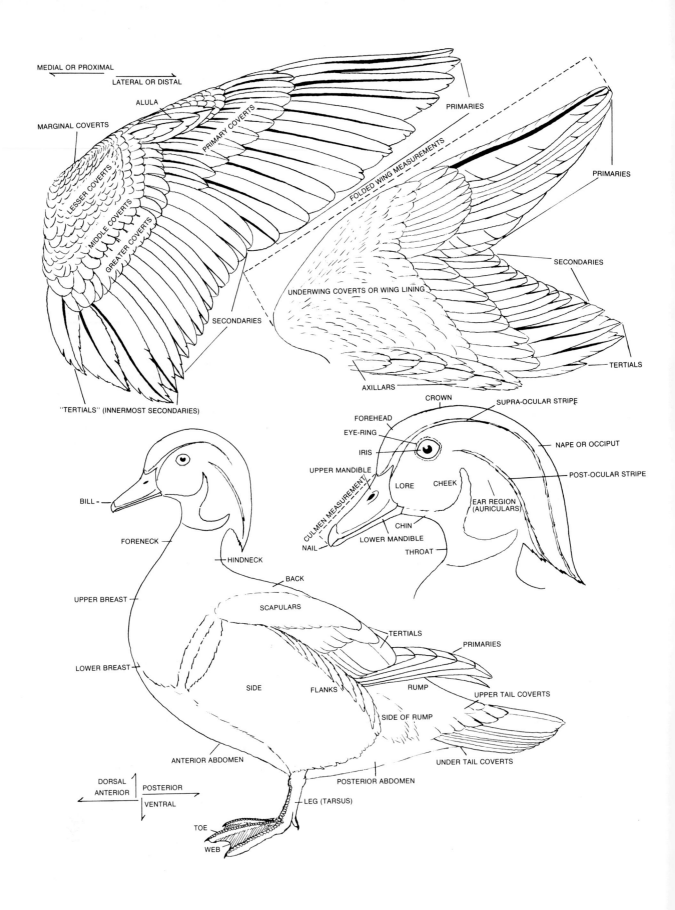

Glossary and Vernacular Name Derivations

Amphipoda: the order of crustaceans that includes the animals called sand fleas and scuds.

Anatidae: the family of birds that includes ducks, geese, and swans.

Anhimidae: the family of birds that includes the screamers.

Anseriformes: the order of birds that includes the families Anatidae and Anhimidae.

Arthropoda: the phylum of animals that includes insects, crustaceans, and other "jointed-legged" invertebrates.

Austral: southern or southerly.

Baer: the Siberian white-eye, or Baer pochard, was named in honor of M. Baer, the 19th-century German ornithologist.

Barrow: the Barrow goldeneye was named in honor of Sir John Barrow (1764–1848), one-time secretary to the British Admiralty.

Bernier: the Madagascan or Bernier teal was named in honor of a ship surgeon associated with a French natural history expedition to Madagascar in the 1830s.

Bewick: the Bewick swan was named in honor of Thomas Bewick (1753–1828), English engraver and naturalist.

Billabong: in Australia, a stagnant backwater, oxbow, or minor branch of a stream.

Bill-down display: a term applied to the post-copulatory display of many pochard species, in which the male swims away from the female in a strongly down-tilted head posture.

Bivalve: a mollusk with two valves, or shells, such as clams.

Boreal: northern or northerly.

Brant: the origin of this vernacular name (which is *brent* in Great Britain) is perhaps from the Welsh *brenig,* or from the Aristotelian *brenthos* or *brinthos,* an unknown bird.

Bridling: a term applied to a male display of some *Anas* species, involving a drawing of the head backward along the scapulars while calling.

Brood: collectively, the young hatched from a single clutch of eggs. Brooding refers to the act of tending a brood; double-brooding refers to the rearing of two broods in a single season.

Bulla: an inflated or bubblelike cavity associated with the syrinx in many male ducks.

Bulrushes: the inclusive name for plants such as tules and three-squares of the genus *Scirpus,* in the family Cyperaceae.

Burping: a term applied to a display of various *Anas* species, consisting of a vocalization accompanied by vertical neck stretching.

Caruncle: a fleshy enlargement of skin, usually on the head or neck.

Cattails: emergent and shoreline plants of the genus *Typha,* family Typhaceae, referred to as cumbungi in Australia.

Cere: a fleshy enlargement of the bill above the nostrils, and the basis for the generic name *Cereopsis.*

Chiloé: the name of the Chiloé wigeon refers to the island of Chiloé, off the coast of Chile.

Cladocera: the order of crustaceans that includes the animals called water fleas (*Daphnia*).

Clutch: the complete number of eggs laid by a single female and incubated simultaneously.

Composite: a member of the plant family Compositae, which includes asters, sunflowers, and similar forms.

Conspecific: a term meaning that two or more populations are or should be considered part of the same species.

Coot: a vernacular name properly restricted to species of the genus *Fulica,* family Rallidae, but sometimes inappropriately applied to scoters.

Copepoda: a subclass of the Crustacea, including many zooplanktonic organisms.

Coscoroba: this swan's vernacular and generic name is derived from the species' typical call.

Cracidae: that family of birds that includes chachalacas, guans, and curassows, within the order Galliformes.

Crèche: an assemblage of flightless young representing several families.

Crustacean: an animal of the class Crustacea, which includes crayfish, crabs, and their relatives.

Decrescendo call: a term applied to a female call of various *Anas* species that is characterized by a series of notes that descend in pitch and volume.

Dimorphism: occurring in two forms (such as sexual dimorphism), and including size and/or color differences.

Display: a term denoting movements and/or vocalizations

that through evolution have come to serve as social signals.

Duck: this vernacular name comes from the Medieval English *duken,* "to dive," and has no taxonomic significance inasmuch as it has been applied to nearly all of the smaller species of Anatidae. It is also sometimes used in the sense of *female,* as in *duck* and *drake.*

Duckweeds: tiny stemless floating plants of the family Lemnaceae, mostly in the genus *Lemna.*

Dump nesting: the laying of eggs in a common nest by two or more females. Among waterfowl, it is difficult to distinguish from "parasitic nesting," the laying of one or more eggs by a female in another's nest, to be incubated by the latter.

Eclipse: the dull, femalelike plumage of male waterfowl assumed after the breeding season in some sexually dimorphic species; actually a highly abbreviated winter or nonbreeding plumage.

Ecotone: an ecological transition zone between two community types.

Eider: the vernacular names for this group of sea ducks is from the Icelandic *ejdar,* used there for the common eider.

Electrophoresis: the separation of a mixture of particles in a fluid medium under the influence of an electric field.

Endemic: a species or other taxon that is native and restricted to a particular area.

Ericad: a plant of the heather family Ericaceae.

Erythristic: a rufous or reddish plumage variation.

Estuarine: associated with an estuary, where a river joins the sea.

Extirpation: the local elimination of a population from an area, as distinct from extinction, the total elimination of a population.

Eyton: the plumed or Eyton whistling duck was named for and by T. C. Eyton, 19th-century English ornithologist (1809–80).

Falcated: sickle-shaped, like the elongated and decurved tertial feathers found on falcated ducks.

Family: a taxonomic category that represents a subdivision of an order and a grouping of related genera, identified by the suffix *idae.*

Feral: existing in a free-living state following escape from captivity or domestication.

Ferruginous: the color of rusty iron.

Fischer: the spectacled, or Fischer, eider was named in honor of J. Fischer von Waldheim, German and later Russian scientist.

Fledging period: the period between hatching and initial flight in birds.

Flightless period: the period between the molting of the flight feathers and their regrowth, during which flight is impossible in adult waterfowl.

Form: a taxonomically neutral term for a species or some subdivision of a species.

Forb: a general term for a broad-leaved herbaceous plant.

Fulvous: dull yellowish brown, tawny.

Fuscous: dark brownish gray to brownish black.

Galliformes: the order of birds that includes pheasants, partridges, quails, and other "gallinaceous" birds.

Genus (plural, genera): a taxonomic category representing a grouping of related species.

Goose: this vernacular name comes from the Medieval English *goos* or *gos; gosling* refers to a baby goose. The term has no taxonomic significance, since it has been applied to the typical geese (*Anser* and *Branta*), but also to the magpie goose and various true ducks such as pygmy geese.

Goosander: this word is probably derived from the Old Norse *Gas* and *ønd,* meaning goose-duck.

Grunt-whistle: a term applied to a display of some *Anas* species, involving a scooping upward of water by the bill, usually accompanied by a vocalization.

Halophytic: refers to plants adapted to life in highly saline soils.

Hartlaub: the Hartlaub duck was named in honor of Dr. Gustav Hartlaub, 19th-century German ornithologist.

Head-throw: a term applied to a display of some pochards and sea ducks, involving a backward tossing of the head while calling.

Head-up-tail-up: a term applied to a display of some *Anas* species, involving a simultaneous stretching of the neck and cocking of the tail while calling.

Herbaceous: a term applied to nonwoody plants, or herbs.

Hectare: an area of 10,000 square meters, equivalent to 2.47 acres.

Heterozygotic: of mixed genetic origin; carrying two alleles at the same locus on a pair of homologous chromosomes.

Home range: an area occupied by but not necessarily defended by a pair or family during a particular period or throughout the year.

Hydrophyte: a plant adapted to growing in water.

Immature: the age class in birds that follows the juvenile period but precedes sexual maturity; used in this book for species that do not become sexually mature in their first year of life and thus usually have a distinct subadult plumage.

Inciting: functional or ritualized threatening movements and/or calls of female ducks, associated with the formation and maintenance of pair bonds.

Incubation: the application of heat to an egg by an adult bird; the incubation period is the period between the start of incubation and hatching.

Insular: having an island distribution.

Intergrade: to exhibit a gradual rather than discontinuous transition in traits of adjoining populations.

Isolating mechanism: properties of individuals that prevent successful interbreeding with individuals belonging to different populations.

Isopoda: an order of crustaceans that includes both terrestrial and aquatic forms.

Jaeger: raptorial gull-like birds of the family Stercorariidae.

Jheel: in India, a marsh, pool, or lake, particularly one left after inundation.

Juvenile: the age class in birds during which the juvenal plumage is carried, and in which initial flight (fledging) occurs.

Kinked-neck call: a term applied to a display of pochards, involving calling while stretching and bending the neck.

Lacustrine: associated with a lake.

Legumes: plants of the family Leguminoseae, such as peas and beans.

Littoral: the tidal zone of the ocean; also applied to the shallow edges of a lake.

Maccoa: the vernacular name of the maccoa duck probably stems from the Afrikaans *kacaauw*, later *makou*, a term used for a kind of duck (*makou-eend*) and also the spur-winged goose (*makougans*).

Mantle: feathers of the back, including the scapulars and interscapulars.

Mast: acorns and similar nutlike fruits from trees.

Meller: the Meller duck was named in honor of Dr. C. Meller, an Englishman who discovered this species and collected the first specimens.

Mock drinking: a term applied to ritualized drinking; also called false drinking.

Mock preening: a term applied to a ritualized preening of the feathers.

Molt-migration: a migration undertaken prior to the postnuptial molt, to an area where the flightless period is passed.

Monotypic: a term applied to a taxonomic category that has only one unit in the category immediately subordinate to it, such as a tribe with only one genus.

Muskgrass: plants of the genus *Chara*, in the algae family Characeae, sometimes also called stonewort.

Nuptial: refers to the plumage in which pair formation or breeding occurs.

Ochraceous: the color of ochre, earthy yellow.

Order: a taxonomic category that represents a grouping of related families.

Ostracoda: a subclass of the Crustacea, including many small aquatic forms.

Pair bond: a prolonged individual association between a male and female in monogamous species, lasting either for a single breeding season or permanently.

Páramo: the moist, grassy, and shrubby zone above timberline in the northern Andes.

Phyletic: refers to the evolutionary history of a taxon.

Phyllopoda: a group of crustaceans having leaflike swimming feet that also serve as gills, such as fairy shrimp.

Phylum: a taxonomic category representing a major subdivision of the animal kingdom.

Plankton: small animals (zooplankton) and plants (phytoplankton) that float or drift in the water.

Pochard: a vernacular name applied to various species of the genera *Netta* and *Aythya*, probably derived from Low German *poken*, "to poke," or Old French *pochard*, "a drunkard."

Pondweeds: the inclusive name for a group of aquatic plants, primarily of the genus *Potamogeton*, in the family Potamogetonaceae.

Postnuptial: refers to the molt following the breeding season, during which the flight feathers are lost and the body feathers are also replaced.

Puna: the dry, grassy alpine zone of the central Andes.

Race: as used here, equivalent to subspecies; not necessarily the same as "form," a geographically definable population of unspecified taxonomic rank.

Radjah: the vernacular name of the radjah shelduck is apparently a variant of *rajah*, and is based on the specific name given the species by Garnot. The alternative name, *Burdekin duck*, is based on the name of a river in Queensland.

Renesting: a second or later attempt at nesting by a female whose initial nesting effort was a failure. Distinct from multiple brooding, the rearing of more than one brood per season.

Reticulate: having a fine network of scales on the tarsus.

Ross: the Ross goose was named in honor of Bernard Ross (1827–74), chief factor of the Hudson Bay Company.

Salvadori: the Salvadori duck was named in honor of Tomasso Salvadori (1835–1923), director of the Zoological Museum at Turin, Italy, and an authority on the birds of Papua, New Guinea.

Scaup: the vernacular name of this group of pochards is from the Old French *escalope* and the Old Dutch *schelpe*, referring to the mollusks on which the birds often feed.

Scutellate: having a vertically aligned series of scales on the front of the tarsus.

Sheldrake: this vernacular name refers to shelducks and sheldgeese inclusively, and is derived from the Medieval English *sheld*, and *drake*, a male duck.

Sedge: the vernacular name of a group of grasslike plants of the genus *Carex* in the family Cyperaceae.

Sneak: a term applied to a display of some pochards, in which the head and neck are lowered to or nearly to the water.

Species: a "kind" of organism, or more technically, a group or groups of actually or potentially interbreeding populations that are reproductively isolated from all other populations. The term *species* remains unchanged in the plural, and also refers to the taxonomic category below that of genus and above that of subspecies.

Speculum: a pattern of distinctive feather coloration on the wing, sometimes iridescent, usually involving the secondary feathers.

Steller: the Steller eider was named in honor of G. W. Steller (1709–46), a German naturalist who discovered this species.

Step-dance: a term applied a display in which a pair of birds tread water in parallel while variably raising one wing.

Sternum: the breastbone of birds, which in all waterfowl is deeply keeled.

Subadult: refers to a late plumage of immature birds that require more than one year to attain their adult or definitive plumage.

Subfamily: a subdivision of a family composed of one or more genera, and identified by the suffix *inae*.

Subspecies: a group of local populations of a species that occupies part of the species' range and differs taxonomically from other local populations; a geographic race.

Superspecies: two or more species with largely or entirely nonoverlapping ranges and that are clearly derived from a common ancestor but are too distinct to be considered conspecific.

Swan: this vernacular name is akin to the Medieval English *soun* and the Latin *sonore,* meaning to make a noise or sound. It is restricted in use to the species here included in the genera *Cygnus* and *Coscoroba.* The Latin *cygnus* is also the basis for *cygnet,* meaning a young swan.

Synonym: in taxonomy, referring to differing names proposed for the same taxon; in such cases the older, or "senior," synonym represents the valid name, provided that other conditions are met.

Syrinx: the sound-producing structure of the trachea in birds.

Tarn: term used in Scotland and England for a small mountain lake or pool.

Taxon (plural, taxa): a term for any category used in scientific classification (taxonomy), or for any particular example of such a category.

Teal: this vernacular name has no taxonomic significance, and has been applied to a variety of small ducks of varied ancestry; the word's origin is uncertain, but may be from the Dutch *telen* or *tele,* "to produce."

Territory: an area, either fixed (as around a nest) or moving (as around a mate or brood), that is defended by an individual or pair from incursion by other individuals of the species.

Tiaga: the boreal or northern coniferous forest.

Trachea: the windpipe, which extends from the glottis to the junction of the bronchi, where the syrinx of waterfowl is located.

Tribe: a subdivision of a family or subfamily composed of one or more genera and identified by the suffix *ini.*

Triumph ceremony: a behavior pattern of geese and swans that typically involves mutual calling and posturing by members of a pair following a hostile encounter.

Turning the back of the head: a term applied to a male display of various duck species, in which the nape region is oriented toward a female.

Univalve: a mollusk with a single valve, or shell, such as snails.

Vermiculations: fine, wavy pigmentation patterns on feathers that vaguely resemble worm tracks.

Vlei: a term used in South Africa for a temporary lake or marsh.

Water lily: a group of aquatic plants with large floating leaves, including several genera in the family Nymphaeaceae.

Wigeon (or widgeon): This name is now usually applied to three species of *Anas* formerly separated as *Mareca,* but sometimes is used for other species such as the Cape teal. The origin of the word is obscure, but it is related to the Old French *vigeon.*

Wigeon grass: the inclusive name of a genus (*Ruppia*) of plants in the pondweed family Potamogetonaceae.

Xerophytic: refers to plants adapted to life in dry environments.

Index

The following index is limited to the species of Anatidae; species of other bird families are not indexed, nor are subspecies included. However, vernacular names applied to certain subspecies that sometimes are considered full species are included, as are some generic names that are not utilized in this book but which are still sometimes applied to particular species or species groups. Complete indexing is limited to the entries that correspond to the vernacular names utilized in this book; in these cases the primary species account is indicated in italics. Other vernacular or scientific names are indexed to the section of the principal account only.

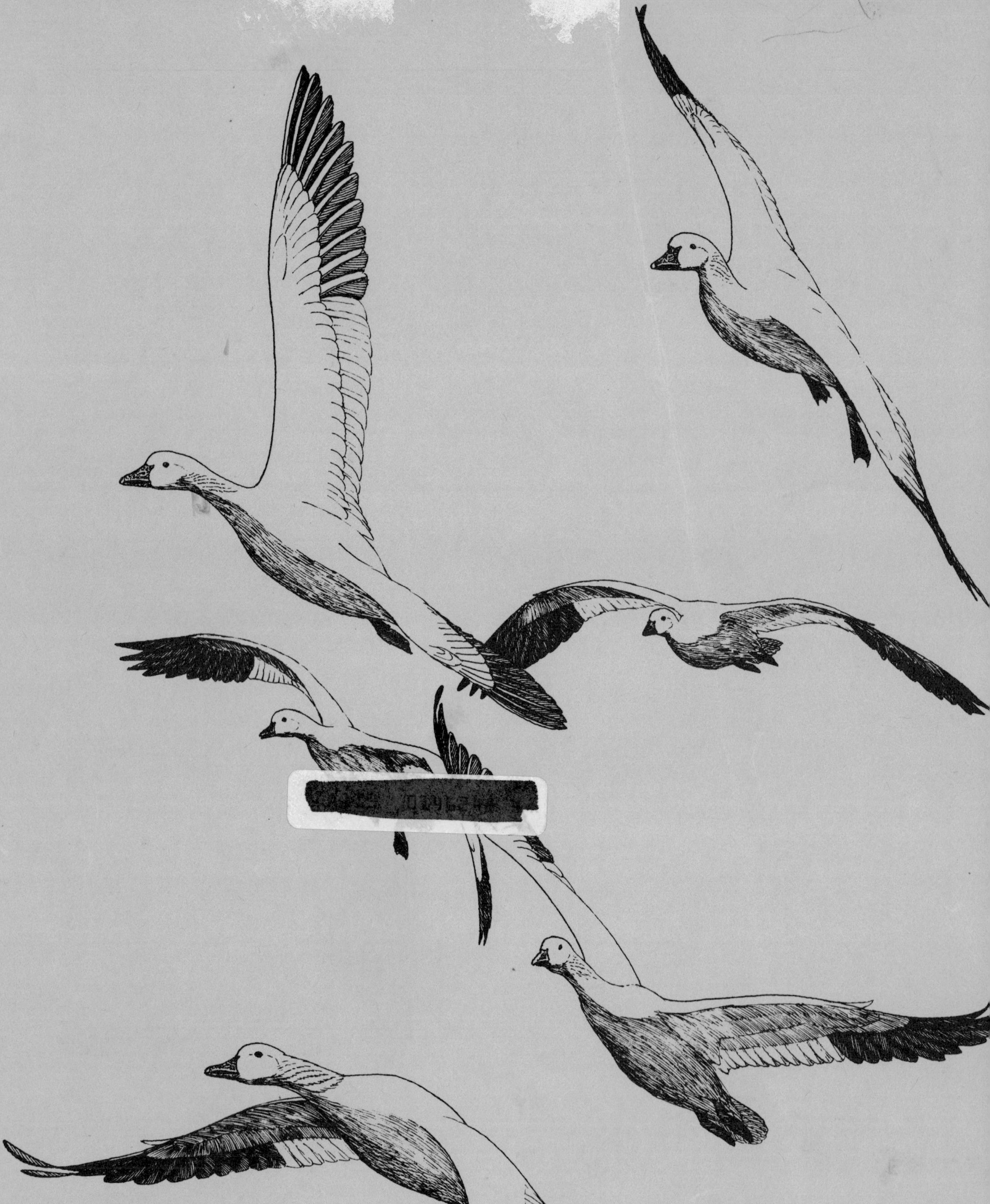